Developments in Soil Science – Volume 31

Digital Soil Mapping

An Introductory Perspective

Developments in Soil Science
Series Editors: A.E. Hartemink and A.B. McBratney

Developments in Soil Science – Volume 31

Digital Soil Mapping
An Introductory Perspective

Edited by

P. Lagacherie
Institut National de la Recherche Agronomique, France

A.B. McBratney
The University of Sydney, Australia

M. Voltz
Institut National de la Recherche Agronomique, France

ELSEVIER

Amsterdam • Boston • Heidelberg • London • New York • Oxford • Paris
San Diego • San Francisco • Singapore • Sydney • Tokyo

Elsevier
Radarweg 29, PO Box 211, 1000 AE Amsterdam, The Netherlands
The Boulevard, Langford Lane, Kidlington, Oxford OX5 1GB, UK

First edition 2007

Library of Congress Cataloging-in-Publication Data
A catalog record for this book is available from the Library of Congress

British Library Cataloguing in Publication Data
A catalogue record for this book is available from the British Library

ISBN-13: 978-0-444-52958-9
ISBN-10: 0-444-52958-6
ISSN: 0166-2481

Printed and bound in The Netherlands

07 08 09 10 11 10 9 8 7 6 5 4 3 2 1

CONTENTS

List of contributors xi

Foreword xvii

Preface xix

Dedication xxi

A. Introduction

1. Spatial soil information systems and spatial soil inference systems: perspectives for digital soil mapping
P. Lagacherie and A.B. McBratney 3

B. Digital soil mapping: current state and perspectives

2. A review of digital soil mapping in Australia
E. Bui 25

3. The state of the art of Brazilian soil mapping and prospects for digital soil mapping
M.L. Mendonça-Santos and H.G. dos Santos 39

4. The soil geographical database of Eurasia at scale 1:1,000,000: history and perspective in digital soil mapping
J. Daroussin, D. King, C. Le Bas, B. Vrščaj, E. Dobos and L. Montanarella 55

5. Developing a Digital Soil Map for Finland
H. Lilja and R. Nevalainen 67

C. Conception and handling of soil databases

6. Adapting soil data bases practices to the proposed EU INSPIRE directive
J. Dusart 77

7. Storage, maintenance and extraction of digital soil data
C. Feuerherdt and N. Robinson 87

8. Towards a soil information system for uncertain soil data
G.B.M. Heuvelink and J.D. Brown 97

9. The development of a quantitative procedure for soilscape delineation using digital elevation data for Europe
E. Dobos and L. Montanarella 107

10. Ontology-based multi-source data integration for digital soil mapping
B. Krol, D.G. Rossiter and W. Siderius 119

D. Sampling methods for creating digital soil maps

11. Optimization of sample configurations for digital mapping of soil properties with universal kriging
G.B.M. Heuvelink, D.J. Brus and J.J. de Gruijter 137

12. Latin hypercube sampling as a tool for digital soil mapping
B. Minasny and A.B. McBratney 153

13. Methodology for using secondary information in sampling optimisation for making fine-resolution maps of soil organic carbon
A. Dobermann and G.C. Simbahan 167

14. Designing spatial coverage samples using the *k*-means clustering algorithm
D.J. Brus, J.J. de Gruijter and J.W. van Groenigen 183

15. Adequate prior sampling is everything: lessons from the Ord river basin, Australia
E.N. Bui, D. Simon, N. Schoknecht and A. Payne 193

E. New environmental covariates for digital soil mapping

16. The use of airborne gamma-ray imagery for mapping soils and understanding landscape processes
J. Wilford and B. Minty 207

17. Visible–NIR hyperspectral imagery for discriminating soil types in the La Peyne watershed (France)
J.S. Madeira Netto, J.-M. Robbez-Masson† *and E. Martins* 219

18. Land-cover classification from Landsat imagery for mapping dynamic wet and saline soils
S. Kienast-Brown and J.L. Boettinger 235

19. Producing dynamic cartographic sketches of soilscapes by contextual image processing in order to improve efficiency of pedological survey
J.-M. Robbez-Masson† 245

20. Conceptual and digital soil-landscape mapping using Regolith-Catenary units
R.N. Thwaites 257

21. Soil prediction with spatially decomposed environmental factors
M.L. Mendonça-Santos, A.B. McBratney and B. Minasny 269

F. Quantitative modelling for digital soil mapping

22. Integrating pedological knowledge into digital soil mapping
C. Walter, P. Lagacherie and S. Follain 281

23. Decomposing digital soil information by spatial scale
R.M. Lark 301

24. Digital soil mapping with improved environmental predictors and models of pedogenesis
N.J. Mckenzie and J.C. Gallant 327

F.i. Example of predicting soil classes

25. A comparison of data-mining techniques in predictive soil mapping
T. Behrens and T. Scholten 353

26. Digital soil mapping: An England and Wales perspective
T.R. Mayr and B. Palmer 365

27. Pedogenic understanding raster classification methodology for mapping soils, Powder River Basin, Wyoming, USA
N.J. Cole and J.L. Boettinger 377

28. Incorporating classification trees into a pedogenic understanding raster classification methodology, Green River Basin, Wyoming, USA
A.M. Saunders and J.L. Boettinger 389

29. Rule-based land unit mapping of the Tiwi Islands, Northern Territory, Australia
I.D. Hollingsworth, E.N. Bui, I.O.A. Odeh and P. McLeod 401

30. A test of an artificial neural network allocation procedure using the Czech soil survey of agricultural land data
L. Boruvka and V. Penizek 415

31. Comparison of approaches for automated soil identification
C. Albrecht, B. Huwe and R. Jahn 425

F.ii. Example of predicting soil attributes

32. Digital mapping of soil attributes for regional and catchment modelling, using ancillary covariates, statistical and geostatistical techniques
I.O.A. Odeh, M. Crawford and A.B. McBratney 437

33. Comparing discriminant analysis with binomial logistic regression, regression kriging and multi-indicator kriging for mapping salinity risk in northwest New South Wales, Australia
J.A. Taylor and I.O.A. Odeh 455

34. Fitting soil property spatial distribution models in the Mojave Desert for digital soil mapping
D. Howell, Y. Kim, C. Haydu-Houdeshell, P. Clemmer, R. Almaraz and M. Ballmer 465

35. The spatial distribution and variation of available Phosphorus in agricultural topsoil in England and Wales in 1971, 1981, 1991 and 2001
S.J. Baxter, M.A. Oliver and J.R. Archer 477

36. The population of a 500-m resolution soil organic matter spatial information system for Hungary
E. Dobos, E. Micheli and L. Montanarella 487

37. Regional organic carbon storage maps of the western Brazilian Amazon based on prior soil maps and geostatistical interpolation
M. Bernoux, D. Arrouays, C.E.P. Cerri and C.C. Cerri 497

38. Improving the spatial prediction of soils at local and regional levels through a better understanding of soil-landscape relationships: soil hydromorphy in the Armorican Massif of western France
V. Chaplot and C. Walter 507

G. Quality assessment and representation of digital soil maps

39. Quality assessment of digital soil maps: producers and users perspectives
P.A. Finke 523

40. Using soil covariates to evaluate and represent the fuzziness of soil map boundaries
M.H. Greve and M.B. Greve 543

41. The display of digital soil data, 1976–2004
P.A. Burrough 555

42. Are current scientific visualisation and virtual reality techniques capable to represent real soil-landscapes?
S. Grunwald, V. Ramasundaram, N.B. Comerford and C.M. Bliss 571

Author Index 581

Subject Index 595

Colour Plate Section to be found at the end of the book

LIST OF CONTRIBUTORS

Albrecht, C. Soil Physics Group, University of Bayreuth, D-95440 Bayreuth, Germany.

Almaraz, R. USDA Natural Resources Conservation Service, Lancaster, CA, USA.

Archer, J.R. 8 Melrose Road, Merton Park, London SW19 3H, UK.

Arrouays, D. Institut National de la Recherche Agronomique – INRA, Infosol, 45160, Ardon, France.

Ballmer, M. USDA Natural Resources Conservation Service, Ventura, CA, USA.

Baxter, S.J. Department of Soil Science, The University of Reading, Whiteknights, PO Box 233, Reading RG6 6DW, UK.

Behrens, T. Institute of Geography, Chair of Physical Geography, Eberhardt-Karls University of Tuebingen, Ruemelinstrasse 19-23, 72070 Tuebingen, Germany.

Bernoux, M. Institut de Recherche pour le Développement – IRD, UR041-SeqC, Labo Most, 34000 Montpellier, France.

Bliss, C.M. Soil and Water Science Department, University of Florida, 2169 McCarty Hall, PO Box 110290 Gainesville, FL 32611, USA.

Boettinger, J.L. Department of Plants, Soils & Biometeorology, Utah State University, Ag Science Building, 4820 Old Main Hill, Logan, UT 84322-4820, USA.

Boruvka, L. Department of Soil Science and Geology, Czech University of Agriculture in Prague, Prague 6 – Suchdol, CZ 165 21, Czech Republic.

Brown, J.D. Institute for Biodiversity and Ecosystem Dynamics, Universiteit van Amsterdam, Nieuwe Achtergracht 166, 1018 WV Amsterdam, The Netherlands.

Brus, D.J. Soil Science Centre, Wageningen University and Research Centre, PO Box 37, 6700 AA Wageningen, The Netherlands.

Bui, E.N. HSSE Academic Group, National Institute of Education, Nanyang Technological University, 1, Nanyang Walk, 637616, Singapore.

Burrough, P.A. 5 Meadow Close, Goring on Thames, Reading RG8 OAP, UK.

Cerri, C.C. Centro de Energia Nuclear na Agricultura – CENA, Universidade de São Paulo, CP.96. 13400-970 Piracicaba, SP, Brazil.

Cerri, C.E.P. Centro de Energia Nuclear na Agricultura – CENA, Universidade de São Paulo, CP.96. 13400-970 Piracicaba, SP, Brazil.

Chaplot, V. IRD, centre IRD d'île de France, 32, avenue Henri Varagnat – 93143 Bondy Cedex, France.

Clemmer, P. United States Department of the Interior – Bureau of Land Management, Denver, CO, USA.

Cole, N.J. USDA Natural Resources Conservation Service, Buffalo, WY, USA.
Comerford, N.B. Soil and Water Science Department, University of Florida, 2169 McCarty Hall, PO Box 110290, Gainesville, FL 32611, USA.
Crawford, M. Faculty of Agriculture, Food & Natural Resources, McMillan Building A05, The University of Sydney, NSW 2006, Australia.
Daroussin, J. Institut National de la Recherche Agronomique – Unité de Science du Sol, Centre de recherche d'Orléans – BP 20619 – 45166 Olivet Cedex, France.
De Gruijter, J.J. Soil Science Centre, Wageningen University and Research Centre, PO Box 37, 6700 AA Wageningen, The Netherlands.
Dobermann, A. Department of Agronomy and Horticulture, University of Nebraska-Lincoln, PO Box 830915, Lincoln, NE 68583-0915, USA.
Dobos, E. Department of Physical Geography and Environmental Sciences, University of Miskolc, 3515 Miskolc-Egyetemváros, Hungary.
dos Santos, H.G. EMBRAPA Solos, Rua Jardim Botânico, 1024, CEP. 22.460-000, Rio de Janeiro, RJ, Brazil.
Dusart, J. European Commission - DG Joint Research Centre, Institute for Environment and Sustainability, TP 262, 21020 Ispra, Italy.
Feuerherdt, C. Kilter Pty Ltd., 536 Hargreaves Street, Bendigo, Victoria 3550, Australia.
Finke, P.A. Dept. of Geology and Soil Science, University of Ghent, Krijgslaan 281, B9000, Ghent, Belgium.
Follain, S. UMR Sol, Agronomie, Spatialisation, Agrocampus Rennes-INRA, 65 rue de St. Brieuc, 35042 Rennes Cedex, France.
Gallant, J.C. CSIRO Land and Water, GPO Box 1666, Canberra, ACT 2601, Australia.
Greve, M.B. Department of Agroecology, Danish Institute of Agricultural Sciences, PO Box 50, DK-8830 Tjele, Denmark.
Greve, M.H. Department of Agroecology, Danish Institute of Agricultural Sciences, PO Box 50, DK-8830 Tjele, Denmark.
Grunwald, S. Soil and Water Science Department, University of Florida, 2169 McCarty Hall, PO Box 110290, Gainesville, FL 32611, USA.
Haydu-Houdeshell, C. USDA Natural Resources Conservation Service, Victorville, CA, USA.
Heuvelink, G.B.M. Soil Science Centre, Wageningen University and Research Centre, PO Box 37, 6700 AA Wageningen, The Netherlands.
Hollingsworth, I.D. Faculty of Agriculture, Food and Natural Resources, Ross Street Building A03, The University of Sydney, Sydney, NSW 2006, Australia.
Howell, D. USDA Natural Resources Conservation Service, Arcata, CA, USA.
Huwe, B. Soil Physics Group, University of Bayreuth, D-95440 Bayreuth, Germany.

Jahn, R. Institute of Soil Science and Plant Nutrition, Martin-Luther-University Halle-Wittenberg, Weidenplan 14, D-06108 Halle, Germany.
Kienast-Brown, S. USDA Natural Resources Conservation Service, Logan, UT, USA.
Kim, Y. Humboldt State University, Arcata, CA, USA.
King, D. Institut National de la Recherche Agronomique – Unité de Science du Sol – Centre de recherche d'Orléans, BP 20619 – 45166 Olivet Cedex, France.
Krol, B.G.C.M. Department of Earth Systems Analysis, International Institute for Geo-Information Science and Earth Observation (ITC), PO Box 6, 7500 AA, The Netherlands.
Lagacherie, P. INRA, UMR LISAH AgroM-INRA-IRD, 2 place Pierre Viala, 34060 Montpellier, France.
Lark, R.M. Environmetrics Group, Biomathematics and Bioinformatics Division, Rothamsted Research, Harpenden, Hertfordshire AL5 2JQ, UK.
Le Bas C. Institut National de la Recherche Agronomique – Unité de Science du Sol – Centre de recherche d'Orléans – BP 20619 – 45166, Olivet Cedex, France.
Lilja, H. MTT Agrifood Research Finland, 31600 Jokioinen, Finland.
Madeira Netto, J.S. Ministerio da Agricultura, Pecuaria e Abasticimento, Empresa Brasiliera de Pesquisa Agropecuaria Embrapa, Parque Estacao Biologica - PqEB s/n, Edificio Sede - Plano Piloto, CEP 70770-901, Brasilia, DF, Brazil.
Martins, E. Embrapa Cerrados, BR 020, km 18., Planaltina, DF, CEP 73310-970, Brazil.
McBratney, A.B. Australian Centre for Precision Agriculture, Faculty of Agriculture, Food & Natural Resources, McMillan Building A05, The University of Sydney, NSW 2006, Australia.
McKenzie, N. J. CSIRO Land and Water, GPO Box 1666, Canberra, ACT 2601, Australia.
McLeod, P. Department of Natural Resources Environment and the Arts Natural Systems Division Goyder Building Palmerston PO Box 30 NT Australia.
Mayr, T. R. National Soil Resources Institute, Cranfield University, Silsoe, Bedfordshire MK45 4DT, UK.
Mendonça-Santos, M.L. EMBRAPA Solos – Centro Nacional de Pesquisa de Solos, Rua Jardim Botanico 1024, 22.460-000, Rio de Janeiro, RJ, Brazil.
Micheli, E. Dept. of Agrochemistry and Soil Science, Szent István University, Gödöllő, Hungary.
Minasny, B. Australian Centre for Precision Agriculture, Faculty of Agriculture, Food & Natural Resources, McMillan Building A05, The University of Sydney, NSW 2006, Sydney, Australia.
Minty, B. Geoscience Australia, GPO Box 378, Canberra, ACT 2601, Australia.

Montanarella, L. Soil and Waste Unit, Institute of Environment and Sustainability, European Commission Joint Research Centre, Ispra, Italy.
Nevalainen, R. Geological Survey of Finland PO Box 1237, FIN-70211, Finland.
Odeh, I.O.A. Australian Centre for Precision Agriculture, Faculty of Agriculture, Food & Natural Resources, McMillan Building A05, The University of Sydney, NSW 2006, Australia.
Oliver, M.A. Department of Soil Science, The University of Reading, Whiteknights, PO Box 233, Reading RG6 6DW, UK.
Palmer, R.C. National Soil Resources Institute, York Science Park, York, Yorkshire YO10 5DG, UK.
Payne, A. Department of Agriculture, Western Australia, Baron-Hay Court, South Perth, WA 6151, Australia.
Penizek, V., Department of Soil Science and Geology, Czech University of Agriculture in Prague, Prague 6 – Suchdol, CZ 165 21, Czech Republic.
Ramasundaram, V. Computer and Information Science and Engineering, University of Florida, USA.
Robbez-Masson, J.M.[†✠]
✠Deceased 23/7/2005, enquiries to Philippe Lagacherie, INRA, UMR LISAH AgroM-INRA-IRD, 2 place Pierre Viala, 34060 Montpellier, France.
Robinson, N. DPI, Primary Industries Research Victoria, 99 Harley Street Strathdale, Victoria 3550, Australia.
Rossiter, D.G. Department of Earth Systems Analysis, International Institute for Geo-Information Science and Earth Observation (ITC) PO Box 6, 7500 AA, The Netherlands.
Saunders, A.M. Utah State University, PO Box 11297 Hilo, HI 96721, USA.
Schoknecht, N. Department of Agriculture, Western Australia, Baron-Hay Court, South Perth, WA 6151, Australia.
Scholten, T. Institute of Geography, Chair of Physical Geography, Eberhardt-Karls University of Tuebingen, Ruemelinstrasse 19-23, 72070 Tuebingen, Germany.
Simbahan, G.C. Department of Agronomy and Horticulture, University of Nebraska-Lincoln, PO Box 830915, Lincoln, NE 68583-0915, USA.
Siderius, W. Department of Earth Systems Analysis, International Institute for Geo-Information Science and Earth Observation (ITC) PO Box 6, 7500 AA, The Netherlands.
Simon, D. CSIRO Land and Water, GPO Box 1666, Canberra, ACT 2601, Australia.
Taylor, J.A. Australian Centre for Precision Agriculture, Faculty of Agriculture, Food & Natural Resources, McMillan Building A05, The University of Sydney, NSW 2006, Australia.

Thwaites, R.N. School of Natural Resource Sciences, Queensland University of Technology, GPO Box 2434, Gardens Point, Brisbane, QLD 4001, Australia.
Van Groenigen, J.W. Soil Science Centre, Wageningen University and Research Centre, PO Box 37, 6700 AA Wageningen, The Netherlands.
Voltz, M. INRA, UMR LISAH AgroM-INRA-IRD, 2 place Pierre Viala, 34060 Montpellier, France.
Vrščaj, B. Agricultural Institute of Slovenia, Centre for Soil and Environmental Research, Hacquetova 17, SI1001 Ljubljana, Slovenia.
Walter, C. UMR Sol, Agronomie, Spatialisation, Agrocampus Rennes-INRA, 65 rue de St. Brieuc, 35042 Rennes Cedex, France.
Wilford, J. Cooperative Research Centre for Landscape Environments and Mineral Exploration, c/o Geoscience Australia, GPO Box 378, Canberra, ACT 2601, Australia.

FOREWORD

Back in the late 1960s when I first began my research career, my aim was to investigate soil variability in the context of soil survey procedures. In those days the approach to soil mapping had changed little over many years, and involved field surveyors spending considerable time observing soils in the field using their knowledge and experience to present what they considered to be the optimum expression of the patterns of soils in the landscape on a map. This ability to observe the nature and patterns of soils in their landscape and to produce a representation of this in the form of a map was a special and often undervalued skill. In part, the relatively low value afforded to these skills was related to the products of their labours, frequently soil series maps, and occasionally simple interpretative maps for properties such as land capability. Once the maps were produced it was considered the end of the surveyor's task, and the interpretation and ancillary knowledge that went with the production of the map was in the surveyor's mind and occasionally in the field notes, but it was rarely available to the general map user. My own research sought to incorporate knowledge of the nature and pattern of soil variability into guiding the process of field surveying and the interpretation of the information presented in the map and presenting this information in the final map and accompanying report. In some respects this research was before its time because there was little enthusiasm from map producers to change the soil mapping procedures, and a lack of awareness on the part of most map users on how the additional information on soil variability could be used to their advantage. How things have changed in the last 30 years! Much of what I could only dream about in the past has now become reality, soil mapping is no longer simply the traditional 'old fashioned' exercise of integrating soil information from field observation and laboratory analysis by placing lines on maps, but is at the forefront of our use of digital technologies to present these data in a wide range of formats and to allow data analysis and interpretations as new soil-related questions are raised.

These changes have occurred because during recent years there has been an amazing increase in the technology available for computation and information technology. In particular there have been tremendous strides in the ability to manage and portray spatial and temporal data. The variability in soils is substantial both in space and time and these technologies are increasingly being applied to soils data. The current databases of soil information are in the whole the work of traditional soil surveyors recording in a format which has changed little over the last 50 years, as such the new techniques of spatial analysis may be applied to them, but their structure frequently imposes considerable restrictions in the manner in which the data can be presented. Similarly, the soil maps that have been produced around the world at a wide variety of scales are valuable documents in their own right but their use is often very restricted. What is needed is a completely new approach to the collection and presentation of the data, but as a first important step the existing soil maps must be digitized so that their data may be more fully utilized. Newly established databases will incorporate field and laboratory observational information, which is spatially and temporally referenced, together with remotely sensed data of increasingly fine spatial resolution, brought together with spatial and non-spatial inference systems. This is the key to Digital Soil Mapping.

Reviewing the scientific literature over the last two decades it is evident that many soil scientists have been working individually on various aspects of what might be broadly termed 'Digital Soil Mapping'. It is evident to these researchers that if the potential of the soil information we have available and those data currently being collected is to be fully realized, the future lies in Digital Soil Mapping. This text brings together a number of leaders in the broad field of Digital Soil Mapping to provide an introduction to many of the established and newly developed procedures and approaches in this field and indicate the potential way forward in the coming years. For the experts in the field the book is a most valuable synthesis of the 'state of the art', for the non-specialist the text provides a reasonably detailed introduction to this rapidly growing and increasingly important field of both soil science research, but perhaps more importantly the manner in which most soil information will be handled and presented in the not too distant future. In addition, the procedures encompassed in this broad field offer new tools and opportunities to soil scientists in other areas of the subject. Many in the Soil Science community complain that in the broad scheme of things too little attention is given to information about soils; in part, this is frequently because of its inaccessibility to the non-specialist. Digital Soil Mapping offers the possibility of responding to the demands of the non-specialist and making our data more accessible to a much wider community of soil users. In particular, the use of new techniques to predict soil attributes is a most valuable tool to be offered to many current and potential new users of soil information.

This is the first book that brings together publications from some of the key workers in the field of this newly developed and recognized area. I commend it to soil scientists and those who wish to use soil science information.

Stephen Nortcliff, University of Reading
Secretary-General
International Union of Soil Sciences (IUSS)

PREFACE

This book is based on scientific contributions to a Global Workshop on Digital Soil Mapping held in Montpellier (Campus Agronomique de la Gaillarde) in September 2004.

Although it has been put into practice for several years by soil scientists through the development of soil databases, soil information systems and the increasing use of numerical techniques in the prediction of soil variability, the concept of Digital Soil Mapping (DSM) has only been introduced recently and this workshop is the first specifically devoted to it. The definition of Digital Soil Mapping certainly needs to be further elaborated, but for the time being we feel that it could be formulated as "the creation and population of spatial soil information systems by numerical models inferring the spatial and temporal variation of soil types and soil properties from soil observation and knowledge and from related environmental variables".

Accordingly, the workshop aimed to gather around the topic of Digital Soil Mapping, soil surveyors interested in numerical techniques and pedometricians willing to extend their approaches to more operational scales. Eighty scientists from 17 different countries attended this meeting, which provided a large overview of the state of the art of this nascent discipline.

The workshop was organized with the financial support of our institutes, namely INRA (Institut National de la Recherche Agronomique) and the University of Sydney. It was hosted by the School of Agronomy of Montpellier (Sup Agro) and received the support of several soil science organizations: the IUSS (International Union of Soil Sciences) Division of Soil in Space and Time, the IUSS Commission on Soil Geography, the French Soil Science Society (AFES) and the IUSS Commission on Pedometrics.

In this book we have compiled the best ideas and methodologies that emerged from this workshop. Rather than being the last word on the subject, this is the first word in book form – hence the subtitle. We envisage significant developments in the coming years. From this perspective a Digital Soil Mapping working group has been created within IUSS. As Digital Soil Mapping is moving from the research phase to the operational production of digital soil maps, we hope it will soon provide a significant contribution to the management of land resources throughout the world.

P. Lagacherie, A.B. McBratney and M. Voltz
Montpellier and Sydney

DEDICATION

Jean-Marc Robbez-Masson giving a presentation at the DSM Workshop, September 2004. (Photograph courtesy of David Howell.)

This book is dedicated to the memory of Dr. Jean-Marc Robbez-Masson who was a major contributor to the Workshop on Digital Soil Mapping and has two chapters in this volume. Tragically Dr. Robbez-Masson lost his life in an accident in the Alps at the end of July 2005.

Jean-Marc taught soil science and geomatics in the National School of Agronomy of Montpellier (sup-agro) from 1986 until his untimely demise. He was also an innovative scientist in Digital Soil Mapping and landscape studies. His PhD thesis, presented in 1994, dealt with the numerical representation and delineation of soilscapes. This research produced a widely distributed computer program called CLAPAS (URL http://sol.ensam.inra.fr/Produits/Clapas/NousGB.asp), which has been extensively applied. He was also involved in several remote sensing research projects mostly applied to the spatial distribution of soil properties. Jean-Marc is the principal author of the website "sol et paysage" (http://sol.ensam.inra.fr/Paysages/) which is a remarkable example of the diffusion of soil knowledge to a wide community. Lastly, his humour, his awareness of others and his generosity made Jean-Marc a precious friend to all those who had the privilege to share a piece of his life, either in or out of his working sphere.

Vale Jean-Marc, we thank you for your contribution to Digital Soil Mapping, your ideas will live on!

The Editors

A. Introduction

Developments in Soil Science, volume 31
P. Lagacherie, A.B. McBratney and M. Voltz (Editors)

Chapter 1

SPATIAL SOIL INFORMATION SYSTEMS AND SPATIAL SOIL INFERENCE SYSTEMS: PERSPECTIVES FOR DIGITAL SOIL MAPPING

P. Lagacherie and A.B. McBratney

Abstract

Given the relative dearth of, and the huge demand for, quantitative spatial soil information, it is timely to develop and implement methodologies for its provision. We suggest that digital soil mapping, which can be defined as the creation, and population of spatial soil information systems (SSINFOS) by the use of field and laboratory observational methods, coupled with spatial and non-spatial soil inference systems, is the appropriate response. Problems of large extents and soil-cover complexity and coarse resolutions and short-range variability representation carry over from conventional soil survey to digital soil mapping. Meeting users' requests and demands and the ability to deal with spatially variable and temporally evolving datasets must be the key features of any new approach.

In this chapter, we present a generic framework that recognises the procedures required. Within quantitatively defined physiographic regions, SSINFOS must be populated and spatial soil inference systems (SSINFERS) must be developed. When combined this will allow users to derive the data they require. Further work is required on the development of these systems, and on the data requirements, the optimal forms of inference and the appropriate representation of the products of digital soil mapping.

1.1 Introduction

With the great explosion in computation and information technology have come vast amounts of data and tools in all field of endeavour. This has motivated numerous initiatives around the world to build spatial data infrastructures (Masser, 1999; Kok and van Loenen, 2004) aiming to facilitate the collection, maintenance, dissemination and use of spatial information. Soil science potentially contributes to the development of such generic spatial data infrastructure through the ongoing creation of regional, continental and worldwide soil databases (see examples in Table 1.1), which have been reviewed recently by Rossiter (2004), and which are now operational for some uses, for example land resource assessment (FAO, 1976; Fischer and Antoine, 1994) and risk evaluation (Lim and Engel, 2003).

Table 1.1. *Examples of (spatial) soil information systems.*

	Entity	Name	URL
Regional			
	Murray-Darling Basin, Australia	MDBSIS	www.brs.gov.au/mdbsis/
	Languedoc-Roussillon, France	BDsolLR	http://sol.ensam.inra.fr/BDSolLR/
National			
	Canada	CanSIS	sis.agr.gc.ca/cansis/
	USA	NASIS	nasis.nrcs.usda.gov/
	Australia	ASRIS (CSIRO)	www.clw.csiro.au/aclep/ASRIS2004.htm
Multinational			
	Europe	European soil database (JRC)	eusoils.jrc.it/ESDB_Archive/ESDBv2
	World	ISIS (ISRIC)	lime.isric.nl/index.cfm?contentid = 218
	World	SOTER (ISRIC/FAO)	www.fao.org/landandwater/agll/soter.stm
	World	Digital soil map of the world (FAO)	www.fao.org/landandwater/agll/dsmw.stm

Unfortunately the existing soil databases are neither exhaustive nor precise enough for promoting an extensive and credible use of the soil information within the spatial data infrastructure that is being developed worldwide. The main reason is that their present capacities only allow the storage of data from conventional soil surveys, which are scarce and sporadically available. The Netherlands has complete coverage at a nominal spatial resolution of 100 m (Table 1.2). In France, in contrast, a highly developed western economy, but with a large land area, only 26% of the country is covered at a nominal spatial resolution of 500 m and 13% at a nominal spatial resolution of 200 m (King et al., 1999). One third of Germany is covered with soil maps at a nominal spatial resolution of 10 m (1:5000) but most of these are not yet digital (Lösel, 2003). However, complete coverage of Germany at coarser resolutions (nominally 100 and 400 m) is available. The situation in larger countries such as Australia and Brazil is much worse. In Australia, for example, prior to Moran and Bui's (2002) work, the Murray-Darling Basin, Australia's most important agricultural region comprising some 14% of the land area, had 50% coverage at 500 m and 3% at 200 m. Brazil is uniformly covered by the Soil Map of Brazil and the Agricultural Suitability Map of Brazil at a nominal spatial resolution of 10 km, exploratory soil maps by the RADAM/EMBRAPA Solos project (1: 1,000,000 or nominally 2 km) and Agroecological Zoning (diagnosis of environmental and agro-socio-economic features, nominally 4 km or 1: 2,000,000).

Table 1.2. *Suggested resolutions and extents of digital soil maps.*

Name	Approx. USDA survey order[a]	Pixel size and spacing[b]	Cartographic scale[b]	Nominal spatial resolution[b]
D1	0[c]	<(5 m × 5m)	>1:5000	<(10 m × 10m)
D2	1, 2	(5 m × 5 m) – (20 m × 20m)	1:5000 – 1:20,000	(10 m × 10 m) – (40 m × 40m)
D3	3, 4	(20 m × 20 m) – (200 m × 200m)	1:20,000 – 1:200,000	(40 m × 40 m) – (400 m × 400m)
D4	5	(200 m × 200 m) – (2 km × 2km)	1:200,000 – 1:2,000,000	(400 m × 400 m) – (4 km × 4km)
D5	5	>(2 km × 2km)	<1:2,000,000	>(4 km × 4km)

[a]Soil Survey Staff (1993), Table 2.1.
[b]Digital soil maps are partly defined by their block size and spacing (refer to Bishop et al. (2001), Figure 3) – which here we equate with pixel size – the cartographic scale is calculated as 1 m/(side length of 1000 pixels), assuming that the smallest area discernible is 1 mm by 1 mm.Conversely, the pixel size (p) of a 1:100,000 conventional map can be calculated as, $p = 1/\chi \times \lambda = 100{,}000 \times 0.001 = 100\,\text{m}$ if we consider the smallest area resolvable on a map, (λ), with representative fraction χ, to be 1 mm by 1 mm.Following notions in microscopy, and the Nyquist frequency concept from signal processing, it may be argued that the minimum resolution is the size of two by two pixels – we define this here as the nominal spatial resolution. Small pixel sizes correspond to fine resolutions, and large pixel sizes correspond to coarse resolutions.
[c]This order was suggested by the late Dr Pierre C. Robert, University of Minnesota, for applications in precision agriculture.

The main reason for this lack of soil spatial data is simply that conventional soil survey methods are relatively slow and expensive. Furthermore, there is a worldwide crisis in collecting new field data in general that leads some to be very pessimistic about future developments in conventional soil surveying (Nachtergaele and van Ranst, 2002). We believe technologies, such as hand-held field spectrometers, will come to the rescue.

To face this situation, we think that the current spatial soil information systems have to extend their functionalities from the storage and the use of digitised (existing) soil maps to the production of soil maps *ab initio*. This is precisely the aim of digital soil mapping, which can be defined as *the creation, and population of spatial soil information systems by the use of field and laboratory observational methods coupled with spatial and non-spatial soil inference systems.*

The development of digital soil mapping methods has been a growing activity for the past decades. It is moving inexorably from the research phase (Skidmore et al., 1991; Favrot and Lagacherie, 1993; Moore et al., 1993) to production of maps for regions, catchments and whole countries. In 1965, G.E. Moore (1965) observed that integrated circuit complexity evolved exponentially. As a consequence of this observation a scaling law was developed in the 1970s, stating that electronic device feature sizes would decrease by a factor of 0.7 every 3 years or the processing power of microchips doubles every 18 months. Although this empirical law has attracted various kinds of critics, this prediction has proved to be accurate enough that it has become well established within the computer and the literature up until the present day we obtain the graph shown in Figure 1.1. We can readily see a information technology industries. Because digital soil mapping is underpinned by information technology one might speculate a relationship between the size of DSM project that might be tackled and time. Taking data on the number of pixels described or predicted from the earlier projects found in exponential growth with time, the fitted line describes a tenfold increase every 7.1 years, or a doubling in the number of pixels every 26 months – slightly slower than Moore's law. The implications of this, for the digital soil mapping of the world, are shown in Table 1.3.

This evolution is contemporaneous with the increasing development of spatial data infrastructures, which provide more and more exhaustive mapping of soil-forming factors, for example relief through Digital Terrain Model (DTM), and organisms for land use from remotely sensed images. Meanwhile, the classical toolbox for observing and characterising soils in the field (codified observations of auger hole and pits and laboratory chemical analysis) is more and

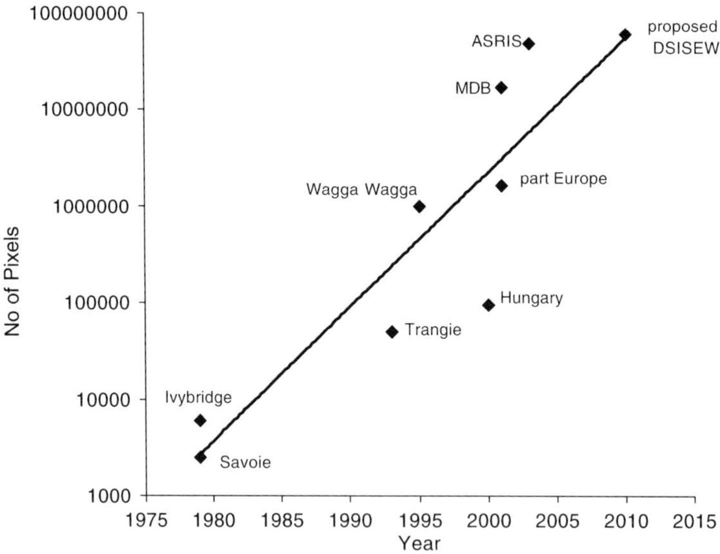

Figure 1.1. *The size of digital soil mapping projects (log_{10}[no. of pixels]) as a function of time with a fitted regression line that has a gradient of 0.14.*

Table 1.3. *Predictions for a digital map of the world on the basis of current rate of progress based on* $\log_{10}$*(no. of pixels) = −272.6977+0.1395053 year AD.*

Resolution (m)	No. of megapixels	Earliest date
1000	149	2015
100	1490	2027
50	5970	2032
10	14,900	2039

more integrated within GIS thanks to new tools such as GPS or PAD observation forms. It is also complemented with new field observation techniques able to hasten and objectify the collection of soil data, for example geophysics (Barrett, 2002; Bishop and McBratney, 2001) and visible–NIR spectrophotometry (Shepherd and Walsh, 2002; Tabbagh et al., 2000). In parallel, a body of research work in geographical information science heralds the evolution from classical raster or vector GIS, tools limited to the collection and storage of all kinds of spatial data, to more sophisticated systems able to represent more complex spatial models dealing with multi-scale variations, uncertainty and variations in time (Burrough and Frank, 1995; Mennis, 2002), and to embed spatial reasoning procedures such as inductive learning (Malerba et al., 2002), or hierarchical reasoning (Van Oosterom and Shenkelaars, 1995; De Bruin et al., 1999). Therefore, a perspective now exists to integrate in modernised GIS packages all the computational work on digital soil mapping, which has been done so far outside the framework of simple raster or vector GIS. In this perspective, we feel it is timely to outline a general intellectual and operational framework for digital soil mapping which can integrate the recent developments in numerical soil mapping techniques reviewed in detail by McBratney et al. (2003) and Scull et al. (2003) in the light of the knowledge on soil cover, which has been accumulated for a century or so by soil surveyors.

This chapter is an attempt to sketch this framework. We first elicit some challenges for digital soil mapping, which sets its agenda. Then we present and detail the concept of a SSINFER, a general framework that incorporates the research material accumulated over the past 25 years in the domain of digital soil mapping. We finally discuss some issues that are associated with the proposed framework.

1.2 From soil surveying to digital soil mapping: old and new challenges

Digital soil mapping aims to create and populate spatial soil information systems (SSINFOS) that could help users to make decisions over territories to deal with agricultural and agro-environmental problems. Although existing SSINFOS concern exclusively regional or national scale (Rossiter, 2004), we think that we must set more ambitious goals, namely the production of SSINFOS able to provide spatial soil information at the D3 level (Table 1.2). In the language of digital soil maps (Bishop et al., 2001), different from that of conventional cartography, scale is a difficult concept and is better replaced by resolution and spacing. D3 survey, which in conventional terms has a scale of 1:20,000 down to 1:200,000, has a block or cell size from 20 to 200 m, a spacing also of 20–200 m and a nominal spatial resolution of 40–400 m (see Table 1.2). The target extents range from 2×2 km to 2000×2000 km, which correspond to sub-catchments, catchments and regions. Such resolutions and extents are compatible with the use of SSINFOS in applications aiming to help decision making at the land parcel level for reaching a common objective defined collectively at the extent of a given region. Examples of such applications are improving water quality over a basin (Burkart et al., 1999) or amelioration of wine quality over a production area (Bodin and Morlat, 2003). In these types of applications, the lack of soil data has often been outlined (Coulibaly et al., 2004).

There are some challenges that we summarise further under four headings. The first two (1.2.1 and 1.2.2) are simply inherited from conventional soil surveying, the importance of the third (1.2.3) has been increasingly recognised over the last few decades, while the fourth (1.2.4) is specific to digital soil mapping.

1.2.1 Large extents and soil-cover complexity

Dealing with large extents increases the chances of investigating complex patterns of soil variation. This was early recognised by soil surveyors as illustrated by the following statement (Burrough, 1993) "because many landscapes are the result of many processes operating both simultaneously and in historical sequence, it is no wonder that patterns of variation on the earth surface are so complex". In parallel, the rare geostatistical studies over large territories have also shown evidence of this complexity (McBratney et al., 1982).

To face the complexity of soil cover over large extents, a number of authors have viewed the soil cover as a hierarchy of soil systems (e.g. Wilding and Drees, 1983, Table 4). This hierarchy has also been taken into account in some modern soil databases (Oldeman and van Engelen, 1993; King et al., 1994). An hierarchical view of the soil cover is also the basis of multi-scale soil survey strategies (Astle et al., 1969; Favrot, 1981; Brabant, 1989), which have been proposed to reconcile fine resolutions and large extents. According to such strategies, a given region is first stratified in landscape units or physiographic units. Each of these units is then sampled by a representative reference area for characterising its own soil pattern by means of a more detailed survey. In at least one of these strategies (Brabant, 1989) the sequence "stratification – characterisation" is repeated, which leads to defining five levels of soil surveying.

This accumulated experience suggests that digital soil mapping also has to deal with this multi-scale view of the soil cover. This means that, apart from very simple soil patterns, digital soil mapping must also be viewed as a combination of several soil mapping procedures operating at nested scales.

1.2.2 Coarse resolutions and short-range variability representation

Owing to limited sample spacing and coarse resolutions, soil surveying and digital soil mapping can deal with only a part of the total soil variability. The unmapped soil variability arises from both the variation in soil properties that cannot be related to a known cause with our current available knowledge – (apparently) random variability (Wilding and Drees, 1983) – and the variation governed by identified factors which operate over distances that are shorter than the resolution of the soil survey, for example hedgerow influences (Dercon et al., 2003). In practice, both of these are assimilated into a single short-range variation that falls below the conventional soil survey or digital soil mapping resolutions. The importance of these short-range variations can be evaluated from the compilation of published variograms made by McBratney and Pringle (1999). For basic soil properties such as clay content and sand content (Fig. 1.2) there is evidence that this short-range unmapped variability is far from negligible. Consequently, what is spatially estimated by soil surveying or digital soil mapping is most often far more complex than a pedon associated with a single

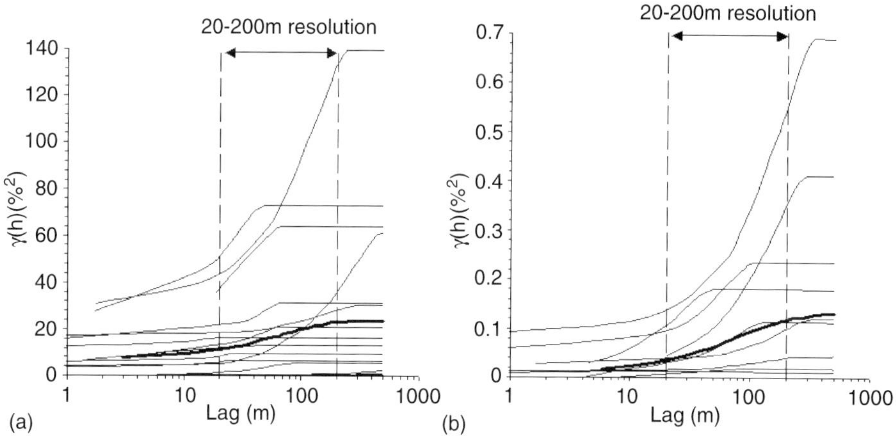

Figure 1.2. *Variograms of (a) clay content (%) and (b) carbon content (dag/kg) collected from published geostatistical studies (McBratney and Pringle, 1999).*

polygon or exact-valued soil properties as conveyed by the well-known notion of representative soil profile.

This problem was early recognised by soil surveyors and dealt through the adaptations of the taxonomic levels to the map scale and, when necessary, the definition of soil associations and soil complexes (e.g. USDA, 1951). Fridland (1972) raised the idea that the within-small-scale-mapping unit variations could be described by soil patterns for which he proposed a complete descriptive framework. Some other authors (e.g. Boulet et al., 1982; Butler, 1982) also attempted to break the nexus between pedon and mapping unit by suggesting the notion of a toposequence. More imprecise but more realistic descriptions of soil mapping units were also proposed using fuzzy logic (McBratney and De Gruijter, 1992; Lagacherie et al., 1994). In spite of these attempts, descriptions of soil mapping units by representative profiles – and the underlying choropleth map model (Burrough, 1993) – have been most often preferred by the users of soil maps for the sake of simplicity, and perhaps visual perception. In these contexts, short-range variation was implicitly taken into account by means of uncertainty measures, for example coefficient of variation (Beckett and Webster, 1971). Although the soil databases offer new possibilities of describing short-range variability – for example descriptions of properties of soil classes by intervals of values, average distance to the nearest correctly mapped soil class (Gaultier et al., 1993; Moran and Bui, 2002) – few attempts have been made to exploit these possibilities (Lagacherie et al., 2000).

We think that digital soil mapping research must address this problem, using the possibilities of information technology to provide the user sound descriptions of the unmapped short-range soil variations. This could allow better understanding of these variations, providing new basis for evaluating the quality of the map. There must be a response to the lack of metadata raised by Rossiter (2004) in his review of current SSINFOS.

1.2.3 Meeting users' requests and demands

In their review of soil science developments over the past 40 years, Mermut and Eswaran (2000) underlined that the demands from the society to the soil science community has dramatically increased, which has led to the emergence of new areas of interest such as land and soil quality, recognition of problems of land degradation and desertification, cycling of biogeochemicals, and soil pollution assessment and monitoring. These have been added to the old topics traditionally investigated by soil science such as soil fertility assessment or land management. This phenomenon has been amplified by the paramount developments of soil databases (Rossiter, 2004), which has attracted new users of soil information. The consequence is that the output of soil survey, and of digital soil mapping, is becoming less and less under the control of soil surveyors themselves, which is not necessarily detrimental. It is simply a symptom of increasing societal demand. Therefore, the soil properties that must be predicted are no longer the few selected by soil surveyors because of their relative accessibility, but can be imposed by other specialists for their own models. Similarly, the geographical support at which soil data must be provided, for example land parcel, watershed, administrative entity, etc., is now defined by the users with respect to the particular problem which they have to solve.

A conventional soil survey, because of its static nature, cannot readily track the evolution of the user's request. A challenge for digital soil mapping is to be flexible enough to satisfy present and future user demands.

1.2.4 Dealing with spatially variable and temporally evolving datasets

Digital soil mapping has some intrinsic difference with conventional soil survey. Conventional soil survey is essentially a solitary activity carried out by small teams with a similar range of skills involved in all the stages of the survey, whereas digital soil mapping involves a number of professionals with different skills (e.g. pedometricians, soil surveyors and GIS specialists), each of them providing its contribution to the spatial prediction of soil properties. In this new scenario, the end-user can also be much more active in the mapping process than he/she used to be, for example by collecting new soil data in the field with simplified guidelines as proposed by Cam et al. (2003) or providing new source of predictive data, for example proximal sensor data (Barnes et al., 2003).

The consequence is that we shift from a scenario in which the available datasets were merely static and uniformly distributed over the study region (as they were fixed at the beginning of conventional soil surveys) to a situation in which the datasets are different from one point to another and are subject to continual updating (according to the current state of the spatial data infrastructure and user activities). In this new context, digital soil mapping must provide the user the best possible map for a given area on a given date. This means that the set of mapping procedures at a given location must be constantly adapted to the current data configuration. The user should have access to all the information and steps of the mapping procedure to be able to modify them in accordance with his objectives and its specific study area.

Current SSINFOS cannot achieve this challenge, since they are static data by nature. More powerful structures must be developed which give ready access not only to the output data but also to all features or steps of the mapping procedures (input data, parameter settings, etc.). Self-updating of the Soil Information System would be then possible by applying these procedures to new areas within the same region as soon as the data configuration allows this application.

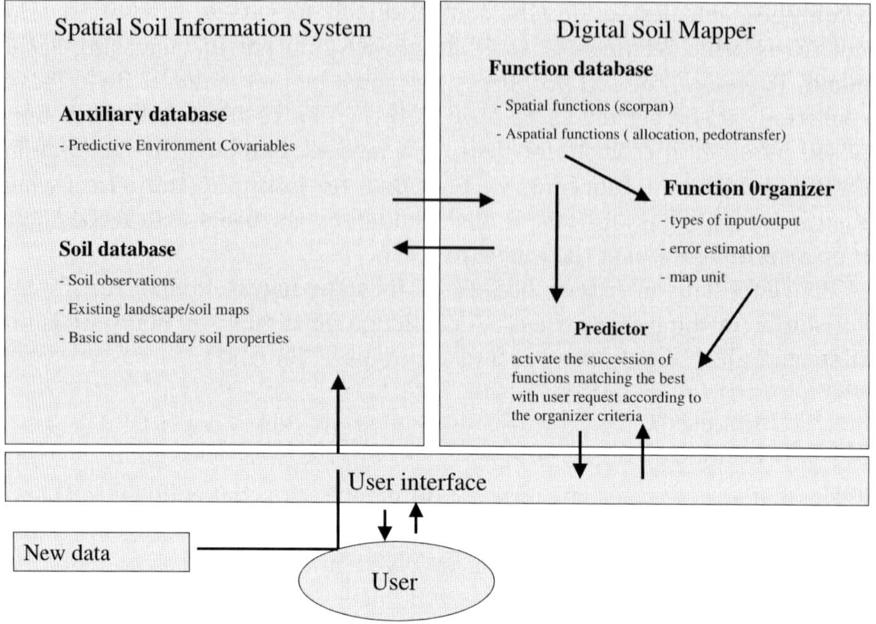

Figure 1.3. *Spatial soil inference system (SSINFERS) as output of digital soil mapping.*

1.3 The spatial soil inference system

As defined by McBratney et al. (2002), a soil inference system takes information on what we more-or-less know with a given level of uncertainty and infers data that we do not know with minimal inaccuracy, by means of properly and logically conjoined functions. We propose to extend to digital soil mapping this concept of soil inference system, which was initially defined for the optimal use of pedotransfer functions (PTFs) (McBratney et al., 2002). A general picture of what could be a SSINFER is presented in Figure 1.3. It incorporates two basic entities within a common user interface: A SSINFOS and a Digital Soil Mapper (DSMAP).

1.3.1 Spatial soil information system

A Spatial Soil Information System includes two components. First it is a georeferenced soil database with various types of soil information: soil observations and laboratory analysis at georeferenced sites, digitised soil maps, images of basic soil properties, for example clay content, pH, etc., and images of secondary soil properties, for example infiltration parameters, field capacity, lime requirement, etc. A number of such soil databases now exist as reviewed by Rossiter (2004). User interfaces for entering new data, formulating queries, and visualising the outputs are also available.

In addition, we propose to link within the SSINFOS the soil database with an auxiliary database of predictive environmental covariables that are available over the area of interest. McBratney et al. (2003, Section 4) provided a detailed inventory of these variables, – that is "the seven scorpan factors" – and of their sources: information on soil themselves either by conventional soil survey expertise or by remote and proximal sensing (*s*), climate variables (*c*), vegetation and land use (*o*), relief (*r*), parent material (*p*), age or elapsed time (*a*) and spatial coordinate alone (*n*).

The Australian Soil Resource Information Systems (ASRIS) (Johnston et al., 2003) can be considered as an example of the proposed SSINFOS, since it associates soil data layers with two scorpan variables used for soil prediction, namely a relief map derived from a digital elevation model (DEM) and associated environmental variables (*r*), and a lithology map (*p*). Looser associations could also be made in the framework of existing regional or national spatial infrastructure.

1.3.2 Digital Soil Mapper

A DSMAP includes a numerical form of the knowledge required to infer new soil data from the one already available in the current SSINFOS.

Three components are identified (Fig. 1.3):

1. A function database that consists in a set of spatial and non-spatial functions for predicting soil classes and attributes.
2. A function organiser that collects, arranges and categorises the functions with respect to different criteria. Some of these criteria are generic in the sense that they can be applied within any inference system. This concerns, for example, the required nature and amount of inputs, the nature of outputs and the uncertainty that is associated with the function (McBratney et al., 2002). Some other criteria could be more specific to the DSMAP. For example, we propose that the organiser associate the soil prediction functions to mapping units following McBratney et al. (2002) who proposed to associate PTFs with the soil types from which they were generated. This allows us to account for the complexity of the soil cover over the targeted extents, allowing variations of the prediction functions according to the pedological context.
3. A predictor, which consists of an inference engine that successively selects and activates the soil prediction functions according to a user request and to the criteria attached to each function. As in the case of the earlier soil inference system proposed by McBratney et al. (2002), the succession of functions attempts to minimise prediction uncertainty, that is to provide the best possible soil map. This can be performed by a set of logical rules that can simply be a collection of "if-then" statements or more likely based on probabilistic Bayesian inference. This can also be a more interactive tool in

which users play an active role in selecting themselves ad hoc prediction functions from their own knowledge.

Such a DSMAP provides the possibility of exploiting any new data which are added to the SSINFOS in a given study area of interest, that is a new scorpan layer, or a set of soil observations provided by a user or by the spatial data infrastructure. As these new data are integrated in SSINFO, DSMAP adds progressively to SSINFO more precise digitised soil maps and images of soil properties that will progressively replace the former ones. In that, spatial soil inference systems (SSINFERS), that is the association of a SSINFOS and a DSMAP, must be understood as a response to the challenge of spatially variable and temporally evolving datasets discussed in the previous section. It also provides a way for describing unmapped soil patterns within complex soil mapping units by means of the set of functions attached to the mapping unit which translate the spatial relations between soil and its forming factors.

In the following section, we detail how the DSMAP could work and what might be the existing procedures that would be integrated in the database functions.

1.4 Various kinds of inference and their associated functions

Figure 1.4 illustrates different inferences that can occur within the DSMAP with different functions. Two basic components can be distinguished: (i) a 'scorpan' component that produces soil class maps and maps of soil properties from soil observations and scorpan layers and (ii) an attribute component that derives new properties from these previously produced outputs.

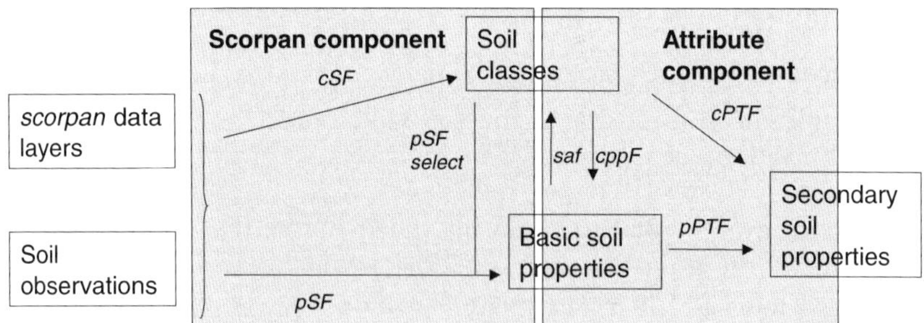

Figure 1.4. *The different inferences within the spatial soil inference system. cSF: class scorpan function; pSF: property scorpan functions; saF: soil allocation function; cppF: class-to-primary properties functions; cPTF: class pedotransfer function; pPTF: property pedotransfert function.*

1.4.1 The scorpan component

A detailed inventory of the scorpan functions that are included in this component has been recently presented by McBratney et al. (2003). All the functions that are used in the spatial inference component refer to the scorpan model, a generalisation and a formalisation of the old Jenny model (1941) that has begun to emerge in number of papers published lately. A general formulation of these scorpan functions was proposed by McBratney et al. (2003) through the equation $S = f(Q)+e$ where S stands for soil class or soil attribute, Q is the scorpan predictor variables included in the auxiliary database (see Section 3.1) and e is the prediction error. The general approach for establishing these functions is to take m observations of S in the field at known locations $[x,y]$ and relate them with some kind of function to a set of pedologically meaningful predictor variables Q which will generally be a set of variables or data layers of size M at locations $[X,Y]$ from the auxiliary database with the $[x,y] \subset [X,Y]$. Once the model is fitted at the m observation points, the predictions can be extended to the M points or cells in the raster layer thereby giving a digital map. The efficiency of the method relies on the fact that hopefully $m \ll M$ and because S is much more difficult and expensive to measure than the Q.

In the context of the proposed SSINFERS, the above-described scorpan functions could be established in training areas in which a given data configuration is available and then applied to other areas as soon as they match the required data configuration. Early examples of the use of such scorpan functions have been already proposed (Favrot and Lagacherie, 1993; Lagacherie et al., 1995) in the framework of the reference area method (Favrot, 1981). This must be developed within a more extensive and generic framework.

Two types of scorpan functions must be distinguished according to the nature of their output (Fig. 1.4): class scorpan functions (cSF), which produce a soil map, and property scorpan functions (pSF), which produce a map (or image) of a continuous soil property. The required output will obviously determine which type of function must be applied. If digitised soil maps are either lacking in SSINFO or not precise enough, cSF will also be applied whatever the required output for first determining the mapping unit that the organiser uses as criteria for selecting a scorpan function (the csf-then-psF-select route in Fig. 1.3). In situations of complex and multi-scaled soil patterns as discussed in Section 1.2, nested delineations of soil classes can be undertaken, involving a succession of several cSF at increasing extents. An alternative to this is to use scale-decomposition functions (Zhu et al., 2004) that can be also stored in the function database.

A lot of cSF and pSF exist in the literature. They can strongly differ according to the inputs they require. They can include scorpan functions using digital layers of environmental variables only for the production of landscape maps, geostatistical functions involving dense sampling of soil observations, functions using intermediate data configurations such as sparse sampling of soil observations, local expert knowledge and existing digitised soil maps. Apart from their inputs, the scorpan functions differ also in the forms they take. Readers are referred to McBratney et al. (2003, Table 5) for a documented and exhaustive list of scorpan functions as well as a description of various methods for describing the functions (McBratney et al., 2003, Table 3).

1.4.2 The attribute component

The attribute component includes the functions that infer new attributes, that is class labels or property values, from the set of attributes provided by either field observations or the scorpan component. Attribute functions have to be distinguished according to their role in the inference system and the nature of their input and output. Three main groups of functions are elucidated further: class-to-primary property functions (cppF), soil allocation functions (saF) and PTFs (cPTF and pPTF).

Class-to-primary properties functions

The cppF's aim to describe the content of pre-defined soil classes with respect to the primary soil properties, that is those determined classically by soil observations. The most current approach for building such function is design-based estimation (Brus and De Gruijter, 1997) by performing additional sampling of primary properties within the soil class. The outputs are (i) an estimate of the mean values of the property of interest over the soil class, (ii) a statistical description of the within-class variability that can be more or less detailed according to the sampling strategy (Chapter 22). In situations where additional sampling is not possible, Cazemier et al. (2001) proposed an alternative involving possibility distributions.

The cppF's can also be considered as a makeshift solution for describing unmapped soil patterns that could occur at short distance as described in the previous section.

Soil allocation functions

The SaF's aim to allocate soil individuals to pre-existing soil classes using a set of soil properties that can be provided by either field observation or a scorpan estimate. This is useful in situations where the soil map that would provide the soil class is lacking and where the soil class is required to apply class-to-secondary property PTFs (Fig. 1.3) or scorpan functions using a soil classification (Chapter 22). Several techniques have been proposed for performing such allocation, for example crisp allocation rules (Baille et al., 1988; Galbraith et al., 1998; Falipou and Legros, 2004), fuzzy k-means with extragrades (McBratney, 1994; Mazaheri et al., 1995) or fuzzy pattern matching (Lagacherie, 2005). A comparison of saF is given in Chapter 31.

Pedotransfer functions (cPTF and pPTF)

Pedotransfer functions (Bouma, 1989) aim to predict hard-to-measure soil properties, which are required by the soil data user, from primary soil properties. They have become a 'white-hot' topic in the area of soil science and environmental research. Reviews on the development and the use of PTFs, particularly for predicting soil hydraulic properties, have been given by Rawls et al. (1991), Wösten (1997), Pachepsky et al. (1999), Wösten et al. (2001) and McBratney et al. (2002). Wösten (1997) recognised two types of PTFs based on the amount of available information, namely class and continuous PTFs (Fig. 1.4). Class PTFs predict certain soil properties based on the class (textural, horizon, etc.) to which the soil sample belongs. Continuous PTFs predict certain soil properties as a continuous function of one or more measured variables. McBratney et al. (2002) proposed a more detailed classification that accounts for the crisp/fuzzy nature of the inputs and outputs. The amount of existing PTFs led McBratney et al. (2002) to propose a soil inference system for optimising their use.

1.5 The way forward

In proposing the framework presented above, we have attempted to demonstrate that the large and apparently heteroclite body of research work that has been carried out for the last quarter of a century in the areas of pedometric and digital soil mapping can converge towards an operational tool that could greatly accelerate soil mapping all over the world. To cement this hope, there are still some important issues that must be addressed. The major ones correspond to three questions and form the basis of much of the discussion presented in the chapters hereafter.

1.5.1 Which input data?

The proposed soil inference system uses two different types of data, namely soil observations, for example profile descriptions, chemical analysis, field measurements and detailed soil survey, and environmental variables, for example DTM and remote sensing. These two types of data have very different status, and thus research problems are also different. Soil observations are scarce and costly, which means that a careful selection of the observed sites is necessary to reduce the cost of digital soil mapping. Therefore, the definition of adequate sampling strategies appears of great importance for the future of digital soil mapping. A discussion about the possible sampling strategies in digital soil mapping is available in McBratney et al. (2003).

McBratney et al. (2003) present in detail the environmental variables that are used in digital soil mapping. They correspond to six of the seven factors or sets of variables in the *scorpan* model, which is an actualised version of the Jenny's model. (The seventh factor, *s*, is mainly a statement about soil properties or classes being able to be predicted from other soil properties or classes or from prior predictions at the same locations.) The aim is to obtain information on all of these factors. It will be a matter of convenience (particularly access to data sources) and scientific contention which variables are used to represent the factors. Indeed, this is an area that has not been studied well enough. The creation of these digital maps of the input environmental variables representing the six factors in the *scorpan* model is seen as an integral part of the digital soil resource assessment approach and a very valuable, environmentally useful, by-product of the new approach. McBratney et al. (2003) reviewed recently the work that has been done in this field.

1.5.2 *How to select the best inference?*

As shown in Figure 1.4, many pathways are possible for a given output. Furthermore within each type of function, represented by an arrow in Figure 1.3, there are many alternatives, for example according to the inputs or to the form of the functions. The organiser (Fig. 1.3) would therefore play a crucial role in selecting the best possible inference . Unfortunately, sufficient and structured expertise is not available that can be put in the organiser for determining a priori, given a study area and the input data, which will be the best inference. However, the criterion of minimisation of error proposed by McBratney et al (2002) could be applied to assess the quality of the output of each possible inference. Further research on uncertainty representation and error propagation is then an important aspect of this objective. Heuvelink (1998) summarised the concepts and main results on this subject.

Although much work has been done in the last 20 years on the different functions that are involved in soil inference system, marginal progress has been made in building functions or chains of functions that better integrate soil surveyor knowledge and numerical procedures. The state of the art of such techniques is provided in Chapter 22.

1.5.3 *How best to represent the results of digital soil mapping?*

The proposed addition of an inference component to the current soil information systems leads to revisit the problem of user interfaces. Although such interface already exists to facilitate soil data diffusion (Rossiter, 2004), progress must be made to help users making a decision, especially in uncertainty visualisation and 3D representation. Examples of progress in this area are given in Chapters 42 and 43.

Somewhat naïvely perhaps, we believe that digital soil mapping will help society to climb the soil data, information, knowledge and wisdom ladder.

References

Astle, W.L., Webster, R., Lawrance, C.J., 1969. Land classification for management planning in the Luangwa valley of Zambia. J. Appl. Ecol. 6, 143–169.

Baille, M., Bourrelly, L., Lagacherie, P., 1988. Modélisation de la connaissance du pédologue: application à la reconnaissance d'unités de sol, applications de l'intelligence artificielle à l'agriculture, l'agrochimie et aux industries alimentaires, Caen, 54–73.

Barnes, E.M., Sudduth, K.A., Hummel, J.W., Lesch, S.M., Corwin, D.L., Yang, C.H., Daughtry, C.S.T., Bausch, W.C., 2003. Remote- and ground-based sensor techniques to map soil properties. Photogramm. Eng. Remote Sens. 69, 619–630.

Barrett, L.R., 2002. Spectrophotometric color measurement in situ in well drained sandy soils. Geoderma 108, 49–77.

Beckett, P.H.T., Webster, R., 1971. Soil variability: a review. Soils Fertilizers 34, 1–15.

Bishop, T.F.A., McBratney, A.B., 2001. A comparison of prediction methods for the creation of field-extent soil property maps. Geoderma 103, 149–160.

Bishop, T.F.A., McBratney, A.B., Whelan, B.M., 2001. Measuring the quality of digital soil maps using information criteria. Geoderma 105, 93–111.

Bodin, F., Morlat, R., 2003. Characterizing a vine terror by combining a pedological field model and a survey of the vine growers in the Anjou region (France). J. Int. des Sciences de la Vigne et du Vin 37, 199–211.

Boulet, R., Chauvel, A., Humbel, F.X., Lucas, Y., 1982. Analyse structurale et cartographie en pédologie. Les cahiers de l'ORSTOM, série pédologie 19, 309–351.

Brabant, P., 1989. La cartographie des sols dans les régions tropicales: Une procédure à 5 niveaux coordonnés. Science du sol 27, 369–394.

Brus, D.J., De Gruijter, J.J., 1997. Random sampling or geostatistical modelling. Choosing between design-based and model-based sampling strategies for soil. Geoderma 80, 1–59 with Discussion.

Burkart, M.R., Gassman, P.W., Moorman, T.B., et al., 1999. Estimating atrazine leaching in the Midwest. J. Am. Water Resour. Assoc. 35, 1089–1100.

Burrough, P.A., 1993. Soil variability: a late 20th century view. Soils Fertilizers 56, 529–562.

Burrough, P.A., Frank, A., 1995. Concepts and paradigms in spatial information: are current geographical information systems truly generic? Int. J. Geogr. Inf. Sci. 9, 101–116.

Butler, B.E., 1982. A new system for soil studies. J. Soil Sci. 33, 581–595.

Cam, C., Vital, P., Fort, J.L., Lagacherie, P., Morlat, R., 2003. Un zonage viticole appliqué, basé sur la méthode des secteurs de référence, en vignoble de Cognac (France). Etude et Gestion des sols 10 (1), 35–42.

Cazemier, D.R., Lagacherie, P., Martin-Clouaire, R., 2001. A possibility theory approach for estimating available water capacity from imprecise information contained in soil databases. Geoderma 103, 115–134.

Coulibaly, L., Labib, M.E., Hazen, R., 2004. A GIS-based multimedia watershed model: development and application. Chemosphere 55, 1067–1080.

De Bruin, S., Wielemaker, W.G., Molenaar, M., 1999. Formalisation of soil-landscape knowledge through interactive hierarchical disaggregation. Geoderma 91, 151–172.

Dercon, G., Deckers, J., Govers, G., Poesen, J., Sanchez, H., Vanegas, R., Ramirez, M., Loaiza, G., 2003. Spatial variability in soil properties on slow-forming terraces in the Andes region of Ecuador. Soil Tillage Res. 72 (1), 31–41.

Falipou, P., Legros, J.P., 2004. Classol. Un Système expert pour aider à classer les sols dans le Référentiel pédologique. Etudes et Gestion des Sols 11, 285–298.

FAO, 1976. A Framework for Land Evaluation. FAO Soils Bulletin 32, Rome.

Favrot, J.C., 1981. Pour une approche raisonnée du drainage agricole en France : la méthode des secteurs de référence. C.R. académie d'agriculture de France, séance du 6 mai 1981, 716–723.

Favrot, J.C., Lagacherie, P., 1993. La cartographie automatisee des sols: une aide a la gestion ecologique des paysages ruraux. Comptes Rendus de L'Academie d'Agriculture de France 79, 61–76.

Fischer, G.W., Antoine, J., 1994. Agro-ecological land resources assessment for agricultural development planning. A case study of Kenya. Making land use choices for district planning. FAO Soils Bulletin 71, Food and Agriculture Organization of the United Nations and International Institute for Applied Systems Analysis, Rome.

Fridland, V.M., 1972. Pattern of the Soil Cover. Israel Program for Scientific Translations, Jerusalem.

Galbraith, J.M., Bryant, R.B., Ahrens, R.J., 1998. An expert system for soil taxonomy. Soil Sci. 163 (9), 749–758.

Gaultier, J.P., Legros, J.P., Bornand, M., King, D., Favrot, J.C., Hardy, R., 1993. L'organisation et la gestion des données pédologiques spatialisées: Le projet DONESOL. Revue de Géomatique 3, 235–253.

Heuvelink, G.B.M., 1998. Error Propagation in Environmental Modelling with GIS. Taylor & Francis, London.

Johnston, R.M., Barry, S.J., Bleys, E., et al., 2003. ASRIS: the database. Aust. J. Soil Res. 41, 1021–1036.

King, D., Daroussin, J., Tavernier, R., 1994. Development of a soil geographic database from the soil map of the European Communities. Catena 21, 37–56.

King, D., Jamagne, M., Arrouays, D., Bornand, D., Favrot, J.C., Hardy, R., Le Bas, C., Stengel, P., 1999. Inventaire cartographique et surveillance des sols en France. Etat d'advancement et exemples d'utilisation. Étude et Gestion des Sols 6, 215–228.

Kok, B., van Loenen, B., 2004. How to assess the success of National Spatial data infrastructure? Comput. Environ. Urban Systems 29 (6), 699–717.

Lagacherie, P., 2005. An algorithm for fuzzy pattern matching to allocate soil individuals to pre-existing soil classes. Geoderma 128, 274–288.

Lagacherie, P., Andrieux, P., Bouzigues, R., 1994. Fuzziness and uncertainty of soil boundaries: from reality to coding in GIS. In: P.A. Burrough and A. Frank (Eds.), Spatial Conceptual Models for Geographic Objects with Undetermined Boundaries. Taylor & Francis, Baden, Austria, pp. 275–287.

Lagacherie, P., Cazemier, D.R., Martin-Clouaire, R., Wassenaar, T., 2000. A spatial approach using imprecise soil data for modelling crop yields over vast areas. Agric. Ecosyst. Environ. 81 (1), 5–16.

Lagacherie, P., Legros, J.P., Burrough, P.A., 1995. A soil survey procedure using the knowledge on soil pattern of a previously mapped reference area. Geoderma 65, 283–301.

Lim, K.J., Engel, B.A., 2003. Extension and enhancement of national agricultural pesticide risk analysis (NAPRA) WWW decision support system to include nutrients. Comput. Electron. Agric. 38, 227–236.

Lösel, G., 2003. Application of heterogeneity indices to coarse-scale soil maps. Abstracts, Pedometrics 2003, International Conference of the IUSS Working Group on Pedometrics, Reading University, Reading, England, September 11–12, 2003.

Malerba, D., Exposito, F., Lanza, A., Lisi, F.A., Appice, A., 2002. Empowering a GIS with inductive learning capabilities: the case of INGENS. Comput. Environ. Urban Systems 27, 265–281.

Masser, I., 1999. All shapes and sizes: the first generation of national spatial data infrastructures. Int. J. Geogr. Inf. Sci. 13, 67–84.

Mazaheri, S.A., Koppi, A.J., McBratney, A.B., 1995. A fuzzy allocation scheme for the Australian Great Soil Groups Classification System. Eur. J. Soil Sci. 46, 601–612.
McBratney, A.B., 1994. Allocation of new individuals to continuous soil classes. Aust. J. Soil Res. 32, 623–633.
McBratney, A.B., De Gruijter, J.J., 1992. A continuous approach to soil classification by modified fuzzy-k-means with extragrades. J. Soil Sci. 43, 159–175.
McBratney, A.B., Mendonça Santos, M.L., Minasny, B., 2003. On digital soil mapping. Geoderma 117, 3–52.
McBratney, A.B., Minasny, B., Cattle, S., Vervoort, R.W., 2002. From pedotransfer functions to soil inference systems. Geoderma 109, 41–73.
McBratney, A.B., Pringle, M.J., 1999. Estimating average and proportional variograms of soil properties and their potential use in precision agriculture. Precision Agric. 1, 219–236.
McBratney, A.B., Webster, R., McLaren, R.G., Spiers, R.B., 1982. Regional variation of extractable copper and cobalt in the topsoil of south-east Scotland. Agronomie 2 (10), 969–982.
Mennis, J.L., 2002. Derivation and implementation of a semantic GIS data model informed by principles of cognition. Comput. Environ. Urban Systems 27, 455–479.
Mermut, A.R., Eswaran, H., 2000. Some major developments in soil science since the mid-1960s. Geoderma 100 (3–4), 403–426.
Moore, G.E., 1965. Cramming more components onto integrated circuits. Electronics 38, 114–117.
Moore, I.D., Gessler, P.E., Nielsen, G.A., Peterson, G.A., 1993. Soil attribute prediction using terrain analysis. Soil Sci. Soc. Am. J. 57, 443–452.
Moran, C.J., Bui, E.N., 2002. Spatial data mining for enhanced soil map modelling. Int. Geogr. Inf. Sci. 16, 533–549.
Nachtergaele, F., van Ranst, E., 2002. Qualitative and quantitative aspects of soil databases in tropical countries. In: G. Stoops (Ed.), Evolution of Tropical Soil Science: Past and Future. Koninklijke Academie voor Overzee Wetenschappen, Brussel, pp. 107–126.
Oldeman, L.R., van Engelen, V.W.P., 1993. A world soils and terrain digital database (SOTER) – An improved assessment of land resources. Geoderma 60, 309–325.
Pachepsky, Y.A., Timlin, D.J., Ahuja, L.R., 1999. The current status of pedotransfer functions: their accuracy, reliability and utility in field- and regional-scale modeling. In: D.L. Corwin, K. Loage, and T.R. Ellsworth (Eds.), Assessment of Non-Point Source Pollution in Vadose Zone-Geophysical Monograph, Vol. 108. American Geophysical Union, Washington, DC, pp. 223–234.
Rawls, W.J., Gish, T.J., Brakensiek, D.L., 1991. Estimating soil water retention from soil physical properties and characteristics. Adv. Soil Sci. 16, 213–234.
Rossiter, D.G., 2004. Digital soil resource inventories: status and prospects. Soil Use & Management 20 (3), 296–301.
Scull, P., Franklin, J., Chadwick, O.A., McArthur, D., 2003. Predictive soil mapping: a review. Progr. Phys. Geogr. 27, 171–197.
Shepherd, K.D., Walsh, M.G., 2002. Development of reflectance spectral libraries for characterization of soil properties. Soil Sci. Soc. Am. J. 66, 988–998.
Skidmore, A.K., Ryan, P.J., Dawes, W., Short, D., O'Loughlin, E., 1991. Use of an expert system to map forest soils from a geographical information system. International J. Geogr. Inf. Sci. 5, 431–445.
Soil Survey Staff, 1993. Soil Survey Manual. USDA, Washington, D.C Handbook No 18.
Tabbagh, A., Dabas, M., Hesse, A., et al., 2000. Soil resistivity: a non-invasive tool to map soil structure horizonation. Geoderma 97, 393–404.
USDA, 1951. Soil Survey Manual. U.S. Dept. of Agriculture. Handbook 18. U.S. Govt. Washington, DC. 503 pp.
Van Oosterom, P., Shenkelaars, V., 1995. The development of an interactive multiscale GIS. Int. J. Geogr. Inf. Sci. 9, 489–507.
Wilding, L.P., Drees, L.R., 1983. Spatial Variability in Pedology in Pedogenesis and Soil Taxonomy. 1 – Concepts and Interactions. Elsevier, Amsterdam.

Wösten, J.H.M., 1997. Pedotransfer functions to evaluate soil quality. In: E.G. Gregorich and M.R. Carter (Eds.), Soil Quality for Crop Production and Ecosystem Health Developments in Soil Science, Vol. 25. Elsevier, Amsterdam, pp. 221–245.

Wösten, J.H.M., Pachepsky, Y.A., Rawls, W.J., 2001. Pedotransfer functions: bridging gap between available basic soil data and missing soil hydraulic characteristics. J. Hydrol. 251 (3–4), 123–150.

Zhu, J., Morgan, C.L.S., Norman, J.M., Yue, W., Lowery, B., 2004. Combined mapping of soil properties using a multi-scale tree-structured spatial model. Geoderma 118, 321–334.

B. Digital soil mapping: current state and perspectives

As digital soil mapping is now moving inexorably from the research phase to the effective production of soil maps, it is destined to play a great role in the development of current and future soil spatial information systems. The way digital soil mapping will be integrated into current soil inventory and soil data acquisition programs has thus to be carefully addressed to ensure an effective benefit to the users. It seems obvious that no single and ideal way can be proposed because the current state of soil data is strongly influenced by the pedological, historical and economic particularities of each region of the world. Indeed, these regions exhibit great differences in their development of digital soil mapping. This section aims to illustrate this by showing four examples of past and current development of Soil Information Systems at the national or regional scale in which digital soil mapping is at different stages of development.

Chapter 2 presents a review of digital soil mapping in Australia where there is a rich tradition of research in this area. DSM approaches was recognised early as a relevant alternative to classical soil inventory programs because of the large size of Australia (relative to its population). In this region of the world, digital soil mapping has truly come of age and is being implemented on a wide scale by land-resource agencies with substantial economical benefits.

Chapter 3 presents the state of the art of soil mapping in Brazil and discusses the future possibilities of digital soil mapping there. In this continental-sized developing country where survey programs have been strongly restrained by government budgets, digital soil mapping is viewed as a credible alternative, although there have been no attempts so far to prove its relevance. The development of spatial information system(s) to store and facilitate the use of the existing soil and environmental data is viewed as a prerequisite. The need for a research and development consortium is stressed in order to overcome organisational problems in undertaking Digital Soil Mapping.

Chapter 4 presents the Soil Geographical Database of Eurasia as the product of the collective expertise of European soil surveyors. In this region of the world where there is a long tradition of environmental studies and soil surveying, digital soil mapping is seen more as a complementary tool to enrich the existing databases and ameliorate their precision by means of a large recourse of local soil data and soil-surveyor knowledge in association with newly available spatial data on soil covariates.

Chapter 5 illustrates this strategy for the particular case of Finland where an ongoing digital soil mapping project aims to produce a 1:250,000 scale georeferenced soil database. It describes a straightforward digital soil mapping approach that exploits a large set of previously collected soil data in association with spatial data on relevant environmental variables. The approach is favoured by relatively simple and well-known soil variations at the national level.

Developments in Soil Science, volume 31
P. Lagacherie, A.B. McBratney and M. Voltz (Editors)

Chapter 2

A REVIEW OF DIGITAL SOIL MAPPING IN AUSTRALIA

E. Bui

Abstract

Australia has a rich tradition of research in methods for land resource assessment; thus, not surprisingly, Australian scientists have been amongst the pioneers of digital soil mapping (DSM) and its underpinning technologies, which have been evolving since the late 1960s. In particular, Australian scientists have been at the forefront of development of methods for terrain analysis and predictive models for soil distribution. This chapter reviews briefly the historical development of DSM and its supporting technologies in Australia in the 1990s and considers the delivery and economics of DSM.

2.1 Introduction

Digital soil mapping (DSM) is the computer-based operationalization of ideas for predicting soil distribution in landscapes that have evolved since the beginning of soil survey (Hewitt, 1993; Lagacherie et al., 1995; McBratney et al., 2003; Scull et al., 2003). It has also been called 'predictive soil modelling' (Hewitt, 1993; Scull et al., 2003) and 'quantitative soil survey' by McKenzie and Ryan (1999). McBratney et al. (2003) argue that it constitutes a shift from the soil-landscape paradigm (Hudson, 1992) that heralds a new paradigm.

DSM relies on advances in computing and information processing that have occurred over the last 30 years: standardised data entry forms, relational databases, geographical information systems (GIS), digital elevation models (DEM) and remote sensing. Faster, more powerful computers have been critical as the datasets involved are very large – many gigabytes or even terabytes in size. These advances have made DSM feasible for soil survey organizations only recently – for most, the last 10 years or less. The reviews by McBratney et al. (2003) and Lagacherie and McBratney (see Chapter 1) cover the development of technology that has enabled the advance of digital mapping around the world.

In this chapter, I will outline the recent historical development of DSM and its supporting technologies in Australia. Previous reviews of soil survey methods in Australia include Taylor (1970), Beckett and Bie (1978), Gibbons (1983) and McKenzie (1991).

2.2 The progression of DSM in Australia

Since the late 1960s, Australian soil scientists have realised that the advent of computers was set to revolutionise soil survey (Stewart, 1968; Moore et al., 1981). Uptake of digital technology started with a move to digitise hard copies of existing soil maps; the creation of geo-referenced databases for new surveys; use of GIS to store data layers and to produce new maps; for details, refer to individual papers in Moore et al. (1981). GIS have now become standard tools in soil survey agencies.

2.2.1 Pre-requisites for DSM

Nevertheless, GIS are not the only or even the major requirement for DSM. More important is the existence of spatial analysis software used to generate data to infer factors of soil formation, tools such as ANUDEM (Hutchinson, 1989; 1996) to derive hydrologically sound DEMs and TAPES-G to derive terrain attributes from DEMs (Gallant and Wilson, 1996; Wilson and Gallant, 2000). To a large degree, I think that it is the development of DEMs and derived terrain attributes that has enabled the application of DSM (e.g. Moore et al., 1993; Odeh et al., 1994). Nevertheless, environmental factors beyond DEMs and terrain analysis have been represented, especially climate and parent material. The representation of climate surfaces has relied on weather monitoring data, DEMs and software such as ANUCLIM to derive climatic surfaces (Houlder et al., 1999). The representation of parent material has relied on re-interpretation of geological data to generate lithology classes that account for likely texture and chemical weathering during pedogenesis (Johnston et al., 2003). Digital remote-sensing data such as bands from the Landsat Thematic Mapper can be used to map vegetation (Lees and Ritman, 1991) and are commonly used by many soil survey agencies even if it is only for the base map (thus in visual interpretation rather than as predictors in modelling).

Nevertheless, lingering limitations in available data for use in DSM in small- to medium-scale soil surveys remain. The price of remotely sensed data is often a limitation when large areas are involved. The extent of high-resolution DEMs (80 m or less) to match the resolution of remote-sensing imagery is limited. Likewise, the coverage of airborne geophysical data which can be useful for DSM (Chapter 16) is patchy across the continent. Currently available geo physical datasets for Australia are best viewed from http://www.ga.gov.au/news/index.jsp#index or searched on http://www-b.ga.gov.au/asdd/tech/zap/basic.html.

One of the underpinning technologies for DSM is the existence of relational databases. These were developed for soil data in Australia in the 1980s–mid-1990s: Queensland's WARIS (Rosenthal et al., 1986) and other State databases; a national standard for database design and exchange protocols that could run on personal computers was developed as PC SITES, Soil Information Transfer and Evaluation System (ACLEP, 1997).

Other critical tools for DSM are the methods and computer-aided models for predicting soil distribution in landscapes. Statistical, including geostatistical, methods have been used to make predictive models of soil properties (e.g. McKenzie and Austin, 1993; Odeh et al., 1994; Gessler et al., 1995; McKenzie and Ryan, 1999). Recently, data-driven machine-learning methods that rely on induction have been applied: for example, Minasny and McBratney (2002) used artificial neural networks to predict soil physical properties; Bui and Moran (2001, 2003) used classification trees to predict soil and surficial geology classes; and Henderson et al. (2005) used decision trees to generate a variety of soil property maps. Certainly, the presence in Australia of Alex McBratney who has been so active in developing pedometric techniques that can be applied to DSM has been instrumental in the uptake of DSM here. Methods for DSM have recently been reviewed thoroughly by McBratney et al. (2000; 2003) and Scull et al. (2003). Personally I like rule-induction methods because they can capture and represent the way people learn from observations in the field.

A consequence of predictive modelling is that it generates an explicitly stated model of the distribution of soils in landscapes. Thus, it remedies one often-cited shortcoming of traditional surveys – the failure to capture the tacit mental model of the surveyors (Bui et al., 1999; McKenzie and Ryan, 1999; Hewitt, 1993). Nevertheless, I admit that large tree models or artificial neural networks can be difficult to interpret.

Quantitative models include a quantifiable level of uncertainty, in terms of model statistics. For example, Gessler et al. (1995) express the probability of occurrence of E-horizons, developed by statistical modelling, as a map. Bagging, which combines the results of several predictions (Breiman, 1996), allows an estimate of the consistency of predictions which is akin to an uncertainty estimate (Bui and Moran, 2003). The use of GIS to render the predictive models into maps then makes it easy to estimate spatial variability quantitatively. For example, in the digital version of the soil-landform map of the Murray–Darling Basin (MDB), there is an estimate of the area proportion of dominant soil types, which is calculated for each map unit and each polygon (Bui, 1998).

2.2.2 DSM comes of age

Australian researchers have been amongst the leaders of DSM. To some extent, this is not surprising since Australia has had a long history of research in land resource assessment (see Stewart, 1968; Beckett and Bie, 1978), quantitative geomorphology (Speight, 1983), remotely sensed image processing for natural resource management (Laut et al., 1977; Graetz et al., 1980) and the development of electronic soil information systems (Moore et al., 1981). Much of the work reported in the collections by Stewart (1968) and Moore et al. (1981) presages DSM. Thus, the leading role of Australians is not a fortuitous accident due to the current presence of key researchers in Australia (e.g. Alex McBratney, Mike Hutchinson, Neil McKenzie, John Gallant; even Peter Burrough did a stint in Australia early in his career).

Throughout the 1980s and 1990s Ph.D. students have been instrumental in advancing DSM (Fig. 2.1); notably Inakwu Odeh who was awarded his Ph.D. in

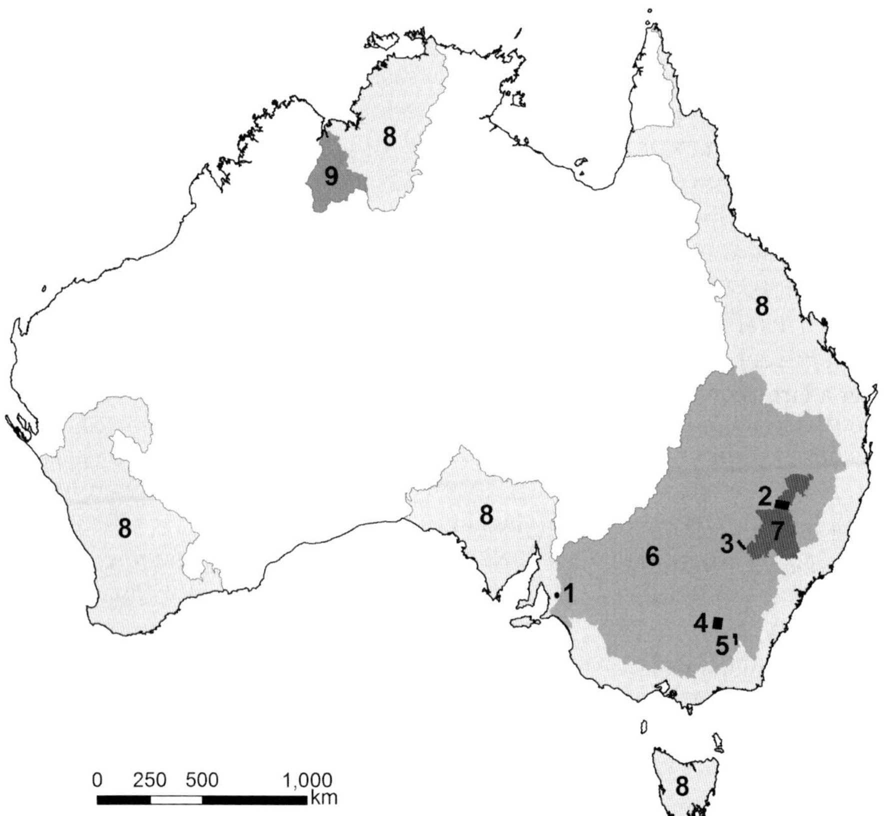

Figure 2.1. *Map of Australia showing location of soil surveys cited. The digital Atlas of Australian Soils (Leahy, 1993; BRS, 2000) covers the whole continent at 1:2 map. 1, Mt. Lofty Ranges (Odeh, 1990); 2, Edgeroi (McGarry et al., 1989; Triantafilis and McBratney, 1993); 3, Trangie (McKenzie, 1992); 4, Wagga Wagga (Gessler, 1996); 5, Bago-Maragle forest (O'Connell, 1998); 6, MDBSIS; 7, southern brigalow belt (Soil Information Systems Unit, 2002); 8, ASRIS; 9, Ord (Chapter 15).*

1990 derived several papers from it subsequently (e.g. Odeh et al., 1992, 1994, 1995). Neil McKenzie used environmental correlation as a guiding principle and was the first to implement Generalised Linear Models for soil property prediction – he was guided by Mike Austin and Henry Nix from the then CSIRO Division of Land Use Research (McKenzie, 1992; McKenzie and Austin, 1993). Paul Gessler received his Ph.D. in 1996 while publishing several classic papers along the way (e.g. Moore et al., 1993; Gessler et al., 1995). For his part, John Triantafilis mapped soil salinity in the Namoi valley, building on the earlier gridded sampling design of McGarry et al. (1989) (e.g. Triantafilis and McBratney, 1993; Triantafilis, 1996; Triantafilis et al., 2001). O'Connell (1998) applied DSM to forest soils (O'Connell et al. 2000). More recent Ph.D. students have worked at the very detailed scale for applications of DSM in precision agriculture (Viscarra-Rossel, 2001; Bishop, 2002).

Many of the pioneering Australian efforts that have led to DSM were incubated at the CSIRO. This organization has had a history of continual re-structuring over its 78 years of existence (CSIRO, 2001) – there have been many incarnations of the current CSIRO Land and Water – and under these circumstances, inevitably, there is re-focusing of effort and a lack of continuity in lines of research. Neil McKenzie remains the only link to the earlier work done by the CSIRO. Therefore in my opinion, the recent emergence of DSM has been borne of a confluence of technology and necessity more than anything – Australia being a large continent with a small population and an economy that still relies heavily on its natural resources. Traditional methods of soil survey take a long time; they are labour-intensive and therefore expensive. To speed up soil survey in this context, one has to look to computers.

For example, MDB covers 1,058,000 km^2, about 14% of Australia (Fig. 2.1). When the MDB Commission wanted a new soil map of the MDB at a scale of 1:250,000, it was willing to invest only $1 M over 3 years into the project. We had a nine-person core team, with only one experienced full-time soil surveyor. Obviously, to deliver the desired product, we had to develop a new method for rapid soil survey. How this was done has been reported in the project final report (Bui, 1998) and in a scattered series of journal papers (Bui et al., 1999; Bui and Moran, 2001; Moran and Bui, 2002) which culminated with Bui and Moran (2003). In a nutshell, we used existing soil maps as training areas to develop landscape association rules that could be extrapolated to neighbouring unmapped areas. We relied heavily on machine-learning algorithms. Luckily, the computer technology was sufficiently advanced to support the amount of data crunching involved! We were working with grids with 250-m resolution which meant that every dataset for the whole MDB was 19×10^6 pixels. Some datasets were several Gbytes. The project certainly pushed our network of Unix workstations to its limit. I may be biased but I think that it is this project more than any other that demonstrates that DSM is feasible on a large scale for soil survey organizations and paves the way forward for integrating old and new data – essentially because of the shear extent of the mapping exercise – because we investigated and used a variety of predictors to represent the factors of soil formation (Moran and Bui, 2002; Chapter 22; Chapter 23) and we mapped soil types. Before the MDBSIS project, most other studies had used predictive modelling to map only soil properties, had relied on new sampling and were much more restricted in their extent of application (Fig. 2.1) – they were still research laboratory prototypes rather than full-scale implementations of DSM.

Nevertheless, the MDBSIS project and its outgrowths, notably the Australian Soil Resources Information System (ASRIS) and Ord (Chapter 15) projects (Fig. 2.1), have highlighted several issues: (1) the fundamental need to implement structured sampling strategies that the researchers have long emphasised (e.g. Webster and Butler, 1976; McGarry et al., 1989; Odeh et al., 1990; Pettitt and McBratney, 1993; McKenzie et al., 2000a); (2) problems arising from the changing basis of classification systems (e.g. Butler, 1958; Northcote, 1979); and (3) problems with within-unit spatial variability in existing maps (e.g. Webster and Butler, 1976).

2.2.3 Technology transfer

Although the recent reviews by McBratney et al. (2000; 2003) and Scull et al. (2003) focus on the methods for predictive modelling in DSM, there is another critical aspect of DSM: the computer-based delivery of information via the Internet or digital communication technology (Chapter 41). One of the first agencies to use Web-based display was the then Victorian Department of Natural Resources and Environment. Victoria Resources on-line became live in 1997: see http://www.dpi.vic.gov.au/web/root/Domino/vro/vrosite.nsf/pages/vrohome. It includes images of maps of soil distribution, soil-landscape descriptions, photographs of representative soil profiles, a metadata directory of published map products, explanations of soil classification, information on key soil properties (pH) and soil management. Similar Web-based map metadata and information delivery is now a common practice for all State soil survey agencies. Exceptionally in 1997 – because for many agencies there were issues of intellectual property around soil maps, Geosciences Australia enabled on-line production of new maps derived from their GIS data layers by remote users (Chopra, 1997). Since then, other agencies have followed suit; see for instance, http://www.nrme.qld.gov.au/era/projects/wwwburn.html.

Surprisingly, the uptake of DSM by Australian soil survey organizations (e.g. Goldrick et al., 2001; Rampant, 2000; Slater and Grundy, 1999) has only lagged by a few years. More recently, Agriculture WA has also tried predictive modelling methods for mapping soils in the Ord River basin (Schoknecht et al., 2002;

Chapter 15). Given that DSM requires not only soil survey teams with new skills and expensive IT hardware and software but also a paradigm shift (McBratney et al., 2003), I think that the speed of uptake has been rapid. Undoubtedly, the existence of the Australian Collaborative Land Evaluation Programme (ACLEP) since 1992 has facilitated technology transfer but another factor has been serendipitous, due to the demographic profile of soil scientists at individual agencies: the presence of key young individuals trained in IT and willing to experiment, and supportive managers. The upcoming publication of the new edition of the guidelines for conducting soil surveys (McKenzie et al., 2006), known in Australian soil science circles as "the blue book", entails that DSM methods will become mainstream within a few years.

2.3 Historical highlights in DSM in Australia

The development of DSM in Australia from 1990 to 2002 is chronicled in the ACLEP Newsletter (http://www.clw.csiro.au/aclep/newsletters/newsletters.htm). Because of the time lag between the actual carriage of research and the publication of journal articles, the ACLEP Newsletter has been the best means of keeping abreast of new developments in DSM in Australia. Here I highlight some of what I see as milestones in Australia since 1990; many are the production of key datasets that enable DSM, and the review only includes those that have continental coverage.

1990–1991: Digital Atlas of Australian Soils, a digitised version of the maps comprising the Atlas of Australian Soils at a scale of 1:2,000,000 produced by Keith Northcote and colleagues over the period 1960–1968 (Leahy, 1993; BRS, 2000)

1991: 1/40th degree of latitude and longitude (~2.5 km) DEM was calculated from continent-wide coverages of point elevation and stream line data using an elevation gridding technique, which incorporates a drainage enforcement algorithm (Hutchinson and Dowling, 1991).

1991–1992: McKenzie and Hook (1992) produced a look-up table (LUT) to enable interpretations of the digital Atlas of Australian Soils which was revised in 2000 (McKenzie et al., 2000b). In 1996, Spouncer et al. (1996) generated another LUT linked to the digital Atlas of Australian Soils, for making interpretations of saline and sodic soils.

1992–1996: 18-s DEM of Australia, a gridded DEM computed from topographic information including point elevation data, elevation contours, stream lines and cliff lines. The grid spacing was 18 s in longitude and latitude (~500 m). It was first released in 1992.

1993–1996: Expector software development (Simon Cook and Rob Corner, CSIRO in WA). Expector is a software tool to aid soil surveyors in formally defining the relationships between landscape characteristics and soil properties using Bayesian causal networks (Cook et al., 1996).

1995: ACLEP workshop on 'Quantitative Land Resource Assessment'. At this influential workshop, issues pertaining to digital environmental predictors, integration of multiple data layers, survey design and sampling, spatial prediction and scale were presented by invited speakers and illustrated with computer demonstrations tailored around Brendan Mackey's GIS course at the Australian National University, Paul Gessler's data around Wagga Wagga and Simon Veitch's ASSESS software (Bowyer and Veitch, 1994).
1995: GEOIMAGE (http://www.geoimage.com.au) produced a continental ortho-rectified Landsat MSS mosaic.
1996: (1) ACLEP co-sponsored the first of an annual series of Enhanced Resource Assessment (ERA) workshops built around the Queensland ERA project (Slater and Grundy, 1999). (2) ACLEP workshop on DEMs at the Australian National University. The workshop lasted a week and consisted of lectures on theoretical and applied issues in the morning with hands-on computer laboratory exercises in the afternoons. The exercises included the generation of DEMs and the derivation of terrain attributes.
1995–1998: MDBSIS, a commission to produce a new soil map of the MDB that could be used at a scale of 1:250,000. Owing to the 3.5-year timeframe for the project, no new sampling or field mapping was possible. Too few site (point) data were available to be of use because of the lack of any Federal/State government agreements regarding access to databases – this was later addressed under the National Land and Water Resources Audit with ASRIS (see below). Owing to data licensing agreements with the various State agencies – providers of previously mapped information, it was not possible to use most of the existing soil polygons directly; therefore, MDBSIS was more than a soil correlation exercise and it required the development of an entirely new legend. The complete final report (Bui, 1998), the new soil-landforms map and metadata about the ancillary supporting datasets were made available through the Internet.
1997: ACLEP co-sponsors a workshop on digital communication technologies for land resource assessment with the then Victorian Department of Natural Resources and Environment.
1997: Release of 9-s DEM Version 1, a gridded DEM computed from topographic information including point elevation data, elevation contours, stream lines and cliff lines. The grid spacing was 9 s in longitude and latitude (~250 m).
1998–2000: 9-s DEM Version 2.
1999–2001: ASRIS, a collation of the best available information to produce for the catchments that include Australia's agricultural zone (Johnston et al., 2003):

- a national database about soil and land resources, including
  - soil profiles (point database in Oracle);
  - soil and land resources maps with interpretations from LUTs (Carlile et al., 2001a,b);
  - other relevant datasets (lithology, DEM and derived landscape descriptors, climate surfaces, Landsat MSS imagery);

- modelled spatial estimates of key soil properties (pH, organic C, bulk density, %clay, texture, thickness, saturated hydraulic conductivity, available water content, total N, total P and erodibility) and their uncertainties presented at a grid cell resolution of ~1.1 km^2 (0.01°) (Henderson et al., 2001, 2005). The maps are individually described on the Australian Natural Resources Atlas at http://www.nlwra.gov.au/atlas. ASRIS datasets are part of the National Land and Water Resources Audit Australian Natural Resources Data Library and have been made available through the Internet (http://adl.brs.gov.au/).

2002–2004: ACLEP has taken over further developments of ASRIS. ASRIS II will update the point database to 5000 soil profiles (10,000 by 2006) with full-quality assurance. A hierarchy of spatial units with soil depth, water storage, permeability, fertility and erodibility attributes will be available on the Internet and allow comprehensive reporting on land suitability and soil resources. Another update is planned for 2006. The delivery of the information will be via ArcIMS. The technical report relating to the system is available at: http://www.asris.csiro.au/methods.html.

2.4 Economics of DSM

In 1996, the benefit:cost ratio of digitizing the Atlas of Australian Soils in 1990 was estimated at more than 10, in terms of 1993 dollars, with a 7% discount over 30 years (ACIL, 1996). Given that the digital Atlas of Australian Soils continues to be used, this ratio would be bigger now.

If the MDB Commission were to pay for a traditional soil survey of the MDB, it would take 330 person-years and cost tens of millions of dollars. Factoring in the in-kind contribution of partners in the MDBSIS project, the total cost was ~\$3.3 M. This gives an estimated cost of \$3 km^{-2} but does not include the original cost of the surveys used as training data, so this is essentially the cost of value-adding.

The NSW Soil Information Systems Unit estimated that the production of a soil-landscape map for the southern brigalow belt (54,000 ha; Fig. 2.1) by conventional survey methods would have taken 21 person-years while the DSM approach cut down that time to 5 person-years (Soil Information Systems Unit, 2002). The cost of the southern brigalow belt project was estimated as \$9.35 km^{-2} (Soil Information Systems Unit, 2002), which compared favourably with an average cost of \$28.00 km^{-2} for conventional survey as estimated by McKenzie (1991).

WA Agriculture estimated the cost of the Ord mapping (Schoknecht et al., 2002; see Bui et al. in Chapter 15) by conventional methods at \$700,000 whereas the predictive modelling project was costed at \$150,000 – this has increased since more field work has been necessary. On an area basis, this compares as \$12.64 km^{-2} with \$2.71 km^{-2}.

None of the economic estimates above has factored in the infrastructure cost of the computing network and ancillary data sets, or the social cost of DSM – the need for a new generation of soil surveyors with new skills. In countries with a young population this will not be an issue but in countries with an aged work-force, it will slow down the uptake of DSM.

Regardless of the apparent cost savings of DSM, there are several other advantages of this approach that defy monetary valuation:

- it generates an explicitly stated model of the distribution of soils in landscapes;
- it is repeatable with existing data but can be updated as new data come on line and
- it includes a quantifiable, spatially variable level of uncertainty.

So I believe DSM to be potentially better in terms of quality assurance/quality control. However, until DSM and conventional survey can be directly evaluated through a field-based uncertainty analysis, it is impossible to adequately compare their costs and benefits.

2.5 Conclusion

Progress in DSM (and underpinning datasets) in Australia in the 1990s shows that it has truly come of age and can be implemented on a large scale by land resources agencies. Australia now has a national soil and land resources digital database and arrangements for its maintenance and upgrade until 2006. DSM is used commonly where the required predictor data are of sufficient resolution. The inclusion of DSM in the new edition of the guidelines for conducting soil surveys in Australia (McKenzie et al., 2006) means that it will become more firmly established over the next decade.

Acknowledgment

I thank Neil McKenzie for his critical review of the first draft of this manuscript and for supplying references documenting the early years of quantitative soil survey in Australia; and David Simon for making the figure.

References

ACIL Economics and Policy Pty Ltd., 1996. An economic framework for assessing the benefits and costs of land resource assessment in Australia. ACLEP Newsletter 5/4 (December), 1–4.

ACLEP, 1997. Soil Information Transfer and Evaluation System: Version 1.2. ACLEP Technical Report No. 5. CSIRO Land and Water, Canberra, Australia.

Beckett, P.H.T., Bie, S.W., 1978. Use of soil and land-system maps to provide soil information in Australia. CSIRO Division of Soils Technical Paper No. 33.

Bishop, T.F.A., 2002. Unpublished Ph.D. Thesis, University of Sydney, Sydney, Australia.

Bowyer, J., Veitch, S., 1994. ASSESS: A system for selecting suitable sites for a landuse. In: Proceedings OZRI 8, Hobart.

Breiman, L., 1996. Bagging predictors. Machine Learn 24, 123–140.

BRS, 2000. Digital Atlas of Australian Soils. Bureau of Rural Science: Canberra. http://www.affa.gov.au/docs/rural_science/datasets/atlas/index.html

Bui, E.N. (Ed.), 1998. A Soil Information Strategy for the Murray-Darling Basin (MDBSIS). Final Report to Murray Darling Basin Commission. http://www.affa.gov.au/content/output.cfm?ObjectID=D2C48F86-BA1A-11A1-A2200060B0A05768

Bui, E.N., Moran, C.J., 2001. Disaggregation of polygons of surficial geology and soil maps using spatial modelling and legacy data. Geoderma 103, 79–94.

Bui, E.N., Moran, C.J., 2003. A strategy for filling gaps in soil survey over large spatial extents: an example from the Murray-Darling basin of Australia. Geoderma 111, 21–44.

Bui, E.N., Loughhead, A., Corner, R., 1999. Extracting soil-landscape rules from previous soil surveys. Aust. J. Soil Res. 37, 495–508.

Butler, B.E., 1958. The diversity of concepts about soils. J. Aust. Inst. Agric. Sci. 24, 14–20.

Carlile, P., Bui, E., Moran, C., Simon, D., Henderson, B., 2001a. Method used to generate soil attribute surfaces for the Australian Soil Resource Information System using soil maps and look-up tables. CSIRO Land and Water Technical Report 24/01. http://www.clw.csiro.au/publications/technical/

Carlile, P., Bui, E, Moran, C., Minasny, B., McBratney, A.B., 2001b. Estimating soil particle size distributions and percent sand, silt and clay for six texture classes using the Australian Soil Resources Information System point database. CSIRO Land and Water Technical Report 29/01. http://www.clw.csiro.au/publications/technical/

Chopra, P., 1997. Linking a GIS to the World Wide Web–the Australian Geological Survey Organisation (AGSO) introduces on-line customisable maps. ACLEP Newsletter 6/1, 2–6.

Cook, S.E., Corner, R.J., Grealish, G., Gessler, P.E., Chartres, C.J., 1996. A rule-based system to map soil properties. Soil Sci. Soc. Am. J. 60, 1893–1900.

CSIRO, 2001. 75 years of Australian Science. CSIRO National Awareness, Dickson, ACT.

Gallant, J.C., Wilson, J.P., 1996. TAPES-G: a grid-based terrain analysis program for the environmental sciences. Comput. Geosci. 22, 713–722.

Gessler, P.E., 1996. Statistical soil-landscape modelling for environmental management. Ph.D. Thesis, The Australian National University, Canberra, ACT.

Gessler, P.E., Moore, I.D., McKenzie, N.J., Ryan, P.G., 1995. Soil-landscape modelling and spatial prediction of soil attributes. Int. J. Geogr. Inf. Systems 9, 421–432.

Gibbons, F.R. 1983. Soil mapping in Australia. In: Soils: An Australian Viewpoint. CSIRO Publishing, Melbourne, pp. 265–276.

Goldrick, G., Chapman, G.A., Simons, N.A., Milford, H.B., Murphy, C.L., McGaw, A.J.E., Edye, J.A., Macleod, A.P., 2001. New technology and soil landscape mapping in NSW. In: Proceedings of the Geospatial Information and Agriculture Symposium, Sydney.

Graetz, R.D., Gentle, M.R., O'Callaghan, J.F., Foran, B.D., 1980. The application of Landsat image data to land resource management in the arid lands of Australia. Technical memorandum 80/6, CSIRO Division of Land Resources Management, Perth, WA.

Henderson, B., Bui, E., Moran, C., Simon, D., Carlile, P., 2001. ASRIS: continental-scale soil property predictions from point data. CSIRO Land and Water Tech. Rept. 28/01. http://www.clw.csiro.au/publications/technical/

Henderson, B.L., Bui, E.N., Moran, C.J., Simon, D.A.P., 2005. Australia-wide predictions of soil properties using decision trees. Geoderma 124, 383–398.

Hewitt, A.E., 1993. Predictive modelling in soil survey. Soils Fertilizers 56, 305–314.

Houlder, D., Hutchinson, M., Nix, H., McMahon, J., 1999. ANUCLIM version 5.0. User Guide. Centre for Resource and Environmental Studies. The Australian National University, Canberra.

Hudson, B.D., 1992. The soil survey as a paradigm-based science. Soil Sci. Soc. Am. J. 56, 295–309.

Hutchinson, M.F., 1989. A new method for gridding elevation and stream line data with automatic removal of spurious pits. J. Hydrol. 106, 211–232.

Hutchinson, M.F., 1996. A locally adaptive approach to the interpolation of digital elevation models. In: Proceedings of the Third International Conference on Integrating GIS and Environmental Modeling. NCGIA, University of California, Santa Barbara.

Hutchinson, M.F., Dowling, T.I., 1991. A continental hydrological assessment of a new grid-based digital elevation model of Australia. Hydrol. Process. 5, 45–58.

Johnston, R.M., Barry, S.J., Bleys, E., Bui, E.N., Moran, C.J., Simon, D.A.P., Carlile, P., McKenzie, N.J., Henderson, B.L., Chapman, G., Imhoff, M., Maschmedt, D., Howe, D., Grose, C., Schoknecht, N., Powell, B., Grundy, M., 2003. ASRIS: The database. Aust. J. Soil Res. 41, 1021–1036.

Lagacherie, P., Legros, J.P., Burrough, P.A., 1995. A soil survey procedure using the knowledge of soil pattern established on a previously mapped reference area. Geoderma 65, 283–301.

Laut, P., Heyligers, P.C., Keig, G., Loffler, E., Margules, C., Scott, R.M., Sullivan, M.E., 1977. Environments of South Australia. CSIRO Division of Land Use Research, Canberra.

Leahy, S., 1993. Atlas of Australian soils: combined map unit descriptions. National Resources Information Centre, Parkes, ACT, Australia.

Lees, B.G., Ritman, K., 1991. Decision-tree and rule-induction approach to integration of remotely sensed and GIS data in mapping vegetation in disturbed or hilly environments. Environ. Manage. 15, 823–831.

McBratney, A.B., Odeh, I.O.A., Bishop, T.F.A., Dunbar, M.S., Shatar, T.M., 2000. An overview of pedometric techniques for use in soil survey. Geoderma 97, 293–327.

McBratney, A.B., Mendonça Santos, M.L., Minasny, B., 2003. On digital soil mapping. Geoderma 117, 3–52.

McGarry, D., Ward, W.T., McBratney, A.B., 1989. Soil studies in the Lower Namoi Valley – Methods and data 1: the Edgeroi data set. CSIRO, Division of Soils, Melbourne, Australia.

McKenzie, N.J., 1991. A strategy for coordinating soil survey and land evaluation in Australia. Divisional Report No. 114, CSIRO Division of Soils, Canberra.

McKenzie, N.J., 1992. Soils of the Lower Macquarie Valley, New South Wales. Divisional Report No. 117, CSIRO Division of Soils, Canberra.

McKenzie, N.J., Austin, M.P., 1993. A quantitative Australian approach to medium and small scale surveys based on soil stratigraphy and environmental correlation. Geoderma 57, 329–355.

McKenzie, N., Hook, J., 1992. Interpretations of the Atlas of Australian Soils. CSIRO Division of Soils Technical Report 94/1992.

McKenzie, N.J., Ryan, P.J., 1999. Spatial prediction of soil attributes using terrain analysis. Geoderma 89, 67–94.

McKenzie, N.J., Cresswell, H.P., Ryan, P.J., Grundy, M., 2000a. Contemporary land resource survey requires improvements in direct soil measurement. Commun. Soil Sci. Plant Anal. 31, 1553–1569.

McKenzie, N.J., Jacquier, D., Ashton, L.J., Cresswell, H.P., 2000b. Estimating soil properties using the Atlas of Australian Soils, Technical Report 11/00, CSIRO Land and Water, Canberra.

McKenzie, N.J., Webster, R., Grundy, M.J., Ringrose-Voase, A.J., (Eds.) 2006. Australian Soil and Land Survey Handbook: Guidelines for Conducting Surveys, 2nd edition. CSIRO Publishing, Melbourne.

Minasny, B., McBratney, A.B., 2002. The neuro-m method for fitting neural network parametric pedotransfer functions. Soil Sci. Soc. Am. J. 66, 352–361.

Moore, A.W., Cook, B.G., Lynch, L.G., 1981. Information systems for soil and related data. Proceedings of the Second Australian Meeting of the ISSS Working Group on Soil Information Systems. Canberra, Australia, 19–21 February 1980. Pudoc, Wageningen, The Netherlands.

Moore, I.D., Gessler, P.E., Nielsen, G.A., Peterson, G.A., 1993. Soil attribute prediction using terrain analysis. Soil Sci. Soc. Am. J. 57, 443–452.

Moran, C.J., Bui, E.N., 2002. Spatial data mining for enhanced soil map modelling. Int. J. Geogr. Inform. Sci. 16, 533–550.

Northcote, K.H., 1979. A Factual Key for the Recognition of Australian Soils. Rellim Technical Publishers, Glenside, SA.

O'Connell, D.A., 1998. Predicting soil and hydraulic properties in small forested catchments. Unpublished Ph.D. Thesis, Centre for Resource and Environmental Studies, Australian National University.

O'Connell, D.A., Ryan, P.J., McKenzie, N.J., Ringrose-Voase, A.J., 2000. Quantitative site and soil descriptors to improve the utility of forest soil surveys. Forest Ecol. Manage. 138, 107–122.

Odeh, I.O.A., 1990. Soil Pattern Recognition in a South Australian SubCatchment. Unpublished Ph.D. Thesis, University of Adeliade, South Australia.

Odeh, I.O.A., McBratney, A.B., Chittleborough, D.J., 1990. Design of optimal sample spacings for mapping soil using fuzzy-k-means and regionalized variable theory. Geoderma 47, 93–122.

Odeh, I.O.A., McBratney, A.B., Chittleborough, D.J., 1992. Soil pattern-recognition with fuzzy-c-means – application to classification and soil-landform interrelationships. Soil Sci. Soc. Am. J. 56, 505–516.

Odeh, I.O., McBratney, A.B., Chittleborough, D.J., 1994. Spatial prediction from landform attributes derived from a digital elevation model. Geoderma 63, 197–214.

Odeh, I.O., McBratney, A.B., Chittleborough, D.J., 1995. Further results on prediction of soil properties from terrain attributes: heterotopic cokriging and regression-kriging. Geoderma 67, 215–226.

Pettitt, A.N., McBratney, A.B., 1993. Sampling designs for estimating spatial variance-components. Appl. Stat. –J. R. Stat. Soc., C 42, 185–209.

Rampant, P.C. 2000. Spatial prediction of soil types using decision trees. Proceedings of Soil 2000: new horizons for a new century. Australian and New Zealand Second Joint Soils Conference, 3-8 December 2000, Christchurch, New Zealand 2, 239–240.

Rosenthal, K.M., Ahern, C.R., Cormack, R.S., 1986. WARIS: a computer based storage and retrieval system for soils and related data. Aust. J. Soil Res. 24, 441–456.

Schoknecht, N., Payne, A., Bui, E., Simon, D., 2002. A new approach for more effective land unit mapping in the Ord River catchment of WA. Proceedings of Australian Soil Science Society Conference, University of Western Australia, Perth, WA.

Scull, P., Franklin, J., Chadwick, O.A., McArthur, D., 2003. Predictive soil mapping: a review. Prog. Physical Geogr. 27 (2), 171–197.

Slater, B.K., Grundy, M.J., 1999. Enhanced Resource Assessment: integration of innovative technologies by agency scientists. In: Proceedings of Second International Conference on Soil Resources. University of Minnesota, St. Paul.

Soil Information Systems Unit, 2002. Soil Landscape Reconnaissance Mapping – Brigalow Belt South Stage 2. New South Wales Dept of Land and Water Conservation, Sydney, NSW.

Spouncer, L.R., Merry, R.H., Michalski, C.H., 1996. Allocation of acidity profile classes to Atlas of Australian Soils mapping units. CSIRO Division of Soils Technical Report 20/1996.

Speight, J.G., 1983. Field description of landforms for Australian soil and land surveys. Technical memorandum 83/6, 2nd revision. CSIRO Division of Water and Land Resources, Canberra.

Stewart, G.A. (Ed.) 1968. Land Evaluation. Macmillan, Melbourne.

Taylor, J.K., 1970. The Development of Soil Survey and Field Pedology in Australia, 1927–1967. CSIRO, Melbourne.

Triantafilis, J., 1996. Quantitative assessment of soil salinity in the lower Namoi valley. Unpublished Ph.D. Thesis. The University of Sydney, New South Wales, Australia.

Triantafilis, J., McBratney, A.B., 1993. Application of continuous methods of soil classification and land suitability assessment in the lower Namoi valley. CSIRO Division of Soils Divisional Rep. No. 121. p. 172. CSIRO, Melbourne, Australia.

Triantafilis, J., Ward, W.T., Odeh, I.O.A., McBratney, A.B., 2001. Creation and interpolation of continuous soil classes in the lower Namoi valley. Soil Sci. Soc. Am. J. 65, 403–413.

Viscarra-Rossel, R.A., 2001. Development of a proximal soil sensing system for the continuous management of acid soil. Unpublished Ph.D. Thesis, The University of Sydney, Sydney, Australia.

Webster, R., Butler, B.E., 1976. Soil classification and survey studies at Ginnindera. Australian Journal of Soil Research 14, 1–24.

Wilson, J.P., Gallant, J.C., 2000. Terrain Analysis: Principles and Applications. John Wiley, New York.

Developments in Soil Science, volume 31
P. Lagacherie, A.B. McBratney and M. Voltz (Editors)

Chapter 3

THE STATE OF THE ART OF BRAZILIAN SOIL MAPPING AND PROSPECTS FOR DIGITAL SOIL MAPPING

M.L. Mendonça-Santos and H.G. dos Santos

Abstract

In this chapter, we shall discuss about the state of the art of the Brazilian soil survey and mapping, including a brief history of soil surveys in Brazil, a summary description of survey methods and techniques, mapping paradigms, as well as the present-day needs and current challenges. Digital soil mapping is viewed as an opportunity to recover the unaccomplished soil mapping program in Brazil. We also focus on several attempts to make a national soil database, starting at the beginning of the 1980s with *SisSolos*, followed by *SigSolos*, and lately, SigWeb *"Iniciativa Solos br"*, available at http://www.cnps.embrapa.br/soilsbr and the country's new challenges to improve soil mapping, as well as some insights into digital soil mapping. Traditional soil surveys in Brazil have covered almost the whole country; these soil surveys are mainly in small-scale mapping, except for the Amazon region, which is poorly provided with soil surveys. Four main governmental institutions, Brazilian Agricultural Research Corporation (EMBRAPA), Brazilian Institute of Geography and Statistics (IBGE), Agronomic Institute of Campinas (IAC) and Geological Survey of Brazil (CPRM), execute soil surveys at the national and state levels. Private consultants also perform soil surveys, particularly on larger scales, under private contracts. Consequently, there is much dispersed information about soil surveys, but at least the methods and procedures are kept reasonably uniform all over the country. The systematic, governmental-supported soil mapping of the entire country, as initially planned, has been cancelled for a long time, although the demand for soil survey information continues at the same or even higher levels in some regions. At present, complete soil mapping covers 17 states out of 26, and the Federal District, at scales ranging from 1:100,000 to 1:600,000, covers approximately 35% of Brazilian soil, as well as a full uniform cover at scales of 1:1,000,000 and 1:5,000,000. Extensive zones still lack complete soil information at suitable scales and survey levels, needed to face the current problems of use, management, conservation, prevention and recovery of agriculturally and nonagriculturally degraded areas. Nowadays, soil surveys are made only on governmental demands, to support agroecological zonings and evaluation of environmental-impact projects, precision agriculture, degraded-area reclamation, planning of rural settlements and land-use planning, and are always linked to multidisciplinary activities. Soil and environmental data organisation, structuring and availability is imperative to perform digital soil mapping, which will certainly generate demands of quality databases as well as of the necessary tools for institutions involved in soil surveys in Brazil.

3.1 Introduction

Here, we present a brief history of the soil surveys in Brazil, a summary description of survey methods and techniques, mapping paradigms, as well as the present-day needs and current challenges considering digital soil mapping as an opportunity to recover the unaccomplished soil-mapping program in Brazil.

At present, approximately 35% of Brazil, 17 out of 26 states and the Federal District, is covered with soil maps at several intermediate scales (1:100,000–1:600,000) and full coverage of the country is available at exploratory and schematic levels (scales 1:1,000,000 and 1:5,000,000). In Brazil, soil surveys are still necessary, mainly at larger scales, to support evaluation of soil resources for planning, and management of agricultural and environmental projects. Detailed and semidetailed soil surveys are now available in small areas, to support local-specific agricultural and environmental projects. Yet, extensive areas in the whole country lack complete soil information in appropriate scales and survey level to support solutions to the current problems of use, management, conservation, prevention and recovery of agriculturally and nonagriculturally degraded areas. Large territorial extension, regional inequalities, permanent shortage of budgets, lack of material and human resources and insufficient institutional support forced the country to opt for the execution of generalised small-scale surveys.

Nowadays, government-requested soil surveys are made only, in general, to support agroecological zonings and evaluation of environmental impact projects, precision agriculture, degraded-area reclamation, planning of rural settlements and land-use planning, always linked to multidisciplinary activities. Requirements to perform digital soil mapping could help in soil data organisation, structuring and availability, among the institutions involved in soil surveys in Brazil.

3.2 A history of soil survey in Brazil

The first steps

As reported in Santos (1992), the first soil survey according to methods and procedures of the Soil Survey Manual (Soil Survey Staff, 1951) was carried out in the beginning of the 1950s by Mendes and his collaborators, for multipurpose interpretations related to soil fertility, conservation and soil management for agricultural use in Itaguaí County, Rio de Janeiro State (Mendes et al., 1954). In that survey, soil classes (supposedly soil series) were defined according to texture, horizon sequence, soil thickness, soil reaction, consistence, structure and additional characteristics as slope, stoniness and erosion used as soil-mapping-unit phases.

According to Mendes et al. (1954), the so-called soil series had been defined since 1936, without reference to this level of detail. They reported that this level of soil survey had been started by agronomists in charge of soil- and water-use planning and irrigation projects in semiarid northeastern Brazil. Mendes et al. (1954) also mentioned similar surveys in São Paulo State, for soil management and conservation in experimental stations and private farms at the soil-type level by the Agronomic Institute of Campinas (IAC).

Generalised soil studies based on geology, native vegetation cover and main regional crops were conducted by Paiva Neto et al. (1951), for characterisation of broad soil classes of São Paulo State over a period of 15 years. Another kind of soil survey executed at that time was geared towards soil-conservation planning of private farms, ending with a land-capability-use map (Marques et al., 1953).

The national soil-survey program

The very handy pioneer, detailed soil surveys did not proceed due to economical, institutional and technological constraints. It was soon realised that the best option would be to organise a national soil-survey program at the reconnaissance or exploratory levels, which was faster and less costly, to meet the demands for soil information, scarcely available in the country at that time.

Soil surveys in Brazil, supported by an organisational structure of national service, started with the institutionalisation of the Soils Commission in 1947, within the Ministry of Agriculture (Brasil, 1958). In 1953, the Soils Commission went through a reorganisation process and, under a new administration, prepared a reconnaissance soil survey program for the whole country, starting with Rio de Janeiro and São Paulo states in 1954 and 1955, respectively (Brasil, 1958 and 1960). The Soils Commission went through another reorganisation in 1957 and, without interruption, proceeded with the national soil-mapping program at the reconnaissance level (Brasil, 1958).

In the 1960s, the making of the soil map of the northern and central regions marked the beginning of a new period of field training of soil scientists in Brazil and the United States, contributing to a substantial increase in soil survey manpower, made possible through a technical cooperation between United States Agency for International Development (USAID) and Brazil, with full participation of the Soil Conservation Service/United States Department of Agriculture (SCS/USDA), currently known as the Natural Resources Conservation Service (NRCS) (Santos, 1992).

As soil-survey operations continued all over the country, at different levels and scales, but predominantly in exploratory and reconnaissance levels, methods and procedures were updated constantly, with publication of documents such as Bennema & Camargo (1964), Reunião Técnica de Levantamento de

Solos, 5 (1964), Reuniao Tecnica de Levantamento de Solos, 7 (1967), Reunião Técnica de Levantamento de Solos, 10 (1979), Lemos & Santos (1996), Embrapa (1995, 1997, 1999).

The RADAM project

In 1971, the "Departamento Nacional de Produção Mineral" (DNPM) planned and executed a project on remote sensing for the Amazon region using the side looking airborne radar (SLAR), known as the RADAM project, which produced a new picture of Amazonia, resulting in 117 maps and 18 volumes on geology, geomorphology, vegetation, soils and land-use potential. After 1976, the RADAM project extended its survey operations to the rest of the country, as the RADAMBRASIL project, and covered the whole country with about 190 maps, 38 report volumes (4, not published yet), at a uniform scale of 1:1,000,000.

The 1:5,000,000-scale soil map of Brazil

This set of information on small-scale maps summed up with more detailed soil maps and reports at State level, made possible the compilation of the soil map of Brazil at the scale 1:5,000,000 (Embrapa, 1981) considered, so far, a major contribution to increase knowledge of tropical and subtropical soils, as well as the basic information reference for development and improvement of the Brazilian system of soil classification (Embrapa, 1999).

A new soil map of Brazil was compiled in 2001 (Embrapa, 2001), as shown in Plate 3 (see Colour Plate Section), based on field information provided by larger scale soil surveys, done in the period of 1981–2001 with cartographic quality improvements and taxonomically updated, at the order level, according to current Brazilian soil classification system (SiBCS). Plate 3 (see Colour Plate Section) encompasses the main soil orders as defined in SiBCS, showing a marked predominance of Latossolos [Oxisols] (about 40% of the total area of Brazil), followed by Argissolos [Ultisols] (20%) and Neossolos [Entisols] (15%) (Coelho et al., 2002).

The current SiBCS is structured into six categorical levels hierarchically organised on the basis of morphopedogenetic soil attributes to emphasise the constitutive nature of the soils. It shows an increasing amount of taxonomic class information, from higher to lower levels of classification. The six categorical levels are composed of 13 orders, 43 suborders, 179 great groups and 749 subgroups, according to the new revised and updated version of the system, to be republished in 2006. The number of classes at family and series levels is unpredictable, depending on identification and registration of soil classes at these levels all over the country. To follow the progressive review of SiBCS, information is available at http://www.cnps.embrapa.br/sibcs.

3.3 Soil survey methods and techniques

For classification and mapping, soil is considered as a natural body resulting from the interactions of climate, organisms, relief and parent material, acting altogether in variable intensities during a certain period of time (Jenny, 1941). These soil-forming factors define the nature of the soils, their distribution and settings in the landscape. In a profile they are conceptual individuals, while in the landscape they constitute a continuum, having a set of physical, chemical, mineralogical and biological attributes. This concept is linked to the evolution of soil and the patterns of distribution in the landscape, showing where and why certain types of soils occur such as to constitute geographical bodies equivalent to the "pedons" and "polypedons" (Knox, 1965).

The traditional soil mapping method comprises the preliminary interpretation of existing aerial photographs or other remote-sensing materials of the area to be surveyed, followed by preliminary field investigation for soil class identification, soil description and sampling procedures for laboratory analysis aiming the elaboration of a preliminary mapping legend of the area.

Field mapping and continuous updating of a preliminary soil legend are the usual procedures. During field mapping, soil correlations with geology and probable parent material of the soils, vegetation cover and landscape features such as drainage, topography, slope types and grades are made (Embrapa, 1995).

Map units are defined and identified by subdivisions of higher categories into lower levels of soil classification according to selected soil attributes such as texture, base and aluminium saturation, cation exchange capacity, clay activity, sodium saturation, sulphates and soluble salts.

Different survey levels are carried out, depending on survey objectives, extent of area covered and availability of cartographic base material. Four main kinds of soil surveys are made, for different purposes and demands, namely detailed, semidetailed, reconnaissance and exploratory soil surveys.

Exploratory soil surveys are commonly carried out in large areas to provide qualitative generalised information. A wide range of cartographic base material may be used. Scales ranging from 1:750,000 to 1:2,500,000 are commonly used, minimum-size delineation from 22.5–250 km^2 and a field observation density of 0.04 observations per square kilometre. This level of information corresponds approximately to spatial resolutions $>4 \times 4$ km, or pixel size and spacing of $>2 \times 2$ km, according to McBratney et al. (2003).

Reconnaissance soil surveys are usually made with broad objectives, comprising qualitative and semiquantitative appraisal of soil resources, such as agricultural and nonagricultural land-suitability evaluation. Common map scales are 1:100,000–to 1:500,000, the minimum area delineation ranging from 0.4–10 km^2

with field observation densities of 0.04–2.0 observations per square kilometre, corresponding to 400 × 400 m to 4 × 4 km spatial resolutions and pixel size and spacing of 100 × 100 m to 500 × 500 m (McBratney et al., 2003).

Semidetailed soil surveys are made to meet more well-defined objectives, to provide basic information for rural-settlement projects, integrated land-use planning and soil management and conservation projects. Preferable scale of publication is 1:50,000, with minimum size delineations of 10 ha and field average observation density of 0.02–0.2 observations per square kilometre. The corresponding spatial resolution is approximately 40 × 40 m to <400 × 400 m and pixel size and spacing corresponding to 50 × 50 m (McBratney et al., 2003).

Detailed soil surveys are relatively scarce in Brazil and are made to support conservation planners, to provide characterisation and delineation of soil areas in agricultural-experimental stations, and to support irrigation, drainage and precision agriculture executive projects. Publishing scales are ⩾1:20,000, minimum size delineations of 1.6 ha and an observation density of 0.2–0.4 per ha, corresponding approximately to spatial resolution of 10 × 10 m to 40 × 40 m or pixel sizes and spacing of 5 × 5 m – 20 × 20 m (McBratney et al., 2003).

Usually, most reconnaissance soil surveys are carried out at the subgroup categorical level, subdivided in phases of vegetation, relief, rockiness, stoniness, slope, erosion status and other features important to soil use.

A peculiar characteristic of Brazilian soil-survey strategies is to define mapping units designed for scales as small as 1:250,000 or 1:500,000, or even smaller, with a larger amount of information, not always conformable with the survey level (taxonomically detailed and cartographically generalised), but very useful in a country with severe economic and financial resource constraints. Much of the knowledge needed to accomplish this kind of survey is dependent on the experience of soil surveyors working in specific areas (Embrapa, 1995). For this purpose qualitative field correlation is used between soil fertility, parent material and vegetation cover, as well as among soil and landscape relationships, in order to predict soil classes in higher categorical levels. It is common to carry out periodic field-soil correlation and classification discussions to update soil surveys and their interpretations (Reunião de Classificação, Correlação e Aplicação de Levantamentos de Solos, 5, 1998).

In larger scale mapping, more intensive field work comprising point observation and sampling is employed, and soil mapping unit components may reach the family and series categorical levels, although no officially defined and recognised soil series exist anywhere in the country.

Soil-mapping unit boundaries are located by means of aerial photographs and topographic-map interpretation in more detailed soil surveys, and radar and satellite images in smaller-scale soil mapping, using much of the previous knowledge of the area, correlation and inferences. These soil-map units are defined according to different levels of abstraction, from the most generalised schematic soil maps such as exploratory and reconnaissance (low, medium and high intensities) to semidetailed and detailed soil surveys.

In the traditional method of soil mapping, field-mapping delineation are usually transferred to geodetically corrected Universal Transverse Mercator (UTM) bases, and digitised by means of geographic information system (GIS) software.

3.4 Present-day Brazilian soil survey and mapping cover

Besides the full cover at 1:1,000,000 and 1:5,000,000, other soil surveys and mapping covers at different levels and scales were performed in Brazil, as illustrated in Figure 3.1. It can be seen that that north and northwest area of Brazil, including part of the Amazon region, are least endowed with soil survey information due to many factors mainly related to difficult access, governmental polices to forest and biodiversity preservation and soil fertility constraints.

On the other hand, the south and southeast of Brazil are more intensively mapped at different levels and scales, mostly reconnaissance and semidetailed with a few detailed soil surveys. In these intensively cultivated agricultural areas, demands for soil survey were greater than any other region and correspond to the most developed area of Brazil.

Other important soil surveys are located in semiarid northeastern Brazil, mostly exploratory and reconnaissance and many semidetailed and detailed soil surveys, demanded by governmental polices aimed to assist small farmers, regional development and irrigation projects. The western and central regions of Brazil are also less endowed with soil information, apart from a few exploratory maps, due to less-intensive land use in the past, mainly cattle raising and pasture in extensive areas. Currently, there is an increasing demand for soil surveys to meet the information needs for soil management and conservation practices in the savannah areas.

In the IBGE/EMBRAPA records, 1432 soil surveys and maps are listed, at different scales and extents, which are depicted schematically in Figure 3.1. Public institutions have done most of these soil surveys and mapping, but some of them come from private organisations and were undertaken for specific purposes. In fact, Brazil has undertaken more detailed soil surveys than usually known, but many were not made according to standard methodology and also are not always available because they were undertaken by private organisations under contract.

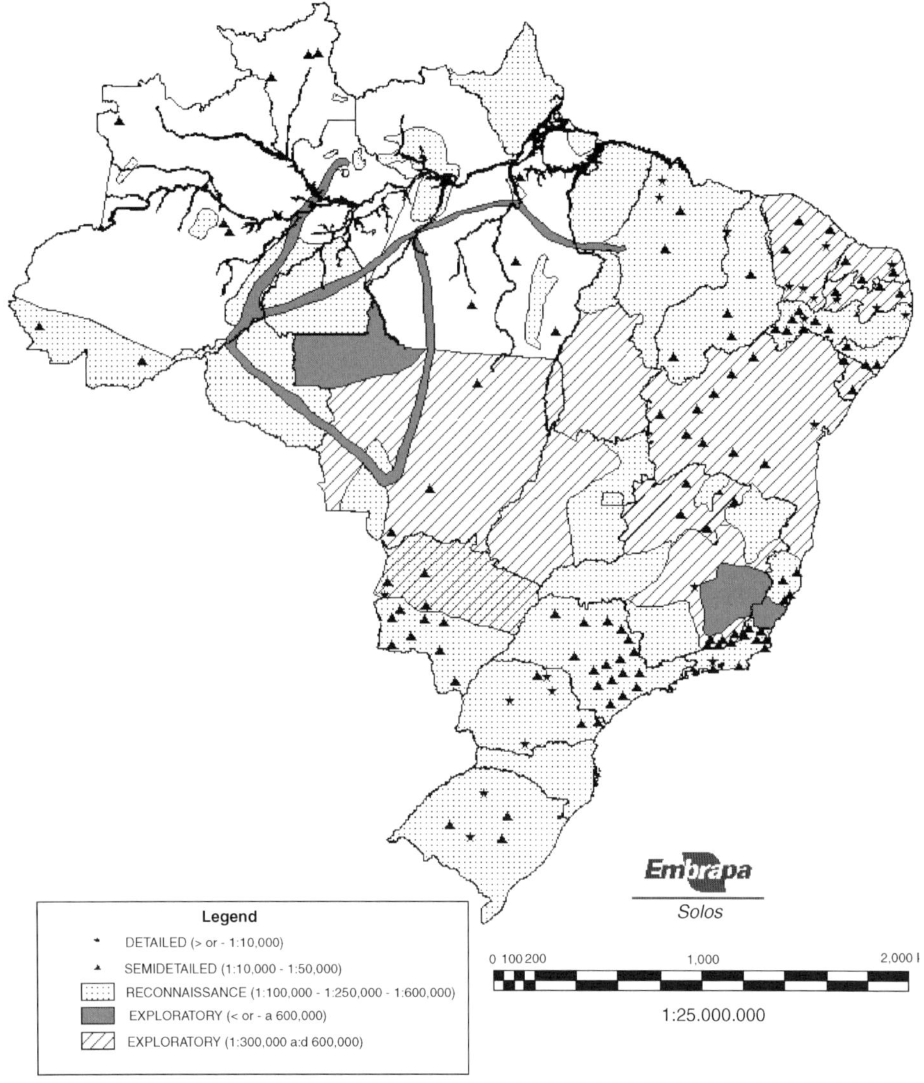

Figure 3.1. *Brazilian soil survey and mapping cover (after Santos, 1992, information updated to 2005).*

3.5 Efforts to build a national soil database

SisSolos: soil information system

The precursor of the Brazilian soil database (BSD) was called *SisSolos*, which was the first attempt to systematise the soil data in Brazil. It was developed at the beginning of the 1980s by Brazilian Agricultural Research Corporation

(EMBRAPA) with the technical consultation of the French Institute of Scientific Research for Development in Cooperation (ORSTOM), currently IRD (Meneguelli et al., 1983). The SisSolos database was written in the COBOL language to be run on mainframes, and the data were stored in magnetic streamers. This database was a pioneer work, and very up-to-date at the time in terms of data entry, organisation, storage, retrieval and even having forms as the user interface. The database, which demanded fairly cumbersome filling-up of the different forms (Meneguelli and Séchet, 1984), was used until the end of the 1980s, when the maintenance as well as the technology became obsolete. By that time, 11,533 soil profiles were stored in SisSolos, covering the regions shown in Figure 3.2 (Meneguelli et al., 1983).

Figure 3.2. *Soil profiles stored in SisSolos until May 1983 (shown in grey, by areas).*

Information technology has changed and the need of a national soil database continues to be the main goal of Embrapa, mainly for Embrapa Solos, the national soil research centre of Embrapa, responsible for soil-data production. Since the end of the 1980s, armed with the new capabilities of information technology, Bhering et al., (1998) started a new soil database, *SigSolos* – georeferenced system of soil information, modelled with ERwin, a relational database, implemented in MS Access (using VBA language) for PCs with the Windows 95 operating system, with a more user-friendly interface for data entry, storage, retrieval and export (Fig. 3.3). Unfortunately, the soil data stored in the SisSolos streams could not be imported into SigSolos. A great effort was then made to enter data from soil reports to SigSolos.

SigSolos is described in Bhering et al. (1998) as a database interface aimed at the organisation and systematisation of georeferenced soil information, using a management database system (SGBD) to deal with alphanumeric data, and a GIS to manage the spatial information. The data-entry interface of SigSolos is very friendly and is based on the current fields of the Brazilian soil-profile description form, including the morphological characteristics, as well as the physical chemical and mineralogical data (Fig. 3.4).

By the end of the 1990s, the SigSolos interface began to present technical problems of maintenance and data exportation, and Embrapa Solos could no longer count on the team that had developed the database. A total of 2292 soil profiles and 6944 soil horizons were stored in SigSolos at that time, which unlike SisSolos, were distributed in almost all Brazilian states (Fig. 3.5).

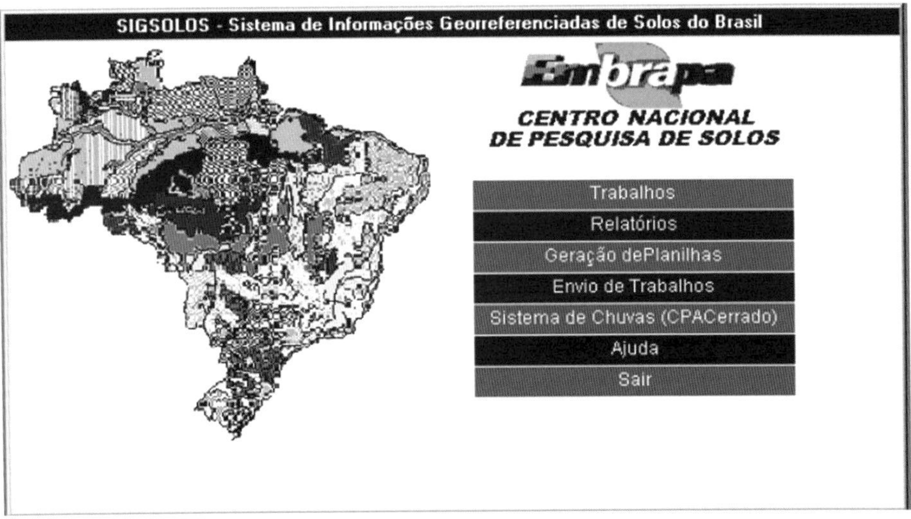

Figure 3.3. *SigSolos opening menu with the main components.*

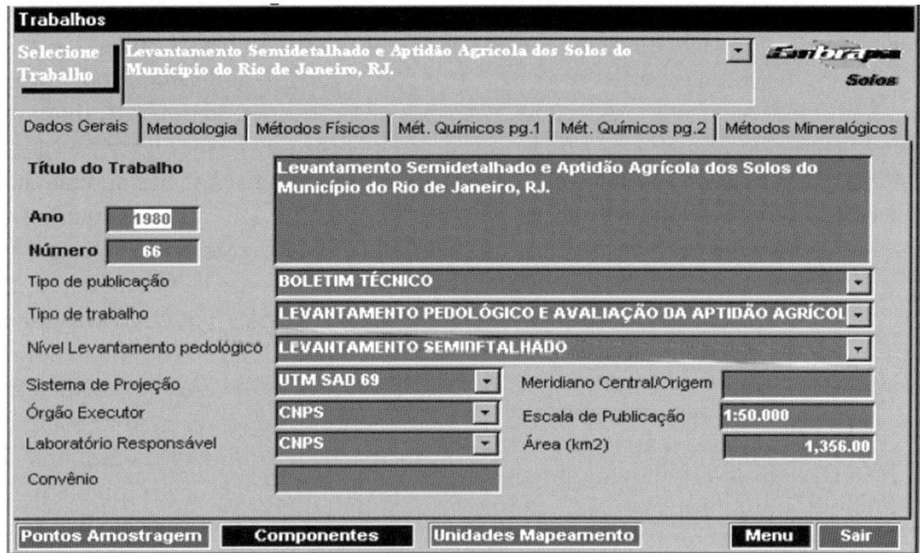

Figure 3.4. *The menu "Trabalhos" gives access to the main soil data, organised by the different soil survey reports, encompassing the soil morphological, physical, chemical and mineralogical characteristics.*

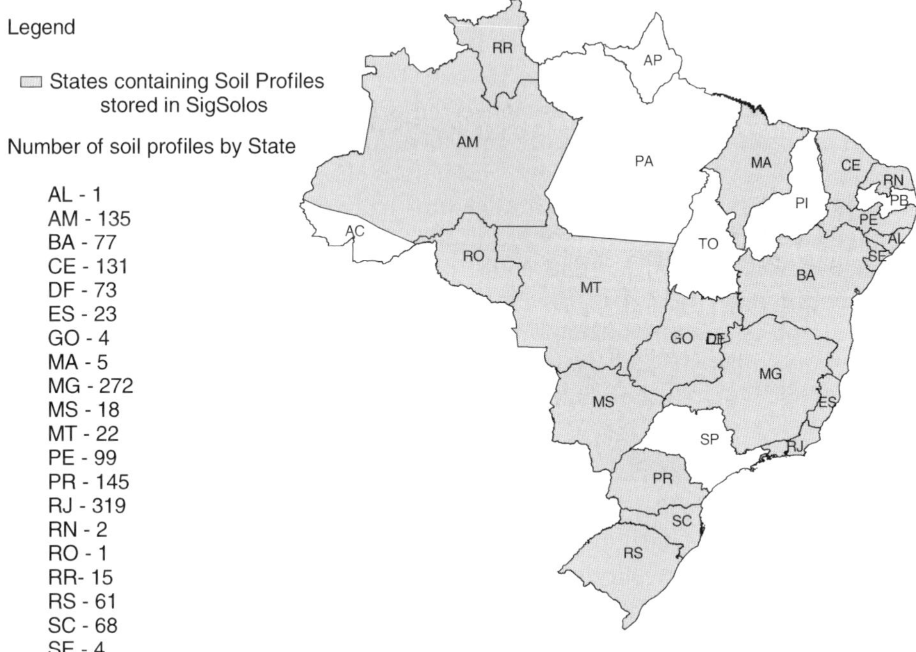

Figure 3.5. *Soil profiles stored in SigSolos—distribution by states in Brazil.*

In both soil databases, most soil profiles do not have associated geographical coordinates.

Iniciativa Solos.br (http://www.cnps.embrapa.br/solosbr/)

The *Iniciativa Solos.br* is a strategic action of Embrapa Solos that has as its main objective to make available and to provide continuous data acquisition, information and knowledge on Brazilian soils in a web-based platform.

Inciativa Solos.br already has three experimental public interfaces:

- **Interface BSB** (soil base of Brazil), for data on physical, chemical and morphological characteristics of some Brazilian soil profiles that can be accessed through different search options.
- **Interface SigWeb**, showing interactive Brazilian soil maps, with cartographic quality and some basic GIS facilities, such as queries to the databases associated with the maps with alphanumeric and graphical answers.
- **Knowledge interface,** which makes available all Embrapa Solos's publications.

3.6 Current challenges of Embrapa Solos

Considering the current state of affairs of soil data, and the continental extent of Brazil and also that conventional soil mapping is a slow and an expensive process, the current challenges are

- to recover, improve and systematise the existing soil data and information into an up-to-date and friendly interface, to be used in knowledge output about soil and its environment. Similar to other countries, the soilsurvey institutions in Brazil, under the leadership of Embrapa Solos, are greatly interested in building a national soil database that could incorporate all the existing soil information that could be used to produce additional soil information other than soil classification per se; and
- to use soil data and information and ancillary data on soil formation factors to model and map soil by using quantitative techniques based on Jenny's model and its upgrades such as the *scorpan* model (McBratney et al., 2003).

These challenges are being undertaken currently through Embrapa Solos's research projects aiming at recovering, auditing and implementing soil data using new database technologies. For Rio de Janeiro state, for example, all available soil data were successfully recovered (more than 800 soil profiles and 3047 soil horizons) and audited, and the missing spatial coordinates were estimated according to georeferenced printed maps. This data then populated a new database interface in Microsoft Access in order to support the objectives of the research project "Inventory of Soil Organic Carbon in Rio de Janeiro State".

Another attempt is being undertaken in a way similar to that of the Rio de Janeiro state, recovering and auditing soil profiles from SigSolos, referring to all other Brazil regions with output in spreadsheets. These soil data spreadsheets will populate the new national soil database interface (most probably in Oracle with an SQL Server) that is under consideration by Embrapa Solos. Therefore, a more comprehensive soil database will be soon available at Embrapa Solos home page, under institutional arrangements for general use.

Another research project on digital soil mapping is ongoing in collaboration with the University of Sydney with the aim of modelling and predicting soil classes and properties (such as soil organic carbon, water-retention capacity, etc.) from soil and ancillary data for the Rio de Janeiro state.

The use of quantitative techniques for spatial prediction in digital soil mapping has improved recently (McBratney et al. 2000; McBratney et al., 2003), benefited by advances in the processing capacity of computers, allowing the use of existing data in conjunction with statistical and mathematical approaches for derivative soil information.

This approach will help to optimise soil mapping and to support the main questions with which Brazil is confronted: food production, management of the hydrological resources and environmental protection, among others, which are influenced by the inappropriate use and management of the natural resource and by social and economic pressures.

3.7 A framework to undertake the Brazilian digital soil mapping

Considering the large geographic area of Brazil, the organisation of soil data and information and the environmental variables in order to reach the digital soil-mapping requirements could be a cumbersome task, involving different governmental and private institutions as well as people and infrastructure. A framework to undertake digital soil mapping in Brazil is illustrated in Figure 3.6.

The first step should be to reinforce EMBRAPA as the leading institution of a National Consortium in the same way it has been done for ecological–economic zoning (EEZ) in Brazil, with Embrapa Solos in charge of assembling existing soil data, systematising and auditing them, and deciding about the need of complementary postsurvey sampling. This will be required in most cases, and it could be done in reference areas (Lagacherie et al., 2001). The result of this initial step will give rise to a more comprehensive and sound BSD. Pedotransfer functions can be used to estimate soil properties not usually found in the traditional soil survey reports (mainly, soil bulk density and soil hydraulic properties).

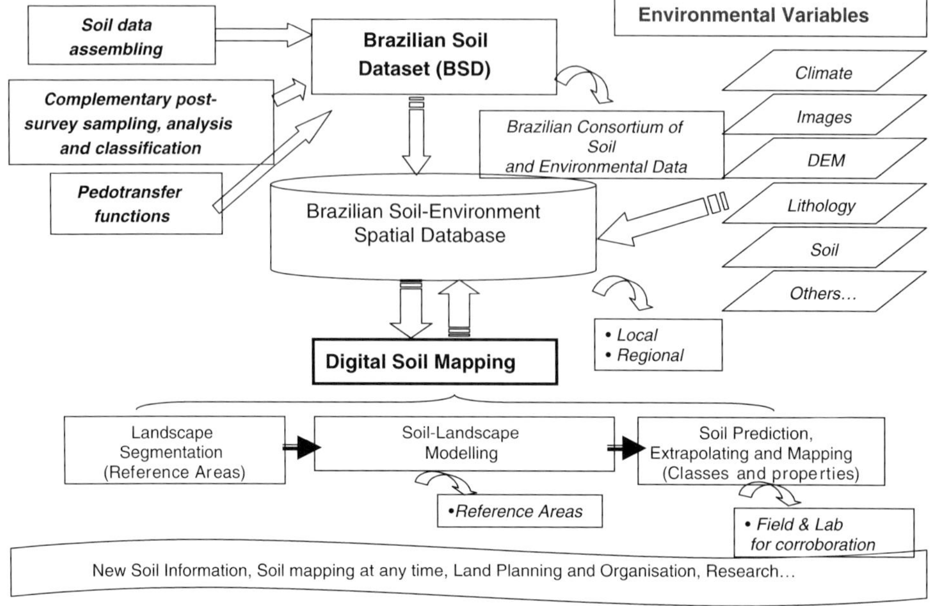

Figure 3.6. *A framework for Brazilian digital soil mapping.*

The Consortium could also facilitate the assembly and use of environmental variables related to soil-forming process, following the *scorpan* model of McBratney et al., (2003), as well all other reliable information. This information, in conjunction with the soil dataset, will constitute the Brazilian soil-environment spatial database (BSESD). Within this background, digital soil mapping can be finally undertaken for further extrapolation of the rules for larger areas.

Under the scope of digital soil mapping, McBratney et al. (2003) propose three preferable spatial resolutions, namely >2 km, 20 m to 2 km and <20 m, corresponding to national-to-global, catchment-to-landscape, and local extents, respectively.

Considering the above framework, the strategy proposed is to use the information stored in the BSESD with spatial-prediction functions (SSPFe) and the spatially autocorrelated errors, to start modelling soil for a "homogeneous" local or regional extent, in order to understand the relationships between soil and landscape, and then to use the rules obtained to predict soil classes and properties at a regional extent. In other words, we suggest starting the soil-landscape modelling in the corresponding cartographic scale of $\leqslant$1:100,000 and then apply the rules in soil prediction to scales of $>$100,000. The predictions and relationships found could be returned as input to the BSESD, which is the main product of the digital soil mapping process. Soil maps could be produced at any time as needed, using a variety of quantitative techniques, as suggested in McBratney et al., (2003).

3.8 Concluding remarks

In this chapter, we report the state of the art of Brazilian soil survey and mapping, as well as some insights into digital soil mapping. Digital soil mapping appears as a new challenge for assembling and organising the existing soil and environmental data, allowing at any time and any scale the mapping of classes and soil properties to meet the current demands of the country, optimising time and resources.

It will be difficult to assemble and organise data process, as well as to accept the new paradigm represented by *scorpan* model with the SSPFe and spatially autocorrelated errors. Concerning institutional aspects, a Consortium will be fundamental. Notwithstanding, acceptance and assimilation of a new soil mapping paradigm will take time, but will certainly find its space in research projects and soil survey executions.

Acknowledgments

The authors would like to thank their colleagues Neli Meneguelli, Jesus Mansilla Baca and Sílvio Bhering for making available some unpublished information on the SisSolos and SigSolos databases, as well as Mario Diamante Áglio and Carolina Xavier de Lima for information on the spatialisation of Brazilian soil surveys and soil profiles stored in SigSolos, respectively. We thankfully acknowledge Philippe Lagacherie for the comments in an earlier draft of this chapter, as well as FAPERJ E-26/171.360/2001 research project that made possible the recovery of the Rio de Janeiro soil data.

References

Bennema, J. and Camargo, M.N., 1964. Segundo esboço parcial de classificação de solos brasileiros; subsídios a IV Reunião Técnica de Levantamento de Solos. Rio de Janeiro, Departamento de Pesquisa e Experimentação Agropecuária, (Mimeo).

Bhering, S.B., Chagas, C.S., Andrade Júnior, O., Carvalho Júnior, W., Baca, J.F.M., Tanaka, A.K., 1998. Base de informação georreferenciada de solos: metodologia e guia básico do aplicativo SigSolos – CD ROM. EMBRAPA-CNPS, Boletim de Pesquisa n. 11.

Brasil, 1958. Ministério da Agricultura. Centro Nacional de Ensino e Pesquisas Agronômicas. Comissão de Solos. Levantamento de reconhecimento dos solos do Estado do Rio de Janeiro e Distrito Federal, contribuição a carta de solos do Brasil. Rio de Janeiro, 350p. (Brasil. Ministério da Agricultura. CNEPA. SNPA. Boletim Técnico, 11).

Brasil, 1960. Ministério da Agricultura. Centro Nacional de Ensino e Pesquisas Agronômicas. Comissão de Solos. Levantamento de reconhecimento do Estado de São Paulo, contribuição a carta de solos do Brasil. Rio de Janeiro, 634p. (Brasil. Ministério da Agricultura. CNEPA. SNPA. Boletim Técnico, 12).

Coelho, M.R., Santos, H.G. dos and Silva, E.F., 2002. O recurso natural solo. In: Manzatto, C.V.; Júnior, E.F.; Peres, J.R.R. Uso Agrícola dos Solos Brasileiros. Rio de Janeiro: Embrapa Solos, 1–11pp.

Embrapa, 1981. Centro Nacional de Pesquisa de Solos (Rio de Janeiro, RJ). Mapa de Solos do Brasil (1:5,000,000). Rio de Janeiro, 1 map.

Embrapa, 2001. Centro Nacional de Pesquisa de Solos (Rio de Janeiro, RJ). Mapa de Solos do Brasil (1:5,000,000). Rio de Janeiro, 1 map.

Embrapa, 1995. Centro Nacional de Pesquisa de Solos (Rio de Janeiro, RJ). Procedimentos normativos de levantamentos pedológicos. Brasília: EMBRAPA-SPI, 101p.

Embrapa, 1997. Centro Nacional de Pesquisa de Solos (Rio de Janeiro, RJ). Manual de métodos de análise de solo. 2nd edition. Rev. Atual. Rio de Janeiro, 212pp.

Embrapa, 1999. Centro Nacional de Pesquisa de Solos (Rio de Janeiro, RJ). EMBRAPA. Centro Nacional de Pesquisa de Solos (Rio de Janeiro, RJ). Sistema brasileiro de classificação de solos: Brasília: Embrapa Produção da Informação. Embrapa Solos, Rio de Janeiro, 412p.

Jenny, H., 1941. Factors of soil formation – a system of quantitative pedology. Dover Publications, New York, 281 pp.

Knox, E.G., 1965. Soil individuals and soil classification. Soil Sci. Soc. Amer. Proc. 29, 79–84.

Lagacherie, P., Robbez-Masson, J.M., Nguyen-The, N., Barthes, J.P., 2001. Mapping of reference area representativity using a mathematical soilscape distance. Geoderma 101, 105–118.

Lemos, R.C.de., Santos, R.D.dos., 1996. Manual de descrição e coleta de solo no campo, 3.ed. Sociedade Brasileira de Ciência do Solo, Campinas, 83p.

McBratney, A.B., Odeh, I.Q.A., Bishop, T.F.A., Dumbar, M.S., Shatar, T.M., 2000. An overview of pedometric techniques for use in soil survey. Geoderma 97, 293–327.

McBratney, A.B., Mendonça-santos, M.L., Minasny, B., 2003. On digital soil mapping. Geoderma 117, 3–52.

Marques, J.Q.A., Grohman, F., Bertoni, J., 1953. Levantamento conservacionista. Levantamento e classificação de terras para fins de conservação do solo. In: Reunião Brasileira de Ciência do Solo, 2., Campinas, 1949. Anais. Rio de Janeiro, Sociedade Brasileira de Ciência do Solo, 651–682pp.

Mendes, W., Cruz-Lemos, P. de O, Lemos, R.C., Carvalho, L.G. de O. and Rosenburg, R.J., 1954. Contribuição ao mapeamento, em séries, dos solos do município de Itaguaí. Rio de Janeiro, CNEPA – Instituto de Ecologia e Experimentação Agrícola, 53p. (Brasil. Ministério da Agricultura. CNEPA, IEAE. Boletim, 12).

Meneguelli, N.A., Assis, D.S., Séchet, P., 1983. SISSOLOS – Manual de Uso. Serviço Nacional de Levantamento e Conservação de Solos. EMBRAPA, Rio de Janeiro (Documentos n. 4). 245p.

Meneguelli, N.A., Séchet, P., 1984. SISSOLOS – guia de entrada. Serviço Nacional de Levantamento e Conservação de Solos. EMBRAPA, Rio de Janeiro (Documentos n. 9). 91p.

Paiva Neto de, J.E., Catani, R.A., Kupper, H., Medina, H.P., Verdade, F.C., Gutmans, M., Nascimento, A.C., 1951. Observações sobre os grandes tipos de solos do Estado de São Paulo. Bragantia. Campinas 11 (7/9), 227–253.

Reunião de Classificação, Correlação e Aplicação de Levantamento de Solos, 5., 1998. Rio de Janeiro. Anais. Rio de Janeiro: EMBRAPA-CNPS, 140p. (Unpublished).

Reunião Técnica de Levantamento de Solos, 5., 1964. Rio de Janeiro. DPEA-DPES, 35p.

Reuniao Tecnica de Levantamento de Solos, 7., 1967. Rio de Janeiro, Sumula. DFPEA-DPFS, 35p (mimeo).

Reunião Técnica de Levantamento de Solos, 10., 1979. Rio de Janeiro. Súmula. Rio de Janeiro: EMBRAPA-SNLCS, 83p. (EMBRAPA. SNLCS. Série Miscelânea, 1).

Santos, H.G. dos, 1992. Country Report: Brazil. In: Soil Survey: Perspectives and strategies for the 21st century, (Zinck ed.) International Workshop for heads of national soil survey organizations, Enschede, The Netherlands (ITC Publication n. 21), 61–67p.

Soil Survey Staff, 1951. Soil Survey Manual. USDA, Handbook no. 18. Washington, D.C.

Developments in Soil Science, volume 31
P. Lagacherie, A.B. McBratney and M. Voltz (Editors)

Chapter 4

THE SOIL GEOGRAPHICAL DATABASE OF EURASIA AT SCALE 1:1,000,000: HISTORY AND PERSPECTIVE IN DIGITAL SOIL MAPPING

J. Daroussin, D. King, C. Le Bas, B. Vrščaj, E. Dobos and L. Montanarella

Abstract
In this Chapter, we describe the steps followed to build and improve the Soil Geographical Database of Eurasia at scale 1:1,000,000 and suggest automatic soil mapping techniques to improve it.

The work started in 1952 with the compilation of materials provided by the contributing countries to publish the Soil Map of the European Communities at scale 1:1,000,000 in 1985. From then on, it was computerised, geo-referenced, structured to form a geographic database, enriched using archives of the original materials, extended geographically and thematically, harmonised over country borders, updated and documented. Despite this streamline of development and enhancements, it has limitations inherent to the initial objective of publishing a paper map and to the wide variety of views on soil mapping that were brought together from the community of contributors.

The next logical step would thus be to improve the Soil Database by using existing detailed soil information and newly European-wide available satellite data and digital elevation models (DEMs). Existing soil maps at larger scale available in some parts of Europe could be used as training references to assess the morphometric (relief) and spectral characteristics of the Soil Mapping Units (SMUs) from DEM and satellite images, and the results extrapolated beyond to the surrounding areas and compared with the present delineation included in the 1:1,000,000 Soil Database. The Limousin Region of France was chosen as a test area but other geomorphologically diverse and representative areas in Europe should be included as well. This work is in progress but more partners will be needed in a collaborative project effort to establish a multi-scale European Soil Information System.

4.1 Introduction

The Soil Geographical Database of Eurasia at the1:1,000,000 scale (SGDBE1 M), in its current state (version 4), is the result of a long construction process and remains the only source of harmonised spatial soil information covering Europe at that scale. The main stages in its development are described in Section 4.2. They all have in common the use of existing soil information and human expertise.

Exhaustive, detailed, digital and easily available sources of information that can characterise some soil-forming factors are now available, for example from DEMs or remote sensing. Soil experts expect them to serve as a source of basic non-soil datasets for improving existing soil databases, or perhaps for deriving new thematic layers for some soil parameters. Section 4.3 presents some ideas for developing methods applicable in the context of the SGDBE1 M that should formalise, quantify and model the relations between the soil-forming factors that constitute the basis of the expertise of soil cartographers.

4.2 History of the Soil Geographical Database of Eurasia: from map to database

4.2.1 Publishing the Soil Map of the European Communities at scale 1:1,000,000

Starting in 1952 (Fig. 4.1), a group of national experts on soil survey from most countries in Europe was sponsored by the Food and Agriculture Organisation of the United Nations (FAO), to prepare a common map. Following guidelines drawn up during their meetings, each contributor came up with draft documents for their country. It is important to note that these documents reflected the state of progress of soil mapping within each participating country. Some of them already had a full coverage at large scale (e.g. Belgium), in which case the materials provided consisted of a synthesis following an up-scaling approach. For others (e.g. France), a completely new map was drawn based on scarce detailed knowledge but significant input from soil experts. Owing to lack of

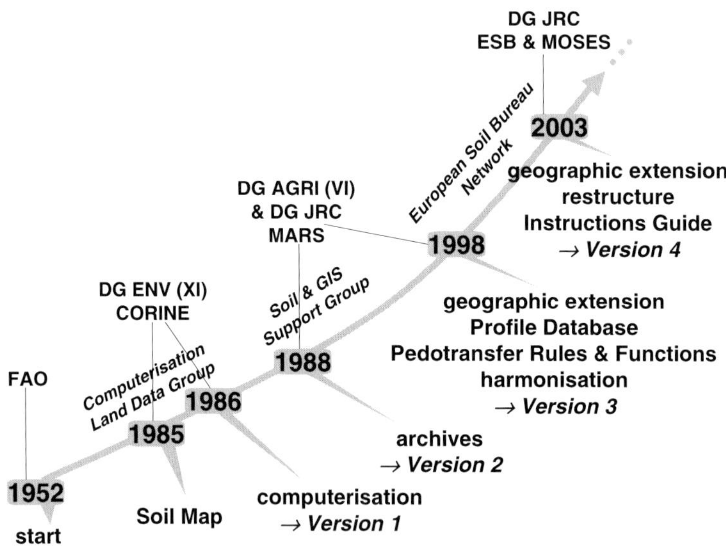

Figure 4.1. *History of the Soil Geographical Database of Eurasia.*

funding, the project was abandoned by the FAO in 1973. In 1978, by a decision of the Working Group on Land Use and Rural Resources of the European Community (EC), the materials for the 12 member countries of the EC were finally brought together. On the basis of these, the Soil Map of the European Communities was published at scale 1:1,000,000 in 1985 together with its accompanying explanatory notes (CEC, 1985). The main objective of that work was to draw a paper map; thus, the effort for harmonising information among countries was limited to the major entry in the legend of the map, that is the soil name. Moreover, simplification was necessary for the sake of map readability; thus, only the dominant soil type within each Soil Mapping Unit (SMU) was shown on the map.

4.2.2 Digitising the Soil Map of the European Communities

The published soil map was computerised under the CORINE project (Co-ordination for the Environment). First, the lines delineating the SMU polygons were digitised. Then the codes for the few soil properties shown on the map for the major soil types were associated with each polygon as attributes: soil name in the FAO-UNESCO 1974 legend (FAO-UNESCO, 1974), texture and slope. The database was then geo-referenced to the co-ordinate system used in the CORINE project, but due to poor geo-referencing quality of the original materials, this step faced some difficulties. It sometimes needed an imprecise technique called rubber sheeting, which can result in locally poor spatial precision. Examples of such problems can be noticed at map sheet and country borders, or when a river is expected to flow within a Fluvisol. Finally, the database was converted to ArcInfo® coverage format.

This resulting Soil Geographical Database of the European Communities at scale 1:1,000,000 was made available to the scientific community in 1986, as version 1.0, through the CORINE project.

4.2.3 Enhancing the Soil Database of the European Communities with the archives of the original Soil Map

The draft documents provided by each contributor to draw the original soil map are archived at the Laboratory of Tropical Pedology of the University of Ghent in Belgium. They are all in hand-written paper form consisting of draft maps showing SMU lines and identifiers, and draft tables describing these SMUs.

At the scale of work and considering the resources available for the project, each soil type that could be identified and described could not always be geographically delineated. As in most small- or medium-scale soil mapping projects, there is a discrepancy between the graphic – the mapping – and the semantic – the describing – levels of resolution. Each SMU is thus composed of one or more Soil Typological Unit (STU), that is soil type description.

As stated above, this richness of information was lost during the publication process by necessity of simplification and readability of the final map. Although database management systems do not have such limitations, the computerised version of the map strictly conformed to the published map.

In 1988, under the auspices of the Monitoring Agriculture with Remote Sensing (MARS) project of the Commission of the European Communities, all the information available from the archives held in Ghent were retrieved, structured and added to the database. The result was version 2.0 of the Soil Geographical Database of the European Communities (King et al., 1994). This version contained all the information provided by the contributors and its general structure is still the one in use today (Fig. 4.2): SMU polygons are depicted in a geographic dataset that is linked to a semantic dataset formed of related SMU and STU tables. All soil characteristics are held as attributes of the STU table.

4.2.4 Updating, harmonising and improving the Soil Database by national expertise

From then on, the database followed a continuous flow of updates and enhancements under the pressure and impulse of both its extending user and contributing communities.

The archives, when available, were again used (1992–1995) to geographically extend the database coverage to all European countries that were involved in the original FAO project, before it was abandoned. These data were then geometrically and semantically checked and corrected by national contributors. When the archives were not available, when up-to-date data could be used or for extension to Russia and Mediterranean countries, national contributors were asked to provide directly soil information following the same structure and coding as those used for the entire database (Jamagne et al., 2002).

A soil profile analytical database was compiled to complement STU description (Madsen and Jones, 1995). Unfortunately, because of the difficulties of collecting geo-referenced data at European level, the data currently stored in the profile database are not comprehensive geographically and have poor replication. Thus, the profile database is difficult to use.

At this stage, in 1998, the Soil Geographical Database of Europe at scale 1:1,000,000 version 3 was released and made publicly available through the Geographical Information System for the Community (GISCO) programme.

A Pedotransfer Rules database built upon the SGDBE1 M allows inference of new soil attributes from those originally supplied. Depth to an obstacle to roots and available water capacity for plants are examples of such inferred properties needed for applied research projects (Le Bissonnais et al., 2005).

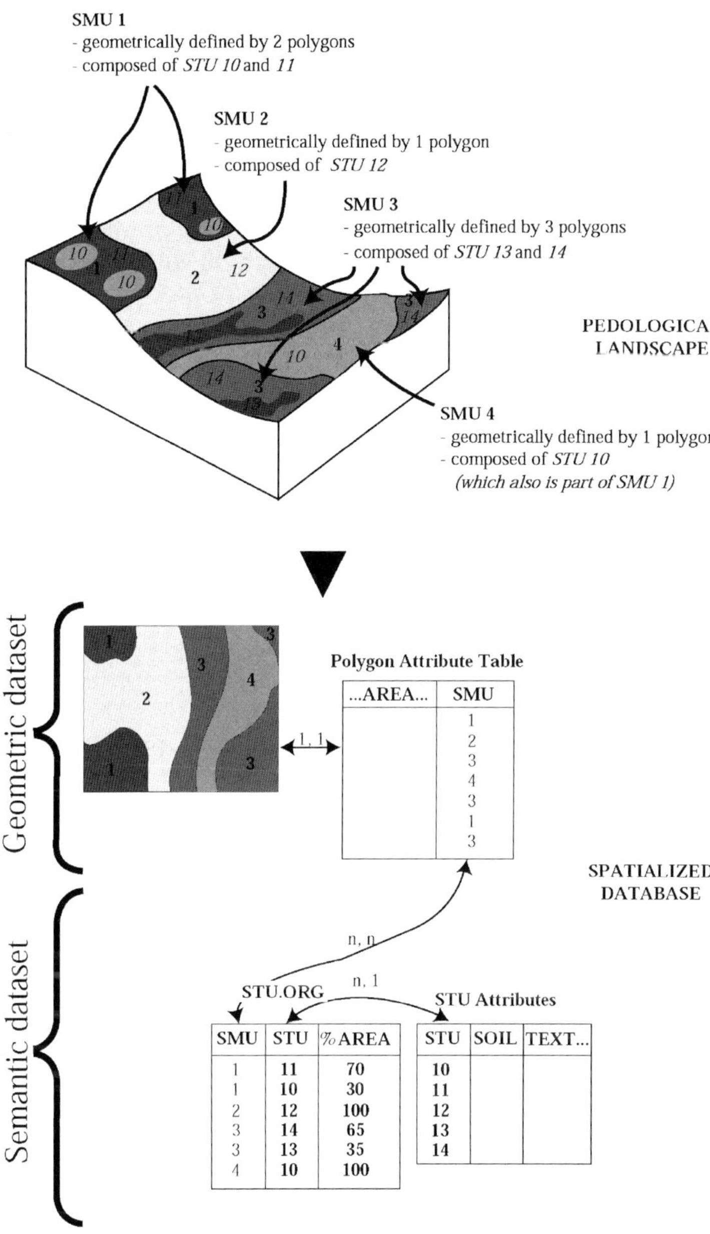

Figure 4.2. *Structure of the soil database.*

For the soil map, harmonisation between countries was limited to the soil name of the dominant soil type within each SMU. This was a shortcoming of the database each time other soil attributes or secondary soil types were needed. A harmonising procedure was set up and meetings were organised between neighbouring countries in an attempt to reduce discrepancies along country border. Unfortunately, its result has been mainly 'cosmetic' and did not resolve major conceptual differences in the mapping strategies between countries. One of the biggest problems with the database has been the difficulty of 'correlation', that is harmonising the mapping and classifying of soils by different experts in different countries. This harmonisation effort called upon the national contributors' knowledge but it has been hampered by the initial strategy to build upon existing soil data rather than setting up a new common methodology for data acquisition such as in the CORINE Land Cover project.

Extension to other European countries, as well as those of the Mediterranean basin, is an ongoing process. To achieve this goal, the structure of the database was revised to make it clearer (Lambert et al., 2003). Moreover, attributes were added to adapt to new classification schemes such as for the World Reference Base (WRB) for Soil Resources (FAO et al., 1998). The result is the current version 4.0 of the Soil Geographical Database of Eurasia at scale 1:1,000,000.

4.2.5 Benefits and criticism of the Soil Database

As an inherent consequence of its history, the SGDBE1 M has several shortcomings and limitations. Its geometric precision is sometimes locally poor. Its semantic information is pedogenetically oriented making it difficult to exploit other than by soil scientists. Despite harmonisation efforts, it is geometrically and semantically heterogeneous from country to country. Its semantic component lacks end user needed information (although this is partly overcome by means of pedotransfer rules) and is often too coarse for applications; for example, soil texture is coded into only five classes, which are too wide for applications, whereas most of the national systems of classifying texture use many more classes.

Nevertheless, the Soil Database represents the only available harmonised soil information system covering Europe and neighbouring areas at the scale 1:1,000,000 (Plate 4 (see Colour Plate Section)). More importantly, the work achieved over the long period of its development has brought together the soil-science community – institutions and experts – from all over the continent and further in the area of the Mediterranean basin and majority of the former Soviet Union countries, in a common effort to provide exhaustive and homogeneous data on soils. It has already proved its usefulness and has raised considerable interest in the development of a comprehensive knowledge base on soils. This is clearly evident from the number and importance of its 'client' projects: crop yield forecasts (Vossen and Meyer-Roux, 1995), soil vulnerability assessment (Jones et al., 2003), carbon sequestration (Jones et al., 2004) and soil erosion risks (Le Bissonnais et al., 2005).

4.3 Digital mapping to improve the existing Soil Database

Availability from remote-sensing imagery of exhaustive, continentally covering, good-quality, detailed, fine-resolution information describing the earth's surface is seen by the soil science community as a major opportunity to improve the quality of existing soil information systems (Bui and Moran, 2003) and to create completely new ones. Indicators about the main soil-forming factors such as relief, vegetation cover, land use and climate can be extracted and interpreted from these non-soil, but soil-related, data sources. Attempts will be made in this direction to build upon the SGDBE1 M, to improve its quality and reduce its identified shortcomings.

4.3.1 Improve the existing Soil Database rather than build a new one

Drawing a soil map or database is traditionally achieved through the view of experts who have an understanding of interactions between soil-forming factors. If successful, implementing digital soil mapping (DSM) methods could be perceived as capturing such expertise and embedding it into models of the soil formation (Fig. 4.3). These models can be developed by correlating existing soil information with the above-mentioned digital data sources and translating this correlation into quantitative equations. Such models would then act as filters or

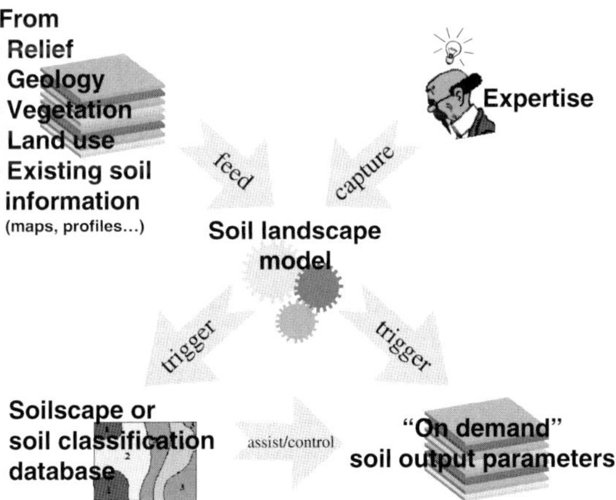

Figure 4.3. *A perception of digital soil mapping (DSM).*

views that would be triggered only on demand upon a database constituted of a pile of layers of easily accessible basic input parameters to produce required output soil landscape and soil parameter layers.

The amount of knowledge and expertise used to produce a soil map is such that no model will be capable of embedding them as a whole. It is thus reasonable to assume that existing soil information systems will still serve as input in the inference process for some time to come. In this perception, the process will be run to improve or enrich the existing soil information system rather than create a new one from scratch.

4.3.2 Improve delineation of Soil Mapping Units

A test area covering the Limousin Region in France was chosen for its wide variety of soil, geologic and relief situations, for the expertise already existing on its soils, and because a regional soil map is in preparation. In the future, more geographically evenly distributed pilot areas in Europe should be chosen to better cover the diversity of geomorphology, soil, land use and climate over the continent. A database was built comprising datasets available easily and freely to cover not only the test area but also the whole of Europe. Two Internet sources were identified: the Shuttle Radar Topography Mission (SRTM) web site (http://srtm.usgs.gov/) from which a digital elevation model (DEM) at 3 arc seconds resolution was downloaded, and the Moderate Resolution Imaging Spectroradiometer (MODIS) web site (http://modis.gsfc.nasa.gov/) from which reflectance and surface temperature datasets at respectively, 500 m and 1000 m resolutions were downloaded. From these databases, a number of relief indicators were derived from slope to potential drainage density index (Dobos, 1998). These indicators allow the characterisation of some soil-forming factors.

Our first objective is to test existing methods – and eventually to propose new ones – which use these indicators to improve the delineation and characterisation of SMUs in the SGDBE1 M. Following Dobos et al. (2001), methods to be tested include supervised classification (maximum likelihood) and unsupervised classification, including fuzzy *k*-means analysis. But however inaccurate they are, we assumed that current SMUs have a strong meaning as a variable in this statistical evaluation. Thus, the principle would be to consider them as a first stratification information when applying the above-mentioned methods rather than to build a completely new delineation of SMUs using only external data.

Another objective is to evaluate the freely available data sources mentioned above with other datasets such as DEMs from the Catchment Characterisation and Modelling (CCM) River and Catchment Database, the French National Geographic Institute (IGN), Landsat 7 images from the Image 2000 database and Spot imagery.

4.3.3 Allocate Soil Typological Units

In the Soil Database, SMUs are generally composed of several soil types (STUs) for which the location is not spatially defined within the SMU polygons. This strongly reduces the significance of SMU delineation because they are a container for a mixture of different soil type objects, which can have very different properties.

A methodology that makes use of high-resolution DEM derived information was set up to spatially reallocate STUs within corresponding SMUs (Bock et al., 2005). This disaggregation method produces a refined delineation in which each new container (SMU) is 'pure', that is containing one and only one soil type (STU) (Fig. 4.4). As a consequence, the soil type object and its spatial definition become more consistent with one another. This refined delineation would

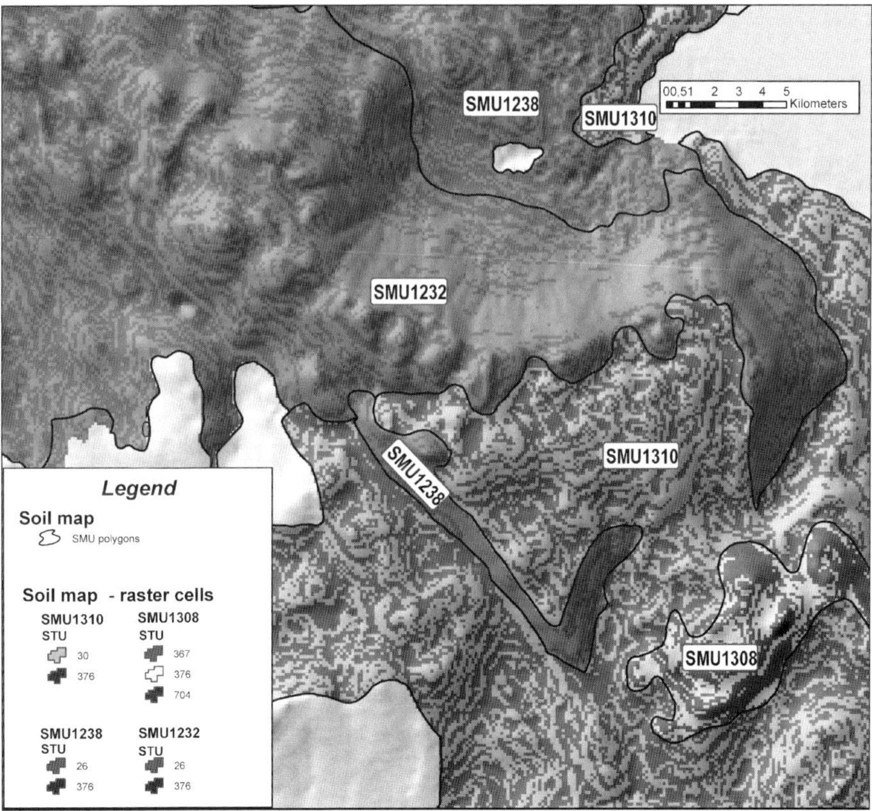

Figure 4.4. *Spatial reallocation of the Soil Typological Units (STUs) within the borders of four adjacent Soil Mapping Unit (SMU) polygons.*

therefore be expected to be much more powerful when serving as a stratifier to seed the multivariate analysis.

The method has been tested with elevation data at resolutions of 100 , 90 and 25 m. In particular, the new SRTM 3 arc seconds (~90 m) resolution DEM offers the opportunity to be used as a European-wide source of detailed topographic information. Some of its constraints and inconvenient have been explored (Vrščaj and Montanarella, 2004, Vrščaj et al., 2006).

4.4 Conclusion

The Soil Geographical Database of Eurasia at scale 1:1,000,000 summarises more than 50 years of collective effort to assemble, improve and harmonise information on soils gathered from almost 40 countries, each one with its own culture and history of soil survey. It is an example of a repository of high-value knowledge. However, it is now time to improve the database taking advantage of state-of-the-art data collection and analysis technologies. Some pre-testing is ongoing but a wider project, involving more partners in a collaborative effort, needs to be set up in the context of refining the European Soil Information System (EUSIS).

References

Bui, E.N., Moran, C.J., 2003. A strategy to fill gaps in soil survey over large spatial extents: an example from the Murray-Darling basin of Australia. Geoderma 111, 21–44.

Bock, M., Rossner, G., Wissen, M., Remm, K., Langanke, T., Lang, S., Klug, H., Blaschke, T., Vrščaj, B., 2005. Spatial indicators for nature conservation from European to local scale. Ecological indicators 5 (4), 322–338.

CEC, 1985. Soil Map of the European Communities at scale 1:1 000 000. CEC, Luxembourg.

Dobos, E. 1998. Quantitative analysis and evaluation of AVHRR and terrain data for small scale soil pattern recognition. Ph.D. Thesis, Purdue University, West Lafayette, IN, USA.

Dobos, D., Montanarella, L., Nègre, T., Michelic, E., 2001. A regional scale soil mapping approach using integrated AVHRR and DEM data. Int. J. Appl. Earth Obs. Geoinf. 3, 30–42.

FAO, ISRIC, ISSS, 1998. World Reference Base for Soil Resources. World Soil Resources Reports No. 84.

FAO-UNESCO, 1974. FAO-UNESCO Soil Map of the World: Vol. 1, Legend. UNESCO, Paris.

Jamagne, M., Daroussin, J., Eimberck, M., King, D., Lambert, J.-J., Le Bas C., Montanarella, L., 2002. Soil geographical database of Eurasia and Mediterranean countries at 1:1 M. 17th World Congress of Soil Science, 14–21/08/2002, Bangkok, Thailand Symposium. 44(494), 1–10.

Jones, R.J.A., Hiederer, R., Rusco, E., Loveland, P.J., Montanarella, L., 2004. The map of organic carbon in topsoils in Europe, Version 1.2 September 2003: Explanation of Special Publication Ispra 2004 No. 72 (S.P.I.04.72). European Soil Bureau Research Report No. 17, EUR 21209 EN. Office for Official Publications of the European Communities, Luxembourg.

Jones, R.J.A., Spoor, G., Thomasson, A.J., 2003. Assessing the vulnerability of subsoils in Europe to compaction: a preliminary analysis. Soil Tillage Res. 73, 131–143.

King, D., Daroussin, J., Tavernier, R., 1994. Development of a Soil Geographical database from the Soil Map of the European Communities. Catena 21, 37–26.

Lambert, J.-J., Daroussin, J., Eimberck, M., Le Bas, C., Jamagne, M., King, D., Montanarella, L., 2003. Soil Geographical Database for Eurasia and The Mediterranean: Instructions Guide for

Elaboration at scale 1:1,000,000, version 4.0. EUR 20422 EN. Office for Official Publications of the European Communities, Luxembourg.
Le Bissonnais, Y., Daroussin, J., Jamagne, M., Lambert, J.J., Le Bas, C., King, D., Cerdan, O., Léonard, J., Bresson, L.M., Jones, R.J.A., 2005. Pan-European soil crusting and erodibility assessment from the European Soil Geographical Database using pedotransfer rules. Adv. Environ. Monitor. Model. 2, 1–15.
Madsen, H.B., Jones, R.J.A., 1995. Soil profile analytical database for the European Union. Danish J. Geogr. 95, 49–57.
Vossen, P., Meyer-Roux, J., 1995. Crop monitoring and yield forecasting activities of the MARS Project. In: D. King, R.J.A. Jones, and A.J. Thomasson (Eds.), European Land Information Systems for Agro-Environmental Monitoring. EUR 16232 EN. Office for Official Publications of the European Communities, Luxembourg, pp. 11–29.
Vrščaj, B., Montanarella, L., 2004. Possible improvements of Vector Soil Maps with DEM data, a GIS method. Seminar. EC Joint Research Centre, Institute for Environment and Sustainability, European Soil Bureau, Ispra, Italy, 25/03/2004.
Vrščaj, B., Daroussin, J., Montanarella, L., 2006. SRTM as a possible source of elevation information for soil landscape modelling. In: G. Jordan, R.J. Peckham (Eds.), Digital Terrain Modelling: Development and Application in a Policy Support Environment. Springer Verlag (in press).

Developments in Soil Science, volume 31
P. Lagacherie, A.B. McBratney and M. Voltz (Editors)

Chapter 5

DEVELOPING A DIGITAL SOIL MAP FOR FINLAND

H. Lilja and R. Nevalainen

Abstract

This chapter describes our methods and experiences of making a 1:250,000 georeferenced soil databases for Finland with digital soil mapping systems. We use the following source materials: maps of quaternary deposits, geophysical low-altitude flight measurements, various GIS – materials from different sources and old soil surveys. We are aware that our pretty simplified way of digital soil mapping, mostly based on a parent material (lithogenic) approach, is vulnerable to international criticism and that is why we explain why this is possible in Finland, which consists of young soils (<12,000 years).

5.1 Introduction

Soils of Finland (~338,000 km^2) were formed quite recently (<12,000 years BP) after the Weichselian glaciation. Owing to their rather weak development, the soils have been nationally classified and mapped according to texture and content of organic matter, and little attention has been paid to pedogenic classification. Therefore, Finland is quite broadly presented in the Soil Map of the World and in the Soil Map of Europe. The country is covered mainly by maps of Quaternary deposits at different scales (1:20,000–1:1,000,000). The Quaternary deposits of about one third of the area have been mapped at scale 1:20,000 while the 1:1,000,000 map is the only one to cover uniformly the whole country. Over the last 10 years, the Finnish national soil classes have been related to the Food and Agriculture Organisation (FAO) and World Reference Base (WRB) systems and to Soil Taxonomy. The soil classes occurring in Finland are closely connected to the parent material and the wetness of the soil. In 2000, we started to compile a new nationwide map and database of the Quaternary deposits and soils at scale 1:250,000 to conform to the requirements of the Georeferenced Soil Database for Europe. The schema of this objective is presented in Figure 5.1. The work is planned to be completed by the end of 2009.

The Geological Survey of Finland (GSF) (soil map layers, aerial geophysics box in Fig. 5.1) maps the soil in the surface layer and at the depth of 1 m as polygons with minimum size 2–6 ha according to Finnish soil classification

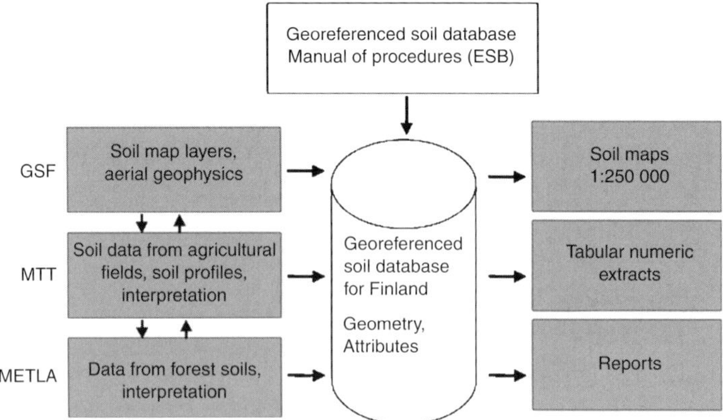

Figure 5.1. *Schema of the procedure for implementing the georeferenced soil database for Finland.*

system. This Finnish system is based mostly on the content of organic matter and soil texture. Geologists verify the results generated this way mainly by using data on soil types obtained from soil testing of agricultural land and from forest soil surveys and from limited amount of fieldwork. The MTT Agrifood Research Finland, as a member of the European Soil Bureau (ESB) Network, derives FAO and WRB names for the soil polygons and builds the geometric and semantic dataset to the database that will conform to the manual of ESB (European Soil Bureau, 2002). The Finnish Forest Research Institute (METLA) produces and makes available information on forest soils. The results of soil samples analysed for the agricultural soil mapping and forest soil inventories are used as a source of data included in the database. About 100 soil profiles have been analysed during the on-going project.

Thus, the geometric part of Quaternary geology map 1:200,000 and soil map 1:250,000 are both based on the same basic data; polygons, which have determined and verified top- and bottom-soil texture according to the Finnish qualification system. The GSF will produce 1:200,000 Quaternary geology maps on the basis of that data and MTT bases the geometrical part of 1:250,000 soil database on the same data.

5.2 Available soil data in Finland

5.2.1 History of soil mapping in Finland

In Finland, systematic geological mapping started more than one hundred years ago. Several stages can be seen in the mapping history, reflecting both the increasing knowledge of the geological history as well as the changing needs of society. In short, combined geological maps from the Precambrian crystalline bedrock and the overlying loose Quaternary deposits were first produced in the scale of 1:200,000. Later on, during the first half of the 20th century, geological maps in the scale of 1:400,000, separately from bedrock and Quaternary deposits, were produced to give a complete coverage of Finland. This period was followed by a more detailed mapping programme at the 1:100,000 scale including separate maps from both the bedrock and the Quaternary formations. Finally, a new mapping programme of the Quaternary deposits in a scale of 1:20,000 was started in cooperation between the GSF, the National Land Survey (NLS) and the Agricultural Research Centre of Finland (MTT) in the beginning of 1980s. The programme, accompanied with a renewal of the Finnish base map system, focused mainly on areas of intensive land use, including agricultural areas in southern Finland.

In the late 1990s, the GSF started to develop methods for a project to produce a new countrywide map and database of the quaternary deposits of Finland at scale 1:200,000, using modern GIS techniques and utilising previous soil maps of different scales and numerous other data sources.

5.2.2 MTT data

MTT has three different datasets

The first dataset consist of 36 measured profiles (Yli-Halla et al., 2000). The properties of soils in this dataset have been separately determined for surface soils (Ap-horizon), changed bottom soil (B-horizon) and unchanged bottom soil (C-horizon).

The second one is an agricultural soil dataset. This dataset consists of texture determinations of some 18,000 samples and plenty of basic chemical soil data, which have been put into digital form. The properties of soils in the dataset have been separately determined for surface soils to a depth of 20 cm and for changed bottom soil (samples taken between 15 and 80 cm).

The third dataset contains only surface soils from agricultural lands, but the number of samples is very high (about 122,000). This dataset contains topsoil texture determinations classified by Finnish agriculture soil classification system and a lot of basic chemical data like pH, Ca, Mg, K, Na and P as mg/l of soil.

5.2.3 METLA data

METLA is so-called national forest inventory data, 'National Forest Invent', and is made once every 5 years. This data contains information on thickness of organic and mineral layers from 70,000 sampling sites. The soil classification is made according to Finnish classification system.

5.3 Methodology

5.3.1 Quaternary mapping and mapping of surface layer

In Finland moraine is the most common substrate covering 53% of land area, whereas glaciofluvial deposits cover 17% and organic post-glacial peat deposits 16%. Bedrock outcrops cover 14%. With the methods used in the new 1:200,000 mapping programme glaciofluvial deposits, till formations, sedimentation basins and geomorphologic features are delineated using a digital elevation model (DEM) with 20-m resolution, NLS Topographic Database, gravel deposit maps (1:20,000) together with observation and field data from earlier investigations and projects in Finland. Other data includes remote sensing data and digital aerial photographs.

The details of the Finnish terrain and the country's built environment have been incorporated in the NLS Topographic Database, the source of the most accurate information about Finland's topography. The topographic database covers the whole of Finland, except for the northernmost parts of Lapland. This database is used as very accurate digital maps together with global positioning system (GPS) in the field mapping work. The following area representations are also taken from NLS Database: rock exposures, boulders, mineral resources extraction and paludified areas.

The mapped moraine landforms contain drumlins, flutings, hummocky moraines and end moraines. Glaciofluvial accumulations include formations like eskers and deltas. Other sorted depositional features consist of beach deposits, fluvial deposits and aeolian deposits. All these are very easily distinguished from DEM and topographic data. Also sedimentation basins of post-ice-age sea- and lake-stages are outlined using elevation data and DEMs to discover areas where deposition of fine-grained sediments like silt and clay has taken place.

5.3.2 Airborne geophysical data

An important development during the mapping project has been the classification of airborne geophysical data for the purpose of the interpretation. Especially the use of radiometric data components, potassium (K), thorium (Th) and uranium (U) and their ratios has proved to be useful in mapping soil deposits (Lerssi et al., 2003; see also Chapter 16). The use of ratios removes the effect of moisture changes, so the K/Th-ratio can be used to distinguish coarse- and fine-grained sediment formations (Fig. 5.2), (Hyvönen et al., 2002). The usefulness of airborne data is, in contrast, strongly dependent on variations of gamma radiation or conductivity in bedrock, which can obscure soil effects. In some areas it is therefore difficult to use geophysical data for soil mapping, which increases the need of fieldwork in these areas.

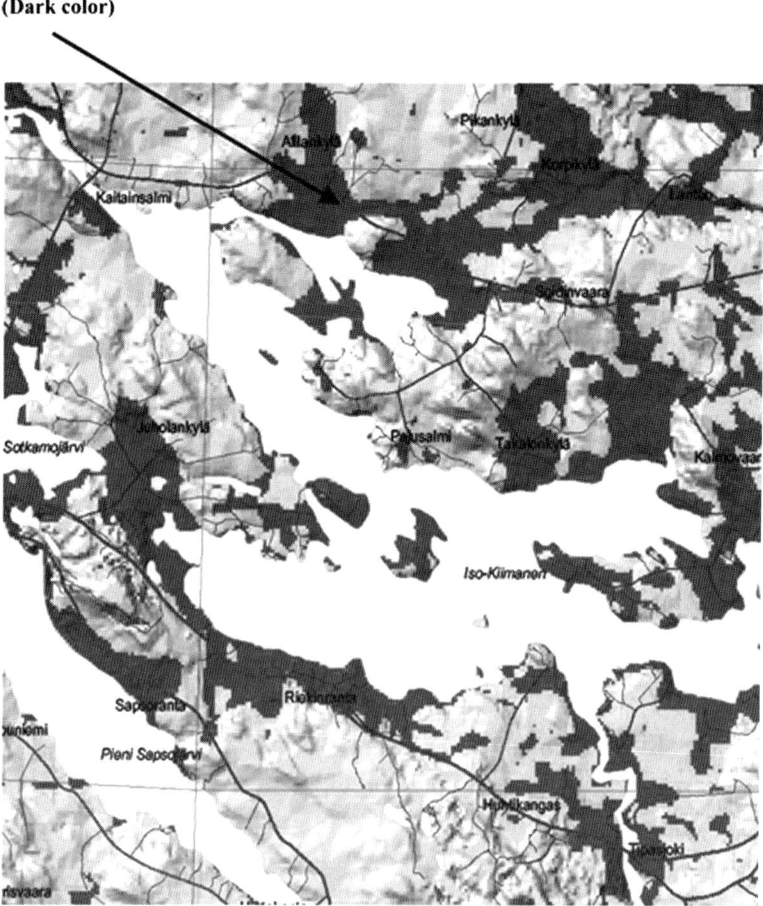

Figure 5.2. *Delineating fine-grained sediments with classified airborne electromagnetic imaginary component*

Peat deposits are drawn using topographic maps (1:20,000), aerial photographs and peat database of GSF. The thickness of mires is varying very much. Some are very thin, especially in the Western Finland (Ostrobothnia), and the mires are commonly deeper but smaller where the topography fluctuates more. Using classified airborne radiometric potassium (K) data with peat-drilling data the thickness of peat deposits can be separated into two classes, that is (1) thinner than 0.6 m and (2) thicker than 0.6 m. Wet soils, mostly in low topographic positions, can also be identified (Plate 5) (see Colour Plate Section).

After interpretation the geological soil data polygons at a depth of 1 m are combined to a single layer using GIS techniques and ArcMap software. The surface soil polygons are mapped mainly by interpreting airborne geophysical data together with topographic maps using ERMapper software and image processing techniques. Surface layers consist of polygons of thin peat (<60 cm), wet soils, thin till over unweathered bedrock, outcrops and boulder fields. To check the interpretation, soil data from MTT and METLA (Finnish Forest Research Institute) is used as a reference. There are about 300,000 topsoil texture observations from agriculture fields and 70,000 topsoil texture observations from forest areas. The amount of the fieldwork is quite limited; about 1 week is reserved for two people to check an area of 2000 km^2 in the field. Finally the revised map is stored in a database and published in the form of a Quaternary Geological Map at 1:200,000. Polygons of different soil layers (raw data) are delivered to MTT for deriving FAO/WRB classification (Plate 5).

5.3.3 Food and Agriculture Organisation and World Reference Base classifications

The allocation to FAO and WRB classifications is based on over 50 classified profiles, which were classified according to these systems. Our studies on soil profiles (e.g., Mokma et al., 2000; Yli-Halla and Mokma, 1999; Yli-Halla and Mokma, 2002) indicate that clay eluviation/illuviation has not occurred to the extent that argic horizons would occur. A relatively well-drained mineral parent material commonly results in a given FAO/WRB soil class, as follows: clay soils → Vertic Cambisols, loamy soils → Eutric Cambisols, silt soils → Eutric Regosols, sandy soils and glacial till → Haplic Podzols, young sandy fluvial and aeolian deposits → Arenosols. Wet soils, mostly in low topographic positions, can be identified by geophysical low flight measurements, currently covering the whole of Finland. Depending on the thickness of the organic topsoil, Dystric Gleysols and Dystric/Eutric Histosols can be identified. We applied this approach by incorporating digital elevation, slope, aspect, soil wetness, and soil depth and soil organic matter as covariates. The soil depth is treated as a covariate because thick and thin peat will get different FAO classifications. Soil organic matter is included because of its significant role in Finnish Soil Classification system.

5.3.4 The procedure to characterise individual soil properties

The properties of individual soils are estimated statistically from MTT's datasets. The properties of soils in the first dataset have been separately determined for surface soils (Ap-horizon), changed bottom soil (B-horizon) and unchanged bottom soil (C-horizon). From the second dataset 2854 profiles were selected for statistics. The following properties were estimated statistically: pH, estimated CEC on the basis of Ca content, top and bottom heights of a profile and the texture percentage. Statistics from the third dataset was calculated per county. In these calculations we had to estimate that surface soil describes well enough the distribution of soils at a county level. The following properties were statistically estimated: pH, Ca, Mg, K, Na and P as mg/l of soil.

5.4 Results

Our system to produce a geometric and semantic dataset has been proved to work and it is now in working condition. The necessary computer applications have been developed and they have been tested to work. We have achieved about 35% coverage of Finland in the geometric part of the database. Owing to the uniqueness of our approach we have a need to test the reliability of our new soil maps. For that purpose, we will select test areas from different topographical regions. These areas should also represent different soil regions. In those test areas we will also investigate methods like classification and regression trees, fuzzy systems, neural networks and limited field surveys to discover the reliability of the maps.

5.5 Conclusions

There is an increasing need for soil data for environmental impact assessment. The authorities are often faced with the fact that sufficiently detailed soil information (particularly in digital form) is non-existent in the area of interest. Better supply of soil data is required also as a consequence of the EU Thematic Strategy for Soil Protection and the probable Soil Framework Directive.

Computerised methods for data handling and advanced techniques for preparing soil maps and databases are now available, and they can effectively be used for the production of soil databases, if sufficiently verified and supported by measured data. The recently launched project to produce a map and database of soils and Quaternary deposits at scale 1:250,000, conforming to the manual of the ESB and expected to be completed by the end of 2009, will make soil data of Finland nationally and internationally more available.

In the validation (ground truthing) we have in Finland quite good and continuously updated reserve. In our country a farmer must take a sample from his field segments once in a 5-year period to get agro-environmental subsidies. Laboratories analyse the samples, and the texture of surface soil is also determined. A farmer can give EU-registration codes of his field segments to a laboratory. Giving this code is not compulsory, however. There are about a million field segments in the Finnish field segment registry system. If the registry code of the soil sample and the code in field registry match, we can get a point, which should represent a texture of topsoil in the segment. According to our experiences about 30% of analyses can be made to match. This means about 300,000 determined topsoil texture points for ground truth in area of Finland.

This opportunity has raised some discussions of soil monitoring possibilities. Combined with National Forest Invent data we could have about 370,000 points with determined topsoil texture covering practically the whole area of Finland. From that basis we could produce a topsoil map for the whole country, which could be updated every 5 years. An interesting option to be investigated for soil monitoring issues.

References

European Soil Bureau, 2002. Georeferenced Soil Database for Europe, Manual of Procedures Version 1.1, European Soil Bureau, Scientific Committee, 172pp.

Hyvönen, E., Turunen, P., Vanhanen, E., Sutinen, R., 2002. Airborne gamma-ray surveys in Finland. Geological Survey of Finland, Special Paper 36.

Lerssi, J., Hyvönen, E.,Väänänen, T., 2003. Airborne geophysical surveys assessing the general scale Quaternary mapping project in Finland. 9th Meeting of Environmental and Engineering Geophysics, Prague, 2003.

Mokma, D.L., Yli-Halla, M., Hartikainen, H., 2000. Soils in a young landscape on the coast of southern Finland. Agr. and Food Sc. in Finland 9, 291–302.

Yli-Halla, M., Mokma, D.L., 1999. Classification of soils of Finland according to soil taxonomy. Soil Surv. Horizons, 40, 59–69.

Yli-Halla, M., Mokma, D., 2002. Problems encountered when classifying the soils of Finland. European Soil Bureau Research Report 7, pp. 183–189.

Yli-Halla, M., Mokma, L., Peltovuori, T., Sippola, J., 2000. Suomalaisia maaprofiileja. Abstract: Agricultural soil profiles and their classification. Maatalouden tutkimuskeskus. Sarja A 78, 104pp.

C. Conception and handling of soil databases

Soil information systems (databases) have been developed and improved since the mid-1970's. Most countries now have national databases of soil profiles and properties. Some are better populated than others and many have poor geographic referencing which is a problem for Digital Soil Mapping. Also, there is no particular standard for these databases. Increasingly, international databases are required for solving large-scale environmental problems or looking at food-security issues. The future development of soil databases should not be performed without consideration of the spatial infrastructures that are being developed everywhere in the world, and Digital Soil Mapping must take in consideration the standards that are proposed, and interact with the data that proposed spatial infrastructure could provide. Part C illustrates some of these aspects.

Chapter 6 discusses the problem of standardising spatial soil information systems and harmonising them between countries. These issues are illustrated for the European Union through the advent of the proposed INSPIRE directive which is a framework for minimum requirements for spatial data infrastructure for members states.

Chapter 7 outlines the development and adoption of a new database structure for the storage, maintenance and retrieval of soil data in the Australian state of Victoria, which facilitates the improved distribution of data to state, national and international communities.

Uncertainty estimation is a key feature of DSM. This begins with the uncertainty estimates of soil data themselves. Chapter 8 presents a framework to facilitate the storage of information about soil data quality, including the uncertainties associated with soil data within a conventional database design. Characteristic uncertainty models are defined for different soil 'data types', classified according to their attribute scale and their space-time variability.

Chapter 9 describes a quantitative procedure for creating a harmonised soilscape cover for the whole of Europe by using digital elevation data. The procedure is based largely on the SOTER (soil and terrain digital data base) approach (ISRIC, 1995). It may be seen as a useful first step towards the development of soil information systems.

In Chapter 10, the problem of semantic matching when multi-source data of different origin need to be integrated in a unique prediction procedure or in a common soil information system is raised. This issue has been rarely addressed in contrast to spatial merging of data of different resolution and sampling density. The chapter explores how the specification of the conceptualisation, so-called ontology, behind each data set can help the semantic data integration.

The challenge will be to adapt these aspects to heavily-populated spatial soil information systems.

Developments in Soil Science, volume 31
P. Lagacherie, A.B. McBratney and M. Voltz (Editors)

Chapter 6

ADAPTING SOIL DATA BASES PRACTICES TO THE PROPOSED EU INSPIRE DIRECTIVE

J. Dusart

Abstract
Harmonised and standardised spatial reference data are increasingly essential to support environmental policies (e.g. Soil Thematic Strategy). However, soil databases often appear heterogeneous, lack common geographical references or are poorly documented (metadata). Analysis based on such data sets may generate inconsistent results that will not benefit the users. The opportunity to review and adapt soil-mapping practices came with the initiative taken by the European Commission to position the establishment and operation of an infrastructure for spatial information in Europe under a legal framework, the INSPIRE Directive. The proposed Directive includes minimum requirements for harmonising, documenting and disseminating spatial data. Member States will be asked to establish and operate infrastructures for different spatial themes in accordance with a set of technical guidelines still to be defined and approved. Over the last 10 years, the European Soil Bureau Network has gained much experience in establishing harmonised soil data for Europe and can, therefore, provide an ideal area of experimentation for the practical implementation of the recommended technical guidelines. This chapter reviews some of the changes to be made to the 1:1,000,000 scale European Soil Database to fulfil the INSPIRE requirements.

6.1 Introduction

In April 2002, the European Commission approved a Communication entitled 'Towards a Thematic Strategy for Soil Protection' (CEC, 2002). This communication recognises eight major threats to the soils of Europe and proposes strategies for preventing soil degradation and the integration of soil protection into other policy areas. Following the publication of this innovative text, several technical working groups (TWG) were established to make recommendations on initiatives to be taken for fulfilling those objectives (CEC, 2004a).

Of particular interest is statement from the TWG on monitoring the urgency to co-ordinate efforts among EU Member States to implement appropriate soil monitoring systems. This analysis has been confirmed by a GMES study on building a European information capacity for environment and security (CEC,

2004b). In order to support the development and implementation of an effective and integrated policy on soil protection, information is needed on a number of different issues that include soil types and their distribution, soil properties and soil degradation.

However, the history of soil survey in different countries has been very different (Jones et al., 2005) and, as a consequence, the methods and principles of soil classification and description have varied. Considerable problems have arisen in developing consistent databases on soils even if several stakeholders have initiated efforts of harmonisation. One might quote among many initiatives the successful completion of the European Soil Database at the scale 1:1 million (King et al., 1994; Lambert et al., 2003).

6.2 Tackling the problems identified: the INSPIRE initiative

With the increasing interest in the spatial dimension of environmental policies and the need for more detailed reporting, Commission services and numerous stakeholders involved in the use of spatial data have started to debate the possible development of a co-ordinated framework for spatial data infrastructure. From this long and tenuous process, a proposal for a new Directive, the INSPIRE Directive, has been approved recently by the Commission and should be submitted to the European Parliament and to the Council in the coming 2 years (CEC, 2004c). Once approved, the Member States will have to transpose the Directive into their national legislation and start implementing it.

The Directive's main objective is to contribute to the formulation, implementation, monitoring and evaluation of Community policies at all levels, from the EU to the local scale. The Directive includes minimum requirements for harmonising, documenting and disseminating spatial data. Among the spatial data themes targeted, soil and subsoil should be characterised according to depth, texture, structure and content of mineral particles and organic material, stoniness and, where appropriate, mean slope and anticipated water storage capacity.

Member States will be expected to establish and operate infrastructures for those spatial themes in accordance with a set of implementing rules (still to be defined and approved) before the Directive might be transposed and implemented. The guidelines include, for example, common coordinate reference systems for uniquely reference geographical information, a harmonised multi-resolution grid net, standardised metadata collection and access, and industry standards for web dissemination.

The European Soil Database (ESBSC, 2003) is a good example of a pan-European dataset that offers room for improvement according to those guidelines. Crossing the recommendations and requirements issued by the above-mentioned working groups with the INSPIRE principles helps to identify actions to be taken to improve the potential use of the database and ensure cross-compliance between soil thematic strategy objectives and the Directive (Table 6.1). INSPIRE principles and their implementing guidelines may be seen as filters that guarantee the building of a harmonised and distributed infrastructure for spatial data.

Table 6.1. *Soil thematic strategy requirements and INSPIRE principles.*

Soil strategy spatial data requirements	Inspire principles	Actions to be taken
Establish a common EU-wide soil inventory (baseline) containing general soil parameters and specific parameters and explore the possibility of achieving a stronger EU coordination of soil monitoring activities through a EU Soil Conservation Service	Spatial data should be stored, made available and maintained at the most appropriate level relevant for European (avoid duplicates)	Central EU Portal for Pan-European data relevant for European policies and additional integration at European level through ESB network
Select a minimum set of common parameters to be monitored (...) which should be part of the existing soil monitoring systems at national level	Combining spatial data from different sources across the EU seamlessly and sharing them between several users and applications must be possible	Common soil classification (WRB), common reference system and geometric consistency with reference layers
Promote the adoption of standardised methods and procedures for the measurement of the selected common parameters	Sharing spatial data collected at one level of public authority between all public authorities must be possible	Use of same methods and exchange formats at different scales and use of a common reference grid for versions of soil database
Establish a regular reporting procedure for the selected parameters from the Member States to the European Commission	Spatial data needed for good governance should be available under conditions that do not restrict their extensive use	Provide access to soil parameters of interest to the identified threats respecting ownership and property rights
Organise regular quality control/quality assurance procedures	Discovering available spatial data, evaluating their fitness for purpose and knowing the conditions applicable to their use should be easy	Document data with metadata (ISO 19115) that include QC/QA info, ownership, access rights and provide searching tools for access to metadata based on standards

Among the elements of the present version of the European Soil Database that could be improved, we can identify the geometric and thematic inconsistencies between countries, the non-standard projection system, the use of multiple reference classifications, the proprietary vector format not fitted for database modelling and the absence of ISO compliant metadata.

Therefore, for a practical point of view, it is somehow difficult to compare the pedogenesis of pan-Europe regarding the soil types because the level of semantic and geometric detail is not the same for each country, and because the mapping processes are not the same, depending on the soil classifications used.

In the coming paragraphs, the special requirements outlined above and the actions to be taken to fulfil them are discussed. Actions will have to be taken on the basis of agreed implementing rules that will complement the Directive.

6.3 The right data at the right level: towards a central EU Portal

Reducing redundancy of data or use of inappropriate data at the wrong level is a key INSPIRE principle. At the European level, the building up of a common EU-wide soil inventory containing general and specific soil parameters will constitute the baseline for European policy implementation. In order to achieve this, the Institute for Environment and Sustainability is setting up a soil portal where users can access all sorts of soil information of relevance for EU-level policies: http://eusoils.jrc.it (last accessed: 10. 07. 2005). The data accessible through the portal will essentially come from the European Soil Database at the 1:1 million scale and derived products.

In the coming years, Member States are expected to build similar access points with data relevant for national or regional scopes. An EU Soil Conservation Service (similar to the one existing in the United States) could have as its main mandate the coordination of soil-monitoring activities throughout Europe and the hierarchical compatibility between levels. The European Soil Bureau Network has largely demonstrated the benefits of a coordinated approach (Montanarella et al., 2005).

This strategy has already been encouraged by the European Environment Agency (EEA, 2001) through its proposal for a comprehensive framework for monitoring, assessing and reporting on soil conditions in Europe. The European Environment Information and Observation Network, EIONET, has largely contributed to the building of a common and harmonised infrastructure but restricted to the political duties of the Agency, such as the publication of the Environment Action Programme. The proposal for an INSPIRE Directive recognises that "without a harmonised framework at Community level, the formulation, implementation, monitoring and evaluation of national and Community policies that directly or indirectly affect the environment will be hindered by the barriers to exploiting the cross-border spatial data needed for policies which address problems with a cross-border spatial dimension" (CEC, 2004c).

6.4 Harmonisation

The proposal for an INSPIRE Directive accords with existing initiatives, meaning that in no case will the Directive substitute national or regional initiatives (concept of subsidiarity) neither will go beyond what is requested by the Directive (concept of proportionality). However, Member States are expected to harmonise their existing soil-monitoring initiatives with the objective to make them exchangeable and spatially continuous (seamlessness).

Harmonising spatial datasets of soil properties supposes that there will be agreement between all stakeholders on the use of common guidelines. The European Environment Agency has released a draft version of guidelines (EEA, 2005) that will evolve within the coming years in parallel with the execution of the working programme of the INSPIRE preparatory phase (CEC, 2004d).

First of all, the highest priority will go to the definition of Annex 1 harmonising and interoperability rules. Regarding Annex 3 (that includes soil themes), priority will be given to geometric consistency, as the perspective of using a common data model goes beyond the objectives of the Directive. Member States have to ensure that their thematic data integrate in a broad infrastructure. However, the success of the European manual of procedures (Lambert et al., 2003; ESBSC, 2003) or the adoption of a common classification scheme (World Reference Base for Soil Resources) demonstrates that specific rules including a data model can be agreed by a large number of stakeholders.

The manual of procedures will be updated in order to reflect the adoption of common geodetic systems (ETRS89 and Lambert Azimuthal Equal Area, "ETRS–LAEA"), and common hierarchical reference grids based on those reference systems. Guidelines will include metadata documentation and discovery, network data services accessible through Internet and updating/uploading data flows.

6.5 Data sharing and re-use

The proposal for an INSPIRE Directive will include measures for increasing the potential of re-use of spatial data sets and network services between public authorities and with third parties. By using a common reference grid instead of the traditional vector-mapping approach as a baseline for soil mapping and monitoring, one can expect that data prepared at lower levels (regional or local scale) can be re-used at higher levels (national or European). The use of a common reference grid facilitates the integration of different types of source data such as digital elevation models, satellite images and land cover datasets, even with different resolutions for the purpose of digital soil mapping.

In addition, permanent soil monitoring will be facilitated through this approach. The French RMQS (*Réseau de Mesures de la Qualité des Sols*) demonstrates some of the benefits of a common reference grid for soil monitoring on a regular basis: http://gissol.orleans.inra.fr/programme/rmqs/rmqs.php (last accessed: 11. 07. 2005). Adopting a common pan-European reference grid will contribute to prepare the ground for a proposal for soil monitoring legislation.

However, the use of a common reference grid supposes that the local level is interested and receives the necessary support for producing highly detailed datasets that might be aggregated at higher levels. The sampling techniques used for collecting soil properties require revision, approach that is far from obtaining the unanimity within the Soil Community. The next LUCAS survey (land use/cover area frame statistical survey) based on the proposed pan-European grid could be an opportunity for testing soil properties collection on a regular grid with stratified sampling.

An example is currently being applied in the preparation of an Eco-pedological map for the Alpine Territory: http://eusoils.jrc.it/projects/alpsis/Main-Alpine.html (last accessed: 10. 07. 2005). A common reference grid snapped on the Lambert Azimuthal Equal Area projection and available at different resolutions (1 and 10 km) constitutes the basis for delivering soil parameters following guidelines adopted in the manual of procedures. More detailed soil information available at regional or national scale will be used for updating the 1:1 million scale European Soil Database.

Another area of improvement for the exchange of data and their re-use resides in the publishing of services based on agreed OpenGIS specifications such as web-mapping services and web feature services. Data services linking opens the door to added value analysis by combining map services. For example, overlaying a web service based on the European Soil Database with the IMAGE 2000 map service used for deriving the CORINE Land Cover 2000 (Nunes de Lima and Peedell, 2004) can be used for the revision of the limits of the soil-mapping units in future releases of the European Soil Database.

6.6 Data access and restrictions

The INSPIRE legal process will affect the types of right, access and use of spatial data between public authorities. This legal framework prolongs initiatives adopted elsewhere such as the re-use of Public Sector Information Directive or the 1998 Aarhus Convention on Access to Information, Public Participation in Decision Making and Access to Justice in Environmental Matters.

However, soil mapping/monitoring often involves particular rights of use regarding privacy. Ownership and copyright concerns of the different contributors have as well to be respected. The complex schema of the European Soil Database demonstrates how restrictions of use can hamper access to soil data. Further processing, such as the rasterisation of the vector data or their publication as map services, has been used for facilitation of access to data. The EUSOILS portal currently contains a raster version of many of the attributes (soil properties) stored in the European Soil Database at a 10 km resolution. All the soil properties layers at higher resolutions can be viewed and queried but cannot be downloaded.

6.7 Metadata

In article 8 of the proposal for an INSPIRE Directive, Member States are requested to ensure that metadata for spatial data sets and services are created and that those metadata are kept up to date. The Directive specifies the minimum information to be included in the metadata and expects that Member States take all the necessary measures to ensure that metadata are complete and of high quality. A calendar fixes the creation of metadata for spatial data sets such as soil to be ready within 6 years after the Directive takes force.

Documenting spatial data are essential parts for establishing spatial data infrastructure portals as they encourage visibility of the data, evaluation of their fitness for purpose and conditions of use. Completeness and high quality of metadata can only be achieved by referring to commonly agreed standards such as ISO 19915:2003 metadata standard. Each thematic community must specify its own profile of the standard. For example, the European Environmental Agency recommends the use of ISO 19115 for environment spatial data documentation by providing a specific profile: http://eionet.eu.int/gis/docs/EEA-MSGI_-v1_1a.doc (last accessed: 10. 07. 2005). The Soil Community might consider using this profile or defining its own specific one.

Metadata have to be prepared carefully, as the existing standards are still evolving. The next metadata implementation specification ISO 19139 will produce an XML schema defining how metadata conforming to ISO 19115 should be stored. When ISO 19139 is finalised, existing metadata will have to be migrated to the new XML format specified by that standard.

Among the remaining issues to tackle, one can point to the following topics: multilinguism, certification/quality for the described resources, thesauri/gazetteer and updating policy of metadata. An INSPIRE core profile will be delivered with a set of rules on how to extend this core profile to specific needs of thematic communities.

6.8 Conclusions

Beyond the different topics discussed in this chapter, there are obviously specific issues that must be tackled such as the organisational, data policy and funding aspects. To better assess the benefits of this substantial effort, awareness has to be raised within the major stakeholders by involving them directly in preparing the implementation measures.

By adopting a proactive attitude, the Soil Community will be able to contribute usefully to the definition of the common implementation rules. Involvement and feedback from thematic communities will guarantee that user needs are taken into account, and the whole approach will provide an ideal testing ground for INSPIRE.

References

CEC, 2002. Communication of 16 April 2002 from the Commission to the Council, the European Parliament, the Economic and Social Committee and the Committee of the Regions: Towards a Thematic Strategy for Soil Protection. [COM(2002) 179 Final]. URL: http://europa.eu.int/eur-lex/lex/LexUriServ/site/en/com/2002/com2002_0179en01.pdf; last accessed: 11. 07. 2005

CEC, 2004a. Soil Thematic Strategy. Working Group on Monitoring: Executive Summary and Final Report. URL: http://forum.europa.eu.int/Public/irc/env/soil/library?l=/reports_working/reports_monitoring/exsumm-150504-finaldoc/_EN_1.0_&a=i; last accessed: 07. 07. 2005

CEC, 2004b. Building a European information capacity for environment and security. A contribution to the initial period of the GMES Action Plan (2002–2003). EUR 21109 EN 150pp. Office for Official Publication of the European Communities, Luxembourg.

CEC, 2004c. Proposal for a Directive of the European Parliament and of the Council establishing an infrastructure for spatial information in the Community (INSPIRE), [COM(2004) 516 Final].

CEC, 2004d. INSPIRE Work Programme Preparatory Phase 2005–2006. Final Draft version 4.5.3. 78pp. European Commission. Brussels. URL: http://inspire.jrc.it/reports/rhd040705WP4A_v4.5.3_final-2.pdf; last accessed: 08. 07. 2005

EEA, 2001. Proposal for a European soil monitoring and assessment framework. Technical report 61 EN, 47pp. European Environment Agency. Copenhagen.

EEA, 2005. Guide to geographical data and maps. Draft version 1.3. *EEA operational guidelines*. 55pp. European Environment Agency. Copenhagen. URL: http://eionet.eu.int/gis/docs/EEA_GIS-guide_2005_June.DOC; last accessed: 07. 07. 2005

ESBSC, European Soil Bureau Scientific Committee, 2003. Georeferenced Soil Database for Europe: Manual of Procedures Version 1.1. EUR 18092 EN, 184pp. Office for Official Publication of the European Communities, Luxembourg.

Jones, R.J.A., Houšková, B., Bullock, P., Montanarella, L. (Eds.) 2005. Soil Resources of Europe, 2nd edition. European Soil Bureau Research Report No. 9, EUR 20559 EN, 420pp. Office for Official Publications of the European Communities, Luxembourg

King, D., Daroussin, J., Tavernier, R., 1994. Development of a soil geographical database from the soil map of the European Communities. Catena 21, 37–56.

Lambert, J.J., Daroussin, J., Eimberck, M., Le Bas, C., Jamagne, M, King, D., Montanarella L., 2003. Soil Geographical Database for Europe & The Mediterranean: Instructions Guide for Elaboration at Scale 1:1 000 000. Version 4.0. EUR 20422 EN, 64pp. Office for Official Publications of the European Communities, Luxembourg.

Montanarella, L., Jones, R.J.A., Dusart, J., 2005. The European Soil Bureau Network. In: R.J.A Jones, B. Houšková, P. Bullock, L. Montanarella (Eds.), Soil Resources of Europe, 2nd edition. European Soil Bureau Research Report No. 9, EUR 20559 EN, 420pp. Office for Official Publications of the European Communities, Luxembourg.

Nunes de Lima, V., Peedell, S., 2004. IMAGE 2000 – A European Spatial Reference. Proceedings, 10th EC GI & GIS Workshop, ESDI State of the Art, Warsaw, Poland, 23–25 June 2004.

Developments in Soil Science, volume 31
P. Lagacherie, A.B. McBratney and M. Voltz (Editors)

Chapter 7

STORAGE, MAINTENANCE AND EXTRACTION OF DIGITAL SOIL DATA

C. Feuerherdt and N. Robinson

Abstract

With the advent of today's technologies more and more soil information is being collected. The increased volume of data collected with these technologies poses questions regarding the storage, maintenance and retrieval of these data. This chapter will outline the database structure adopted and engineered by the Victorian Department of Primary Industries (DPI) from the Australian Soil Resource Information System (ASRIS). The database structure, its collection mechanisms, storage, maintenance and retrieval of data will be discussed, focussing on the improved distribution of data by DPI to state, national and international communities.

7.1 Introduction

There are vast amounts of spatial datasets in existence today. This data is housed in a variety of data libraries ranging from simple file-based systems through to complex database structures. The concept of information domains can help bring order to otherwise disparate datasets.

All data relating to natural resources fit into one of three broad information domains:

1. Climate – incorporates data relating to temperature, rainfall, radiation, etc.
2. Surface – includes land use, surface water, biodiversity, etc.
3. Earth – comprising soil, groundwater, regolith, geology, etc.

Previously, each dataset has been seen as a separate entity. This has made it difficult to analyse different datasets even from the same domain because they have been treated heterogeneously. The adoption of information domains for classifying data has benefits for data analysis, as similar data is stored in a logical, consistent structure.

There are many organisations across Australia that are collecting, storing and utilising soil data. At the national level, the National Heritage Trust (NHT) has undertaken the National Land and Water Resources Audit (NLWRA), which is an attempt to collate and assess a variety of biophysical assets. The data collated as part of this project has been sourced from relevant authorities in each Australian state.

In Victoria there are two state departments that collect and maintain biophysical data, the Department of Sustainability and Environment (DSE) and the Department of Primary Industries (DPI). Primary Industries Research Victoria (PIRVic) is a division of DPI focussing on research and development.

The NLWRA commissioned a team of experts to create a nationally consistent soil database from various states and national sources. A whole of earth domain approach was adopted during the development of the database structure allowing, potentially, any data from the earth domain to be stored in a single database, providing users to access data in ways previously unthought. The resulting database, Australian Soil Resource Information System (ASRIS), provides a structure capable of storing a range of biophysical data, but is currently focussed on soil data. Since its inception, the data structure has been refined to less than 12 tables and contains more than 16,000 fully characterised soil profiles from the agricultural regions of Australia (Johnston et al., 2003).

Previously, a substantial amount of time and effort was put into maintenance of the Victorian Soil Site Database (VSSD), which comprised some 40 distinct tables. This cumbersome structure required a high level of knowledge relating to soils to input, update and extract data.

Rather than go through the process of developing a new database structure, the adoption of the existing ASRIS database was more efficient and would provide future benefits. Prior to the implementation of the new data structure, the VSSD contained over 2000 fully described soil sites. All soil information collected in the field was recorded on paper sheets; a soil technician then manually entered these into a Microsoft Access database. This was a laborious process, with only several points being entered per week. This lack of currency reduced the value of the entered data and resulted in occasional use by soil scientists.

This chapter outlines the adoption of a new database structure for the storage, maintenance and retrieval of soil data in Victoria. The alterations made to the ASRIS structure are discussed in the context of the issues raised by the creators of the structure and the benefits they provide to soil scientists in the DPI.

7.2 Methods and data

In order to allow seamless data migration between PIRVic and the federal authorities it was decided to adopt the ASRIS data structure. In order for the database to be useful in PIRVic it would need to be accessible and useable across all aspects of the business chain (Fig. 7.1). The current ASRIS database is focused on obtaining resource data rather than explicit data collection. The alteration of the ASRIS data structure to increase its applicability across the breadth of the business chain was the ultimate focus of the project.

Prior to the existing data from the VSSD data being migrated into the new structure, thorough crosschecking of the code tables was necessary to compare and document the missing, different and additional codes and measures. This resulted in the creation of a master code table (Fig. 7.2 – Classification codes) containing all codes from ASRIS and the VSSD.

Once the comprehensive codes tables were created and additional tables incorporated, existing data were ready for migration. A separate table was

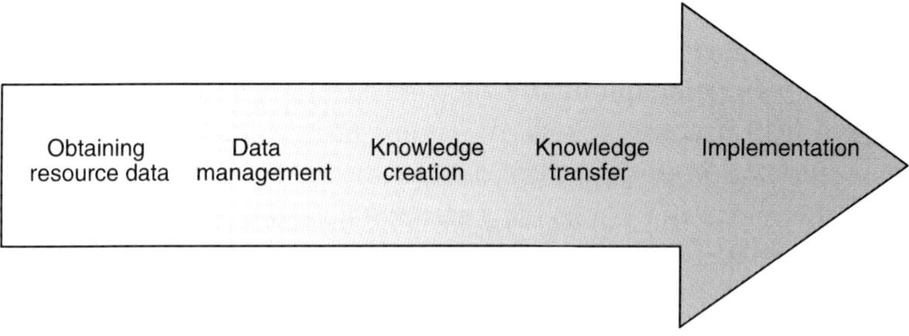

Figure 7.1. *Business chain.*

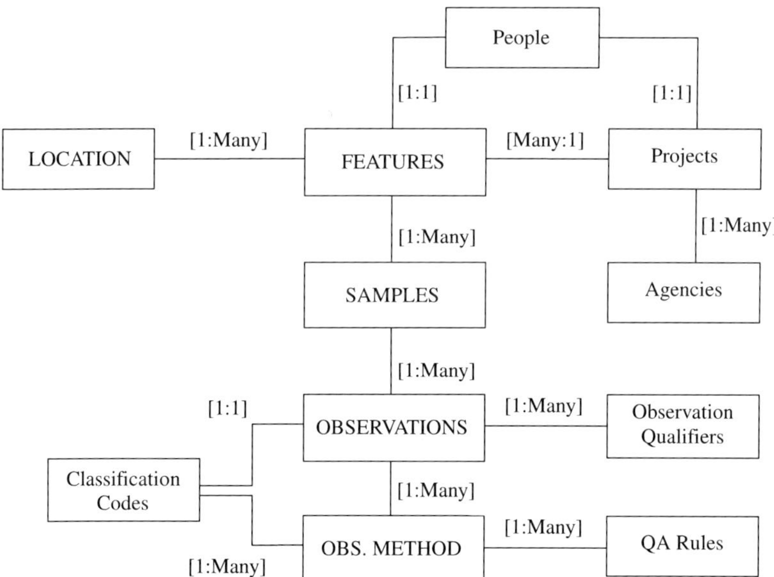

Figure 7.2. *Adopted database structure.*

generated for each measure existing in the VSSD. Each of these individual tables, and their associated location tables were then merged into the new structure shown in Figure 7.2.

In the report prepared by Johnston et al. (2003) as part of the ASRIS development, a list of data quality issues were outlined. PIRVic addressed some of these issues by making alterations to the ASRIS structure. In order to adopt a whole of earth domain approach, the 'Observation qualifiers' table (Fig. 7.2) was added, potentially allowing each measurement or observation to be attributed with the purpose for which it was collected, that is groundwater monitoring, incorporation into national soil network, etc. This table provides intelligence around each observation and can be used to query the data contained in the database from holistic approach rather than site by site, thereby solving two issues listed by Johnston et al. (2003).

The alterations carried out by PIRVic were additions rather than alterations to the existing ASRIS structure. These alterations were communicated to the ASRIS team for consideration in the future development of the database structure.

Some of the remaining issues are perennial across the soil domain; however, PIRVic is looking at the issues in the following ways.

- Inconsistencies in the way soil horizons are named and described. The inconsistencies reflect the evolutions in soil and pedological description in Victoria. The current standard published in the Australian soil and land survey field handbook (McDonald et al., 1990) provides a means of consistency in recent soil description. Past soil descriptions, where appropriate, will be updated to reflect this; however, limitations in data availability may make this task difficult.
- Non-existent or inconsistent taxonomic descriptions. Numerous taxonomic systems have existed and evolved throughout Australian soil and field survey. Most recently the Australian Soil Classification (Isbell, 1996) has been adopted. All taxonomic descriptors have different soil observation requirements for classification; therefore, limiting collation of sites using 'one system' [possibly the Principal Profile Form (Northcote, 1979)] offers the best solution.
- Differences in methods used for measurement of specific properties. Standardisation of methods is difficult without further research to develop statistically significant relationships between tests and methodologies. This has been an issue with methods of the ASRIS database (e.g. carbon measurement). Another factor that will influence method standardisation is the possible distribution of this database to other agencies that collect soil data. It is conceivable that other industries (agronomic, building site soil testers, etc.) could input data into the database, increasing the coverage of soil information. Inclusion of methods that cover all possibilities would be seen as beneficial.

- Failure to distinguish between a null field and a zero result. The approach adopted by Victoria is that if there was no record in the database, there was no observation made. If an observation was made then there should be an appropriate entry into the database. Historical data, however, did not make this distinction, and some data will require interpretation during translation.
- Typographical or other errors in data entry. This issue is perennial with any data entry task. In order to minimise errors made during data entry rules can be attached to each reading type that specify value ranges (i.e. 0–14 for pH) or specific national codes.

The user interface is rudimentary with the addition of some time-saving features. All of these features are located on the 'profile and measurements' form and implemented as buttons utilising VB script and macros. The buttons include:

- The 'add measures' button – by clicking this button the relevant measures are added to the list. This reduces the time taken to make entries. The measures added are dependent on which soil profile you are, that is the surface or any other profile.
- The 'remove nulls' button – once measurement values have been added there may be particular measurement types with no entries (null values). Rather than store these as superfluous records in the database they can be removed by selecting this button.
- The 'add multiples' button – it is possible to have more than one measurement for some measurement types, that is segregations. This button adds these measures, saving the user the tedious task of doing it manually.

The inclusion of these time-saving tools makes it easier for new users to input data in an efficient manner.

7.3 Results

Although the process of adopting a new data structure was tedious, multiple benefits have been realised.

7.3.1 User interface

In order to streamline data entry a simple interface was developed in Microsoft Access. The creation of this entry interface allows the data entry to be easily distributed, both within DPI and to external departments and contractors. The interface allows all soil data to be collected in a homogeneous manner.

The user interface comprised two main forms with four additional forms used for entering primarily descriptive data. The forms allow users to add projects, soil sites and profiles. Once these data are entered, individual measures can be added against each profile (including the surface) using drop-down lists.

These lists are created from the master code table and eradicate data entry errors ensuring that only correct codes are entered.

7.3.2 Data collection protocols

The current interface makes it possible to take a portable computer into the field for data collection. The benefits of collecting data directly into the database include:

- Negating the need to enter paper records upon return to the office.
- Reducing coding errors as only values defined in the database can be used.
- Ability to run simple queries to validate data collected.
- Streamlined process decreasing the time from data collection to data availability.

7.3.3 Data from external sources

The development of an interface in Microsoft Access allows users to collect soil information according to national standards. A wide variety of people are collecting soil data and each has their own requirements, that is soil tests for building sites, soil fertility tests, etc. The ASRIS database has the functionality to store a wide range of measures that cover all perceivable tests. If the database were adopted as a standard collection tool for the various users, the data collected would be of benefit to all users of soil information.

Data from multiple databases can be easily merged back into the master database held in Microsoft SQL by running update queries.

7.3.4 Data retrieval

Currently there is no method of extracting data from the central repository without knowledge in the use of databases and SQL. It is envisaged that data query and extraction functionality will be delivered by means of an SQL client residing on the soil scientist's desktop. This client would allow read-only access to all the data in Microsoft SQL, allowing queries to be performed and results extracted for further analysis.

This concept could be expanded to include a web client allowing anyone with appropriate privileges (i.e. agronomists, etc.) the ability to query and visualise the results.

7.3.5 Information product creation

The predominant purpose of soil data is to provide input into a wide variety of applications including resource allocation and modelling. Rather than provide clients with the detailed soil data stored in the database, generalised information products (Plate 7 (see Colour Plate Section) and Fig. 7.3) are used to highlight key characteristics of particular soil sites and soil units.

Robinson et al (2003) Corangamite Land Resource Assessment DPI Victoria

Component / *Proportion of soil-landform unit*	1 / 25%	2 / 72%	3 / 3%
CLIMATE			
Rainfall (mm)	Annual: 760		
Temperature (°C)	Minimum 9, Maximum 19		
Precipitation: less than potential evapotranspiration	October–March		
GEOLOGY			
Age and lithology	Quaternary basalt and minor scoria, Recent clay, sand and gravel		
Geomorphology	6.1.3 Volcanic Western Plains with poorly developed drainage		
LAND USE	Uncleared: Nature conservation Cleared: Cropping; beef cattle and sheep grazing; dairy		
TOPOGRAPHY			
Landscape	Gently undulating rises		
Elevation range (m)	125–164		
Local relief (m)	20		
Drainage pattern	Radial		
Drainage density (km/km²)	1.9		
Landform	Undulating rises and plains		Drainage depressions
Landform element	Broad gentle crests and upper slopes	Long gentle slopes and plain	Drainage depressions
Slope and range (%)	1 (0–4)	1 (0–3)	2 (1–5)
Slope shape	Convex	Straight	Concave
NATIVE VEGETATION			
Ecological Vegetation Class	Plains Grassy Woodland (0.9%)		
SOIL			
Parent material	Scoria, ash, basalt	Scoria, ash, basalt	Scoria, ash, basalt and colluvium
Description (Corangamite Soil Group)	Shallow red and black friable loams (40)	Shallow red and black friable loams (40)	Shallow red and black friable loams (40) and mottled black texture contrast soils (37)
Soil type sites	SW26, SW96, BD5	SW26, SW96, BD5	SW26, SW25, BD5
Surface texture	Clay loam	Clay loam	Clay loam, light clay
Permeability	High	High	Low to moderate
Depth (m)	<0.7	<1	<1.5
LAND CHARACTERISTICS, POTENTIAL AND LIMITATIONS	Uniform loam, low to moderate water holding capacity and high nutrient holding capacity. Friable, but often stony. Highly permeable, especially shallower profiles.	Uniform loam, low to moderate water holding capacity and high nutrient holding capacity. Friable, but often stony. Highly permeable, especially shallower profiles.	Uniform loam and some texture contrast, low to moderate water holding capacity and nutrient holding capacity in upper soil, higher in subsoil. Friable, but often stony. Higher permeability on shallower profiles. Some compaction and nutrient decline possible in upper part of texture contrast soil.

Figure 7.3. *Example page of a soil landform unit description.*

Currently, the creation of these information products is an intensive process requiring data to be collated from disparate sources including photographic repositories, soil sites database, other spatial data layers and a significant amount of expert knowledge.

The adoption of the new database structure will streamline the creation of these information products. By using templates and the mail merge function in Microsoft Word, it will be possible to extract data into a standard format while making sure it abides to protocols. Manual editing will still be required to format pages appropriately, but in time, it is hoped that this manual editing can be minimised by creating custom tools.

7.3.6 Additional benefits

The process undertaken in Victoria will provide input into the future development of the ASRIS database. A comprehensive report will be produced documenting all alterations and additions made. This could be used to incorporate further improvements into the ASRIS database.

The streamlined interface has allowed relatively inexperienced users to input a field sheet at the rate of 3–4 per h. This has resulted in an increased volume of data being consolidated into digital format, allowing collation of vast amounts of historic and recently collected field data. Having this data in a readily accessible, digital format adds value to irreplaceable soil data collected decades earlier.

7.4 Conclusion

The adoption of the ASRIS data structure has refined the collection, storage, maintenance and extraction of soil data in PIRVic. The database provides a cheap, robust and easily distributed means of collecting soil data and simplifies the process of providing updates into the national database.

The simple structure of the database greatly simplifies the process of inputting and extracting data and therefore increases its use. A field-based application allowing soil scientists to directly input field data is currently being tested. This simple entry application alleviates the requirement for a data entry person to interpret handwritten notes. It also ensures that data are collected to a minimum standard and utilise common nomenclatures. The application also provides the soil scientist with the freedom to edit, alter or update data as further work is carried out.

Although this data model has been tested and adopted for the collection of soil point data, the structure can easily accommodate other data. Attributes of soil landform units, land capability units or even groundwater monitoring data could be collected and stored in the same database with links to Geographical

Information System (GIS) layers representing their spatial extent. The storage of this varied, but inter-related data in one database allows complex querying and extraction of many data layers, potentially improving our understanding of the relationships between these related datasets.

References

Isbell, R.C., 1996. The Australian Soil Classification. CSIRO Publishing, Melbourne.

Johnston, R.M., Barry, S.J., Bleys, E., Bui, E.N., Moran, C.J., Simon, D.A.P., Carlile, P., McKenzie, N.J., Henderson, B.L., Chapman, G., Imhoff, M., Maschmedt, D., Howe, D., Grose, C., Schoknecht, N., Powell, B., Grundy, M., 2003. ASRIS: The database. Aust. J. Soil Res. 41, 1021–1036.

McDonald, R.C., Isbell, R.C., Speight, J.G., Walter, J., Hopkins, M.S., 1990. Australian Soil and Land Survey Field Handbook. Inkata Press, Melbourne.

Northcote, K.H., 1979. A Factual Key for the Recognition of Australian Soils. Rellim Technical Publication Pty. Ltd., Adelaide.

Developments in Soil Science, volume 31
P. Lagacherie, A.B. McBratney and M. Voltz (Editors)

Chapter 8

TOWARDS A SOIL INFORMATION SYSTEM FOR UNCERTAIN SOIL DATA

Gerard B.M. Heuvelink and James D. Brown

Abstract

Understanding the limitations of soil data is essential for both managing environmental systems effectively and encouraging the responsible use of soil data. Explicit assessment of the uncertainties associated with soil data, and their storage in a soil database are therefore important. In practice, users will not want to separate 'uncertain data' from 'certain data' and will therefore require a single database that meets all the requirements of a conventional database, as well as the ability to handle uncertain data specifically. This chapter presents a framework that facilitates the storage of information about soil-data quality, including the uncertainties associated with soil data, in a conventional database design. It comprises a methodology for classifying data according to their attribute scale, which influences the structure of an uncertainty model, and their space–time variability, which determines the need for autocorrelation functions in describing uncertainty. In terms of the former, the key distinctions are among real numbers on a continuous domain, real or integer numbers on a discrete domain, categorical data and narrative data. In terms of the latter, the key distinctions are among data that are constant in space and time (e.g. universal constants); data that vary in time, but not in space; data that vary in space but not in time; and data that vary both in time and space. Thus, we distinguish 13 'data types' to which individual datasets may be assigned. This simplifies the process of assessing uncertainties about soil data because characteristic uncertainty models can be defined for each data type. In general terms, an uncertain soil variable is completely specified by its probability distribution function (pdf). However, the complexity of a pdf varies with the 13 'data types' identified. For example, the (cumulative) pdf of an uncertain numerical constant is simply a nondecreasing function on the real line. The database stores this function or some parameters of it, such as the mean and variance. Other data types are associated with more complex pdfs. For example, an uncertain categorical soil map requires the probability of each soil type occurring at any location to be defined (local uncertainty), as well as the spatial dependencies between these probabilities at multiple locations (spatial uncertainty). In practice, the complexity of the joint pdf will make it difficult or impossible to identify. Assumptions (such as a stationarity assumption) are therefore required to reduce the number of model parameters.

8.1 Introduction

Decisions about the exploitation and management of environmental systems require information about 'environmental variables' for which an understanding of data uncertainties is important. In this context, 'environmental variables' include social and economic indicators, such as 'wealth', 'employment' and 'voting intentions', as well as 'natural' indicators, such as 'terrain', 'river discharge' and 'fish stocks'. In this chapter, however, we concentrate on soil-related data, such as 'soil type', 'soil organic matter content' and 'weathering rate'. In practice, our knowledge of these variables is always limited because instruments cannot measure with perfect accuracy, samples are not exhaustive and abstractions and simplifications are useful when resources are limited. Similarly, knowledge of these variables may be contested within and between groups of scientists due to preferred instruments, preferred sample designs and debated approaches to abstraction and simplification.

While soil data are rarely certain or 'error free', these errors may be difficult to quantify in practice. Indeed, the quantification of error (defined here as a 'departure from reality') implies that the 'true' state of the environment is known. In the absence of such confidence, we are uncertain about the 'true' state of the environment. Uncertainty is an expression of confidence about our knowledge and is therefore subjective. Different people can reach different conclusions about how uncertain something is, depending on their own personal experiences and world-view, as well as the amount and quality of information available to them (Cooke, 1991; Heuvelink and Bierkens, 1992; Brown, 2004).

In recent years, a distinct spectrum of methods, not altogether statistical, has emerged for dealing with situations of 'imperfect knowledge' in scientific research (see Ayyub, 2001 also). These methods include ranked and unranked scenarios (Von Reibnitz, 1988), rough sets, fuzzy sets, qualitative expressions of uncertainty (Funtowicz and Ravetz, 1990) and probability distribution functions (pdfs).

For situations in which pdfs can be estimated 'reliably' (see Brown, 2004), they confer a number of advantages over nonprobabilistic techniques (i.e., in a collective sense). For example, pdfs include methods for describing interdependence or correlation between uncertainties, methods for propagating uncertainties through environmental models and methods for tracing the sources of uncertainty in environmental data and models (Heuvelink, 1998). Notwithstanding these advantages, and the current popularity of stochastic methods in environmental research, there are a number of ongoing challenges for the successful application of pdfs to environmental data. In particular, there is a need to support the identification and estimation of pdfs in specific cases, as well as their storage in environmental databases.

This chapter provides a general framework for characterising uncertain soil variables with probability models. Probability models are developed for uncertain soil variables following a classification by measurement scale and space–time variability. In order to apply these 'general pdfs' to specific cases, they must be tractable to different types and degrees of information on uncertainty, for which a range of simplifying assumptions is introduced and discussed. These include assumptions for representing statistical dependence, both within and between soil variables (auto- and cross-dependence, respectively).

8.2 A taxonomy of uncertain soil variables

Since probability models are influenced by the characteristics of an uncertain variable, it is useful to develop a 'taxonomy of uncertain soil variables'. In this context, it is useful to distinguish between: (1) the space–time variability of a soil variable; and (2) the measurement scale of a soil variable.

Four classes of measurement scale are distinguished, namely:

1. Soil variables measured on a continuous numerical scale (e.g. nitrate concentration of the soil, bulk density and horizon depth);
2. Soil variables measured on a discrete numerical scale (e.g. the number of worms per metre square in the topsoil);
3. Soil variables measured on a categorical scale (e.g. soil type or soil colour);
4. Narrative soil variables, which involves a textual description of a soil (such as a description of its genesis or a soil-profile description).

In addition, four classes of space–time variability are distinguished, namely:

A. Soil variables that are constant in space and time. Arguably there are no constant soil variables, but for some applications the space–time variability of a soil variable may be *assumed* constant;

B. Soil variables that vary in time, but not in space. In practice, these will be soil variables for which the spatial variability is negligible for some practical purpose. Soil variables with a high degree of temporal versus spatial variability might be assumed constant in space for all practical purposes;

C. Soil variables that vary in space, but not in time (apply B to time);

D. Soil variables that vary in time and space. These include soil variables for which temporal variability and spatial variability are both important for some practical purpose.

In practice, it is not helpful to classify the space–time variability (A–D) of narrative attributes (4), although a pdf may be defined for the credibility of this material. Thus, the combination of attribute scale (1–4) and space–time variability (A–D) leads to 13 classes of uncertain attributes (see Table 8.1).

Table 8.1. *Categories for guiding the application of uncertainty models to soil variables.*

Space–time variability	Measurement scale			
	Continuous numerical	Discrete numerical	Categorical	Narrative
Constant in space and time	A1	A2	A3	4
Varies in time, not in space	B1	B2	B3	
Varies in space, not in time	C1	C2	C3	
Varies in time and space	D1	D2	D3	

8.3 Defining pdfs

When all possible outcomes of an uncertain event are known and their associated probabilities are quantifiable, uncertainties may be described with a pdf. In order to represent uncertainty with a pdf, it must be specified completely, which may be done by parameterising it to a chosen shape and estimating the parameters of it (see below). Some or all of the parameters of the pdf, such as the mean and standard deviation, may vary in space and time, or with the size of the measured variable. Furthermore, individual soil variables, and the uncertainties associated with them, may be statistically dependent on space and time, or other attributes, for which a joint distribution must be defined. In order to manage this complexity, it is useful to establish general pdfs for each category of uncertain data. Subsequently, it is useful to provide groups of simplifying assumptions for estimating these pdfs. In this section, we present the general pdf for each category.

Category A1: numerical constant defined on a continuous scale

An uncertain continuous numerical constant A is completely characterised by its (cumulative) marginal pdf:

$$F_A(a) = P(A \leq a) \qquad a \in \Re \tag{8.1}$$

Here, A is a random variable representing the uncertain constant, whereas a represents a possible value for A. The pdf F_A must be a continuous, nondecreasing function whose limit values are $F_A(-\infty) = 0$ and $F_A(+\infty) = 1$. When F_A is continuously differentiable, its derivative $f_A(a)$ exists and is known as the probability density function.

Category A2: numerical constant defined on a discrete scale

Let an uncertain discrete numerical constant B assume a value from the set $\{b_1, b_2, \ldots, b_m\}$ where the b_i are real or integer numbers. Extension to a countable infinite number of classes is allowed. Then B is completely characterised by:

$$F_{\mathrm{B}}(b_i) = P(\mathrm{B} = b_i) \qquad i = 1, \ldots, m \tag{8.2}$$

Each of the $F_{\mathrm{B}}(b_i)$ should be nonnegative and the sum of all $F_{\mathrm{B}}(b_i)$ should be equal to 1.

Category A3: constant defined on a categorical scale

Apply category A2 where the real numbers b_i are replaced with categories c_i.

Category B1: numerical time-series defined on a continuous scale

An uncertain continuous numerical variable $A(t)$ that varies with time is completely characterised by its joint pdf:

$$F(a_1, t_1, \ldots, a_n, t_n) = P(A(t_1) \leq a_1, \ldots, A(t_n) \leq a_n) \tag{8.3}$$

F must be known for any combination of the t_i and a_i, while n may assume any integer value. In other words, the joint pdf must be known for any finite set of arbitrary chosen times t_i ($i = 1, \ldots, n$).

Category B2: numerical time-series defined on a discrete scale

A discrete numerical time-series is completely characterised by:

$$F(b_1, t_1, \ldots, b_n, t_n) = P(\mathrm{B}(t_1) = b_1, \ldots, \mathrm{B}(t_n) = b_n) \tag{8.4}$$

where the b_i are real numbers. F must be known for any combination of the t_i and b_i, while n may again assume any integer value.

Category B3: categorical time-series

Apply category B2, where the real numbers b_i are replaced with categories c_i.

Category C1: numerical spatial variables defined on a continuous scale

Apply category B1 with spatial coordinates x_i, y_i and z_i rather than a temporal coordinate t_i.

Category C2: numerical spatial variables defined on a discrete scale

Apply category B2 with spatial coordinates x_i, y_i and z_i rather than t_i.

Category C3: categorical spatial variables

Apply category B3 with spatial coordinates x_i, y_i and z_i rather than t_i.

Category D1: numerical space–time variables defined on a continuous scale

Apply category B1 with space–time coordinates x_i, y_i, z_i and t_i.

Category D2: numerical space–time variables defined on a discrete scale

Apply category B2 with space–time coordinates x_i, y_i, z_i and t_i.

Category D3: categorical space–time variables

Apply category B3 with space–time coordinates x_i, y_i, z_i and t_i.

Category 4: narrative attributes

In some cases, it may be useful to describe the 'credibility' of narrative information with a (discrete) pdf.

8.4 Simplifying pdfs

Section 8.3 presented some general pdfs for the various categories of uncertain soil variables. In principle, and provided their assumptions are satisfied, these 'general pdfs' completely characterise the quantitative uncertainty of a soil variable (conceptual uncertainties are not included). In practice, however, they usually contain too many degrees of freedom for reliable estimation or storage within a soil database. Thus, the general pdfs need to be simplified in order to make them estimable in practice and tractable to storage within a soil database. Simplification implies that the pdfs in Eqs (8.1)–(8.4) can be adequately described with a relatively small number of parameters, as demanded by a particular application.

There is no unique way in which the simplification needs to be carried out. It will depend on the data category in question (although some categories will require similar or even the same simplifications, see below) and on the type and amount of information available to estimate the parameters of the pdf (assuming a parametric pdf is available). For example, as the information available to estimate a pdf declines, the simplifications required to make it estimable will usually increase. However, it is important to minimise the number and strength of these assumptions because their impacts may be difficult to establish *a priori* (e.g. when uncertain data are propagated through a model).

In this section, we explore the possible simplifications for the various categories of uncertain soil data (Table 8.1). Our propositions are not exhaustive and should be treated as a first-order solution. Further research is necessary to explore the broad range of simplifications that might be introduced for each category of uncertain data, but the simplifications themselves must always be implemented for individual cases and thus cannot be universal in their value or applicability.

8.4.1 Parameterisation of the pdf

The pdf of a numerical or categorical constant may be simplified by describing the uncertainty with a characteristic shape function, for which a small number of parameters must be estimated. Rather than specifying the entire pdf it is therefore sufficient to define the shape function and to estimate its parameters. For example, measurement error in a continuous numerical attribute is often assumed to follow a normal distribution (Heuvelink, 1998). This implies that the pdf F_A is reduced to only two parameters, namely the mean and standard deviation, which describe the bias and average magnitude of uncertainty in the soil attribute, respectively. Similarly, it may be reasonable to assume that the number of stones in a volume of soil is Poisson distributed, for which the discrete pdf F_b is reduced to only one parameter.

However, it will not always be acceptable to represent the pdf of an uncertain soil constant by a characteristic shape function. For example, the uncertainty about the concentration of pollutants in a contaminated soil may follow an unconventional or 'non-parametric' shape. Here, the pdf may be represented by a finite set of its percentiles, which are linearly interpolated. In most cases, the pdf of a categorical soil variable will not satisfy a common shape, implying that the probability of each possible outcome c_i, $i = 1, \ldots, m$ must be specified. For example, the uncertainty about a 'soil type' at some point in space and time is captured by the probabilities of this point occupying any of the m possible soil types, and generally these probabilities will not follow a common-shape function. These probabilities may be estimated with a 'people-driven' approach (i.e. expert elicitation, see Ayyub, 2001) or 'data-driven' approaches (e.g. through a confusion matrix computed on a comparison of observed and predicted soil types; see below).

8.4.2 Statistical independence in space or time

A rigorous simplification for soil variables that vary in space or time is achieved by assuming temporal and spatial independence in their associated uncertainty. This reduces the joint pdfs in Eqs. (8.3) and (8.4) to a set of n marginal pdfs, as described in Eqs. (8.1) and (8.2). These marginal pdfs may be estimated using people-driven and data-driven approaches where, additionally, a stationarity assumption (see below) may be employed. Stationarity implies that the parameters of the n marginal pdfs are constant in space and time.

8.4.3 Stationarity assumption for continuous numerical data

A common assumption for spatially distributed continuous numerical variables, and one that is equally applicable to temporal variables, is to assume (second-order) stationarity (Goovaerts, 1997). In this case, spatial dependence between the marginal pdfs is allowed, but the mean $\mu(x,y)$ and variance $\sigma^2(x,y)$ are assumed constant, and the spatial autocorrelation is assumed to depend only on the distances h_x and h_y between the two points. Spatial autocorrelation is usually described with a 'variogram' function (Goovaerts, 1997; Goovaerts, 2001), which plots (half) the variance of two values as a function of the distance between them (for all pairs of values in the dataset). The variogram function can be estimated from point observations. The same approach may be used to quantify temporal correlation, although in time series analysis it is customary to quantify temporal correlation directly by means of a correlogram or indirectly through autoregressive moving average models (i.e., Box, Jenkins and Reinsel, 1994).

8.4.4 Stationarity assumption for categorical data

In assessing the accuracy of a discrete numerical or categorical variable, an error or 'confusion' matrix may be constructed from a sample of more accurate data and assumed valid for a wider population. The confusion matrix reports the probability p_{kl} that a variable recorded in category c_k actually belongs to class c_l. These probabilities are computed from the sample and are assumed to be valid for the entire domain in space or time; that is, a stationarity assumption is made. This approach originates from the classification of land cover with remote-sensing imagery. Spatial dependence between uncertainties is not included in the confusion matrix.

One approach to deriving the spatial dependence in a discrete numerical or categorical variable is indicator geostatistics (Goovaerts, 1997; Finke, Wladis, Kros, Pebesma, & Reinds, 1999; Kyriakidis and Dungan, 2001). For example, Finke et al. (1999) used indicator variograms and cross-variograms to quantify the uncertainty in categorical soil and landcover maps, and used indicator simulation to generate spatially correlated realisations of these maps for use in an uncertainty propagation analysis. Other techniques for simplifying and estimating the spatial or temporal pdf of a discrete numerical or categorical uncertain soil variable include conditional probability networks (Kiiveri and Cacetta, 1998) and Bayesian maximum entropy (D'Or and Bogaert, 2003).

8.4.5 Markov property

The Markov property states that, given the present, the past and future are independent. In the one-dimensional time domain, the pdf Eq. (8.4) is then fully characterised by the transition probabilities from one time step to the next. For a discrete numerical or categorical soil variable that has m possible values, it is necessary to estimate the $m \times m$ matrix with transition probabilities, which can be derived from an observed time series. Storage of the transition matrix in the soil database is straightforward, provided a stationarity assumption is introduced, whereby the transition probabilities are assumed constant over time.

Extension of the Markov property to two dimensions or more (as required for spatial and space–time data) increases the complexity because 'future' and 'past' are connected through various 'presents'. An example of the Markov-random-field approach is given by Norberg et al. (2002). One example where the Markov property has been exploited in a real-world case study on soil salinity mapping is reported by Kiiveri and Cacetta (1998).

8.5 Future work

The work reported here requires further elaboration on the types of simplifying assumption that might be introduced for each class of uncertain data. For example, both the stationarity assumption and the Markov property aim at reducing the complexity of the general pdf. The differences and similarities between these two approaches and the development of guidelines about which approach is most suitable in a given situation need much further research. Useful simplifications must satisfy two conditions. First, the simplified pdfs must be estimable in practice, as well as tractable to storage within a soil database. Secondly, they must approximate the uncertainty in a soil variable sufficiently for their intended application. Among others, the elaboration and subsequent storage in the database must include the following aspects:

- Uncertainty is subjective. The database must allow the opinions of different 'experts' to be stored.
- Uncertainty information is very sensitive to the support size of the data items (Heuvelink and Pebesma, 1999). Here 'support' refers to the volume, magnitude and length of the entity described. Support size (in time and space) should always be specified in a database.
- The uncertainty in a particular variable may well be statistically dependent on the uncertainty in another variable. Statistical dependencies (and cross-correlations) between uncertain variables can have a marked influence on how uncertainties propagate in a modelling study.

Acknowledgements

The present work was carried out within the project 'Harmonised Techniques and Representative River Basin Data for Assessment and Use of Uncertainty Information in Integrated Water Management (HarmoniRiB)', which is partly funded by the EC Energy, Environment and Sustainable Development programme (Contract EVK1-CT-2002-00109).

References

Ayyub, B.M., 2001. Elicitation of Expert Opinions for Uncertainty and Risks. CRS Press, Florida.

Box, G.E.P., Jenkins, G.M., Reinsel, G.C., 1994. Time Series Analysis: Forecasting and Control. Prentice–Hall, Englewood Cliffs, NJ.

Brown, J.D., 2004. Knowledge, uncertainty and physical geography: towards the development of methodologies for questioning belief. Trans. Inst. British Geographers 29, 367–381.

Cooke, R.M., 1991. Experts in Uncertainty: Opinion and Subjective Probability in Science. Oxford University Press, Oxford.

D'Or, D., Bogaert, P., 2003. Continuous-valued map reconstruction with the Bayesian Maximum Entropy. Geoderma 112, 169–178.

Finke, P.A., Wladis, D., Kros, J., Pebesma, E.J., Reinds, G.J., 1999. Quantification and simulation of errors in categorical data for uncertainty analysis of soil acidification modelling. Geoderma 93, 177–194.

Funtowicz, S.O., Ravetz, J.R., 1990. Uncertainty and Quality in Science for Policy. Kluwer Academic, Dordrecht.

Goovaerts, P., 1997. Geostatistics for Natural Resources Evaluation. Oxford University Press, New York.

Goovaerts, P., 2001. Geostatistical modeling of uncertainty in soil science. Geoderma 103, 3–26.

Heuvelink, G.B.M., 1998. Error Propagation in Environmental Modelling with GIS. Taylor and Francis, London.

Heuvelink, G.B.M., Bierkens, M.F.P., 1992. Combining soil maps with interpolations from point observations to predict quantitative soil properties. Geoderma 55, 1–15.

Heuvelink, G.B.M., Pebesma, E.J., 1999. Spatial aggregation and soil process modelling. Geoderma 89, 47–65.

Kiiveri, H.T., Cacetta, P., 1998. Image fusion with conditional probability networks for monitoring the salinization of farmland. Digital Signal Process 8, 225–230.

Kyriakidis, P.C., Dungan, J.L., 2001. A geostatistical approach for mapping thematic classification accuracy and evaluating the impact of inaccurate spatial data on ecological model predictions. Environ. Ecol. Stat. 8, 311–330.

Norberg, T., Rosén, L., Baran, A., Baran, S., 2002. On modelling discrete geological structures as Markov random fields. Math. Geol. 34, 63–77.

Von Reibnitz, U., 1988. Scenario Techniques. McGraw-Hill, Hamburg.

Developments in Soil Science, volume 31
P. Lagacherie, A.B. McBratney and M. Voltz (Editors)

Chapter 9

THE DEVELOPMENT OF A QUANTITATIVE PROCEDURE FOR SOILSCAPE DELINEATION USING DIGITAL ELEVATION DATA FOR EUROPE

E. Dobos and L. Montanarella

Abstract
Until recently, manual methods were used for delineating soilscapes. The use of digital data sources, such as digital elevation models (DEMs) and satellite data, can speed up the completion of digital soil databases and improve the overall quality, consistency and reliability of the database. Our approach uses DEM for soilscape delineation based on the terrain classification system of the SOTER "Procedures Manual" and Edwin Hammond's landform classification methods, published in 1954.

In this study, the goal was to use a quantitative method to derive terrain classes that match the criteria of the "Georeferenced Soil database for Europe (GSDBE)" procedural Manual and to create a DEM-derived polygon (soilscape) system for Europe. Four terrain attributes were used to define the soilscape: hypsometry (combination of elevation and relief intensity), slope percentage (SP), relief intensity (RI) and dissection (potential drainage density, PDD). The SRTM30 database was used as a base DEM and for the derivation of the SP, RI and PDD layers. Erdas Imagine 8.5, Arc/Info and ArcView 3.2 software were used to handle the data.

We concluded that no major modification is required for the procedures to incorporate information that is derived quantitatively from digital data sources. The resulting database will have all the advantages of quantitatively derived databases, including consistency, homogeneity and reduced data generalisation and edge-matching problems.

9.1 Introduction

Soil information is needed for a wide range of environmental and agricultural applications. Knowledge of soils, combined with climatic and ecological data, is essential for understanding recent and future changes in ecosystems, managing natural resources and estimating crop yields. Climatic variations, population growth, urbanisation, pollution and changes in land management put stresses on natural ecosystems as well as on managed agricultural and forest lands. Soil and terrain information can improve decision-making to respond to such stresses and to evaluate ways to decrease negative impacts on the world economy and environmental quality.

9.1.1 Global soil databases

Until recently, there has been only one small-scale soil map with global coverage, the Soil Map of the World, at a scale of 1:5,000,000. (In this work, the term "small scale" refers to all the continental- and global-scale databases at spatial scales of less than 1:500,000.) The Food and Agriculture Organisation (FAO) of the United Nations and the United Nations Educational, Social and Cultural Organisation (UNESCO) (FAO, 1988) produced this map between 1960 and 1980. The map contains polygons representing soil associations. Attributes provided for each mapping unit include the dominant soil type, the list of associated soils, the textural class and the slope class of the soil association. FAO prepared a map of World Soil Resources at a scale of 1:25,000,000 and also created a generalised version at a scale of 1:100,000,000 (FAO, 1993). Both the Soil Map of the World and the World Soil Resources maps are available from the FAO in digital format.

Since the completion of the FAO soil map, many new data have been documented and new approaches are being used. Numerous new field observations are available, and many national databases, previously under internal embargo, have become accessible to the global community. Insufficient detail, lack of quality control and disparity among existing soil classification systems suggest that a new globally compatible soils database is needed to manage global resources, including geology, hydrology, land use, land cover, climatic parameters and other components of the Earth system (Baumgardner, 2000).

To improve upon the FAO map, in 1986, the International Society of Soil Science (ISSS) initiated an effort to create a SOil and TERrain Digital Database, called SOTER (ISRIC, 1995). Other international organisations have joined the project and supported the idea of having a global soil and terrain database capable of a series of applications. The SOTER mapping approach is based on polygons representing homogeneous physiographic and lithologic units of the Earth's surface. Soil information is assigned to the polygons as soil associations. A SOTER mapping procedure was developed and tested for some pilot areas and found to be useful and informative for representing soil properties. Until recently, there has been only limited progress towards the targeted global coverage. More than half of the terrestrial area, mainly the developed part of the world, is still missing from the SOTER database. There is an ongoing project to complete the SOTER database for Europe. This work needs a lot of existing data and a procedure capable of harmonising the data from different origins to provide a spatially and thematically consistent database for Europe.

Europe has adapted the SOTER mapping approach and incorporated its procedure into the "Georeferenced Soil Database for Europe" (GSDBE) (European Soil Bureau, Scientific Committee, 2001) to create a soil database for the European

Union (EU) and neighbouring countries. It began under an EU program known as Monitoring Agriculture by Remote Sensing (MARS), and focused initially on developing a harmonised and corrected geographical database for the soil cover of 18 European nations, at a scale of 1:1,000,000. Work is ongoing in the near term to extend this database to 26 West, Central and East European countries, mainly those who are members of the EU and neighbouring countries. Recent intentions are that the project will extend its database development activities to include all Eurasian countries and northern Africa (King and Thomasson, 1996; Jamagne et al., 1993).

9.1.2 The aim of the study

The SOTER procedure is based on existing databases. Although a tremendous amount of geographic and attribute soil and terrain data has been collected in the last few decades, the availability or reliability of those data is often limited. One difficult part of the database compilation is the conversion, translation and reprojection of the data to reach a common geographic and attribute coding platform required by a consistent, uniform database. These problems are often difficult to overcome and limit the use of the compiled database. For example, major discrepancies exist between the mapping units in the border areas of neighbouring countries, or the spatial resolution is inconsistent over the mapped area, owing to the varied quality and resolution of the base data used for the compilation.

Our previous studies have demonstrated the usefulness of digital data sources, like DEM and satellite data for creating maps of large areas (Dobos, 1998; Dobos et al., 2000, 2001, 2005b). Worstell (2000) demonstrated the potential use of coarse spatial resolution digital elevation model (DEM) data to delineate physiographic units at regional and global scales. Low-resolution satellite and DEM data (SRTM30) are available at no cost for most of the world's land surface. Scattered soil pedon data exist throughout the world, which can be supplemented with data from higher resolution soil maps. Using these digital data sources, we can develop a quantitative method for SOTER database compilation. The aim of this study was to implement the scientific results of these studies and set up a quantitative methodology for soilscape delineation supporting the GSDBE and SOTER database creation. The procedure is incomplete because the lithology information has not been incorporated. Although the major objective of this study was to develop a quantitative approach for the SOTER procedures, it also demonstrates the potential problems and limitations of the use of digital data sources for soil or terrain characterisation. A similar approach using high-resolution SRTM data with 90 m resolution was also implemented and reported in another paper by Dobos et al. (2005b).

9.2 Materials

9.2.1 The study area

The study area covers the Continent of Europe and represents a great variety of soils, landscape types, landuse, climate, topography and vegetation.

9.2.2 SRTM30 Global Elevation Data

SRTM30 database is an improved version of the GTOPO30. GTOPO30 was completed in late 1996, following development over a 3-year period through a collaborative effort led by staff at the USGS's EROS Data Center (Gesch et al., 1999). Its documentation is available on the following homepage: http://edcdaac.usgs.gov/gtopo30/README.asp#h10.

The GTOPO30 has a spatial resolution of 30 arc-s (~1 km). Produced for use in large-area studies, these global DEM data have been generated at a resolution that is compatible with the AVHRR sensor. The quality of the dataset varies depending on the original datasets used to compile GOTO30. The areas with higher relief are in general of a good quality, while the data of low-relief areas have significant artifacts that limit its use for many type of hydrologic modelling.

SRTM30 was a great step forward in small-scale digital terrain modelling. It is a near-global DEM, covering the Earth surface between the latitudes 60° south and north. The data come from the combination of GTOPO30 and the Shuttle Radar Topography Mission (Farr and Kolbrick, 2000), flown in 2000. The basic product has a resolution of 1 arc-s, but it is not publicly available. These data were first generalised to 3 arc-s by averaging a 3 by 3 cell area, and then to 30 arc-s by a 10 by 10 grid averaging of the 3 arc-s product. There are small (0.15% in total) gaps in the dataset due to the shadowing effect of high-relief areas on the radar beam. These gaps were filled with data from GTOPO30. The data for the areas outside the 60° latitudes are completely from the original GTOPO30.

The data were transformed from geographical coordinates (latitude/longitude) to the standard Lambert Azimuthal Equal-Area projection. Numerous terrain derivatives were created and added to the database.

9.3 Methods

9.3.1 Soil Mapping Unit (Soilscape) delineation

The soilscape delineation is based on two primary phenomena: terrain and lithology. Each soilscape represents a unique combination of terrain and soil characteristics. The two major differentiating criteria are applied in a step-by-step manner, leading to a closer identification of the land area under consideration. Physiography is the first differentiating criterion to be used to characterise a soilscape. The term physiography is used in this context as the description of the landforms of the Earth's surface. It can best be described as identifying and quantifying as far as possible the major landforms, on the basis of the dominant gradient of their slopes and their relief intensity (RI). This process, in combination with a hypsometric (absolute elevation above sea level) grouping and a factor characterising the degree of dissection, can make a broad subdivision of an area and delineate it on the map. Further subdivision of the soilscape according to the lithology (parent material) needs to be done to complete the delineation procedure.

Until recently, manual methods were used for delineating soilscapes. The availability of DEMs makes it feasible to use a quantitative approach. Worstell (2000) has proposed an approach using DEMs for soilscape delineation based on Edwin Hammond's landform classification methods (Hammond, 1954). Worstell has adapted and modified his methods to create a quantitative procedure to classify landforms on a regional scale.

In this study, the goal was to use quantitative methods to derive terrain classes that match the criteria of the GSDBE Manual. This classification is identical to the SOTER methodology to delineate the SOTER Mapping Units. Thus, the soilscapes of the GSDBE and the SOTER Mapping Units are identical and used in parallel in this chapter. Four terrain attributes are used to define the soilscape: hypsometry (elevation), slope percentage (SP), RI and dissection (potential drainage density, PDD). The class limits within these attributes are defined quantitatively in the Chapter 6 of the manual. The quantitatively existing class limits were adapted without any modification. The dissection class limits were derived from the PDD layers (Dobos et al. 2000, Dobos and Daroussin, 2005) through an empirical approach. The SRTM30 database was used as a base DEM and for the derivation of the SP, RI and PDD layers. Erdas Imagine 8.5, Arc/Info and ArcView 3.2 software were used to handle the data (Erdas, 1995; ESRI, 1997).

The adapted classification for the four terrain layers is given in the four sections below.

Slope

The slope layer was derived using the slope command of the ArcView Spatial Analyst module. Following the GSDBE slope classification, six classes were formed with slope intervals of 0–2, 2–5, 5–8, 8–15, 15–30 and 30–60%. However, this kind of classification rarely results in distinct borders between the classes, necessary for defining polygons of practical size. Thus, a different approach was taken to refine the classes. Instead of using the slope values of each of the pixels, we calculated the frequency of the slope classes within a 50-pixel area circle (with a 4-pixel radius) and thus formed five frequency layers, one for each class. These five layers were

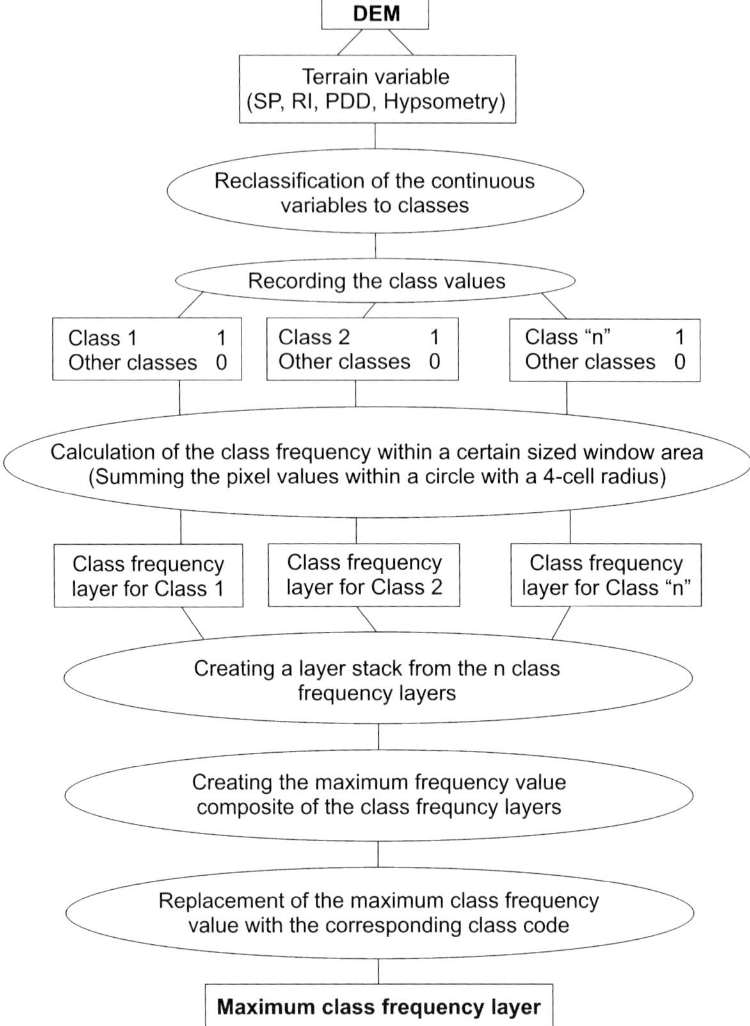

Figure 9.1. *Flowchart of the "maximum class frequency" approach.*

then overlaid, and a new layer was formed having each of its pixels assigned to the slope class with the highest frequency among the five frequency layers. This we called the "maximum class frequency approach" and will be used under this name in this chapter. Figure 9.1 shows the flowchart for deriving it.

Relief intensity

The RI is the difference between the maximum and minimum elevations in a specified size area. Owing to the low spatial resolution of the DEM, the RI values could not be calculated on a 1 km, or slope unit basis. A 4-pixel-radius circle (4-km radius, $\sim$50 km^2) was used instead. Worstell (2000) used a similar range and found the 11 by 11 km area representative for small-scale characterisation. In this study, we tried several circle sizes. Although, the 11 by 11 km window area was found to be appropriate for this purpose, the smaller circular window was chosen to better characterise the thinner, linear features of the terrain. Four classes were formed on the basis of the SOTER criterion: 0–50, 50–100, 100–600, 600–1600 m/50 km^2.

The derivation of the RI-class layer followed the same maximum-class frequency approach as the one used for deriving the slope classes.

Dissection (PDD)

A terrain dissection layer was also generated from the SRTM-30. The PDD function was used to represent the level of terrain dissection. This function was developed by Dobos (2000) and Dobos and Daroussin (2005). The data are processed in two steps. In the first step, a DEM-based drainage network is derived by thresholding the flow-accumulation values (Please refer to ArcInfo® flow-accumulation command for further explanation of the term). A threshold flow-accumulation value (also known as contributing area or catchment area for each cell) expressed in cell counts is selected. Cells having flow-accumulation values higher than the threshold are considered drainage ways and are assigned a value of 1, while the rest of the pixels are set to zero. In the second step, a size for a moving window is selected, and the sum of the drainage-way pixel values for this window is assigned to the central cell. The higher the PDD value, the more dissected is the terrain. For a 1-km resolution DEM, Dobos (1998) suggested setting the flow-accumulation value to 10 and the moving window to a 4-pixel-radius circle, which were used in this project also.

Only three dissection classes are distinguished in the "Manual of procedure" on an area basis: 0–10, 10–25 and over 25 (km lengths of drainage ways)/km^2. In this study, the use of the same class limits would have been meaningless, due to the different approach and resolution. After several empirical trials, the following two classes were set up: 0–7 values for the non-dissected or slightly dissected areas and above 8 for the dissected ones.

The derivation of the PDD-class layer followed the maximum-class frequency approach used for deriving the slope classes and RI classes as described above. The land areas with an RI $>$ 600 were masked out from the PDD layer in order to be consistent with the manual.

Hypsometry (elevation)

The SOTER Procedures Manual use a composite hypsometry variable, which is dependent on the value of RI for its interpretation. The hypsometry classes 1 through 5 are set up for level and slightly sloping lands (RI < 50) and indicates the height above sea level. For areas with RI > 50, the hypsometric classes 6 through 8 are used to indicate the height above the local base, the local RI. The SOTER hypsometry classes 9 through 12 are for steep and sloping lands (RI > 600), and were not used in this study due to the lack of membership. For our purposes, the first step of the class definition involved partitioning the scene into two classes, those with RI < 50 and those with RI > 50. The areas with RI of less than 50 were classified into four classes: 0–300, 300–600, 600–1500 and 1500–3000 (the fifth class had no membership in this area). The derivation of the level-area-hypsometry-class layer followed the maximum class frequency approach. For those areas with RI higher than 50, classes 6 through 10 were defined on the basis of RI: 50–200 (low hills), 200–400 (medium hills), 400–600 (high hills), 600–1500 (low mountain), 1500–3000 (medium mountain) and 3000–5000 (high mountains). These class limits are somewhat different from the limits in the original procedure. The derivation of the sloping-area-hypsometry-class layer has followed the maximum class frequency approach as well. Finally, the two hypsometry layers were united.

9.3.2 Database compilation

The soilscape delineation was based on the classes of the four source images, namely the SP, RI, PDD and hypsometry. These images were vectorised and thematic polygon coverages were generated. These coverages were overlaid and united. Polygons with an area of less than 25 km^2 were treated as sliver polygons and were merged to the neighbouring polygons with which they share the longest border. Plate 9 (see Colour Plate Section) shows the final polygon system for the Carpathian Basin.

The last phase was to assign terrain information to the polygons. The DEM, hypsometry, SP, RI and PDD grid coverages were used as information sources. The minimum, maximum and mean elevation, RI, dissection and slope values were calculated and assigned to each of the polygons.

9.4 Conclusions

Until recently, there has been only one soil map with global coverage, the FAO Soil Map of the World at a scale of 1:5,000,000. This FAO map is not reliable for many applications. This database had to be revised based on emerging input soil data and the state-of-the-art digital mapping technology, which have become available in the last few years. The SOTER project was initiated in 1986 to replace the FAO map and make use of the technological development and new soil and terrain data sources. The mapping procedure uses terrain and parent material characteristics to delineate the mapping units. Soil associations appear only at the attribute level. The objective of this chapter was to introduce a quantitative methodology to derive the physiographic information required for the terrain unit delineation.

The dataset derived through this quantitative manner characterises and classifies the landscape based on its physiographic features. The classification criteria for the slope, hypsometry and relief layers are directly adapted from the SOTER Manual. The terrain dissection classification needed a new approach, which has been presented here. The class criteria were defined on an empirical manner by comparing different setups and their matches with real physiographic features. The objective was to "translate" the original SOTER terrain classification system into a quantitative, DEM-based procedure, keeping the original class limits. We did not address questions on class limits or on the general terrain characterisation idea of SOTER. The questions asked here were (1) whether the SOTER-mapping approach defined in the Procedures Manual could be implemented in a digital, DEM-based framework or not and (2) whether the results of the digitally derived polygons fit into the original SOTER theory or not.

The answer for the first question is definitely positive: the SOTER procedure could be quantified and implemented in a DEM-based digital procedure. The terrain criteria used by the SOTER system can be derived from DEMs. The class limits and thresholds can be kept for three out of the four parameters. The terrain dissection was the only parameter that needed an analogue approach to be developed within the digital framework. The PDD index was used to estimate the terrain dissection. Although the PDD of the digital approach and the dissection – the total length of the drainage ways (as the SOTER procedures measures it) – have the same meaning, the class limits of the original SOTER dissection parameters could not be used directly for the PDD due to the different derivation techniques. PDD had to be calibrated to match the SOTER dissection theory by adjusting the threshold values for the PDD procedure and the class limits for the PDD classes. The calibration/verification process was done empirically using a trial-and-error approach, deriving different PDD layers and class limits. The different PDD layers were overlaid on a higher resolution DEM of the Carpathian Basin, where the authors used local knowledge/expertise to interpret the results and choose the best setup for the PDD.

The potential way to answer the second question would have been a comparison of a "hand-made" or traditional SOTER polygon structure and the digitally derived one. This approach has two limitations or difficulties. The major problem we identified with the traditional SOTER is that the delineation of the polygons system is very much dependent on the person who draw the lines. This fact is very evident when an international SOTER database is tested. Despite the similar environmental conditions, the polygons along the political borders almost never match and even the border themselves are evident and visible on the database. Thus, it is difficult to decide which approach is more accurate and could be used for comparison because there is no exact measure of accuracy available for this complex and necessarily generalised terrain unit system. Therefore, the same approach described above was used to validate and conclude the results. The DEM-derived polygon system was overlaid on a digital terrain map for the Carpathian Basin – where our local knowledge and expertise was considered to be appropriate – and three different types, low-, medium-, and high-relief areas and the transitions between them were visually tested and validated. Although this procedure is not a quantitative verification, the resulting data set was found to be realistically characterising the terrain and the delineated terrain polygons represented real geomorphologic phenomena.

However, potential problems and limitations are still present in the dataset. One major source of its inaccuracy comes from the original DEM. Low-spatial resolution DEMs tend to overgeneralise the terrain features values and decrease their data ranges. This factor should be taken into consideration, and some modification of the original DEM or adjustment of the SOTER physiographic classes may be needed to overcome this problem.

Another limitation of the procedure is the handling of linear features, like the alluvial areas along the major rivers, which should appear as a contiguous mapping unit. On the low-relief areas only the PDD can identify some of these features, but even these are not contiguous, elongated geometric features, but appear as a chain of smaller polygons along the major rivers.

A new, quantitative method for creating a SOTER terrain polygon system has been demonstrated in the chapter. It is designed for mapping large areas of the world quickly and cost-effectively. The resulting SOTER database will have all the advantages of quantitatively derived databases, including consistency, homogeneity, reduced data generalisation and edge-matching problems. Although the results from the above procedures are believed to be accurate enough to serve as a basis for global and regional studies, they should be checked and further revised by local and regional experts to ensure quality. Research should continue to improve the procedures, augment the profile data with new field sampling and incorporate new images and DEM data sources.

Acknowledgements

This study was supported by the European Commission, by the Hungarian National Science Foundation (OTKA, 34210) and by the Bolyai Foundation.

References

Baumgardner, M.F., 2000. Soil databases. In: M.E. Sumner (Ed.), Handbook of Soil Science. CRC Press, Boca Raton, FL.

Dobos, E., Daroussin, J., 2005. Potential Drainage Density Index (PDD). In: E. Dobos, J. Daroussin, and L. Montanarella (Eds), An SRTM-based procedure to delineate SOTER Terrain Units on 1:1 and 1:5 million scales. European Commission, Joint Research Centre, Ispra, Italy EUR 21571 EN.

Dobos, E., Daroussin, J., Montanarella, L., 2005. The development of a quantitative procedure for building physiographic units for the European SOTER database. In: E. Dobos, J. Daroussin, and L. Montanarella (Eds), An SRTM-based procedure to delineate SOTER Terrain Units on 1:1 and 1:5 million scales. European Commission, Joint Research Centre, Ispra, Italy EUR 21571 EN.

Dobos, E., Micheli, E., Baumgardner, M.F., Biehl, L., Helt, T., 2000. Use of combined digital elevation model and satellite radiometric data for regional soil mapping. Geoderma 97, 367–391.

Dobos, E., Montanarella, L., Negre, T., Micheli, E., 2001. A regional scale soil mapping approach using integrated AVHRR and DEM data. Int. J. Appl. Earth Obs. Geoinf. 3, 30–42.

Erdas, 1995. Erdas Field Guide, 3rd edition. Erdas, Atlanta, Georgia.

ESRI, 1997. ARC/INFO Online Manual. Environmental Systems Research Institute, Redlands, CA.

European Soil Bureau, Scientific Committee, 2001. Georeferenced Soil Database for Europe. Manual of Procedures. European Commission, Joint Research Centre, Ispra, Italy. EUR 18092 EN

FAO, 1988. Soil Map of the World, Revised Legend. World Soil Resources Report 60. Food and Agriculture Organisation of the United Nations, Rome.

FAO, 1993. World Soil Resources: An explanatory note on the FAO World Soil Resources Map at 1:25 000 000 scale. Food and Agriculture Organisation of the United Nations, Rome.

Farr, T.G., Kolbrick, M., 2000. Shuttle Radar Topography Mission produces a wealth of data. EOS, Trans. Am. Geophys. Union, 81, 583–585.

Gesch, D.B., Verdin, K.L., Greenlee, S.K., 1999. New land surface digital elevation model covers the earth. EOS Trans. Am. Geophys. Union, 80, 69–70.

Hammond, E.H., 1954. Small scale continental landform maps. Ann. Ass. Am. Geogr. 44, 33–42.

ISRIC, 1995. Global and National Soils and Terrain Digital Databases (SOTER). In: V.W.P. Van Engelen, T.T. Wen (Eds), Procedures Manual (revised edition). ISRIC, The Netherlands.

Jamagne, M., Le Bas, C., King, D., Berland, M. (1993). A geographical database for the soils of Central and Eastern Europe. European Commission, Joint Research Centre, Ispra, Italy.

King, D., Thomasson, A.J., 1996. European soil information policy for land management and soil monitoring, Report EUR 16393 EN. European Commission, Joint Research Centre, Ispra, Italy.

Worstell, B., 2000. Development of soil terrain (SOTER) map units using digital elevation models (DEM) and ancillary digital data. M.S. Thesis, Purdue University, Indiana, USA.

Developments in Soil Science, volume 31
P. Lagacherie, A.B. McBratney and M. Voltz (Editors)

Chapter 10

ONTOLOGY-BASED MULTI-SOURCE DATA INTEGRATION FOR DIGITAL SOIL MAPPING

B. Krol, D.G. Rossiter and W. Siderius

Abstract
There is a need for cheap methods for digital soil mapping on intermediate scales that make optimal use of existing multi-source datasets on both general and detailed scales. Apart from the spatial challenges that often have to be faced in map integration the semantics of datasets have to be well understood for successful data integration. So-called "ontology-based" approaches for the semantic integration of multi-source geographical datasets may be used to give a firm conceptual basis to digital soil mapping from multi-scale, multi-source geographic data. This chapter explores the use of ontologies in semantic data integration. A first version of an approach for ontology-based data integration for soil-landscape mapping is presented consisting of semantic factoring, ontology definition, reference model construction and data integration. This approach is illustrated with the semantic integration of a small-scale (1:400,000) soil-geomorphic map with a geological map at 1:50,000 scale of the Antequera area in Spain. Comparison of the results of our approach to semantic integration with an Antequera geo-pedological legend designed to map soil-landforms in this area at 1:50,000 scale shows a clear correspondence of ontologies.

10.1 Introduction

Planning and decision-making on the multi-functional use of space often takes place at municipal to regional level, thereby requiring information on soil properties and behaviour at intermediate map scales (1:50,000 to 1:100,000). Soil maps at these scales are often not available, especially in developing countries where soil maps over most of the territory are both categorically and cartographically general. However, other thematic maps may be available on appropriate scales. The experienced soil surveyor who understands the soil-landscape relations in the region may well be able to combine detailed thematic information (e.g. topography, physiography, land use and stream networks) with general soil maps to infer the distribution of soil classes or properties; this is referred to as spatial disaggregation (Bui and Moran, 2001). More generally, soil mappers have long used secondary data sources to prepare predictive maps, also called concept maps (Bartsch et al., 1997), to guide field campaigns. These secondary sources present two problems: the obvious one of scale, or more generally cartographic conceptualisation, and the less obvious one of semantics. The first issue has been more thoroughly studied and presents only a limited set of problems, whereas semantics has been little studied and is much more difficult.

Semantic integration has been addressed in the context of topographic mapping by Uitermark (2001) and Uitermark et al. (2005), who consider semantic matching (the finding of corresponding object instances) as the key issue in geographical dataset integration. But this requires a clear identification of objects, which in turn reflects on how we conceptualise the world (Smith and Mark, 2003). This is the motivation for recent interest in the geographic sciences in so-called "ontology-based" approaches for the semantic integration of multi-source geographical datasets; examples are Uitermark et al. (1999), Kavouras and Kokla (2000; 2002), Uitermark (2001) and Fonseca et al. (2002). It seems that these approaches might be used to give a firm conceptual basis to predictive digital soil mapping from multi-scale, multi-source geographic data.

This chapter presents some preliminary work on the use of ontologies in data integration for digital soil-landscape mapping on intermediate scales. We only discuss semantic aspects of data integration and leave for later work the question on how to simultaneously attack both semantic and spatial integration, since the spatial definition of an object can be considered one aspect of its semantics.

10.2 Definitions

The term "ontology" has its origin in philosophy where it refers to the study of the nature of being and existence. This term has been adopted by information science to refer to the being and existence of objects. This somewhat pretentious reuse of a well-established philosophical term has been accepted as useful and even as a signal of a paradigm shift in geoinformation science (Winter, 2001). In this context Gruber (1993) defines an ontology as "a specification of a conceptualization", where the conceptualisation is a view of some reality to be modelled. Similarly, Uitermark et al. (2005) define an ontology as "a structured, limitative collection of unambiguously defined concepts," which thus allows the unambiguous classification of (geographic) concepts. Operationally, ontologies are defined as formal structures that describe entities, classes, properties and functions that correspond to a certain view of the world (Fonseca et al., 2002). Thus, ontologies are used to specify what is being modelled, both conceptually and operationally. There may well be several conceptualisations of the same reality (e.g. by different data producers); if these are formalised as ontologies, the incompatibilities between them should be clear, and provide a starting point for semantic integration.

Ontologies are often discussed at two levels (Fonseca et al., 2002):

- A *domain* ontology describes the concepts related to a certain discipline, for example soil-geomorphic mapping.
- An *application* ontology describes those concepts from a domain ontology that correspond to a particular (geographical) dataset.

These are related by *abstraction rules* (Uitermark et al., 1999) from the various applications to the domain. This is not just a subsetting relation; the concepts from a given dataset are not exactly the same as those of the related domain ontology, although there must be enough similarity to formulate the abstraction rule.

Finally, a *reference model* presents the subset of concepts from the domain ontology relevant to the geographical area where dataset integration is being carried out. This is then used to actually combine the datasets.

We now show how this is applied in practice for soil-landscape predictive mapping. First we outline the approach and then apply it to a particular data integration problem.

10.3 An approach to ontology-based data integration for soil-landscape mapping

The approach has four steps: (1) semantic factoring, (2) ontology definition, (3) reference model construction and (4) data integration.

10.3.1 Semantic factoring

Semantic factoring is the process of transforming object classes from a dataset to be integrated into a set of decomposed fundamental classes. This is a phase in the formal concept analysis method as proposed by Kavouras and Kokla (2000; 2002). It is based on the user's knowledge about the geographical concepts underlying the thematic fields that are considered. Naming conflicts are resolved, equivalent classes are specified and overlaps between classes are identified. These overlapping classes are split into sub-classes. The result of semantic factoring is a list of decomposed classes, as corresponding object instances.

10.3.2 Ontology definition

This involves the definition of application ontologies from the input geographical datasets, and the selection of domain ontology to integrate these. The domain ontology must contain enough concepts to enable the integration of the application ontologies. Each dataset to be integrated will have its own application ontology, but there must be only one domain ontology. This is often an existing conceptual framework. For example several national soil classification systems could be application ontologies for soil maps from the respective countries, and the World Reference Base for Soil Resources could be the domain ontology for an integrated map (Joisten et al., 2002).

10.3.3 Reference model construction

A reference model is constructed as a subset of the domain ontology, including only those concepts from the domain ontology relevant to the geographical datasets to be integrated. Two abstraction rules apply to structure the model (Uitermark et al., 2005): (1) there is a hierarchy of sub-classes and super-classes, and (2) classes are organised in a composite–component relation.

The reference model has two kinds of semantic relationships (Uitermark et al., 1999). First, there are the relationships between the domain ontology concepts and the application ontology classes for one dataset. These can be:

- *Equivalent classes* between the domain ontology and the application ontology;
- Two or more classes from the domain ontology that define an *aggregated or composed* class in the application ontology.

Secondly, there are the relationships between object classes from different geographical datasets, represented by different application ontologies:

- Object classes are *semantically equivalent* if they refer to the same class from the domain ontology;
- They are *semantically related* if they refer to classes in the domain ontology that are sub-class and super-class from each other;
- They are *semantically relevant* if object class A is equivalent to a class in the domain ontology that refers to an aggregated class in object class B, or vice versa.
- Other object classes are not related to each other; they are *incompatible.*

The result of these steps is a reference model that relates the application ontologies for the various datasets to be integrated to the domain ontology and to each other. Hence, the reference model enables reasoning about terrain situations in the study area that are consistently represented in both input datasets (Uitermark, 2001) in the next step and data integration.

10.3.4 Data integration

The actual data integration involves the establishment of explicit relationships between *corresponding instances* (Uitermark et al., 2005) in the datasets involved. These are instances that are from *corresponding classes* and also share a similar location. Here corresponding classes are defined as those classes of the application ontologies that are either semantically equivalent, semantically related or semantically relevant. In practice data integration starts by overlaying the input maps. This results in a matrix of overlapping polygons as candidate instances. Next, the semantic relationships between application ontologies as represented in the reference model are used to select those instances that are also semantically related.

10.4 An example of data integration: soil-geomorphology of Antequera

We illustrate this generic procedure with a case study from the Antequera area of Andalucía, Spain, a long-time fieldwork area for the soil survey course at ITC (Siderius, 2001). The aim is to make a predictive 1:50,000 soil-geomorphic map of the area corresponding to the Antequera map sheet of the 1:50,000 series of the topographical map of Spain (55,000 ha). To illustrate the above-mentioned ideas we selected two input geographic datasets of different themes, at different scales, and produced by different organisations. The first is a 1:400,000 soil-geomorphic map; this is the desired theme but at too coarse a scale. The second is a 1:50,000 geological map, which has the desired scale but only a related theme. They are considered as application ontologies that share the same domain ontology (which is defined as part of the process) and can be integrated on that basis.

Given the large differences in map scale between the two datasets strict spatial integration by map intersection would result in objects that, in many cases, do not at all or only partly share the same location and with substantially different boundary generalisations. Therefore, the application of spatial specialisation–generalisation and component–composite hierarchies will not work in achieving effective data integration. However, these same abstraction mechanisms can be used while concentrating on the semantic integration between the datasets.

10.4.1 Geographical datasets

Source: Exploratory soil-geomorphic map of Andalucía, 1:400,000

This map was published by the Junta de Andalucía (de la Rosa and Moreira, 1987) and covers the entire province; for this exercise we extracted the sub-map corresponding to the study area. The survey methodology is based on the Australian 'land systems' approach (Christian and Stewart, 1968). The legend is a two-level hierarchy: (1) classes of uniform morphogenesis; and (2) within these, sub-classes defined by both physiography and geomorphic dynamics. Map delineations are defined by this hierarchical legend and also by lithology. The delineation description also names the dominant soils at the Soil Taxonomy 1975 (Soil Survey Staff, 1975) order level; this is not used to define the delineation, but to provide information about its major soils. The sub-map contains seven geomorphic map units and five complex lithology classes. These combine into 34 polygons covering the sub-map. The average size delineation is 1617 ha, which corresponds to an optimum size delineation for a 1:320,000 scale map (Forbes et al., 1982). So, this map has the right theme but the wrong scale for our purpose; the categorical level is correspondingly too coarse (Soil Taxonomy orders). A spatial selection corresponding to an area of 3 by 4 km in the Antequera study area as covered by the soil-geomorphic sub-map is presented in Figure 10.1. The corresponding legend classes are shown in Table 10.1.

Source: Geological map of Spain, 1:50,000 series, Antequera map sheet

This map is published by the Instituto Geológico y Minero de España (IGME, 1986). The 39 map units define areas of relatively homogeneous stratigraphy and lithology. Map units of Quaternary stratigraphy are also defined by their geomorphology; the legend is thus a hybrid, depending on the geological age of the strata. The map area corresponds to the study area. Its 55,000 ha is divided into 1272 delineations with an average size delineation of 43.2 ha, very close to the optimum size delineation (40 ha) for this map scale (Forbes et al., 1982). So, this map has the right scale but the wrong theme for our purpose. A spatial selection corresponding to an area of 3 by 4 km in the Antequera study area as covered by the geological map is presented in Figure 10.1. The corresponding legend classes are shown in Table 10.2.

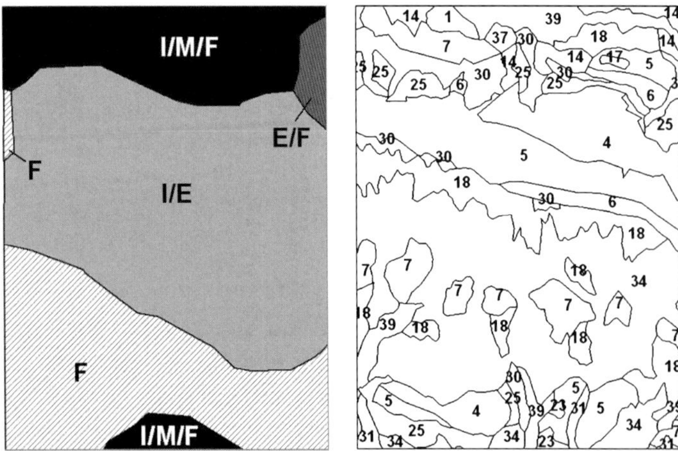

Figure 10.1. *Spatial selection of object instances from the soil-geomorphic sub-map (left) and geology map (right). The spatial subsets shown correspond to the same area of 3 by 4 km in the Antequera study area. Class labels refer to Tables 10.1 and 10.2, respectively.*

Table 10.1. *The legend classes corresponding to a spatial selection in Antequera study area as covered by the soil-geomorphic sub-map.*

Morphogenesis	Physiography	Label	At polygon level combined with lithology class
Fluvio-colluvial	Detritic covers in general (glacis, talus)[a]	F	Quaternary colluvium of conglomerates, sands, clays, marls and gravels of limestone origin, dolomites and other Betic rocks (class D)
Denudational	Hills with moderate structural influence, medium stable	I/M/F	Upper Miocene and Pliocene calcareous sandstones, nodular limestones, molasses, sandy marls, sandstones and calcareous marls (class C)
	Hills with moderate structural influence, medium instable	E/F	Triassic multi-coloured marls and gypsum (class A)
Structural	Mountainous alignment, medium stable[b]	I/E	Jurassic and Cretaceous calcareous marls, limestones and marls (class B)

[a]In Spanish: 'coberturas detriticas en general (glacis, derrubios de ladera)'.
[b]In Spanish: 'alineaciones montañosas. Medios estables'.

Clearly, these two maps provide a wealth of information on soil properties and distribution at the target scale of 1:50,000. In traditional soil survey methodology they would be used as references for the surveyor when developing a soil-landscape legend and finding tentative boundaries between map units. But this would be on an *ad hoc* and intuitive basis. What we attempted to do was formalise this using ontologies into an automated, transparent and documented approach.

10.4.2 The geo-pedological system as a domain ontology

A domain ontology for dataset integration must contain sufficient concepts to allow for the semantic integration of the application ontologies corresponding to the datasets to be integrated (Uitermark, 2001). For this case study, we selected the geo-pedological system of Zinck (Zinck, 1988; Zinck and Valenzuela, 1990) for systematic structuring of soil map legends. This multi-level hierarchical system for soil-geoform identification and classification is an example of the so-called 'geomorphologic approach' to soil survey (Wielemaker et al., 2001) and is also an example of the knowledge-based approach to soil survey (Bui, 2004). It conceptualises geoforms in taxonomy with six hierarchical levels, each with a corresponding categorical level of soil types: (1) *geostructure*, (2) *morphogenetic*,

Table 10.2. *Selected object classes from the geological map. Classes and labels correspond to the spatial selection of object instances from the geological sub-map shown in Figure 10.1 (translated from Spanish).*

Label	Chronostratigraphy	Lithology
Triassic	Multi-coloured clays, marls, sandstones and gypsum	1
Jurassic	Dolomites and dolomitic breccias	4
	Limestones	5
	Limestones, calcareous marls with flint and green marls	6
	Calcareous sandstones with flit and white marls	7
Tertiary	Sands, bioclastic sandstones, conglomerates and grey marls	14
Quaternary	Rounded limestone boulders encased in a cemented marl–sandstone matrix (terrace)	17
	Sub-angular boulders of limestone and bioclastic sandstone in a soft powdery calcareous matrix (glacis)	18
	Crusts, clays and red sands (surfaces)	23
	Breccious crust (colluvium)	25
	Clays and sands with boulders (colluvium)	30
	Clays and red sands with sub-angular boulders (alluvial fan)	31
	Clays and red sands with boulders (slope glacis)	34
	Sandy clays with boulders (alluvial fan)	37
	Sandy clays and boulders (alluvium, valley bottom)	39

environment, (3) *landscape*, (4) *relief/moulding*, (5) *substratum* and (6) *landform*. This system was designed to be applied at multiple cartographic and categorical generalisation levels and so allows for easy generalisation–specialisation relations.

10.4.3 Semantic factoring of the exploratory soil-geomorphic map with respect to the domain ontology

Application ontology classes from the exploratory soil-geomorphic map were compared with domain ontology concepts from the geo-pedological system; these classes must be factored (decomposed) if necessary to fit within the selected domain ontology. *Equivalent* classes are simply listed as such. *Overlapping* classes are decomposed into sub-classes. For example:

- *Detritic covers in general (glacis, talus) physiography* is equivalent to the *piedmont landscape*. In addition, enclosed in its definition are two sub-classes of the landscape level at the *relief/moulding* level. Hence, this class is decomposed into three (sub-)classes:

- *piedmont landscape* (level 3);
- *glacis relief/moulding* (level 4) and
- *talus relief/moulding* (level 4).

In Table 10.3, the decomposition of this example is shown together with the relationships of the resulting application ontology classes and the domain ontology concepts.

Lithology information is presented as lithological complexes in each spatial delineation of the soil-geomorphic map. In the geo-pedological system, lithological classes are presented as substrate classes in a user-defined rock classification system, which is not itself defined as part of the geo-pedological system. As a consequence, the lithology classes in the application ontology classes will by definition overlap with domain ontology classes. Decomposition of the lithological complexes would result in an unmanageably large number of rock

Table 10.3. *Examples of class decomposition and semantic relationships between application ontology classes and domain ontology concepts (x = relationship).*

		Domain ontology concepts				
		Depositional morphogenetic environment	Piedmont landscape	Glacis relief/molding	Talus relief/molding	Alluvial fan relief/molding
Soil-geomorphic classes	**Fluvio-colluvial morphogenesis**	x				
	1. Detritic covers in general (glacis,talus) physiography		x			
	1.1. Detritic covers in general: Glacis			x		
	1.2. Detritic covers in general: Talus				x	
Geology classes	Glacis (unit 18)			x		
	Alluvial fan (unit 31)					X
	Alluvial fan (unit 37)					X

classes. Hence, the lithology is only considered when establishing semantic relationships with the other application ontology that is the Antequera geological map.

10.4.4 Semantic factoring of the geological map with respect to the domain ontology

The lithology classes in this application ontology also overlap with domain ontology classes. The legend classes of the geological map that correspond to the Quaternary period (units 16–39) also correspond to geomorphic classes. These are compared with domain ontology concepts. For example (see also Table 10.3):

- The following geological units represent classes that are identified as subclasses of the *piedmont landscape*:
  - unit 18 represents *glacis*; it is considered as component class of *glacis relief/moulding* and
  - units 31 and 37 represent *alluvial fans* that can be considered as component class of *alluvial fan relief/moulding*.

10.4.5 The reference model

The object class hierarchies, in the domain ontology, for the two application ontologies can be combined by taking their set union, resulting in a refined domain ontology for the study area. This reference model specifies the relationships between the domain ontology classes and the application ontology classes.

The main function of the *substratum* level in this particular reference model is to assess the semantic relationships between object classes of the two datasets. Note that in some cases the *relief/moulding* level (in the domain ontology) could not be specified from the information in the application ontologies. In Table 10.4, reference model constructs are presented for the domain ontology concepts that correspond to the spatial subset shown in Figure 10.1.

10.5 Data integration

Once the semantic correspondence is made via the reference model, the two maps can be integrated. This has two aspects: semantic and spatial. Semantic integration is possible at map unit (object class) level, by assigning semantic information on geoforms to legend classes of polygons of the geological map that is the finer-scale map. The semantic relationships between the two datasets are assessed by evaluation of the relationships between lithology classes and units of the soil-geomorphic and geological maps, respectively.

Table 10.4. *Reference model constructs for the spatial subset of 3 by 4 km map shown in Figure 10.1 ('classes' relate to soil-geomorphic map; 'units' correspond to geological map).*

Morphogenetic environment	Landscape	Relief/moulding	Lithology (substratum)
Structural	Isolated mountain	Undifferentiated	Class B, units 4, 5, 6
Denudational	Hilland	Undifferentiated	Class A, unit 1
		Undifferentiated	Class B, unit 7
		Undifferentiated	Class C, unit 14
		Surfaces	Units 17, 23
		Valley bottom	Unit 39
		Undifferentiated	Units 25, 30
Depositional	Piedmont	Glacis	Class D, units 18, 34
		Talus	Class D
		Alluvial fan	Class D, units 31, 37
		Undifferentiated	Class D, units 25, 30

As an example, following is the relationship between a selection of lithology categories from the two datasets. "Classes" are from the soil-geomorphic map, and "units" are from the geologic map (see also Table 10.4):

- Class A: *Triassic multi-coloured marls and gypsum* is semantically similar to
  - unit 1: *Triassic multi-coloured clays, marls, sandstones and gypsum.*

Candidate object instances that are also semantically similar in this way will be mapped *as structural morphogenetic environment, isolated mountain landscape and undifferentiated relief/moulding.*

- Class D: *Quaternary colluvium of conglomerates, sands, clays, marls and gravels of limestone origin, dolomites and other Betic rocks* is semantically relevant to
  - unit 18: *Quaternary limestone and bioclastic sandstone boulders in a soft powdery calcareous matrix (glacis).*
  - unit 34: *Quaternary clays and red sands with boulders (slope glacis).*

Candidate instances that are also semantically similar in this way will be mapped as *depositional morphogenetic environment, piedmont landscape and glacis relief/moulding.*

A first test of semantic data integration resulted in 50% predicted object classes (Fig. 10.2). Reasons for unsuccessful integration at the level of object classes include differences in geometric definition, the quality of thematic representation of real world situations and many-to-many correspondences from composite class–component class relationships. The simplest approach to spatial integration is a polygon intersection where each result polygon is evaluated separately before it receives a label from the reference model. However, this is

Figure 10.2. *Test result of semantic integration of the soil-geomorphic sub-map and geology map (the spatial subset shown corresponds to an area of 3 by 4 km in the Antequera study area). Predicted objectclasses (50% of subset area) are shown at (domain ontology) landscape level: isolated mountain (black); hilland (dark grey); piedmont (light grey); no semantic integration (white).*

really only applicable with input maps on similar scales, with similar levels of boundary generalisation. By definition this integration must be per polygon as each one has a different spatial extent. A further step is automatic sliver removal based on purely geometric considerations. At contrasting scales (as in this case) it may be better to keep the polygons of the finer-scale map and re-label them with the integrated information from the reference model.

10.6 Validation

We compared the results of our approach to semantic integration with a geo-pedological legend independently designed at ITC (Siderius, 2004) to map soil landforms at 1:50,000 in the Antequera area (Table 10.5). This purpose-built legend uses the same domain ontology that is the geo-pedological system, as our approach for this study area. It identifies in the Antequera area four landscapes (*isolated mountain and ridge, plateau, piedmont and hilland*) on four different sub-strata, and two additional landscapes (*alluvial and lacustrine plains*). There is a clear correspondence of ontologies, although of course the reference model, combining as it does two independent sources with different purposes, is not as 'clean' as the purpose-built model.

Table 10.5. *Selection of legend classes at the landscape, substratum, and relief/moulding level as extracted from the Antequera geo-pedological legend to soil map 1:50,000 (Siderius, 2004).*

Landscape	Substratum[a]	Relief/moulding
Ridge	J	Hogback
	Colluvium (J)	Talus
Piedmont	Colluvium–alluvium (K)	Glacis
	Colluvium–alluvium (J)	Glacis
Hilland	T	Hill-system (sharp summits)
	Colluvium–alluvium (T)	Vale
		Swale
		Foot plain
	J	Hill-ridges
	Colluvium–alluvium (J)	Vale
		Swale
		Foot plain
	M	Hill-ridges
	Colluvium–alluvium (M)	Vale
		Swale
		Foot plain
	S	Hill-system (smooth summits)
	Colluvium–alluvium (S)	Vale
		Swale
		Foot plain

T: multi-coloured marls and gypsum with dolomites and dolomitic limestones (Triassic).
K: gypsum and multi-coloured marls with dolomites and dolomitic limestones (Triassic).
J: dolomites and dolomitic breccias, (dolomitic) limestones, white marls, calcareous sandstones (Jurassic).
M: marls and grey calcareous sandstones (lower Tertiary).
S: bioclastic sandstones and conglomerates (upper Tertiary).
Colluvium–alluvium (*x*): colluvium and alluvium from these rocks reworked in the Pleistocene/Holocene.
[a]The letters refer to the following sources of parent material:

10.7 Discussion and conclusion

The fundamental advantage of an ontology-based approach is its emphasis on concepts that is the semantics of the data. In the simplest case, two sources (application ontologies) may simply be using different words for the same concept (domain ontology), but even where the concepts do not exactly match, we can clearly see the sources of terminological conflict and attempt to resolve them by sub-classing. However, there must be enough commonality to find a common underlying domain ontology.

Another advantage is transparency: the process of integration provides documentation of how names were assigned to polygons in the integrated lap; this is lineage for the legend.

An open question is a combination of semantic and spatial integration: modifying the geometry (polygon splitting or merging) based on expert understanding of the probable soil landscape as expressed in the domain ontology. A simple example, not found in our case, is where two application ontologies use different terms for the same class in a reference model, and the polygons from the two maps are representing the same geographic feature, but with slightly different geometry. The solution here is to use the most precise boundary (perhaps from the finer-scale map) and discard the other. But most cases are not so easy. Still, we expect that the ontology-based approach to semantic integration can be extended to combined semantic–spatial integration.

References

Bartsch, H.U., Kues, J., Sbresny, J., Schneider, J., 1997. Soil information system as part of a municipal environmental information system. Environ. Geol. 30, 189–197.

Bui, E.N., 2004. Soil survey as a knowledge system. Geoderma 120, 17–26.

Bui, E.N., Moran, C.J., 2001. Disaggregation of polygons of surficial geology and soil maps using spatial modelling and legacy data. Geoderma 103, 79–94.

Christian, C.S., Stewart, G.A., 1968. Methodology of Integrated Surveys. In: G.A. Stewart (Ed.), Land evaluation: Papers of a CSIRO Symposium, organized in cooperation with UNESCO, Canberra 26–31 August 1968. Macmillan Company of Australia, South Melbourne, pp. 233–280.

de la Rosa, D., Moreira, M.J., 1987. Evaluación Ecológica de Recursos Naturales de Andalucía, Servicio de Evaluación de Recursos Naturales, A.M.A. Junta de Andalucía (4 maps 1:400,000 and explanatory report). Map: mapa geomorfoedáfico., Sevilla.

Fonseca, F., Egenhofer, M., Davis, C., Camara, G., 2002. Semantic granularity in ontology-driven Geographic Information Systems. Ann. Math. Artif. Intell. 36, 121–151.

Forbes, T.R., Rossiter, D., Van Wambeke, A., 1982. Guidelines for Evaluating the Adequacy of Soil Resource Inventories. SMSS Technical Monograph 4. Cornell University Department of Agronomy, Ithaca, NY, 51pp.

Gruber, T.R., 1993. A translation approach to portable ontologies. Knowledge Acquisit. 5, 199–220.

IGME, 1986. Mapa Geológico de España 1:50 000, No. 1023 (Antequera); Mapsheet and explanatory text, Instituto Geologico y Minero de España.

Joisten, H. et al., 2002. The first international joint soil map of Germany, the Czech Republic and Poland in a scale of 1:50 000, Sheet Zittau, 17th World Congress of Soil Science. International Union of Soil Sciences, Bangkok, CD-ROM, paper 2049.

Kavouras, M., Kokla, M., 2000. Ontology-Based Fusion of Geographic Databases, Spatial Information Management, Experiences and Visions for the 21st Century. International Federation of Surveyors, Commission 3-WG 3.1, Athens.

Kavouras, M., Kokla, M., 2002. A method for the formalization and integration of geographical categorizations. Int. J. Geogr. Inf. Sci. 16, 439–453.

Siderius, W., 2001. Soils of the Antequera area, Malaga province Spain, ITC, Enschede, The Netherlands.

Siderius, W., 2004. Antequera descriptive legend to soil map 1:50 000 (internal draft document), ITC, Department of Earth Systems Analysis, Enschede, The Netherlands.

Smith, B., Mark, D.M., 2003. Do mountains exist? Towards an ontology of landforms. Environ. Plan. B Plan. Design 30, 411–427.

Soil Survey Staff, 1975. Soil taxonomy: A Basic System of Soil Classification for Making and Interpreting Soil Surveys. Agricultural Handbook 436. US Department of Agriculture Soil Conservation Service, Washington, DC, 754pp.

Uitermark, H.T., 2001. Ontology-Based Geographic Data Set Integration. Ph.D. Thesis, Twente University, Enschede, The Netherlands, 139pp.

Uitermark, H.T., van Oosterom, P.J.M., Mars, N.J.I., Molenaar, M., 1999. Ontology Based Geographic Data Set Integration. In: M.H. Boehlen, C.S. Jensen and M.O. Scholl (Eds.), International Workshop on Spatio Temporal Database Management STDBM'99. Lecture Notes in Computer Science. Springer, Berlin, Edinburgh, Scotland, UK.

Uitermark, H.T., van Oosterom, P.J.M., Mars, N.J.I., Molenaar, M., 2005. Ontology-based integration of topographic data sets. Int. J. Appl. Earth Obs. Geoinf. 7, 97–106.

Wielemaker, W.G., de Bruin, S., Epema, G.F., Veldkamp, A., 2001. Significance and application of the multi-hierarchical landsystem in soil mapping. Catena 43, 15–34.

Winter, S., 2001. Ontology: buzzword or paradigm shift in GI science? Int. J. Geogr. Inf. Sci. 15, 587–590.

Zinck, J.A., 1988. Physiography & Soils. ITC Lecture notes (SOL41), Enschede, The Netherlands.

Zinck, J.A., Valenzuela, C.R., 1990. Soil geographic database: structure and application examples. ITC J. 3, 270–293.

D. Sampling methods for creating digital soil maps

There has been little work so far on designing or modifying sampling designs for creating digital soil maps. The methodology is more advanced for soil properties than for soil classes, where geostatisitcal theory has been developed for model-based sampling. Chapters 11–15 highlight the main issues, without necessarily addressing all of them.

Chapter 11 presents a generalisation of the geostatistical approach to designing sampling for soil properties to include covariates. This attempts to minimise the average universal kriging variance. Chapter 12 presents a new approach based on the quantiles of covariates that can be used for properties or classes. Chapter 13 shows how to create strata that can be used to optimise sampling by constrained spatial simulated annealing. Chapter 14 presents a straightforward approach to spatial coverage sampling designs by clustering the spatial locations of a raster. Chapter 15 highlights the generic problems of dealing with legacy soil data and having adequate sampling for fitting prediction models.

From the somewhat limited work presented that the following provisional sampling methodology may be proposed for digital soil mapping that depends on the amount of prior information:

- If there is no variogram or covariates, use *k*-means (optimising coverage of the geographic space)
- If there are covariates, use Latin hypercube sampling (optimising coverage of the covariate space).
- If there is a variogram and covariates with a known trend, then minimise universal kriging variances to optimise coverage of the geographic and covariate space.
- If there is prior data (legacy soil data), then a procedure is required to evaluate its adequacy.

Digital soil mapping will probably be more successful with covariates. Further work needs to be done on all of these approaches particularly the Latin hypercube and the universal kriging approaches and comparative tests need to be done. In many cases, there will be legacy soil data available; methods are required urgently to deal with these. Particular questions that need addressing are, how adequate is the prior sampling and where do we need to make further observations. Answers to these questions could be used to reconcile purposive-with-design or model-based approaches to sampling.

Developments in Soil Science, volume 31
P. Lagacherie, A.B. McBratney and M. Voltz (Editors)

Chapter 11

OPTIMIZATION OF SAMPLE CONFIGURATIONS FOR DIGITAL MAPPING OF SOIL PROPERTIES WITH UNIVERSAL KRIGING

Gerard B.M. Heuvelink, Dick J. Brus and Jaap J. de Gruijter

Abstract

Digital soil mapping makes extensive use of auxiliary information, such as that contained in remote sensing images and digital elevation models. However, it cannot do without taking samples of the soil itself. Therefore, methods and guidelines need to be developed that assist users in designing spatial sample configurations for use in digital soil mapping. Existing geostatistical methods are insufficient because these typically have been developed for situations in which there is no auxiliary information. In this chapter, we explore how the existing methods may be extended to the case in which the auxiliary information is spatially exhaustive and where soil mapping is done using universal kriging. We develop and illustrate a methodology that optimizes the spatial configuration of observations by minimizing the spatially averaged universal kriging variance. The universal kriging variance incorporates trend estimation error as well as spatial interpolation error. Hence, the optimized sample configuration strikes a balance between an optimal distribution of observations in feature and geographic space. The results show that optimal distribution in feature space prevails over optimal distribution in geographic space when the stochastic component of the universal kriging model is weakly spatially autocorrelated. It also prevails when the total number of observations is small. In all other cases, the optimal configuration is close to that obtained with minimization of only the spatial interpolation error. Application to a variety of real-world cases with multiple predictors and different spatial dependence structures is needed to support and generalise these preliminary findings.

11.1 Introduction

Digital soil mapping aims at spatial prediction of soil properties by combining soil observations at points with auxiliary information, such as contained in remote sensing images, digital elevation models and climatological records. Direct observations of the soil are important for two main reasons. First, they are used to establish the character and strength of the relationship between the soil property of interest and the auxiliary information. Second, they are used to improve the predictions based on the auxiliary information, by spatial interpolation of the differences between the observations and predictions. The twofold use of soil observations is nicely illustrated in universal kriging, which is commonly used for digital soil mapping when the auxiliary information is spatially exhaustive (Heuvelink and Webster, 2001; McBratney et al., 2003). Universal kriging treats the soil property of interest as the sum of a deterministic trend, which is taken as a regression on the auxiliary variables, and a spatially autocorrelated stochastic residual. The soil observations are used to estimate the regression coefficients as well as to interpolate the residual.

Fieldwork and laboratory analyses are labour-intensive and costly. It is therefore important that a sampling strategy is employed in which the available resources are used effectively. This means that the twofold use of soil observations in digital soil mapping should be incorporated in choosing an optimal sample configuration. However, the two uses of the soil observations generally impose conflicting requirements on the sample configuration. Estimation of the relationship between the soil property and the auxiliary information benefits from a large spread of the observations in feature space, while spatial interpolation of the differences between observations and predictions gains from a uniform spreading of the observations in geographic space (Lesch et al., 1995; Müller, 2001; Hengl et al., 2003). As yet, it is unclear how these two requirements should be weighed and how the weighing depends on the characteristics of a particular case. Existing methods for optimization of sample configurations have mainly focused on situations without auxiliary information, in which case the soil observations are only used for spatial interpolation. We review these methods briefly.

The first option for selecting sample locations in the context of spatial interpolation is sampling on a regular grid. Commonly used grid shapes are triangular and square lattices. These grid shapes can be evaluated by computing the maximum or average kriging variance, and choosing the shape for which the chosen criterion attains it minimum, given the number of sample points. Yfantis et al. (1987) show that for variograms with a small relative nugget, the maximum kriging variance is minimal for a triangular grid. Besides the configuration, we must also decide on the grid spacing (i.e., sampling density). Obviously, the smaller the grid spacing, the smaller is the average or maximum kriging variance. If a minimum precision is required and a reasonable variogram can be postulated, the variogram can be used to calculate the grid spacing required to achieve this minimum precision (McBratney and Webster, 1981; Olea, 1984; Christakos and Olea, 1992). So, for a range of grid spacings we may calculate the average or maximum kriging variance, plot these values against the grid spacings in a graph and determine the required grid spacing from this graph. Lark (2000) showed how a fuzzy set of required grid spacings can be calculated in situations where we are uncertain about the variogram.

In practice, regular grid sampling may be suboptimal for several reasons. Sampling may be hampered by enclosures that are inaccessible for sampling. The study area may have an irregular shape or the dimensions of the area may be small compared to the grid spacing. The latter may result in prediction error variances that are relatively large near the boundaries of the area. Also, we may want to use previously made observations at locations that cannot be matched with the grid. In these situations, minimization of the kriging variance criterion will yield a sample with irregular spacing. The criterion may be simplified in terms of the distance between sample points and prediction points, leading to the so-called spatial coverage samples, also referred to as space-filling samples (Royle and Nychka, 1998). Brus et al. (Chapter 14) propose the mean of the squared shortest distance as a minimization criterion, and minimize it with the well-known k-means algorithm from cluster analysis. Alternatively, we can define the minimization criterion in terms of the kriging variance. Sacks and Schiller (1988) and van Groenigen et al. (1999) minimized the average and maximum kriging variance using a simulated annealing algorithm.

Little work has been done on optimization of sample configurations when the observations are used both for estimating regression coefficients and for spatial interpolation. Lesch et al. (1995) selected sample sites for multiple linear regression. The fitted regression model was then used to predict the target variable at all locations with observations of the auxiliary variables. Also, the sample locations were required to be 'spatially representative of the entire survey area'. They proposed a site selection algorithm that starts from a sample that is closest to the optimum response surface (RS) design calculated from experimental design theory. The second and third best RS designs are also calculated. Then points of the best RS design are swapped with corresponding points in the second and third best RS design. Swaps are accepted if they improve the spatial coverage. Hengl et al. (2003), struggling with the same problem, proposed an 'equal range' design. In this procedure, the study region is stratified on the basis of the frequency distribution of the auxiliary variables. Stratification limits are set at equal distances in feature space. From each stratum an equal number of sample points is selected randomly, thus ensuring that the entire sample has uniform spacing in feature space. Many samples are generated in this way, and the one with the best spatial coverage is retained. Müller (2001, Sections 5.4 and 5.5) avoids heuristic approaches by introducing information matrices and applying standard gradient algorithms to design measures defined on these matrices. However, approximations were needed to be able to apply this method in a situation where both a trend and a spatially autocorrelated residual are present.

In this chapter, we address the same problem but aim for solutions that are not heuristic and do not involve approximations. This requires that we make some assumptions. We assume that we have only one target soil property that is measured on a numerical scale, that the auxiliary information is spatially exhaustive, and that the relationship between the target variable and the auxiliary variables takes the form of a linear regression equation, with known structure but unknown coefficients. We also assume that the variogram of the residual is known. Especially, the first and last assumptions are quite restrictive, but in the 'Conclusions' section we outline how these might be relaxed for further development of the method. Under the above assumptions, the problem boils down to minimization of the maximum or spatially averaged universal kriging variance, which can be solved using numerical simulation. We illustrate the theoretical results with synthetic examples.

11.2 Universal kriging

We consider the following model (Christensen, 1990):

$$Z(\mathbf{s}) = \sum_{j=0}^{m} \beta_j \cdot g_j(\mathbf{s}) + \varepsilon(\mathbf{s}), \tag{11.1}$$

where $Z(\mathbf{s})$ is the target soil property; $\mathbf{s} = (x \;\; y)'$, a two-dimensional spatial coordinate; $g_j(\mathbf{s})$ explanatory variables (note that $g_0(\mathbf{s}) \equiv 1$ for all $\mathbf{s}$); β_j regression coefficients and $\varepsilon(\mathbf{s})$ a normally distributed residual with zero mean and constant variance $c(0)$. The residual ε is possibly spatially autocorrelated, as quantified through an autocovariance function or variogram.

In what follows, it will be convenient to use matrix notation, so that Eq. (11.1) may be rewritten as

$$Z(\mathbf{s}) = \mathbf{g}'(\mathbf{s}) \cdot \boldsymbol{\beta} + \varepsilon(\mathbf{s}), \tag{11.2}$$

where $\mathbf{g}$ and $\boldsymbol{\beta}$ are column vectors of the m+1 explanatory variables and m+1 regression coefficients, respectively. The universal kriging prediction at an unobserved location s_0 from n observations $Z(s_i)$ is given by

$$\tilde{Z}(\mathbf{s}_0) = (\mathbf{c} + \mathbf{G}(\mathbf{G}'\mathbf{C}^{-1}\mathbf{G})^{-1}(\mathbf{g}_0 - \mathbf{G}'\mathbf{C}^{-1}\mathbf{c}))'\mathbf{C}^{-1}\mathbf{z}, \tag{11.3}$$

where $\mathbf{G}$ is the $n \times (m+1)$ matrix of predictors at the observation locations; $\mathbf{g}_0$ the vector of predictors at the prediction location; $\mathbf{C}$ the $n \times n$ variance–covariance matrix of the n residuals; $\mathbf{c}$ the vector of covariances between the residuals at the observation and prediction locations; and $\mathbf{z}$ the vector of observations $z(s_i)$. $\mathbf{C}$ and $\mathbf{c}$ are derived from the variogram of ε.

The universal kriging prediction-error variance at s_0 is given by

$$\begin{aligned}\sigma^2(\mathbf{s}_0) &= c(0) - \mathbf{c}'\mathbf{C}^{-1}\mathbf{c} \\ &\quad + (\mathbf{g}_0 - \mathbf{G}'\mathbf{C}^{-1}\mathbf{c})'(\mathbf{G}'\mathbf{C}^{-1}\mathbf{G})^{-1}(\mathbf{g}_0 - \mathbf{G}'\mathbf{C}^{-1}\mathbf{c})\end{aligned} \tag{11.4}$$

Although Eqs. (11.3) and (11.4) present well-known results (Christensen (1990, Section VI.2); Müller (2001, Section 2.2)), it is instructive to take a closer look. The universal kriging variance incorporates both the prediction error variance of the residual (first two terms on the right-hand side of Eq. (11.4), i.e., $c(0)–\mathbf{c}'\mathbf{C}^{-1}\mathbf{c}$) as well as the estimation error variance of the trend (third term on the right-hand side of Eq. (11.4)). Comparison of these two components of the prediction variance yields insight into what the main source of error is in a specific situation. If the first component is much larger than the second, then trend estimation has a negligible effect on the prediction error. If, on the other hand, the second component is much larger than the first, then uncertainty about the trend coefficients will be the main contributor to the universal kriging variance. Although it cannot be easily deduced from Eq. (11.4), the contribution of the second component tends to be small when the number of observations is large compared to the number of predictors. Note also from Eq. (11.4) that the universal kriging prediction variance does not depend on the data values themselves, but merely on the spatial pattern of the predictors and the covariance structure of the residual. This enables computation of the prediction variance prior to collecting the data, provided that the predictors and the variogram of the residual are known. This attractive property will be used in the next section to compute optimal sample configurations.

Universal kriging with externally defined predictors is sometimes also referred to as *kriging with external drift* (Goovaerts, 1997; Bishop and McBratney, 2001). In fact, some authors reserve the term universal kriging exclusively for those cases where the trend functions $g_j(s)$ are polynomials in the coordinates (Deutsch and Journel, 1998; Wackernagel, 1998). *Regression kriging* (Odeh et al., 1995; Hengl et al., 2004) uses, as a starting point, the same model as given in Eq. (11.1), and provided the same variogram of residuals is used to estimate the trend and krige the residuals, it yields the same results as presented in Eqs. (11.3) and (11.4).

11.3 Calculation of the optimal sample configuration

Given the model defined through Eq. (11.1), we now want to optimize the sample configuration. For this we first need to define a criterion, which we will take as the spatially averaged universal kriging variance. Thus, given the sample size, our objective is to locate the sample points such that the average universal kriging variance is minimized. In theory, it is very easy to compute the optimal configuration. For each possible configuration, simply compute the criterion and select the one that has the smallest value from all the configurations. In practice, the problem is that there is a huge (or infinite) number of possible configurations, meaning that evaluation of all possible configurations becomes impossible. Instead, some efficient search algorithm will have to be employed. Here, we will use spatial simulated annealing, following van Groenigen et al. (1999, 2000).

Simulated annealing is an iterative procedure that works as follows. Starting from an arbitrary initial configuration of sample points for which the criterion has been computed, a small perturbation in the configuration is brought about by moving one sample point in a random direction. The magnitude of the movement is random as well but it cannot be larger than a chosen value, which is gradually decreased as the iteration continues. The criterion is recomputed for the new configuration. If it is smaller than the previous criterion then the movement is accepted and a new perturbation is made, using the new configuration as a starting point. If, on the other hand, the movement has led to an increase in the criterion's value then there is still a chance that the new (worse) configuration is accepted. This is done to be able to escape from local minima. The probability of accepting a worsening configuration is also gradually decreased as the iteration continues. The iterative procedure is repeated until a maximum number of iterations is reached or until acceptance of new configurations has become very rare. In practice, thousands of iterations or even more are needed to obtain satisfactory results.

There is no guarantee that the optimal configuration will be obtained but confidence can be gained if redoing the iteration with different starting configurations yields approximately the same solutions. Also, the algorithm may be checked by running simplified examples for which it is known that the optimal configuration must have certain symmetry properties.

11.4 Synthetic examples

We use synthetic examples to analyse how the balance between optimization in feature and geographic space works out in practice. We examine how the balance is affected by three types of parameter settings. These are the number of observations, the structure of the trend and the degree of spatial autocorrelation of the stochastic residual. The number of observations is taken as 4, 9 and 16. In practice, usually there will be more observations, but a limited number of observations are used here because it facilitates the interpretation of the results. Three different trends are evaluated (see Eq. (11.1)):

(1) $m = 0$

(2) $m = 1$, $g_1(\mathbf{s}) = x$

(3) $m = 2$, $g_1(\mathbf{s}) = x$ and $g_2(\mathbf{s}) = x^2$

In other words, we consider a constant trend, a trend that is linear in the x-coordinate and a trend that is quadratic in the x-coordinate. Of course, in digital soil mapping one would use 'real' auxiliary variables, but here we take polynomials of a spatial coordinate as an example, again in order to facilitate the interpretation of the results.

Three variograms for the residual ε are distinguished:

(1) exponential variogram with zero nugget variance and spatial range equal to the diagonal of the spatial domain (case of strong spatial autocorrelation),
(2) exponential variogram with 50%-nugget variance and spatial range equal to one quarter of the diagonal of the spatial domain (case of weak spatial autocorrelation),
(3) pure nugget variogram (case of no spatial autocorrelation).

In total, this gives 27 cases. The spatial domain is taken as a square area, with the x-axis in the horizontal and the y-axis in the vertical. The sample configurations resulting from running the simulated annealing optimization are given in Figures 11.1–11.3.

We also computed the gain in precision that is obtained by including trend estimation in the optimization of the sample configuration. We did this by

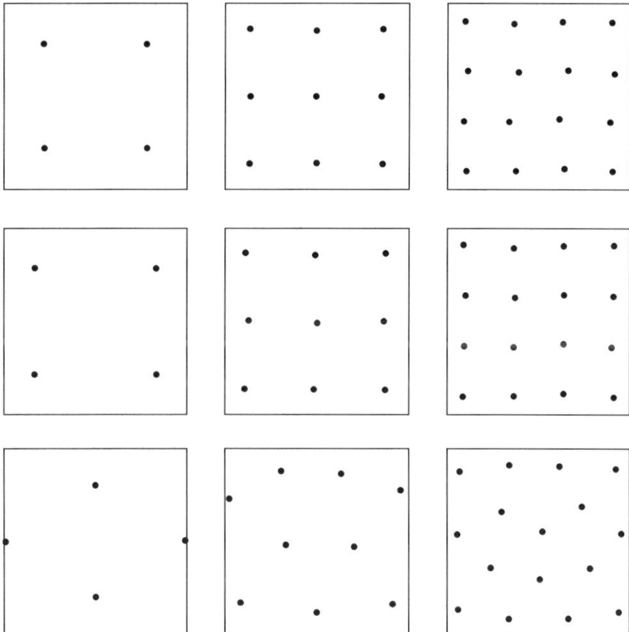

Figure 11.1. *Optimized sample configurations in the case of strong spatial autocorrelation. Cases considered are constant trend (top row), linear trend in x-coordinate (middle row) and quadratic trend in x-coordinate (bottom row). Sample sizes are 4 (left column), 9 (middle column) and 16 (right column).*

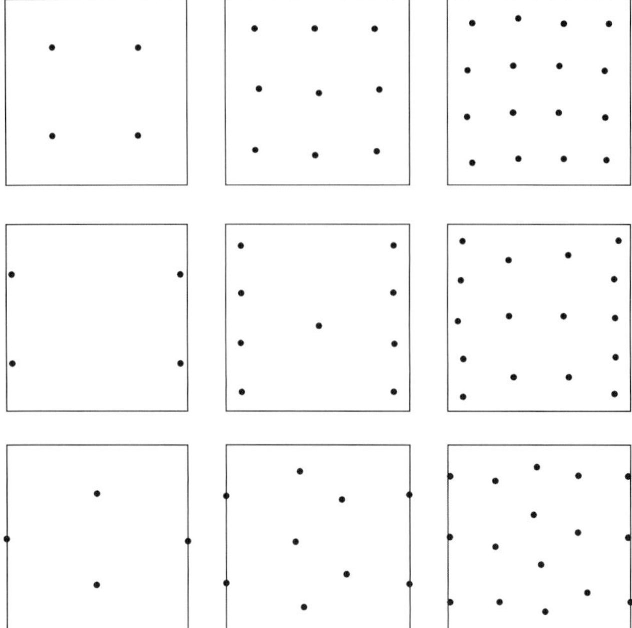

Figure 11.2. *Optimized sample configurations in the case of weak spatial autocorrelation. Cases considered are constant trend (top row), linear trend in x-coordinate (middle row) and quadratic trend in x-coordinate (bottom row). Sample sizes are 4 (left column), 9 (middle column) and 16 (right column).*

comparing the average universal kriging variance of the optimized configuration with that obtained with the configuration optimized for when there is no trend. The results are presented in Tables 11.1–11.3.

11.5 Discussion of results

Optimization of the sample configuration in case of a constant trend and a spatially autocorrelated residual (top rows of Figures 11.1 and 11.2) yields a regular square grid. It is square and not triangular, which can be explained from the fact that the numbers of observations chosen are squares of whole numbers. Note that the iterations have not fully converged because there are some irregularities in the configurations, notably for the 16 observation cases. Increasing the number of iterations would resolve this problem, but the changes made to the configurations would only be marginal. Comparison of the strong and weak spatial autocorrelation case (top rows of Figures 11.1 and 11.2 again) shows that for the former, observations are placed somewhat closer to the boundaries of the spatial domain. The 'optimized' configuration for the case of a constant trend

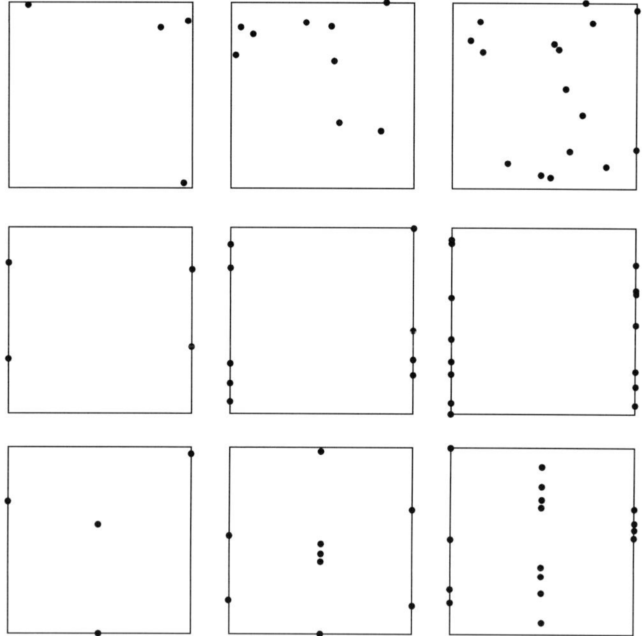

Figure 11.3. *Optimized sample configurations in the case of no spatial autocorrelation. Cases considered are constant trend (top row), linear trend in x-coordinate (middle row) and quadratic trend in x-coordinate (bottom row). Sample sizes are 4 (left column), 9 (middle column) and 16 (right column).*

Table 11.1. *Gain in precision by including trend estimation error in sample configuration optimization. Case of strong spatial autocorrelation. The gain is expressed as the ratio of the average kriging variance obtained with the 'ordinary kriging sample configuration' (i.e. top row of Figure 11.1) and that obtained with the 'universal kriging sample configuration' (i.e. appropriate row of Figure 11.1).*

Trend model	$n = 4$	$n = 9$	$n = 16$
Constant (no trend)	1.0000	1.0000	1.0000
Linear in x-coordinate	1.0282	1.0029	1.0008
Quadratic in x-coordinate	∞	1.0339	1.0020

and no spatial autocorrelation (top row of Figure 11.3) yields a random pattern. This is because, in this case, it really does not matter where the sample points are located. All configurations yield the same value for the average kriging variance and are therefore equally good. The configurations shown in the top row of Figure 11.3 are therefore completely arbitrary.

Table 11.2. *Gain in precision by including trend estimation error in sample configuration optimization. Case of weak spatial autocorrelation. The gain is expressed as the ratio of the average kriging variance obtained with the 'ordinary kriging sample configuration' (i.e. top row of Figure 11.2) and that obtained with the 'universal kriging sample configuration' (i.e. appropriate row of Figure 11.2).*

Trend model	$n = 4$	$n = 9$	$n = 16$
Constant (no trend)	1.0000	1.0000	1.0000
Linear in x-coordinate	1.1722	1.0391	1.0121
Quadratic in x-coordinate	∞	1.1711	1.0469

Table 11.3. *Gain in precision by including trend estimation error in sample configuration optimization; case of no spatial autocorrelation. The gain is expressed as the ratio of the average kriging variance obtained with the 'ordinary kriging sample configuration' (i.e. top row of Figure 11.3).*

Trend model	$n = 4$	$n = 9$	$n = 16$
Constant (no trend)	1.0000	1.0000	1.0000
Linear in x-coordinate	1.1324	1.0935	1.0516
Quadratic in x-coordinate	2.0454	1.1319	1.0935

Comparison of the resulting configurations for the constant trend and linear trend case shows that including trend estimation in the optimization causes the sample points to move towards the extreme values of the x-coordinate (i.e., the left and right boundaries of the spatial domain). The effect is not as strong for the strong autocorrelation case as for the weak autocorrelation case. This is because the spatial interpolation error increases more strongly in the strong autocorrelation case when sample points are moved to the boundaries of the spatial domain. A balance is sought between reduction of trend estimation error and spatial interpolation error, and this turns out to yield a configuration closer to the constant trend configuration when the regression residual is strongly autocorrelated. If the regression residual is spatially uncorrelated (Figure 11.3), then all efforts can be directed to the minimization of the trend estimation error variance because the spatial interpolation error is the same, regardless of which configuration is used. For the trend that is linear in the x-coordinate (middle row of Figure 11.3) this yields solutions in which half of the sample points are placed at locations with the minimum value for x and half at locations with the maximum value for x. This is the minmax D-optimal design (Hengl et al., 2003). The y-coordinates of these locations have no effect on the criterion and are again arbitrarily chosen. It is interesting to observe that in the case of weak spatial autocorrelation, including a linear trend distorts the structure of the optimal configuration in the case of 9 and 16 observations. The rectangular grid configuration is no longer optimal and is replaced by a configuration in which more sample points are placed at extreme values of x. For the strong autocorrelation case the grid structure remains intact but is only stretched in the x-direction.

The results of the quadratic trend case confirm those obtained for the linear trend case. Again a balance is sought between optimization in feature and geographic space, whereby optimization in geographic space is more important when spatial autocorrelation is strong. The only difference with the linear trend case is that in this case, optimization in feature space aims at placing half of the sample points at locations with a central x-value, one quarter at locations with a minimum x-value, and one quarter at locations with a maximum x-value. This is most evident from the configurations presented in the bottom row of Figure 11.3, where optimization in feature space is the only consideration. This also explains why the optimal configuration in case of a quadratic trend and four observations (strong and weak spatial autocorrelation case) yields a diamond-shaped configuration instead of a rectangle. The advantage of the diamond-shaped configuration is that it satisfies the optimal conditions for trend estimation, while still maintaining a fairly uniform spreading in geographic space. Note also that these results are in agreement with well-known results from the theory of experimental design (Atkinson and Donev, 1992; Hengl et al., 2003).

The figures in Tables 11.1–11.3 present the gains in precision. A value equal to 1 means that there is no gain in precision at all. This is the case for all constant trend models because in these cases there is no trend and the two optimum sample configurations are the same. These 'gains' are included in the tables for completeness only. The tables show that the gain is relatively large when the spatial autocorrelation is weak and when the sample size is small. Note that the gain is infinitely large for the case of a quadratic trend and sample size 4. This can be explained as follows. The square configurations as given in the top left of Figures 11.1 and 11.2 yield only two effective observations for trend estimation because two observations that have the same x-coordinate provide the same information as one observation. Thus, two effective observations are available, and these are used to estimate a trend with three unknown regression coefficients. This yields an ill-posed problem, leading to infinite estimation variance. Hence, the gain in precision by employing the configurations as given in the bottom left of Figures 11.1 and 11.2 is infinitely large.

11.6 Conclusions

The synthetic examples presented in this chapter show that taking trend estimation into account can have a marked effect on the optimized sample configuration. However, the effect decreases as the number of sample points increases. Here, we used only 16 sample points as a maximum, which is in fact a small number for real-world situations. In future work, it should be analysed how a further increase in the number of sample points will affect the optimized sample configuration and the associated precision gain. Gathering from the results presented here, one might suspect that in case of many sample points, little gain is to be expected from including trend estimation in optimizing the sample configuration. After all, gains in precision that are 5% or less are not very interesting, and extrapolation of the gains presented in Tables 11.1–11.3 suggests that for larger datasets (tens or hundreds of sample points) the gain will be much smaller than that. However, we cannot be certain about this until we have done the analysis because there are also reasons that suggest that the gain in precision may still be substantial in real-world situations. First, it is important to realise that in many practical cases, much of the variation in the target soil property is explained by the predictors. As a result, the residual will often be weakly spatially autocorrelated. We have seen in the synthetic examples that the effect on the optimal sample configuration and the precision gain is larger in case of weak spatial autocorrelation. Second, we should also bear in mind that in the practice of universal and regression kriging for digital soil mapping, it frequently occurs that a large number of predictors are used. Here we used a maximum of two predictors, but real-world cases typically use between five and ten predictors or even more (e.g. Bishop and McBratney (2001); Hengl et al. (2004)). Including a large number of predictors implies that more regression coefficients need to be estimated, making trend estimation more important. In addition, using more predictors also increases the risk of running into multicollinearity problems. How multicollinearity affects the optimization of the sample configuration is unclear and needs to be investigated as well.

It is remarkable that the optimal sample configuration is not dependent upon the strengths of the relationships between the predictors and the target soil variable. Intuitively, one would expect that predictors that explain a large portion of the variation in the target variable should have a stronger influence than predictors that have a small predictive power. This turns out to be not the case. Even when a predictor does not explain a single bit of variation in the target soil variable, it will influence the sample configuration as much as a predictor with a strong explanatory power. This can be explained as follows. Once we have decided to include a predictor in the universal kriging model, we need to estimate the associated regression coefficient. If we do a poor job then we might end up with an estimate that is far off from its true value. In the example of a predictor that has no explanatory power, we might end up with an estimate that differs substantially from zero while the true (but unknown) regression coefficient is in fact zero. This is potentially as harmful as wrongly estimating a regression coefficient that is in reality nonzero. Thus, in both cases, we should choose the sample configuration such that the estimation error variance is as small as possible (of course also taking the effect on the spatial interpolation error into account).

In order to avoid heuristic or approximate solutions, we made several assumptions that simplified the problem of optimizing the sample configuration for digital soil mapping. First, we assumed that the auxiliary information is spatially exhaustive and used in a universal kriging procedure. If the auxiliary information is not spatially exhaustive then other methods will have to be employed to take advantage of the auxiliary information. One possibility would be cokriging (see also McBratney and Webster (1983)). Optimization of the sample configurations of the primary and secondary variables used in cokriging is not fundamentally different from the optimization problem tackled in this chapter, although some adaptations would be required. Second, we also assumed that we had only one target soil variable. In practice, many soil properties must usually be mapped in the same project. Different soil variables will yield different optimal sample configurations. A reasonable solution to come up with a single configuration seems to be to find a compromise between these configurations, or perhaps better to use a criterion that is a combination of the criteria used for the separate soil variables. In our analysis, we also assumed that the variogram of the stochastic residual of the universal kriging model is known. This is not a very realistic assumption. One solution to circumvent this assumption is to first conduct a preliminary fieldwork that is specifically aimed at estimation of the variogram. The result is then used to develop an optimal sample configuration for minimization of the universal kriging variance. Note that the algorithm presented in this chapter can easily optimize sample configurations subject to fixed locations of previously collected observations. Another solution would be to include the uncertainty in the variogram parameters in the optimization. This would make an elegant solution, but it would require the use of so-called model-based geostatistics (Diggle et al., 1998), which is, as yet, not easy. Sample configuration optimization in combination with model-based geostatistics may also be used in the case where the target soil variable is measured on a categorical scale, where the regression is nonlinear or when the residual follows a non-Gaussian distribution.

In this chapter, we used the average universal kriging variance as a criterion, but other criteria may also be used. A frequently used alternative is the maximum kriging variance. Quite different criteria, such as trend estimation and variogram parameter estimation criteria, or weighed combinations of multiple criteria (Müller, 2001), may also be considered. This can all be done with little modification to the methodology. In the synthetic examples, we used trends that are linear or quadratic functions in the x-coordinate, but it should be stressed that the methodology presented works for any number and any type of trend functions. We now need to apply the methodology to real-world situations with trend functions that are derived from relevant auxiliary information. The tool is there and is ready to be used. It will help us to gain more insight into how the optimal sample configuration depends on the characteristics of a particular situation and to design optimal sample configurations for practical applications, thus encouraging the efficient use of fieldwork resources.

References

Atkinson, G.L., Donev, A.N., 1992. Optimum Experimental Design. Clarendon Press, Oxford.

Bishop, T.F.A., McBratney, A.B., 2001. A comparison of prediction methods for the creation of field-extent soil property maps. Geoderma 103, 149–160.

Christakos, G., Olea, R.A., 1992. Sampling design for spatially distributed hydrogeologic and environmental processes. Adv. Water Resour. 15, 219–237.

Christensen, R., 1990. Linear Models for Multivariate, Time, and Spatial Data. Springer, New York.

Deutsch, C.V., Journel, A.G., 1998. GSLIB: Geostatistical Software and User's Guide, 2nd Ed.. Oxford University Press, New York.

Diggle, P.J., Tawn, J.A., Moyeed, R.A., 1998. Model-based geostatistics. Appl. Stat. 47, 299–350.

Goovaerts, P., 1997. Geostatistics for Natural Resources Evaluation. Oxford University Press, New York.

Hengl, T., Heuvelink, G.B.M., Stein, A., 2004. A generic framework for spatial prediction of soil variables based on regression-kriging. Geoderma 120, 75–93.

Hengl, T., Rossiter, D.G., Stein, A., 2003. Soil sampling strategies for spatial prediction by correlation with auxiliary maps. Australian J. Soil Res. 41, 1403–1422.

Heuvelink, G.B.M., Webster, R., 2001. Modelling soil variation: past, present, and future. Geoderma 100, 269–301.

Lark, R.M., 2000. Designing sampling grids from imprecise information on soil variability, an approach based on the fuzzy kriging variance. Geoderma 98, 35–59.

Lesch, S.M., Strauss, D.J., Rhoades, J.D., 1995. Spatial prediction of soil salinity using electromagnetic induction techniques 2. An efficient spatial sampling algorithm suitable for multiple linear regression model identification and estimation. Water Resour. Res. 31, 387–398.

McBratney, A.B., Mendonça Santos, M.L., Minasny, B., 2003. On digital soil mapping. Geoderma 117, 3–52.

McBratney, A.B., Webster, R., 1981. The design of optimal sampling schemes for local estimation and mapping of regionalized variables. II Program and examples. Comp. Geosci. 7, 335–365.

McBratney, A.B., Webster, R., 1983. Optimal interpolation and isarithmic mapping of soil properties. 5. co-regionalization and multiple sampling strategy. J. Soil Sci. 34, 137–162.

Müller, W.G., 2001. Collecting Spatial Data: Optimum Design of Experiments for Random Fields, 2nd Ed.. Physica-Verlag, Heidelberg.

Odeh, I., McBratney, A.B., Chittleborough, D., 1995. Further results on prediction of soil properties from terrain attributes: heterotopic cokriging and regression-kriging. Geoderma 67, 215–226.

Olea, R.A., 1984. Sampling design optimization for spatial functions. Math. Geol. 16, 369–392.

Royle, J.A., Nychka, D., 1998. An algorithm for the construction of spatial coverage designs with implementation in SPLUS. Comp. Geosci. 24, 479–488.

Sacks, J., Schiller, S., 1988. Spatial designs. In: S. Gupta and J. Berger (Eds.), Statistical Decision Theory and Related Topics IV, Vol. 2. Springer Verlag, New York.

van Groenigen, J.W., Pieters, G., Stein, A., 2000. Optimizing spatial sampling for multivariate contamination in urban areas. Environmetrics 11, 227–244.

van Groenigen, J.W., Siderius, W., Stein, A., 1999. Constrained optimisation of soil sampling for minimisation of the kriging variance. Geoderma 87, 239–259.

Wackernagel, H., 1998. Multivariate geostatistics: an introduction with applications, 2nd Ed.. Springer Verlag, Berlin.

Yfantis, E.A., Flatman, G.T., Behar, J.V., 1987. Efficiency of kriging estimation for square, triangular and hexagonal grids. Math. Geol. 19, 183–205.

Developments in Soil Science, volume 31
P. Lagacherie, A.B. McBratney and M. Voltz (Editors)

Chapter 12

LATIN HYPERCUBE SAMPLING AS A TOOL FOR DIGITAL SOIL MAPPING

B. Minasny and A.B. McBratney

Abstract
Prediction of soil attributes (properties and classes) in digital soil mapping (DSM) is based on the correlation between primary soil attributes and secondary environmental attributes. These secondary attributes can be obtained relatively cheaply over large areas. In the presence of these environmental covariates, a strategic sampling design needs to ensure the coverage of the full range of environmental variables. This could enhance the full representation of the expected soil properties or soil classes. This chapter presents the Latin hypercube sampling (LHS) as a sampling strategy on existing data layers. LHS is a stratified-random procedure that provides an efficient way of sampling variables from their multivariate distributions. It provides a full coverage of the range of each variable by maximally stratifying the marginal distribution. This method is illustrated with examples from DSM of part of the Hunter Valley of New South Wales. Comparison is made with other methods: random sampling, equal spatial strata and principal component (PC). Results showed that the LHS is the most effective way to replicate the distribution of the variables.

12.1 Introduction

Soil sampling strategies have been developed for various purposes:

- Estimating the variogram (Lark, 2002; Pettitt and McBratney, 1993).
- Estimating soil variables within a field using spatial interpolation. This includes method that minimise the kriging variance (McBratney and Webster, 1981; van Groenigen et al., 1999) or filling the spatial coverage (Royle and Nychka, 1998).
- Establishing prediction model from ancillary or secondary variables (Hengl et al., 2003; Lesch et al., 1995).

Brus and de Gruijter (1997) discussed the two fundamental approaches to soil sampling: design-based, which follows the classical survey, and model-based, arising from geostatistical analysis. Design-based sampling is mainly based on probability theory, which allows for unbiased estimates of the statistical parameters. Model-based approaches are more suited where spatial variation and its prediction are important. Most sampling designs mainly have the purpose to provide a good spatial coverage of the area, or to cover the spatial variation. In digital soil mapping (DSM), prediction of soil attributes (properties and classes) is based primarily on the correlation between primary soil attributes and secondary soil and environmental attributes (Dobermann and Simbahan, 2006, Chapter 13). These secondary attributes can be obtained relatively cheaply over large areas, for example digital elevation model (DEM) and its derivatives, and satellite images. Sampling in the presence of these environmental covariates has the purpose of calibrating models for spatial soil prediction. The environmental covariates are assumed to have some relationship with the soil variables.

There are few studies focussing on sampling for the purpose of spatial prediction based on environmental attributes. Lesch et al. (1995) developed an algorithm for calibration of the electromagnetic induction (EMI) data. The sampling strategy not only covers a range of EMI values but also ensures spreading the location of the sampling units. It is questionable from a design-based point of view that a geographic constraint is necessary. Gessler et al. (1995) provided a scheme that randomly selects sampling units along the whole range of compound topographic index (CTI) values. McKenzie and Ryan (1999) used terrain attributes, climatic and geological data to stratify an area into classes, and then random selection of sampling units within each class. McBratney et al. (1999) introduced the variance quad-tree method, which sampled sparsely in uniform areas and more intensively where variation is large. The aim is to maximise efficiency of the sampling scheme while ensuring that the variability within the sampling area is characterised effectively. Hengl et al. (2003) proposed sampling along the PCs of the environmental covariates; the number of sampling units taken from each of the components is the proportion of the total variance described by each of the PCs.

In this book, there are two emphases on sampling: sampling in the geographical space and sampling in the character or feature space. Brus et al. (2006, Chapter 14) use k-means clustering to partition the area of interest into geographically compact subregions, from which one sampling unit is selected purposively or randomly. This ensures a good geographical coverage of the area, especially for regions with irregular shapes. Brus (personal communication) has also suggested that the clustering method also can be performed in the feature space. Heuvelink et al. (2006, Chapter 11) designed a sampling scheme for the purpose of spatial prediction using ancillary variables. This is done by minimising the universal kriging variance. However, the requirements are: known form of the model and the spatial structure of the residuals of the model. This information is usually not available for areas that have never been sampled before. However, this technique could be useful for improving prediction on previously sampled areas or some reasonable assumptions could be made following the ideas of McBratney and Pringle (1999).

The aim of sampling for soil mapping is to construct predictive equations for the soil attributes of interest in terms of the environmental covariates; it would seem wise to span the range of values of each covariate so that the prediction model will not be required to extrapolate beyond its bounds. The soil-environmental covariates form a hypercube in the attribute space, and an efficient sampling should cover this hypercube. It will be desirable to be able to sample the points that can represent the whole range of each of the environmental covariates. One possibility is to use Latin hypercube sampling (LHS) (McKay et al., 1979), a constrained Monte-Carlo sampling scheme.

Latin hypercube sampling is a stratified-random procedure that provides an efficient way of sampling variables from their distributions. It has been used in soil science and environmental studies for assessing the uncertainty in a soil prediction model (Minasny and McBratney, 2002) and the simulation of Gaussian random fields (Pebesma and Heuvelink, 1999; Zhang and Pinder, 2004). LHS selects n different values from each of k variables $X_1, \ldots X_k$ in the following manner. The range of each environmental variable X is divided into n non-overlapping intervals on the basis of equal probability. One value from each interval is selected at random with respect to the probability density in the interval. The n values thus obtained for X_1 are paired with the values of X_2 in a random manner or based on some rules to preserve the correlation with other variables. The values for each of the variable are paired until x k-tuplets are formed. We can search through the data and find the locations that are taxonomically most similar to the combination of values chosen, or find points that match the intervals in the various variables. In either case, we will then have a set of x spatial coordinates (locations) at which we can observe the soil attribute(s) of interest.

12.2 Digital soil mapping–Latin hypercube sampling

In LHS of a multivariate distribution, a sample size n from multiple variables is drawn such that for each individual variable the sample is marginally maximally stratified. A sample is maximally stratified when the number of strata equals the sample size n and when the probability of falling in each of the strata is n^{-1} (Everitt, 2002). For sampling of existing environmental soil-covariates, the procedure is then to select n sample points out of the N locations of the soil-environmental covariates (X), so that the sample points x form a Latin hypercube, or the multivariate distribution of X is maximally stratified. This becomes an optimisation problem, where a search procedure is required; the search can be random or based on some prescribed rules combined with an annealing schedule (Metropolis et al., 1953).

We propose the DSM–LHS algorithm as follows:

(1) Divide the quantile distribution of X into n strata, and calculate the quantile distribution for each of the variables, $q_j^i, \ldots, q_j^{n+1}$, for $j = 1, \ldots k$, and $i = 1, \ldots n+1$. Calculate **C**, correlation matrix for X.
(2) Pick a sample of size n units at random from N,. x ($i = 1, \ldots n$) is the sample and y ($i = 1, \ldots N\text{–}n$) is the reservoir.
(3) Calculate the objective function:

For continuous variables:

$$O_1 = \sum_{i}^{n} \sum_{j=1}^{k} \left| \eta\left(q_j^i \leq x_j < q_j^{i+1}\right) - 1 \right|$$

where $\eta(q_j^i \leq x_j < q_j^{i+1})$ is the number of x_j that falls in between quantile q_j^i and q_j^{i+1}.

For discrete variables or classes:

$$O_2 = \sum_{j=1}^{c} \left| \frac{\eta(x_j)}{n} - K_j \right|$$

where $\eta(x_j)$ is the number of x that belongs to class j in sampled data and K_j the proportion of class j in X.

To ensure that the correlation of the sampled variables will replicate the original data, we add another objective function

$$O_3 = \sum_{i=1}^{k} \sum_{j=1}^{k} \left| \mathbf{C}_{ij} - \mathbf{T}_{ij} \right|$$

where **C** is the correlation matrix of X and **T** correlation matrix of x.

The overall objective function is:

$$O = O_1 + O_2 + O_3$$

(4) Perform an annealing schedule: *Metro* $= \exp[\frac{-\Delta O}{T}]$, where ΔO is the change in objective function, and T is a cooling temperature (between 0 and 1) which is decreased at each iteration.
(5) Generate *rand* a uniform random number between 0 and 1, if *rand* $<$ *Metro*, accept the new values, otherwise discard changes.
(6) Try to perform changes: generate *rand*

If *rand* < 0.5,

Pick a sample randomly from x and swap it with a random point from reservoir y.

Else,

Remove the sampling unit(s) from x that has the largest $\eta(q_j^i \leq x_j < q_j^{i+1})$ and replace it with random sampling unit(s) from the reservoir y.

End if.
(7) Go to step (3).

Repeat the loop until the objective function value falls beyond a certain stopping criterion or until a defined number of iterations have been completed.

The attractiveness of this technique is that categorical as well as continuous variables can be incorporated. The sampling is based on the empirical distribution of the original data; thus, it is non-parametric. In addition, spatial coordinates can be incorporated to ensure a good spread of the sampling points if this is required. However, we reiterate that from a design-based point of view this is not necessary. Additional constraints can be imposed on the objective function, for example distance from road and field boundaries. The final sample will represent a hypercube of the attribute space whereby the distribution and multivariate correlation will be preserved.

12.3 Applications

A part of the Pokolbin area in the Hunter Valley is used for this study. The Hunter Valley is situated approximately 140 km North of Sydney, New South Wales, Australia. The area of this study is about 13.8 km^2; a GIS of the area comes from a variety of different data sources. A DEM for the area was produced from digitised topographic maps, originally produced from ground surveying and aerial photograph interpretation. The DEM forms the basis of the landform attributes. Digital data captured on 19th August 1999 by the Landsat (LS) 7 satellite were acquired for the study area. The LS 7 ETM+ measures eight bands of the electromagnetic spectrum, encompassing the visible, near, middle and thermal infrared regions. A soil-landscape unit describing the landform characteristics and geology was obtained from a 1:250,000 soil and landscape survey by the NSW Department of Land and Water Conservation (Kovac and Lawrie, 1991). There are four soil-landscape units in the study area, coded as follows: 300 (Pokolbin), 400 (Branxton), 600 (Mt. View) and 904 (Ogilvie); the Pokolbin unit occupies 86% of the survey area. All data layers were interpolated onto a common grid of 25 m, a total of 22,125 pixels. We wish to sample 100 points in this area that covers the following soil-environmental variables: soil-landscape units (4 classes), CTI, slope angle, LS band 5 (mid-infrared, spectral range 1550–1750 nm) and normalised difference vegetation index calculated from the LS band 3 (visible red) and band 4 (near infrared): NDVI = (NIR−Red)/(NIR+Red).

We compare four sampling methods:

- LHS with five variables: soil-landscape unit, CTI, slope, LS band 5 and NDVI.
- LHS incorporating spatial coordinates: Easting, Northing, soil-landscape unit, CTI, slope, LS band 5 and NDVI.
- Stratified spatial sampling (Brus and de Gruijter, 1997; Brus et al., 2006), where the whole area is divided into n polygons of equal area and the sampling units are at the centre of each polygon. This method ensures a good spatial coverage of the whole area.
- Simple random sampling, where n points are randomly picked from X.
- Sample along PCs (Hengl et al., 2003), where the data X are transformed into PCs and the sampling units are distributed among the PCs according to the proportion of the total variance explained. In each PC, the distribution is divided into strata and sampling units were selected randomly within each strata.

12.4 Results and discussion

The evolution of the objective function with the DSM–LHS algorithm is shown in Figure 12.1. Starting from random sampling this algorithm searches through the data to find points that can represent a Latin hypercube. The objective function value (O) decreases with increasing iterations with perturbations (iterations $<$ 4000) introduced by the annealing process. For this example we use 20,000 iterations to search for an optimal solution.

The sampling unit locations chosen by LHS and stratified spatial sampling are shown in Plate 12 (see Colour Plate Section). The distribution of the variables

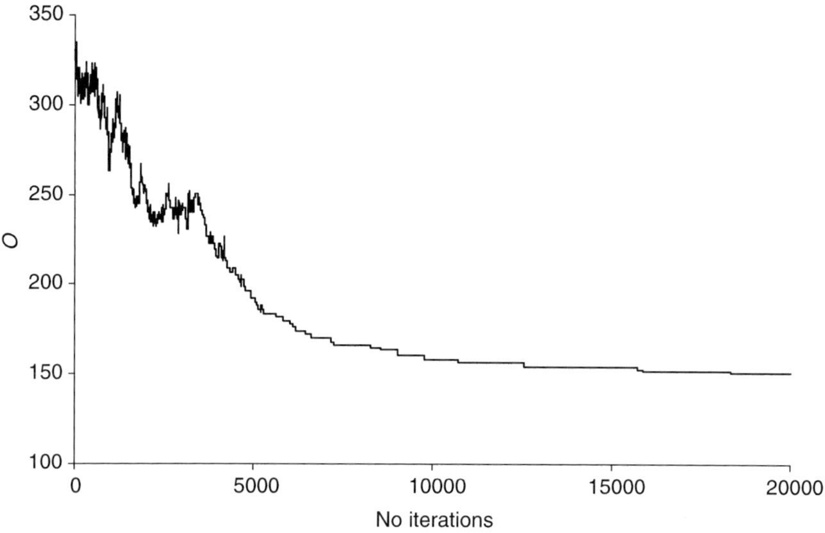

Figure 12.1. *Evolution of the objective function (O) with number of iterations.*

for different sampling techniques is presented in Figure 12.2, and the correlation matrix in Table 12.1. LHS of the target variables mimics the original distribution the best with a good matching correlation matrix. The expense of adding spatial location is the reduction in the range of the values (especially LS band 5). The stratified spatial and random sampling methods are slightly biased in the selection of the values (esp. NDVI). The general perception that a good sampling requires a geographical spread of the points is not well founded. As can be seen an approximate grid in the equal spatial strata, which has a good geographical coverage does not represent the distribution of the variables adequately. Since the objective of the sampling is for calibration purposes, sampling for a good coverage of the environmental attributes is the most appropriate. As seen in Plate 12(a), the spatial distribution of the sample points without considering spatial location is not clustered in the landscape.

Sampling along the PC as suggested by Hengl et al. (2003) shows the worst result by far. When transformed into the orthogonal space (in PC analysis) sampled values from orthogonal space do not have the same probability as prescribed in the original space, which is due to rotation in the attribute space (Minasny and McBratney, 2002).

The ability to pick a sample that represents a hypercube of the original data in LHS–DSM algorithm enables us to build a model to predict soil classes or soil attributes. This model then can be interpolated into the whole area.

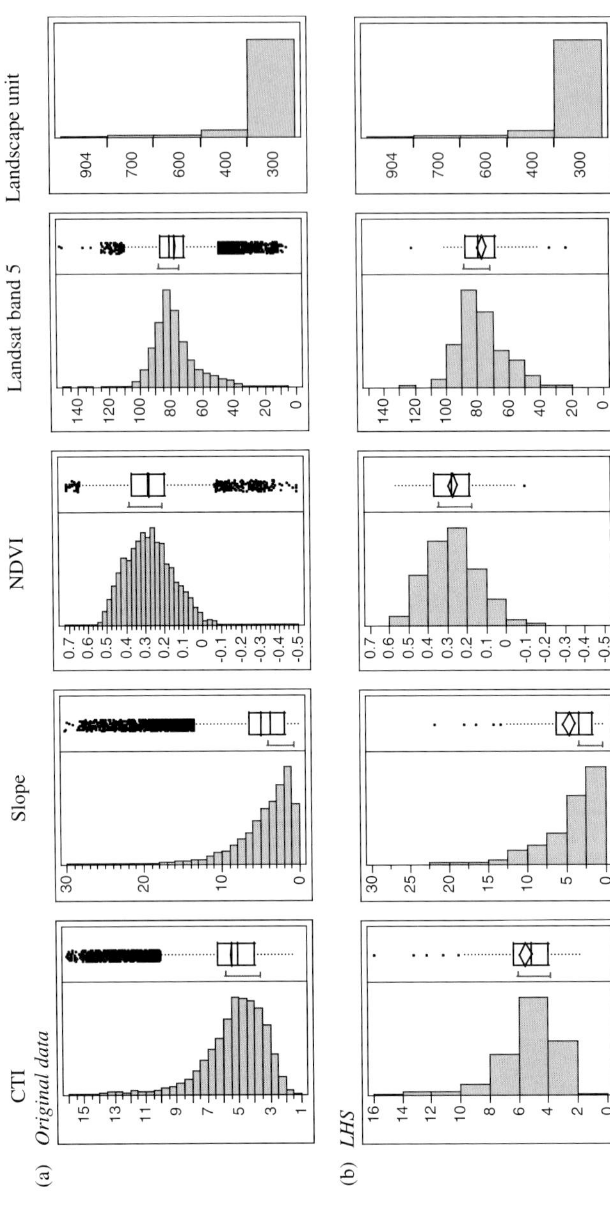

Figure 12.2. *Histogram of the variables (a) original data, (b) Latin hypercube sampling, (c) Latin hypercube sampling with spatial coordinates, (d) equal spatial strata, (e) simple random sampling and (f) sample along principal component.*

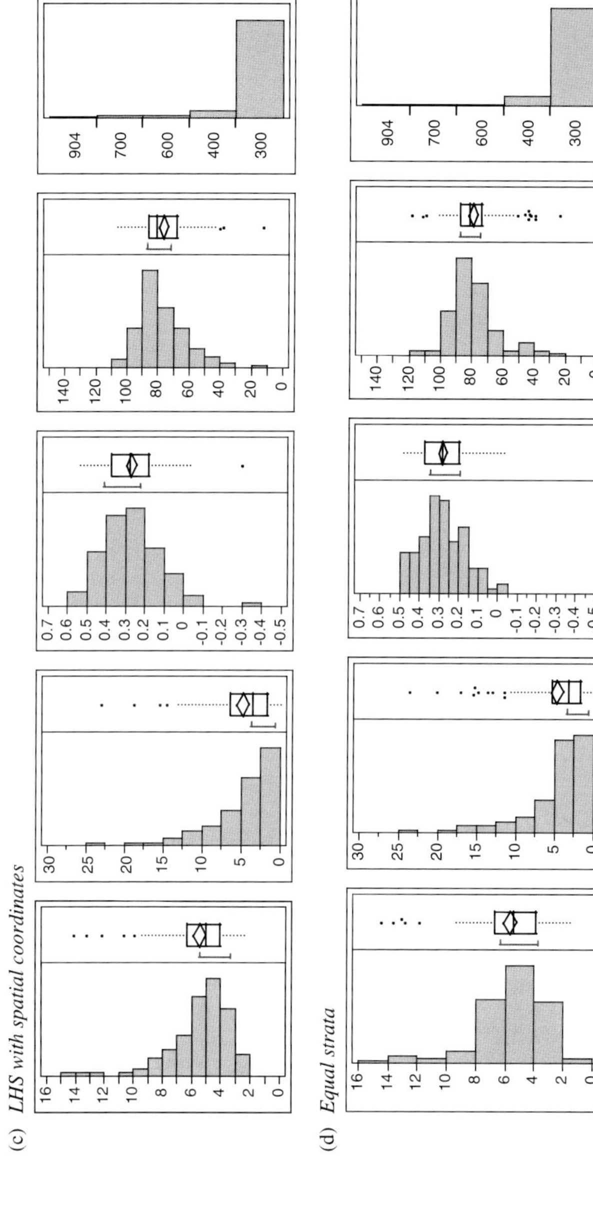

Figure 12.2 *(Continued)*

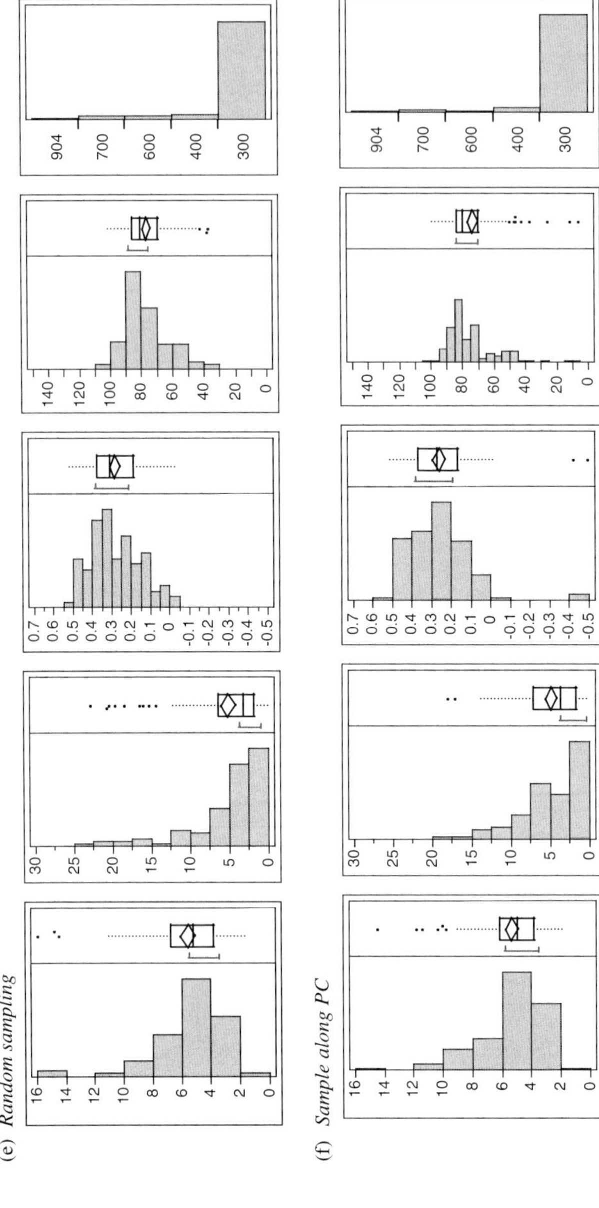

Figure 12.2 *(Continued)*

Table 12.1. *Correlation matrix of the environmental covariates.*

		CTI	Slope	NDVI	MIR
Original data	*CTI*	1.000	−0.491	0.044	−0.096
$N = 22{,}125$	Slope	−0.491	1.000	0.038	−0.212
	NDVI	0.044	0.038	1.000	−0.092
	MIR	−0.096	−0.212	−0.092	1.000
LHS	CTI	1.000	−0.489	0.034	−0.098
$n = 100$	Slope	−0.489	1.000	0.030	−0.168
	NDVI	0.034	0.030	1.000	−0.091
	MIR	−0.098	−0.168	−0.091	1.000
LHS spatial	CTI	1.000	−0.462	−0.004	−0.014
$n = 100$	Slope	0.462	1.000	0.036	−0.236
	NDVI	−0.004	0.036	1.000	−0.060
	MIR	−0.014	−0.236	−0.060	1.000
Equal strata	CTI	1.000	−0.574	0.010	0.095
$n = 100$	Slope	−0.574	1.000	0.054	−0.365
	NDVI	0.010	0.054	1.000	−0.315
	MIR	0.095	−0.365	−0.315	1.000
Random	CTI	1.000	−0.481	0.069	−0.048
$n = 100$	Slope	−0.481	1.000	0.107	−0.226
	NDVI	0.069	0.107	1.000	−0.297
	MIR	−0.048	−0.226	−0.297	1.000
PC sampling	CTI	1.000	−0.581	0.039	−0.028
$n = 100$	Slope	−0.581	1.000	0.001	−0.175
	NDVI	0.039	0.001	1.000	0.247
	MIR	−0.028	−0.175	0.247	1.000

12.5 Conclusion

Sampling in DSM is selecting a limited number of sampling units in an area where secondary soil-environmental attributes are available. This has the purpose of calibration or fitting a model to predict soil classes or soil attributes, where the relationship is usually unknown. An efficient sampling method therefore needs to cover the whole range of values and the values should be maximally stratified. LHS conditioned by the environmental data offers an attractive solution to this sampling problem. The conditioned LHS becomes an optimisation process that is select x (n points) from X (N data) (where $n << N$) so that x forms a Latin hypercube of X. We approach this method by using swapping rules combined with an annealing schedule. This optimisation then can be used to sample not only continuous but also categorical variables, and other criteria can also be added to the objective function. This approach should produce a reasonably efficient way of sampling the soil and its environment so that the full range of conditions are encountered, ensuring a good chance of fitting relationships if they exist. Further work is required to discover whether the quality of the predictions made using this method is superior to those of its rivals.

Acknowledgement

This work is funded by an Australian Research Council Discovery Project on Digital Soil Mapping.

References

Brus, D.J., de Gruijter, J.J., 1997. Random sampling or geostatistical modelling? Choosing between design-based and model-based sampling strategies for soil (with Discussion). Geoderma 80, 1–59.

Brus, D.J., de Gruijter, J.J., van Groenigen, J.W., 2006. Chapter 14. Designing spatial coverage samples using the k-means clustering algorithm. This book.

Dobermann, A., Simbahan, G.C., 2006. Chapter 13. Methodology for using secondary information in sampling optimisation for making fine-resolution maps of soil organic carbon. This book.

Everitt, B.S., 2002. The Cambridge Dictionary of Statistics, 2nd edition. Cambridge University Press, Cambridge.

Gessler, P.E., Moore, I.D., McKenzie, N.J., Ryan, P.J., 1995. Soil-landscape modelling and spatial prediction of soil attributes. Int. J. Geogr. Inf. Sci. 9, 421–432.

Hengl, T., Rossiter, D.G., Stein, A., 2003. Soil sampling strategies for spatial prediction by correlation with auxiliary maps. Aust. J. Soil Res. 41, 1403–1422.

Heuvelink, G.B.M., Brus, D.J., de Gruijter, J.J., 2006. Chapter 11. Optimization of sample configurations for digital mapping of soil properties with universal kriging. This book.

Kovac, M., Lawrie, J.W., 1991. Soil Landscapes of the Singleton 1:250 000 Sheet. Soil Conservation Service of NSW, Sydney.

Lark, R.M., 2002. Optimized spatial sampling of soil for estimation of the variogram by maximum likelihood. Geoderma 105, 49–80.

Lesch, S.M., Strauss, D.J., Rhoades, J.D., 1995. Spatial prediction of soil salinity using electromagnetic induction techniques 2. An efficient spatial sampling algorithm suitable for multiple linear regression model identification and estimation. Water Resour. Res. 31, 387–398.

McBratney, A.B., Pringle, M.J., 1999. Estimating average and proportional variograms of soil properties and their potential use in Precision Agriculture. Precision Agric. 1, 219–236.

McBratney, A.B., Webster, R., 1981. The design of optimal sampling schemes for local estimation and mapping of regionalized variables. Comput. Geosci. 7, 331–334.

McBratney, A.B., Whelan, B.M., Walvoort, D.J.J., Minasny, B., 1999. A purposive sampling scheme for precision agriculture. In: J.V. Stafford (Ed.), Precision Agriculture '99. Sheffield Academic Press, Sheffield, pp. 101–110.

McKay, M.D., Beckman, R.J., Conover, W.J., 1979. A comparison of three methods for selecting values of input variables in the analysis of output from a computer code. Technometrics 21, 239–245.

McKenzie, N.J., Ryan, P.J., 1999. Spatial prediction of soil properties using environmental correlation. Geoderma 89, 67–94.

Metropolis, N., Rosenbluth, A., Rosenbluth, M., Teller, A., Teller, E., 1953. Equation of state calculations by fast computing machines. J. Chem. Phys. 21, 1087–1092.

Minasny, B., McBratney, A.B., 2002. Uncertainty analysis for pedotransfer functions. Eur. J. Soil Sci. 53, 417–430.

Pebesma, E.J., Heuvelink, G.B.M., 1999. Latin hypercube sampling of Gaussian random fields. Technometrics 41, 303–312.

Pettitt, A.N., McBratney, A.B., 1993. Sampling designs for estimating spatial variance components. Appl. Stat. 42, 185–209.

Royle, J.A., Nychka, D., 1998. An algorithm for the construction of spatial coverage designs with implementation in SPLUS. Comput. Geosci. 24, 479–488.

van Groenigen, J.W., Siderius, W., Stein, A., 1999. Constrained optimisation of soil sampling for minimisation of the kriging variance. Geoderma 87, 239–259.

Zhang, Y., Pinder, G.F., 2004. Latin-hypercube sample-selection strategies for correlated random hydraulic-conductivity fields. Water Resour. Res. 39(8), SBH 11.

Developments in Soil Science, volume 31
P. Lagacherie, A.B. McBratney and M. Voltz (Editors)

Chapter 13

METHODOLOGY FOR USING SECONDARY INFORMATION IN SAMPLING OPTIMISATION FOR MAKING FINE-RESOLUTION MAPS OF SOIL ORGANIC CARBON

A. Dobermann and G.C. Simbahan

Abstract

We describe an approach for digital mapping of soil properties based on dual use of secondary information. In the first stage, multivariate secondary information is used to stratify a landscape for sampling. A new clustering algorithm was developed for this, which allows conducting spatially constrained cluster analysis using mixed categorical and continuous secondary attributes. The stratification is used to optimise sampling by constrained spatial simulated annealing (CSSA). Sampled values and correlated secondary attributes are then used in simple kriging with varying local means (SKLMs) to perform final mapping. We applied this method to mapping of soil carbon stock (CS) in the top 0.3 m of soil. Using secondary information greatly reduced the root mean square error (RMSE) of CS prediction as compared with ordinary kriging (OK) of sampled CS alone. When sampling intensity was decreased, RMSE increased more with OK than by using SKLM for local prediction. Among the ancillary variables, on-the-go sensed soil electrical conductivity was most useful for CS mapping. Further research should focus on developing a generic framework for thematic soil mapping at different scales. More standardised procedures are required for processing of secondary information, optimising sampling designs and performing local prediction to minimise cost and achieve high precision.

13.1 Introduction

'Hybrid' geostatistical procedures that account for environmental correlation have become increasingly popular in recent years because they allow utilising available secondary information and often result in more accurate local predictions (Goovaerts, 1999; McBratney et al., 2000). Methods such as co-kriging, kriging with external drift, simple kriging with varying local means (SKLMs) or regression kriging differ in the assumptions about the form of relationships between primary and ancillary variables and how the secondary information is utilised in estimating the primary variable at unsampled locations (Goovaerts, 1997; Hengl et al., 2004; McBratney et al., 2003).

Numerous ancillary variables with potential for digital soil mapping (DSM) are available. At the field scale, examples include categorical attributes derived from digitised soil surveys, attributes derived from digital elevation models (DEMs), remotely sensed images of soil reflectance or transmission, or measurements of soil properties obtained with on-the-go sensors that are driven over a field (Adamchuk et al., 2004; McBratney et al., 2003). While potentially useful, such secondary information is mostly only indicative of the spatial distribution of the primary variable of interest. Which ancillary variables provide the most accurate estimates of a primary variable at unsampled locations must be evaluated across different environments, and there is a need to develop generic methodologies for specific applications.

In recent studies on field-scale mapping of soil organic carbon stock (CS), we have shown that utilising multivariate secondary information in SKLMs resulted in relative improvements in map precision that ranged from 16 to 38% as compared with ordinary kriging (OK) of CS (Simbahan, 2004). This work was done at high sampling intensity, with no attempt to utilise the secondary information for optimising sampling. An issue is how secondary information can be used twice in the mapping process: (i) to optimise sampling without prior knowledge of the spatial variation structure of the primary variable of interest and (ii) to serve as ancillary variables in the local prediction of soil properties using hybrid techniques such as regression kriging.

Most approaches aiming at optimising sampling schemes require prior knowledge or make assumptions about the spatial correlation structure of the primary variable of interest so that optimisation can be based on minimising the kriging variance or similar criteria (McBratney et al., 1981; van Groenigen et al., 1999). Alternatively, sampling optimisation can also be based on stratification and distance criteria to optimise spatial coverage (Brus et al., Chapter 14). More recently, emphasis has shifted towards sampling in the presence of auxiliary information so that principles of environmental correlation can be exploited in the mapping process. Heuvelink et al. (Chapter 11) present a concept in which auxiliary information is used in regression and local prediction, with sampling configurations optimised by minimising the universal kriging variance. However, they assume that the variogram of the residuals of predicting the target variable is known, an assumption that is quite restrictive for practical use. Minasny and McBratney (Chapter 12) show how Latin Hypercube Sampling can be used to ensure that sampling schemes fully represent the multivariate space of correlated variables, which is of importance for establishing good predictive models for use in DSM.

However, prior knowledge of the spatial variation of the primary variable of interest is often not available or gathering it significantly increases the sampling cost. The objective of this chapter is to describe a methodology for using fine-resolution auxiliary information in sampling stratification and optimisation of sampling locations for making fine-resolution maps of soil properties. We will use mapping of soil organic carbon as an example, but the principles should apply to other soil properties in a similar manner.

13.2 An approach for thematic soil mapping using secondary information

We assume that numerous spatial information sources exist that provide indirect information about the spatial distribution of a primary variable of interest. Utilising secondary information in our DSM approach involves five major steps (Fig. 13.1):

1. Acquire the available secondary information, evaluate its usefulness for mapping of the primary variable and process it to a common spatial resolution. Key criteria are (i) availability and cost, (ii) spatial resolution and (iii) a meaningful relationship with the primary variable of interest. Multivariate secondary information with fine spatial resolution is preferred, which should represent soil information with different vertical and lateral resolution measured independently at different times and with different methods. For example, a soil map represents patterns of vertical and lateral soil distribution based on quantitative data and expert judgment, a remotely sensed image mainly represents soil surface properties that affect light reflection, whereas a map of soil electrical conductivity represents a mixture of inherent and management-induced variation in mainly soil chemical properties measured for an intermediate soil depth.
2. Conduct a spatially constrained classification to stratify the area of interest. Two assumptions are made: (i) no direct prior information about the spatial variation of the primary variable is available and (ii) one or more of the available ancillary variables are likely to be spatially correlated with the primary variable. Thus, a stratification based on multivariate secondary information is likely to account for a significant proportion of the variability of the primary variable. A new cluster analysis algorithm is used that utilises both categorical and continuous attributes and results in spatially contiguous clusters (Simbahan, 2004). Categorical variables may include discrete maps of soil types or other landscape features. The user can assign different weights to the categorical variable(s) used, and semivariograms of the continuous secondary variables are used to weight the dissimilarity between individuals.
3. The stratification obtained from the cluster analysis of secondary information serves as a basis for optimising sampling through constrained spatial simulated annealing (CSSA). It is assumed that no direct information about the primary variable to sample is available. Sampling locations are distributed within a field to honour the proportional area of each classified patch as well as to optimise the estimation of semivariograms and achieve minimum distance among locations.

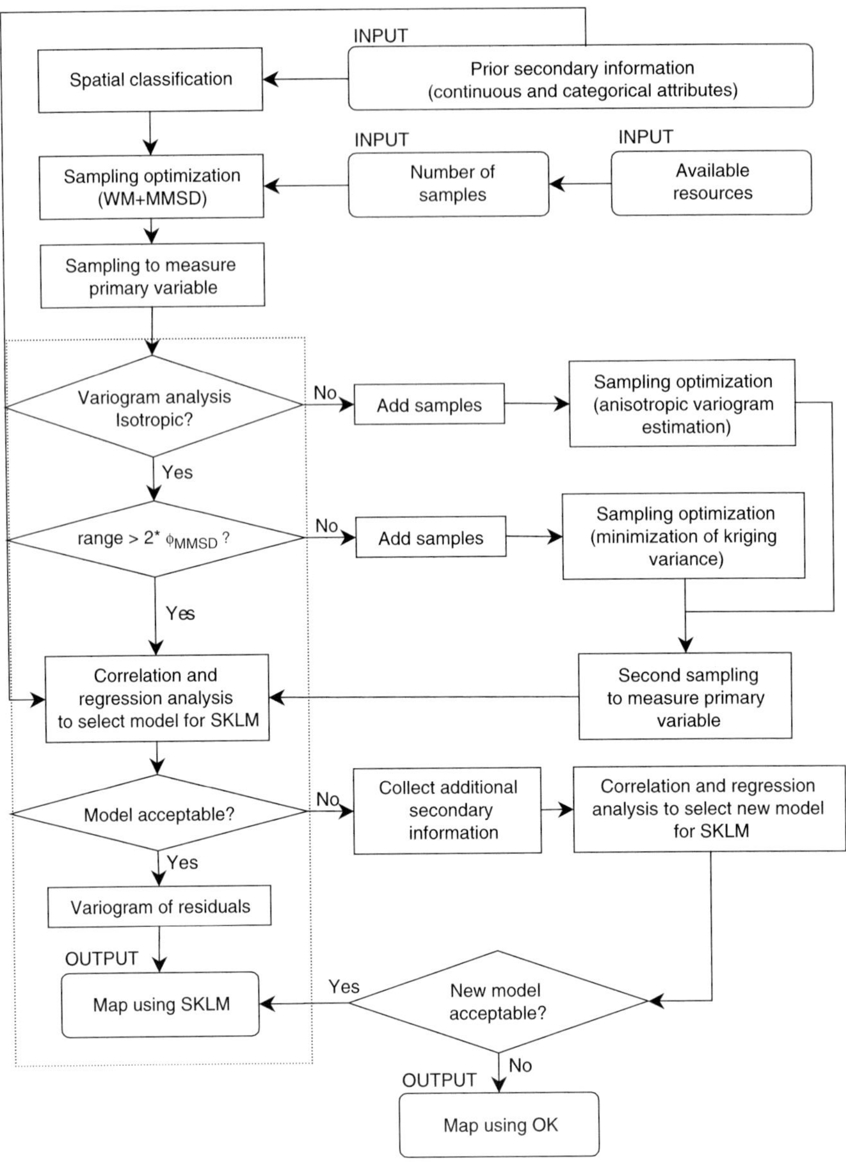

Figure 13.1. *Proposed framework for digital soil mapping (DSM) with dual use of available secondary information. If suitable secondary information is available and correlated with the primary variable of interest, mapping is done using simple kriging with varying local means (SKLMs). If not, ordinary kriging (OK) is used for local prediction.*

4. Sampling to measure the primary variable of interest. If variogram analysis indicates poor structure of the semivariogram or presence of strong anisotropy, additional sampling may be required (Fig. 13.1).
5. Local prediction. The sampled values of the primary variable are used in combination with secondary information to obtain a detailed map of the primary variable. If the correlation between primary and ancillary variables can be modelled reliably, local prediction is done by SKLM. If no correlation exists, OK of the primary variable alone is used for local prediction.

13.3 Case study: mapping of soil organic carbon stock at the field scale

We illustrate the procedure outlined in Figure 13.1 with a case study conducted in a 52-ha agricultural field near Mead, Nebraska, where the objective was to map the amount of soil organic carbon in 0–0.3 m depth (CS, Mg C ha^{-1}). Precise and cost-efficient maps of CS are important in site-specific management of fertilisers and other crop inputs, as well as in programmes that aim to quantify the terrestrial sequestration of atmospheric carbon.

Five layers of secondary data were acquired (Plate 13 (see Colour Plate Section)), assuming that these variables were likely to be related to CS and readily available at fine spatial resolution and a relatively low cost. Soil series was obtained from a digitised soil map (USDA-NRCS, 2004) and represented by four nominal categories in the field: Yutan (Mollic Hapludalfs), Tomek (Pachic Argialbolls), Filbert (Vertic Argialbolls) and Filmore (Vertic Argialbolls). Relative elevation (EL) and slope (SL) were derived from DEM measured by real-time kinematic laser survey. Apparent soil electrical conductivity in 0–0.3 m depth (EC) was measured on-the-go using a Veris-2000 sensor (Veris Technologies, Salina, KS). Field surface reflectance (REF) was derived from a multispectral satellite image obtained in May 2000 (4-m resolution), when surface coverage by plants was minimal. All ancillary variables were pre-processed to a common 4 m × 4 m grid and standardised to zero mean and unit variance. Intensive soil sampling was conducted in April 2001 to obtain an exhaustive sample of CS, which was used to validate the approach shown in Figure 13.1. Three sets of samples were collected: (a) a 68-m triangular grid, (b) a second set comprising 12 stratified transects at 0-, 2.5-, 5-, 10- and 20-m progression and (c) a third set consisting of 25 randomly selected sampling locations. Total number of samples was 202 (Plate 13, top left). At each location, two cores were collected and combined into one sample, for which soil organic C and bulk density were measured to obtain CS.

13.3.1 Spatially constrained classification

A detailed description of the spatial classification algorithm is provided elsewhere (Simbahan, 2004). Our goal was to develop a procedure for spatially constrained classification that would (i) allow using mixed categorical and continuous variables and (ii) results in spatial clusters ('patches') with little clutter or noise. The latter is of particular interest for stratification aiming at optimising sampling.

Let **X** be a vector of N individuals (here: 4 m × 4 m pixels). Further, each individual i has a vector $\mathbf{x}_i[x_1, x_2, \ldots, x_{n_r}]$ comprising n_r continuous attributes, and a vector $\mathbf{c}_i[c_1, c_2, \ldots, x_{n_c}]$ comprising n_c categorical attributes. Conventional non-hierarchical clustering aims to partition the individuals into k disjoint groups such that the similarity of individuals within a group is greater than that to individuals in other groups. The objective is to obtain an $N \times k$ matrix **Y** such that a classification criterion (objective function) is optimised:

$$O = \sum_{i=1}^{N} \sum_{j=1}^{k} y_{ij} \bullet d_{ij} = \mathbf{1}_n^T \mathbf{Y} \mathbf{d}^T \mathbf{1}_n \tag{13.1}$$

where d_{ij} is a metric representing the dissimilarity between individual i and cluster j. Let $\bar{\mathbf{x}}_j$be the vector of means comprising the n_r continuous attributes for cluster j. The Euclidean distance from an individual i to cluster j is then:

$$dn_{ij} = (\mathbf{x}_i - \bar{\mathbf{x}}_j)(\mathbf{x}_i - \bar{\mathbf{x}}_j)^T \tag{13.2}$$

For categorical variables, a simple matching coefficient for the similarity between an individual X_i to the mode of the cluster $\bar{\mathbf{c}}_j$, $j = 1$ to n_c can be defined as:

$$dc_{ij} = \mathbf{w}_c \delta(\mathbf{c}_i, \bar{\mathbf{c}}_j)^T \tag{13.3}$$

where $\mathbf{w}_c$ is a vector of weights assigned to the categorical variables, $\delta(\mathbf{c}_i, \bar{\mathbf{c}}_j)$ a vector of matching coefficients whose elements are 0 if $\mathbf{c}_i = \bar{\mathbf{c}}_j$ and $\delta(\mathbf{c}_i, \bar{\mathbf{c}}_j) = 1$ if $\mathbf{c}_i \neq \bar{\mathbf{c}}_j$. The weights $\mathbf{w}_c$ for categorical variables are added so that expert judgement on the role of the categorical variable can be accounted for in the classification process. The total distance of an individual to a cluster centroid is then calculated as:

$$d_{ij} = dn_{ij} + dc_{ij} \tag{13.4}$$

To add spatial constraints to the Euclidean distance dn_{ij} (Eq. 13.2), the dissimilarity measure of the attribute space is weighted by a non-linear function of the distance in geographic space – the semivariogram. For a continuous variable m, the dissimilarity between two individuals i and l is weighted by (Oliver and Webster, 1989):

$$d_{il}^* = d_{il} \left[\frac{C_{m0}}{C_{m0} + \sum_{i=1}^{n_v} C_{mi}} + \sum_{i=1}^{n_v} \frac{C_{mi}}{C_{m0} + \sum_{i=1}^{n_v} C_{mi}} f_{mi}(h_{il}) \right] \tag{13.5}$$

where h_{il} is the geographic distance (lag) between i and l, C_{m0} represents the nugget effect, C_{mi} the contribution of the ith spatial structure to the semi-variance, $f_{mi}(\mathbf{h})$ a spatial function at lag h and n_v the number of the spatially structured components of the variogram of continuous variable m. This decreases the dissimilarity of points that are close together in geographic space. We extended the method of Oliver and Webster (1989) by using variograms computed for each continuous variable for weighting the similarity between observations. Spatial clustering was then based on minimising the objective function:

$$SS_W = \sum_{i=1}^{N}\sum_{j=1}^{N_c} y_{ij}[(\mathbf{x}_i - \bar{\mathbf{x}}_j)\mathbf{G}(\mathbf{x}_i - \bar{\mathbf{x}}_j)^T + (\mathbf{w}_c)\delta(\mathbf{c}_i, \bar{\mathbf{c}}_j)] \tag{13.6}$$

where $\mathbf{G}$ is the $m \times m$ diagonal matrix whose elements are the spatial weights assigned to variable m and N_c the total number of clusters, subject to other constraints for determining which individuals or clusters can be joined.

In practice, the proposed method consists of an initial allocation to create seed cluster centres, spatial smoothing using a modified majority filter, and iterative re-allocation of individuals on the periphery of clusters, based on their spatially weighted dissimilarities to cluster centres. The spatial smoothing–re-allocation procedure is performed iteratively until no large cluster is split or vanishes. User inputs are the number of spatial clusters (N_k) to be obtained, the minimum size of a cluster and the weight w_c assigned to the categorical variable(s).

In our example, spatial cluster analysis was performed with varying N_k from 18 to 32 and varying w_c from 0 to 5 in small increments. Contour plots of unweighted (S_W) or weighted (wS_W) sum of squares of deviation to the cluster centres as a function of N_k and w_c can be used as a semi-objective method for identifying an optimal combination of N_k and w_c, that is the combination resulting in minimum S_W and wS_W. Both S_W and wS_W increased with greater weight assigned to the categorical variable (soil series), but decreased with increasing number of clusters (Fig. 13.2).

A larger area with minimum S_W (54,000 contour level) and minimum wS_W (33,000 contour) occurred when N_k was about 27–29 and w_c was about 0.7–0.9 (Fig. 13.2). Thus, a solution comprising 28 clusters and a weight of 0.8 was chosen for the final classification, resulting in the map of spatial clusters shown in Plate 13. Clusters were spatially contiguous, free of noise and the stratification captured most of the variation of the four continuous variables as well as that of the categorical variable used in the classification. Relative variance accounted for by the spatial classification ranged from 48% (REF) to 78% (EL). Based on analysis of variance of the 202 collected soil samples, the classification also

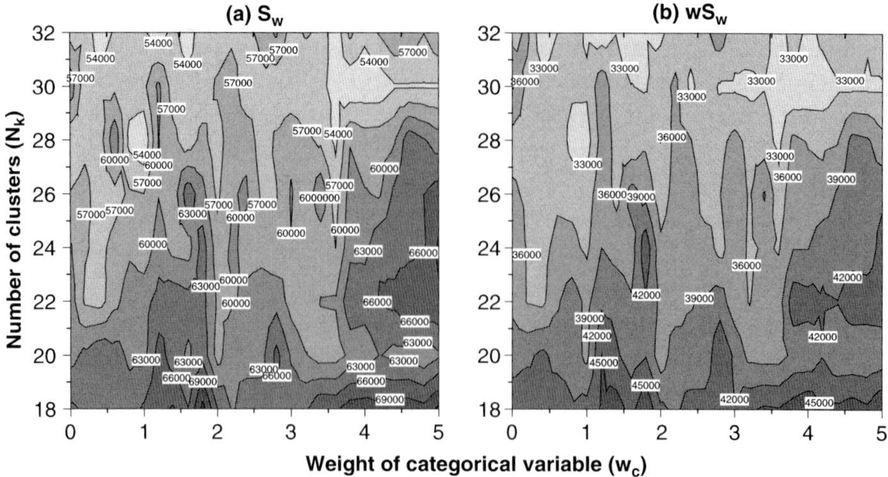

Figure 13.2. *Unweighted (S_w) and weighted (wS_w) total within-cluster error sum of squares as a function of varying number of clusters and weight of the categorical variable (soil series) used in the spatial classification.*

accounted for 68% of the variance of CS, even though this variable was not used in the cluster analysis. In other words, the strata delineated were likely to provide a sound basis for optimising sampling to measure and map CS.

13.3.2 Sampling optimisation

Following the spatial classification, stratified random sampling (SRS) schemes of 50, 100, 150 or 200 sampling units were superimposed on the field, constrained to proportionally allocate samples according to the area of each stratum (patch). The stratified random samples were then further optimised by CSSA, following approaches similar to those used by Van Groenigen and Stein (1998). To preserve the distribution of the sampling units based on the classification, a perturbation of a sample point, x_i, was accepted only if the new location was in the same spatial cluster as the original sample location. To avoid trivial sampling schemes in which pairs of points become very close to each other, a user-defined minimum distance (4 m) between any two points was enforced.

Two criteria were used for sampling optimisation to obtain both good spatial coverage and allow estimation of the semivariogram. In the first approach, the initial sample was optimised by minimising the mean of the shortest distances (MMSDs), subject to the constraints on the acceptable perturbation. MMSD aims for the even spreading of the sampling points over the whole region. The objective function is formulated as the minimisation of the expectation of the distance between an arbitrary chosen point within the study region and its nearest sampling point. If we consider all N points of a finely meshed grid as evaluation points, MMSD aims to minimise the average of the distances of all grid points to their nearest sampling point, that is

$$\phi_{\text{MMSD}}(S) = \frac{1}{N}\sum_{j}^{N} d(x_j, S) \tag{13.7}$$

where x_j is the jth grid point, and $d(x_j,S)$ the distance between grid point x_j and the closest sampling point in sampling set S. Alternatively, samples were also optimised according to the Warrick–Myers (WM) criterion (Warrick and Myers, 1987), which aims to optimise the fit of the realised distribution of point pairs for the experimental variogram to an a priori defined ideal distribution. The criterion solely depends on the distances between sampling points. The objective function to minimise is

$$\phi_{\text{WM}}(S) = \sum_{i=1}^{nc} a \ [(\zeta_i^* - \zeta_i)^2 + b\sigma_i] \tag{13.8}$$

where a and b are user-defined weights that can be used to define the relative importance of the two parts of the function. If $b = 0$, the minimisation function is a simple sum of squares of the deviation between the desired distribution and the realised distribution. Because the sole use of either MMSD or WM may result in non-optimal solutions for mapping (Stein and Ettema, 2003), a third optimisation approach used both criteria jointly. Arbitrarily, two third of the initial stratified sample was first optimised according to MMSD. After convergence of the MMSD algorithm, the remaining one third of the sample was optimised by the WM criterion, using the optimised MMSD samples as prior information.

Figure 13.3 shows the initial stratified sample and optimised sampling schemes based on WM, MMSD and MMSD+WM criteria for 100 samples. Using the WM criterion alone resulted in a sampling scheme that allowed good variogram estimation, but had poor spatial coverage. Other studies have also shown that a straightforward application of the WM criterion causes extreme clustering of sampling points, leading to large kriging variances (Stein and Ettema, 2003; van Groenigen and Stein, 1998). Sampling optimisation using the MMSD criterion alone resulted in a near-triangular grid with even field coverage. At a sampling intensity of 100 samples ϕ_{MMSD} was 28.4 m, which represents a reduction of 8% over a true triangular grid or 12% over a square grid. However, the shortest lag distance class for variogram estimation that was resolved with the MMSD-optimised design was 40–60 m for 100 samples. Sampling optimisation using the combined MMSD (2/3)+WM (1/3) approach provided a good compromise between adequate spatial coverage and estimation of the variogram.

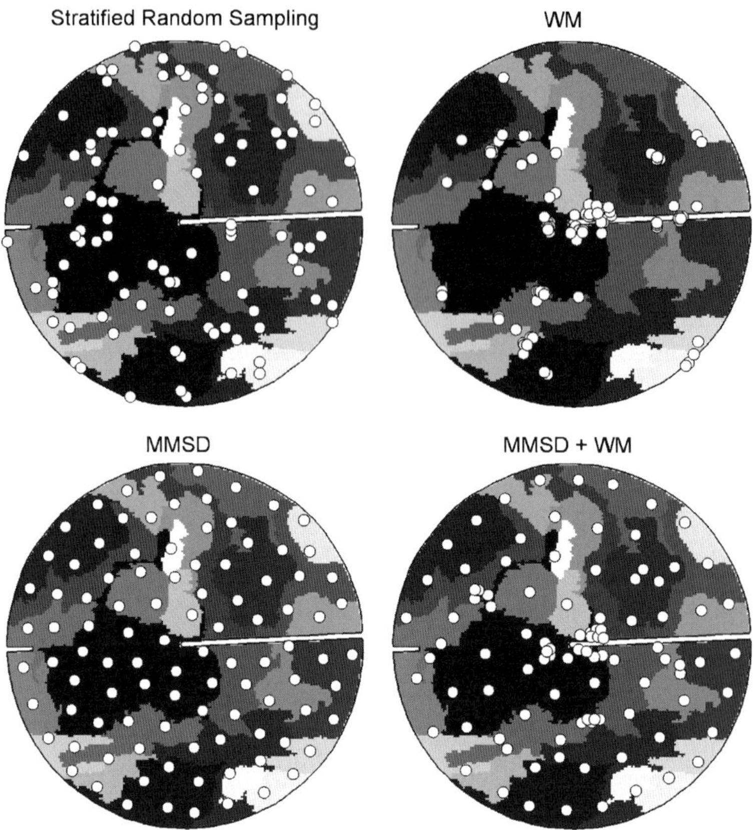

Figure 13.3. *Sampling schemes obtained by utilising different optimisation criteria. In this example, 100 samples were allocated by either using stratified random sampling (SRS, top left), the Warrick–Myers criterion (WM, top right), the minimisation of the shortest distance criterion (MMSD, bottom left) or the combined MMSD+WM approach (bottom right). The underlying stratification (see Fig. 13.2) is indicated by patches with different greyscale colours.*

Compared with the purely MMSD- or WM-optimised sampling schemes, the MMSD+WM-optimised sampling permitted the estimation of the experimental variogram values for all lag classes, including shorter distances (0–20 m lag distance class), and it maintained a small ϕ_{MMSD} of 31.7 m.

13.3.3 Local prediction

The optimised sampling designs were evaluated through repeated sampling of CS values from simulated fields. Sequential Gaussian simulation was used to generate 100 realisations of CS maps, conditional to the measured CS values, the model variogram estimated from these measured CS values and boundary conditions. Each realisation was sampled using (a) SRS, (b) WM-optimised sampling designs and (c) combined MMSD+WM-optimised sampling designs for the 50, 100, 150 and 200 sample densities. Following the procedures outlined in Figure 13.1, each sample was used to map CS by OK and SKLM.

All kriging estimators are variants of the basic linear regression estimator (Goovaerts, 1997):

$$Z*(\mathbf{u}) - m(\mathbf{u}) = \sum_{\alpha=1}^{n(\mathbf{u})} \lambda_\alpha(\mathbf{u})[Z(\mathbf{u}_\alpha) - m(\mathbf{u}_\alpha)] \qquad \alpha = 1, ..., n \tag{13.9}$$

where $n(\mathbf{u})$ is the number of measured data values $z(\mathbf{u}_\alpha)$ used for estimating the attribute Z at location $\mathbf{u}$, $\lambda_\alpha(\mathbf{u})$ the weight assigned to each datum $z(\mathbf{u}_\alpha)$ interpreted as a realisation of the random variable Z and $m(\mathbf{u})$ and $m(\mathbf{u}_\alpha)$ the expected values (means) of $Z(\mathbf{u})$ and $Z(\mathbf{u}_\alpha)$. OK only utilised primary data such as CS at sampled locations $\mathbf{u}_\alpha$ to estimate CS at unsampled locations. Stationarity of the mean is assumed within a local neighbourhood $W(\mathbf{u})$, centred at the location $\mathbf{u}$ being estimated. The mean is deemed to be a constant but unknown value.

Simple kriging with local means (SKLMs) refers to a general group of procedures in which the trend component $m(\mathbf{u})$ in Eq. (13.9) is globally modelled as a function of one or more secondary attributes. These local means, denoted as $m^*_{\mathrm{SK}}(\mathbf{u})$, are assumed to vary smoothly within each local neighbourhood and can be estimated as a function f of secondary attributes y:

$$m^*_{\mathrm{SK}}(\mathbf{u}) = f[y(\mathbf{u})] \tag{13.10}$$

The SKLM estimator is then written as the sum of the regression estimate and the kriged estimate of the spatially correlated residual values at $\mathbf{u}$:

$$Z^*_{\mathrm{SK}} = m^*_{\mathrm{SK}}(\mathbf{u}) + \sum_{\alpha=1}^{n(\mathbf{u})} \lambda^{\mathrm{SK}}_\alpha(\mathbf{u}) \; R(\mathbf{u}_\alpha) \tag{13.11}$$

where $m^*_{\mathrm{SK}}(\mathbf{u})$ is the estimated mean for location $\mathbf{u}$ and $R(\mathbf{u}_\alpha)$ the residuals of the $n(\mathbf{u})$ observation points, $R(\mathbf{u}_\alpha) = Z(\mathbf{u}_\alpha) - m(\mathbf{u}_\alpha)$. In our example, the SKLM approach involved semi-automatic procedures for multiple regression to model the relationship between CS and the four ancillary variables (soil series, EL, EC and REF), using the regression model to predict the local means of CS at sampled and unsampled locations, obtaining the residuals, fitting semivariograms to them, performing the local prediction and obtaining cross-validation statistics. Altogether, for each sampling density (50, 100, 150 or 200) and sampling scheme (SRS, WM and MMSD+WM), we obtained 100 maps of CS by OK and another 100 maps by SKLM, with cross-validation statistics for each.

The MMSD+WM-optimised sampling schemes resulted in higher precision than either the SRS- or the WM-optimised sampling schemes. The root mean square error (RMSE) of prediction was significantly smaller with MMSD+WM-optimised sampling than with either SRS- or WM-optimised sampling, irrespective of sample size and local prediction method (Simbahan, 2004). Using the MMSD+WM sampling schemes, RMSE decreased with increasing number of samples for both OK and SKLM, that is, map precision generally increased with increasing sampling density (Fig. 13.4). However, at all sample sizes, RMSE was significantly smaller with SKLM than with OK and the difference between the two methods increased with decreasing sample size. In other words, in OK, any decrease in sample number caused a considerable loss of information, whereas the relative loss of precision was much smaller with SKLM, particularly when sample size was reduced from 200 to 150 or 100 samples (Fig. 13.4).

Figure 13.5 illustrates these differences for CS maps based on one sampled realisation. At 200 samples per field, maps produced by either OK or SKLM were similar. Reduction of sampling density to 100 samples caused significant smoothing of the OK map, whereas the SKLM map had similar RMSE than with 200 samples and showed more local detail because the relative influence of secondary attributes on the local prediction increased.

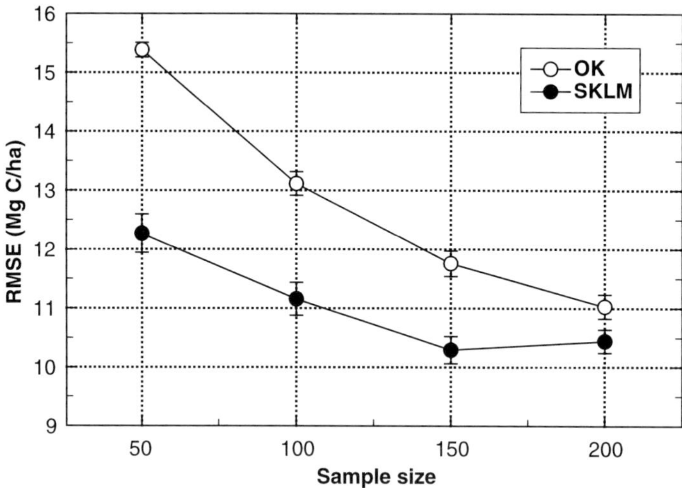

Figure 13.4. *Root mean square error (RMSE) for estimating soil organic carbon stock (CS, Mg C ha^{-1} in 0–0.3 m depth) as a function of sample size and geostatistical prediction method. For each sample number, sampling optimisation was based on the combined MMSD+WM criterion. RMSEs of predicting CS were estimated by cross-validation using either OK or SKLMs. In SKLM, local means of CS were predicted from soil series, elevation, electrical conductivity and soil surface reflectance. Values shown are means (symbols) and standard errors (bars) of 100 map realisations sampled.*

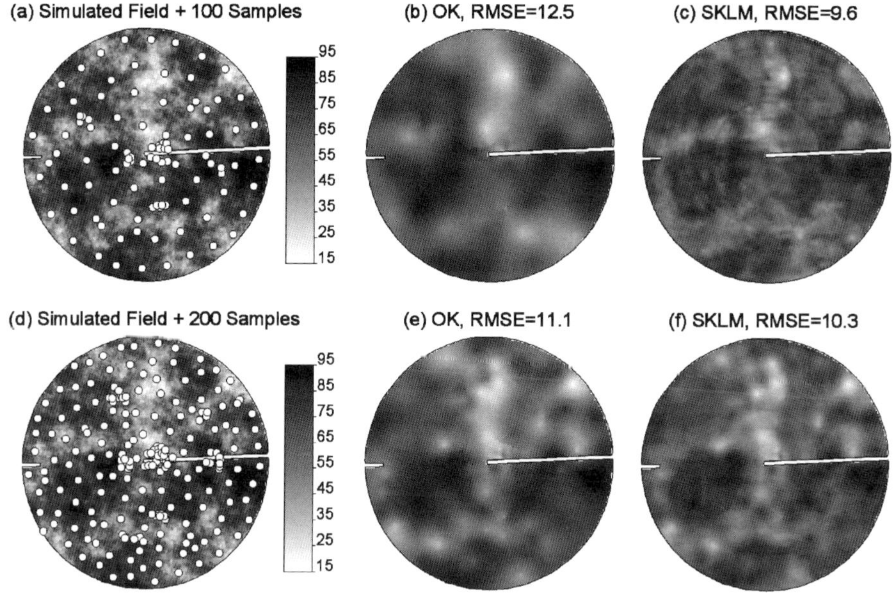

Figure 13.5. *Influence of sampling intensity and local prediction method on the RMSE for estimating soil organic CS (Mg C ha^{-1}). The example shown is one realisation of the simulated CS maps (a, d), which was sampled with either 100 (a) or 200 samples (d) optimised by using the combined MMSD+WM criterion. RMSEs of predicting CS were estimated by cross-validation using either OK or SKLMs. In SKLM, local means of CS were predicted from soil series, elevation, electrical conductivity and soil surface reflectance.*

We have tested this approach in CS mapping at three field sites of similar size (45–65 ha). Results were similar at all sites (Simbahan, 2004), generally suggesting that (i) spatial constraints derived from classification of secondary information improved sampling designs, (ii) the combination of MMSD and WM provided the most suitable sampling schemes for mapping using SKLM techniques and (iii) sample size for mapping of CS using the combined MMSD+WM approach should be about 100 samples per field or 1.5–2.2 samples ha^{-1}. At a density of 50 samples per field, variogram estimation became unreliable, whereas sampling densities of 150 or 200 samples per field would result in relatively little gain in accuracy as compared with the extra cost involved.

13.4 Conclusions

When no or little prior information is available on the spatial variation of a primary variable of interest, correlated secondary information can be used for designing optimal sampling schemes as well as for local prediction by using techniques such as SKLM. In our studies, this resulted in a significant reduction in sample size for mapping of CS and an increase in map accuracy and precision compared with non-optimised sampling and using OK for local prediction. The performance of hybrid geostatistical techniques such as SKLM improves with increasing correlation between primary and secondary variables. No single ancillary variable is likely to be consistently correlated with primary soil properties across different environments. Therefore, to reduce uncertainties, multivariate secondary information should be used to model more stable (deterministic) patterns of soil variation, which are likely to be reproduced to some degree in the different sources of ancillary information used.

The proposed approach is applicable to mapping of soil properties at different spatial scales, but specific procedures, optimisation criteria and constraints, and the available data sources will vary. The algorithm for creating spatially contiguous clusters may have potential for a wider range of applications, not just stratification for sampling. Advantages include the use of mixed variable types, the possibility to assign weights in attribute space, weighting in geographical space and inclusion of other spatial constraints. Optimising sampling schemes based on stratification of secondary information and the combined use of different optimisation criteria are attractive options when no information on the primary variable is available. More research is needed to develop an objective method for determining the appropriate sample size for a given cost of sampling and desired precision of estimates, and for dividing a set number of samples into those that should be optimised by different optimisation criteria. The method proposed could be expanded by utilising other constraints and optimisation criteria to simultaneously account for variation in feature and geographic space as well as sampling cost. The Latin Hypercube Sampling concept discussed in Chapter 12 is one possible addition to the sampling optimisation routine shown in Figure 13.1.

Acknowledgements

This research was supported by the Hatch Act; by the U.S. Department of Energy: (a) EPSCoR program, grant no. DE-FG-02-00ER45827 and (b) Office of Science, Biological and Environmental Research Program (BER), grant no. DE-FG03-00ER62996; and by the Cooperative State Research, Education and Extension Service, U.S. Department of Agriculture under agreements nos. 2001-38700-11092 and 2001-52103-11303.

References

Adamchuk, V.I., Hummel, J.W., Morgan, M.T., Upadhyaya, S.K., 2004. On-the-go soil sensors for precision agriculture. Comput. Electron. Agric. 44, 71–91.

Goovaerts, P., 1997. Geostatistics for Natural Resources Evaluation. Oxford University Press, New York.

Goovaerts, P., 1999. Geostatistics in soil science: state-of-the-art and perspectives. Geoderma 89, 1–45.

Hengl, T., Heuvelink, G.B.M., Stein, A., 2004. A generic framework for spatial prediction of soil variables based on regression kriging. Geoderma 120, 75–93.

McBratney, A.B., Odeh, I.O.A., Bishop, T.F.A., Dunbar, M.S., Shatar, T.M., 2000. An overview of pedometric techniques for use in soil survey. Geoderma 97, 293–327.

McBratney, A.B., Mendonca Santos, M.L., Minasny, B., 2003. On digital soil mapping. Geoderma 117, 3–52.

McBratney, A.B., Webster, R., Burgess, T.M., 1981. The design of optimal sampling schemes for local estimation and mapping of regionalized variables. I. Theory and method. Comput. Geosci. 7, 331–334.

Oliver, M.A., Webster, R., 1989. A geostatistical basis for spatial weighting in multivariate classification. Math. Geol. 21, 15–35.

Simbahan, G.C., 2004. Processing of spatial information for mapping of soil organic carbon. Ph.D. Thesis, University of Nebraska, Lincoln, NE.

Stein, A., Ettema, C., 2003. An overview of spatial sampling procedures and experimental design of spatial studies for ecosystem comparisons. Agric. Ecosyst. Environ. 94, 31–47.

USDA-NRCS, 2004. Soil Data Mart. http://soildatamart.nrcs.usda.gov. United States Department of Agriculture, Natural Resources Conservation Service.

van Groenigen, J.W., Siderius, W., Stein, A., 1999. Constrained optimisation of soil sampling for minimisation of the kriging variance. Geoderma 87, 239–259.

van Groenigen, J.W., Stein, A., 1998. Constrained optimization of spatial sampling using continuous simulated annealing. J. Environ. Qual. 27, 1078–1086.

Warrick, A.W., Myers, D.E., 1987. Optimization of sampling locations for variogram calculations. Water Resour. Res. 23, 496–500.

Developments in Soil Science, volume 31
P. Lagacherie, A.B. McBratney and M. Voltz (Editors)

Chapter 14

DESIGNING SPATIAL COVERAGE SAMPLES USING THE *K*-MEANS CLUSTERING ALGORITHM

D.J. Brus, J.J. de Gruijter and J.W. van Groenigen

Abstract

In situations where we do not want to use available environmental data at the sampling stage of a soil survey because we are uncertain about the correlation with the soil attribute, the best thing we can do is to disperse the points in geographical space. This chapter describes a simple method for selecting such spatial coverage samples. The study area is partitioned into geographically compact subregions from which one point is selected purposively. The partitioning is done by clustering the cells of a fine raster with the k-means clustering algorithm, using the x- and y-coordinates of the midpoints of the cells as classification variables. The centroids of the clusters are used as sample points. The method was tested in two case studies. In the first study, the locations of 23 points were optimised within a square area, and in the second study, 32 points were added to 6 prior points in an irregular shaped area. In both studies, the average kriging variance (AKV) and maximum kriging variance (MaxKV) for the spatial coverage samples were compared with the AKV and MaxKV for the geostatistical samples obtained by directly minimising these criteria with the spatial simulated annealing (SSA) algorithm. The AKV values for the spatial coverage samples and the geostatistical samples obtained with AKV as a minimisation criterion were comparable. If we want to minimise MaxKV, then the SSA procedure is preferable.

14.1 Introduction

Heuvelink et al. (Chapter 11) and Minasny and McBratney (Chapter 12) consider the situation where one has environmental variables, exhaustively known, that are used to design a sample for digital soil mapping. In situations where we have doubts on the strength of the correlation between the environmental variables and the soil variable, we may prefer not to use this prior information at the sampling stage of a survey. The use of weakly correlated environmental variables in designing a sample may lead to suboptimal samples. If after the sampling it appears that there is a moderate or strong correlation, then we still can use this prior information at the prediction (estimation) stage of a survey. If we refrain from using the environmental variables in designing the sample, then the best thing we can do is dispersing the sample points in geographical space. It is well known that, both for mapping and for estimating spatial means of a soil attribute, the precision of the result may be increased by dispersing the points so that they cover the study area as good as possible. A simple method to achieve this is systematic sampling, that is sampling on a regular grid. For mapping a soil variable sampled on a regular grid, the maximum prediction error variance at the centre of the grid cells (squares, triangles or hexagons) is approximately equal everywhere. However, at the border of the area the prediction-error variance (kriging variance) increases considerably when there are no measurements outside the study region that can be used to predict the values near the border. Of course this border effect can be reduced by shifting the points towards the edges, but this goes hand in hand with an increase of the kriging variance at the centres of the grid cells. This raises the question whether we must relax the constraint of a regular configuration of sample points to obtain the best result. This question emerges even stronger when the area is irregularly shaped, or has enclosures that cannot be sampled (build-up areas) or need not be mapped.

A regular configuration of sample points can also be too restrictive when we have measurements in the study region collected in a former survey, which we want to use in the geostatistical interpolation and which do not fit into a regular grid. The previous measurements may have left large spaces unsampled, which we would like to fill in because there the largest gain in precision can be achieved.

When we expect regular grid sampling to be suboptimal under such practical constraints, we must design in some way an irregular configuration of sample points that will lead to more precise spatial predictions than the regular grid. Several methods for optimisation of the locations of sample points have been described in the literature. The methods differ with respect to the criterion that is minimised, and in the way the method searches for the optimal locations of the sample points (optimisation algorithm). In geostatistical samples, a criterion explicitly defined in terms of the prediction error is minimised, usually the average or maximum kriging variance (Sacks and Schiller, 1988; Müller, 2001). This requires knowledge of the variogram, and in many situations this variogram is unknown, or at least uncertain. In spatial coverage samples, a criterion that is defined in terms of the distance between the sample points and the nodes of an interpolation grid is minimised (Royle and Nychka, 1998), and a variogram is not needed.

In this chapter, a simple method is proposed for designing a spatial coverage or space-filling sample. In this method, the area to be mapped is partitioned into compact subregions by clustering the raster cells with the k-means algorithm, using the x- and y-coordinates of the midpoints of the cells as classification variables. The centroids of the clusters are used as sample points.

The aim of this chapter is to demonstrate the proposed method, and to explore the precision of spatial coverage samples obtained with the proposed method in relation to the precision of geostatistical samples.

14.2 Geographical partitioning by *k*-means

K-means is a well-established method of cluster analysis, used to find compact clusters of objects in multivariate property space (Hartigan, 1975). The clusters are represented by their multivariate means, referred to as centroids, and the method aims at minimising the mean squared distance between the objects and their nearest centroid. Here we propose to use this method for a different purpose: partitioning the area into compact subregions. In doing so, the multivariate property space is replaced by the geographical space, with only the *x*- and the *y*-coordinates as properties, and the cells of fine raster represented by their midpoints play the role of objects. In formula, the criterion that is minimised is

$$\mathrm{MSSD} = \frac{1}{N}\sum_{i}^{N} \min_j(D_{ij}^2), \qquad (14.1)$$

where N is the number of raster cells and $\min_j(D_{ij}^2)$the minimum of the squared distances between the ith cell and all cluster centroids.

K-means proceeds in an iterative way, starting with an initial solution, a random configuration of cluster centroids or a random clustering of raster cells, and repeatedly alternating two steps: reallocation of raster cells from one cluster to the other, and recalculation of the coordinates of the centroids. In the reallocation step, each cell is allocated to the nearest centroid. In the recalculation step, each centroid is shifted to the mean of the coordinates of the cells allocated to that centroid. The iteration process is stopped when the mean squared shortest distance (MSSD) cannot be further lowered, or when the improvements become smaller than a user-specified threshold. Recalculation can be done as soon as a cell has been reallocated to a different centroid, or it may be postponed and done when all cells are reallocated.

K-means is a deterministic search technique, which means that, given the raster cells and the stopping rule, all intermediate solutions and the final solution are determined by the initial solution. This is because in each iteration the algorithm calculates the best reallocation of raster cells. The final clustering can be a local instead of a global minimum, and therefore a number of initial solutions should be tried, leading to as many final solutions as the initial solutions tried. The final solution with the lowest value for MSSD is kept as an optimal solution.

Given the optimal clustering, the final step is to select the sample points. For digital soil mapping, the centroids of the ultimate clustering are taken as the sample points. Note that the centroids do not form a probability sample but are selected purposively to minimise MSSD, and therefore cannot be used in design-based inference.

Existing points with measurements on the target variable can be easily accommodated in the *k*-means algorithm, simply by reallocating the cell to both the existing and the new sample points, but limiting the recalculation step to the new sample points. In this way, the existing points remain at their place and only the new points move. There is no problem in using these prior data for mapping, if the support and the measurement method are the same as for the new data and the values have not changed after the collection of the prior data.

For concave areas, one or more of the centroids may move out of the area. Sampling at these outlying centroids is recommendable only when the border is artificial, for example administrative borders, and we expect the same soil conditions outside the study area. If we do not want to sample outside the study area (or when it is impossible to do so), the problem of centroids moving out can be avoided by shifting such points to the nearest grid point after each recalculation. By doing so, the sample points are forced to stay within the area during the optimisation process.

14.3 Case studies

The method will be illustrated by calculating the optimal locations of 23 sample points in a square area. The sample obtained by minimising MSSD with the *k*-means algorithm will be compared with the samples obtained by minimising the AKV or MaxKV with the SSA algorithm of van Groenigen et al. (1999). We used Genstat to optimise the MSSD/*k*-means samples (Genstat committee, 2000). A total of 100 initial samples have been optimised, and the best final solution was kept. For SSA using the AKV and MKV criteria, the optimal locations of the sample points depend on the variogram. We used two variograms, both of the type 'spherical with nugget' with a sill variance of 1 and a range of half the side of the square. The first variogram has a nugget variance of 0.5, and the second has a nugget variance of 0.1. The start temperature for the cooling schedule for SSA was chosen so as to accept 95% of the proposed iterations (van Groenigen and Stein, 1998). The finishing temperature was chosen as 1 billionth of the starting temperature (e.g. from 1250 to 0.000001250), which has proven to be a reliable final temperature. The SSA optimisations ran for 60 h. In most cases, a solution close to the final result was derived within half an hour. In the remaining hours, the criterion value was reduced by only a small percentage.

In practice, the true variogram is unknown, and consequently the variogram used in the optimisation always differs to some extent from the true variogram. We explored the effect of a misspecified variogram on the minimum value for

AKV and MaxKV by using different variograms in the optimisation and in the interpolation (Table 14.1 rows 4, 5, 9 and 10). The AKV and MaxKV on a 400 × 400 grid have been calculated for ordinary kriging, using a neighbourhood of the eight nearest sample points.

Figure 14.1 shows the five optimised samples. The configurations are more or less regular (point symmetric) and can be described as triangular grids distorted at the sides of the square. The configuration obtained with MaxKV/SSA and a variogram with a nugget-to-sill ratio of 0.5 is much less regular (Figure 14.1C). The outer points of the samples optimised with MaxKV are much closer to the sides of the square than those of the samples optimised with AKV or MSSD. The AKV of the MSSD/*k*-means sample is very close to the AKV of the two AKV/SSA-samples, for both variograms (Table 14.1. On the other hand, the MaxKV of the two MaxKV/SSA samples are well below the MaxKV of the MSSD/*k*-means sample. The superiority of the MaxKV/SSA sample over the MSSD/*k*-means sample is retained when the variogram used in the SSA-algorithm is misspecified. When the nugget-to-sill ratio increases, this superiority of MaxKV/SSA over MSSD/*k*-means vanishes.

Figure 14.2 shows the result for an irregularly shaped area (province of Gelderland, the Netherlands) with 32 points added to 6 prior points. The variogram used in the SSA-procedure is spherical with nugget, with a nugget variance of 0.4, a partial sill variance of 0.6 (nugget-to-sill ratio is 0.4) and a range of 50 km.For all three samples, the six prior points fit well in the configuration of the new sample points. The configurations of the MSSD/*k*-means and AKV/SSA sample are more or less comparable. SSA locates points somewhat nearer the borders where the kriging variance is largest, and the AKV can be reduced even more. This effect is much stronger for the MaxKV criterion. The AKV for the optimised AKV/SSA sample equals 0.614, and the MaxKV for the optimised MaxKV/SSA is 0.680. For the MSSD/*k*-means sample the AKV and MaxKV variance are 0.618 and 0.868, respectively.

14.4 Discussion

14.4.1 Maximum kriging variance and k-means

As shown above, k-means is inferior to SSA when MaxKV is chosen as a minimisation criterion. K-means does not account for extrapolation at the borders, whereas SSA does and locates points near the border where the kriging variance is the largest. The same effect can be reached with k-means simply by shifting the borders of the area across a fixed distance, for instance $0.5\sqrt{A/n}$ (i.e. half the side of a square with an area covered, on an average, by 1 sample point), and clustering the raster cells of this enlarged area. For 23 points in a square this results into a MaxKV of 0.860 for a nugget-to-sill ratio of 0.5 (compare with 0.872

Table 14.1. *Average and maximum kriging variance for spatial coverage sample (obtained with k-means) and geostatistical samples (obtained with spatial simulated annealing, SSA) of 23 points in a unit square.*

Minimisation Criterion	Optimisation-algorithm	Variogram used in optimisation	True variogram	Average kriging variance	Maximum kriging variance
MSSD	*k*-means	–	Nug(0.5)+Sph(0.5,0.5)	0.805	1.019
AKV	SSA	Nug(0.5)+Sph(0.5, 0.5)	Nug(0.5)+Sph(0.5,0.5)	0.805	0.988
MaxKV	SSA	Nug(0.5)+Sph(0.5, 0.5)	Nug(0.5)+Sph(0.5,0.5)	0.819	0.872
AKV	SSA	Nug(0.1)+Sph(0.9, 0.5)	Nug(0.5)+Sph(0.5,0.5)	0.805	0.965
MaxKV	SSA	Nug(0.1)+Sph(0.9, 0.5)	Nug(0.5)+Sph(0.5,0.5)	0.813	0.901
MSSD	*k*-means	–	Nug(0.1)+Sph(0.9, 0.5)	0.427	0.793
AKV	SSA	Nug(0.1)+Sph(0.9, 0.5)	Nug(0.1)+Sph(0.9, 0.5)	0.423	0.676
MaxKV	SSA	Nug(0.1)+Sph(0.9, 0.5)	Nug(0.1)+Sph(0.9, 0.5)	0.436	0.535
AKV	SSA	Nug(0.5)+Sph(0.5,0.5)	Nug(0.1)+Sph(0.9, 0.5)	0.424	0.733
MaxKV	SSA	Nug(0.5)+Sph(0.5,0.5)	Nug(0.1)+Sph(0.9, 0.5)	0.445	0.570

Note: MSSD = mean squared shortest distance; AKV = average kriging variance; MaxKV = maximum kriging variance; Nug(c_0) = pure nugget variogram with a nugget variance of c_0; Sph(c,a) = spherical variogram with a sill variance of c and a range of a distance units.

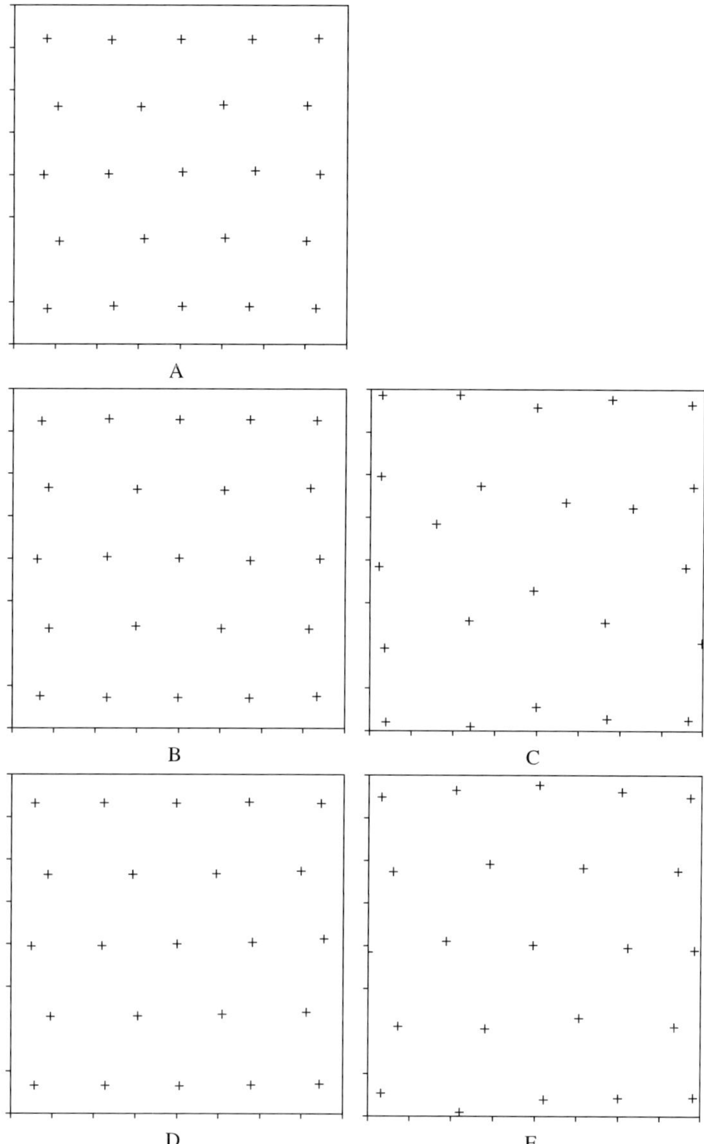

Figure 14.1. *Sample of 23 points in unit square optimised with k-means (A) and spatial simulated annealing with the average kriging variance (B,D) and the maximum kriging variance (C,E) as a minimisation criterion. Variograms used in SSA: Figs B,C: Nug(0.5)+Sph(0.5, 0.5); Figs D,E: Nug(0.1)+Sph(0.9, 0.5).*

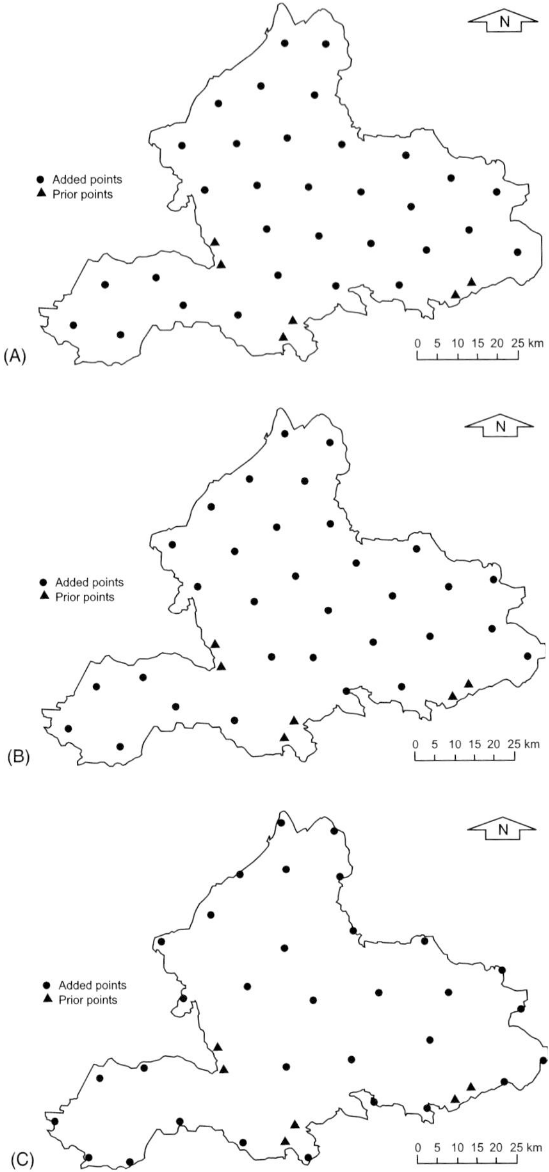

Figure 14.2. *Additional sample of 32 points in province of Gelderland (the Netherlands) with six prior points, optimised with k-means (A) and spatial simulated annealing with the average kriging variance (B) and the maximum kriging variance (C) as a minimisation criterion. Variogram used in SSA: Nug(0.4)+Sph(0.6, 50 km).*

for SSA, Table 14.1 row 3) and 0.541 for a nugget-to-sill ratio of 0.1 (0.535 for SSA, Table 14.1 row 8). For the first variogram, the result obtained with k-means is even slightly better than that with SSA, showing that the rather irregular configuration of Figure 14.1C is a local minimum indeed.

14.4.2 Controlling computing time for k-means

In general, computing time for SSA optimisation is much larger than for *k*-means optimisation due to the random search, and the probability of accepting deteriorations. The above SSA solutions were obtained with 60 h of computing time on a 2.40-Ghz computer, although after half an hour a solution close to the final result was already obtained. The relative fastness of *k*-means allows for a large number of starting points (initial samples), which reduces the chance of ending up with a local minimum. This may outweigh the suboptimality of the MSSD/*k*-means-sample with respect to AKV. In other words, the best result from *k*-means with a large number of starting points could have a smaller AKV than the best result from SSA run for the same CPU time. The number of starting points in the *k*-means procedure must be weighed against the probability of ending up with a local minimum. To get an idea of this probability, we may construct a histogram of the minimised MSSD values based on an arbitrary chosen number of starting points. This histogram can be used to predict the required number of initial samples to be almost sure that a sample close to the global optimum is found.

14.4.3 Upgrading and validation of digital soil-class maps

The proposed method can also be used in situations where one has a digital soil class map, made by classification trees or otherwise, and one wants to sample these soil classes to estimate the class means of soil attributes, that is upgrading digital soil-class maps, or to estimate for each soil class the areal fractions of the soil classes as a measure of map purity, that is validation. For these aims, design-based sampling is appropriate because valid estimation of confidence intervals is then possible. The sample sizes per soil class can be controlled by using the soil classes as strata in stratified random sampling. The precision of the mean (areal fraction) for a given soil class may be increased by dispersing the randomly selected sample points over the spatial extent of that soil class. Again the *k*-means clustering technique can be used to achieve this spreading of the sample points (Brus et al. 1999). Contrary to the mapping situation, we are now not interested in the centroids of the clusters, but in the clusters themselves. These clusters are used as substrata (strata within the soil-class strata).

14.5 Conclusions

Optimisation of sample locations by k-means can be an attractive alternative to optimisation by SSA using the AKV and MaxKV criteria. The AKV of predictions from the spatial coverage sample obtained with k-means is very close to the minimum AKV obtained with SSA. However, with respect to the MaxKV the k-means sample is suboptimal. Using MaxKV as a minimisation criterion, SSA locates points near the borders where the kriging variance is the largest and the strongest reduction of the criterion can be achieved. The same effect can be reached by calculating the centroids of the clusters for a larger area obtained by shifting the border on all sides across a distance $0.5\sqrt{A/n}$.

References

Brus, D.J., Spätjens, L.E.E.M., de Gruijter, J.J., 1999. A sampling scheme for estimating the mean extractable phosphorus concentration of fields for environmental regulation. Geoderma 89, 129–148.

Genstat Committee, 2000. The guide to GenStat. VSN International, Oxford.

Hartigan, J.A., 1975. Clustering Algorithms. Wiley, New York.

Müller, W.G., 2001. Collecting Spatial Data: Optimum Design of Experiments for Random Fields, 2nd ed. Physica-Verlag, New York.

Royle, J.A., Nychka, D., 1998. An algorithm for the construction of spatial coverage designs with implementation in S plus. Comp. Geosci. 24, 479–488.

Sacks, J., Schiller, S., 1988. Spatial designs. In: S.S. Gupta and J.O. Berger (Eds.), Statistical Decision Theory and Related Topics IV, Vol. 2. Springer-Verlag, New York, pp. 385–399.

Van Groenigen, J.W., Stein, A., 1998. Constrained optimization of spatial sampling using continuous simulated annealing. J. Environ. Qual. 27, 1078–1086.

Van Groenigen, J.W., Siderius, W., Stein, A., 1999. Constrained optimisation of soil sampling for minimisation of the kriging variance. Geoderma 87, 239–259.

Developments in Soil Science, volume 31
P. Lagacherie, A.B. McBratney and M. Voltz (Editors)

Chapter 15

ADEQUATE PRIOR SAMPLING IS EVERYTHING: LESSONS FROM THE ORD RIVER BASIN, AUSTRALIA

E.N. Bui, D. Simon, N. Schoknecht and A. Payne

Abstract

Computers are not omniscient, and computer-aided predictive modelling of soil distribution requires adequate representative sampling of landscapes to cover the range of combinations of environmental factors that govern pedogenesis. An example from Australia is presented here that illustrates what can go wrong if prior sampling is not representative. Existing soil mapping and targeted new field work were used as training data for classification tree models of soil distribution, with geology, climate, remote sensing and digital terrain data as predictors. Although good models were generated directly over training areas, extrapolation proved impossible over long distances because of inadequate coverage of the regional environment by the training data.

15.1 Introduction

The Ord river basin in northern Australia covers 55,400 km^2 and straddles the northern territory (NT) and western Australia (WA) (Fig. 15.1). The NT part of the basin has recent land unit mapping at a nominal scale of 1:50,000–1:100,000 as the result of a lengthy mapping program. Because of developmental pressure in the area, land management in the WA part of the Ord river basin would benefit from land unit mapping similar to that available in the NT. Traditionally, soil mapping requires extensive fieldwork and is thus expensive. Funds to conduct land unit mapping using these techniques are not available in WA, and a joint project was initiated with Australia's Commonwealth Scientific and Industrial Research Organisation (CSIRO) Land and Water as part of the Ord–Bonaparte Program to put into trial a new technique for land unit mapping. This technique uses existing mapping (land systems – Aldrick et al., 1978; Stewart et al., 1970; land units – de Salis, 1993; and geology), climate data, remotely sensed data, digital elevation models (DEM) and targeted field work in computer-aided modelling.

The project's objective was to use available soil and land resource information, with targeted field work, to improve the quality of land resource mapping

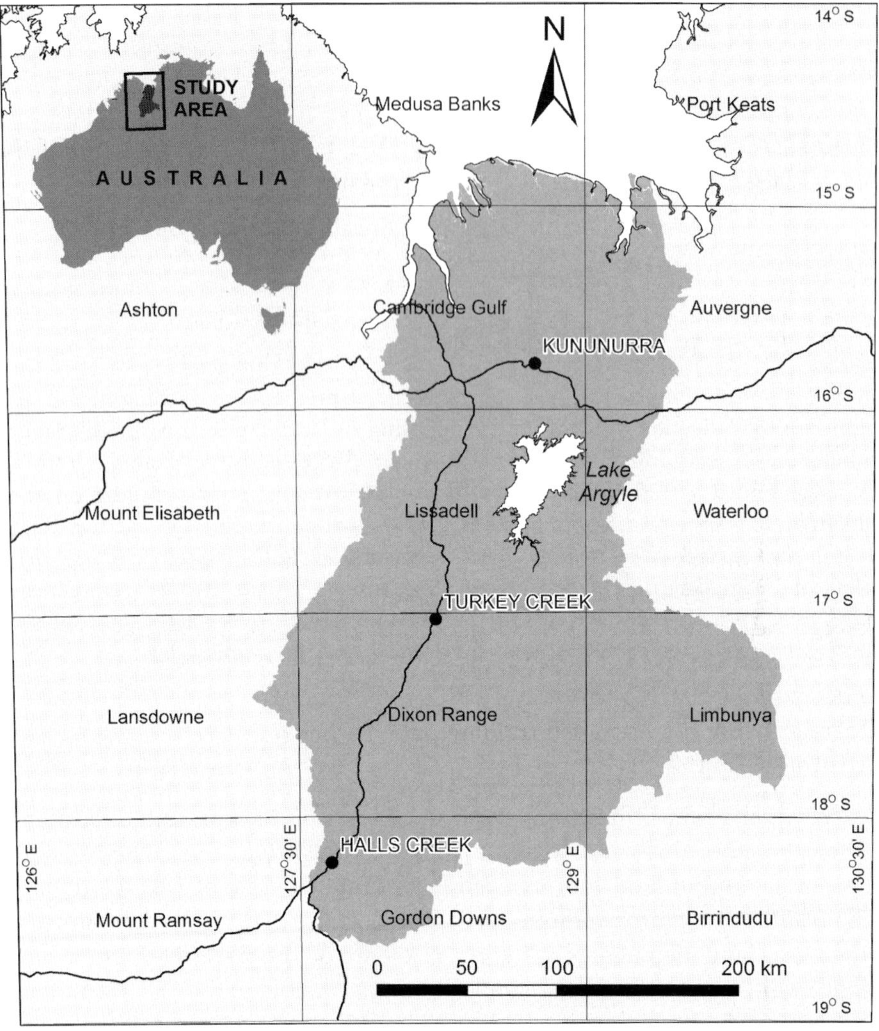

Figure 15.1. *Location of the Ord river basin in Australia with overlay of 1:250,000 map sheets.*

for the part of the Ord river basin and Keep river catchments in Western Australia. This chapter focuses on the problems encountered with sampling to provide training data for modelling.

15.2 Data and methods

An initial attempt at digital soil mapping of the WA at a nominal scale of 1:250,000 was made using classification trees with existing NT land systems

(Aldrick et al., 1978) and Ord regeneration mapping (de Salis, 1993) as training data with a 9-s (~250-m) grid DEM, Landsat multispectral scanner (MSS), climate variables and lithology (1:1,000,000) as predictors. The modelling was based on techniques developed for the Murray–Darling Basin (Moran and Bui, 2002; Bui and Moran, 2003). This initial model predicted land units to a reasonable degree over and adjacent to the training data but was not a good predictor of land units elsewhere in the basin. The first model indicated that the approach could work, and a set of random 10 × 10 km training data regions and sampling points were generated (Fig. 15.2). New field work was undertaken; however, it did not coincide exactly with the randomly selected regions and concentrated on alluvial areas, entirely neglecting some regions across the central and western Ord (Fig. 15.2).

The final coverage of training data represented about 15% of the river basin's area (Fig. 15.2); some land-unit mapping from the NT (Van-Cuylenburg et al., in prep.) was also used. All available land resource mapping was labelled using a concatenated four-part label that described the landform, geology, soil and vegetation of the land unit. This was based on the system adopted by the NT in their land unit mapping work. For example, land unit label'1aktCTHP' where character one (1) describes the landform, character two (a) describes the geology, characters three and four (kt) describe the soil and characters five to eight (CTHP) describe the vegetation type.

Together with more detailed training data, higher resolution predictors were also obtained. Thirty-four predictor datasets across the entire study area were assembled from four categories of digital biophysical datasets.

15.2.1 Biophysical dataset 1: digital elevation models

A total of 12 predictor datasets were derived from the best available DEM for the area. Two DEMs were available:

- A 1-s (~30-m grid) DEM derived from 1:50,000 Army topographic data was available for the majority of the area. (see Fig. 15.3 for coverage)
- A 9-s (~250-m grid) DEM for the remainder of the Ord catchment (mainly the Gordon Downs 1:250,000 map sheet on Fig. 15.1) (http://www.ga.gov.au).

Predictor datasets derived from the DEMs (refer to Moran et al. (2001) for definitions):

1. Slope (percent)
2. Relief (10-s radius window)
3. Flow accumulation
4. Flowline distance to the nearest stream

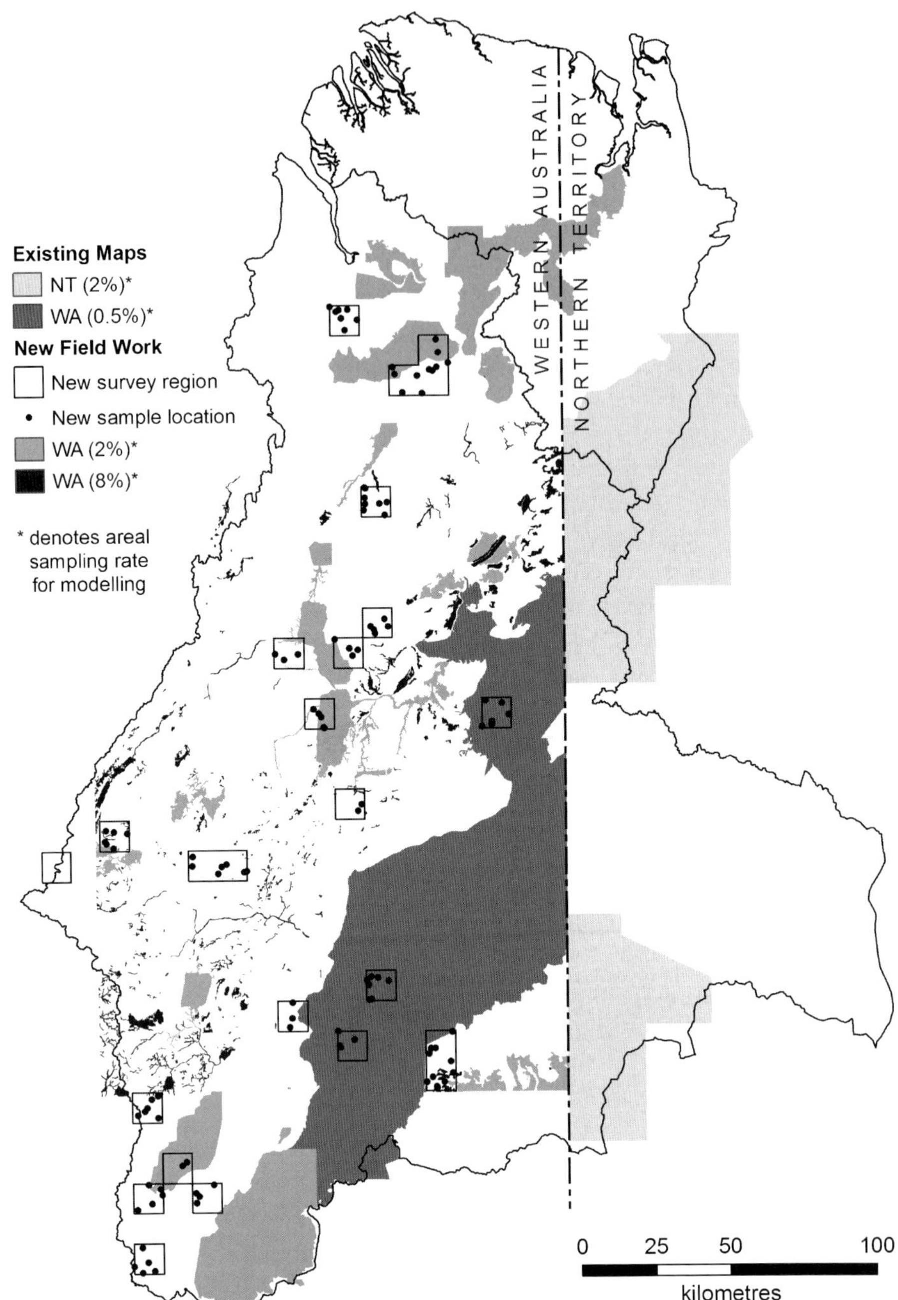

Figure 15.2. *Location of training data from WA and NT.*

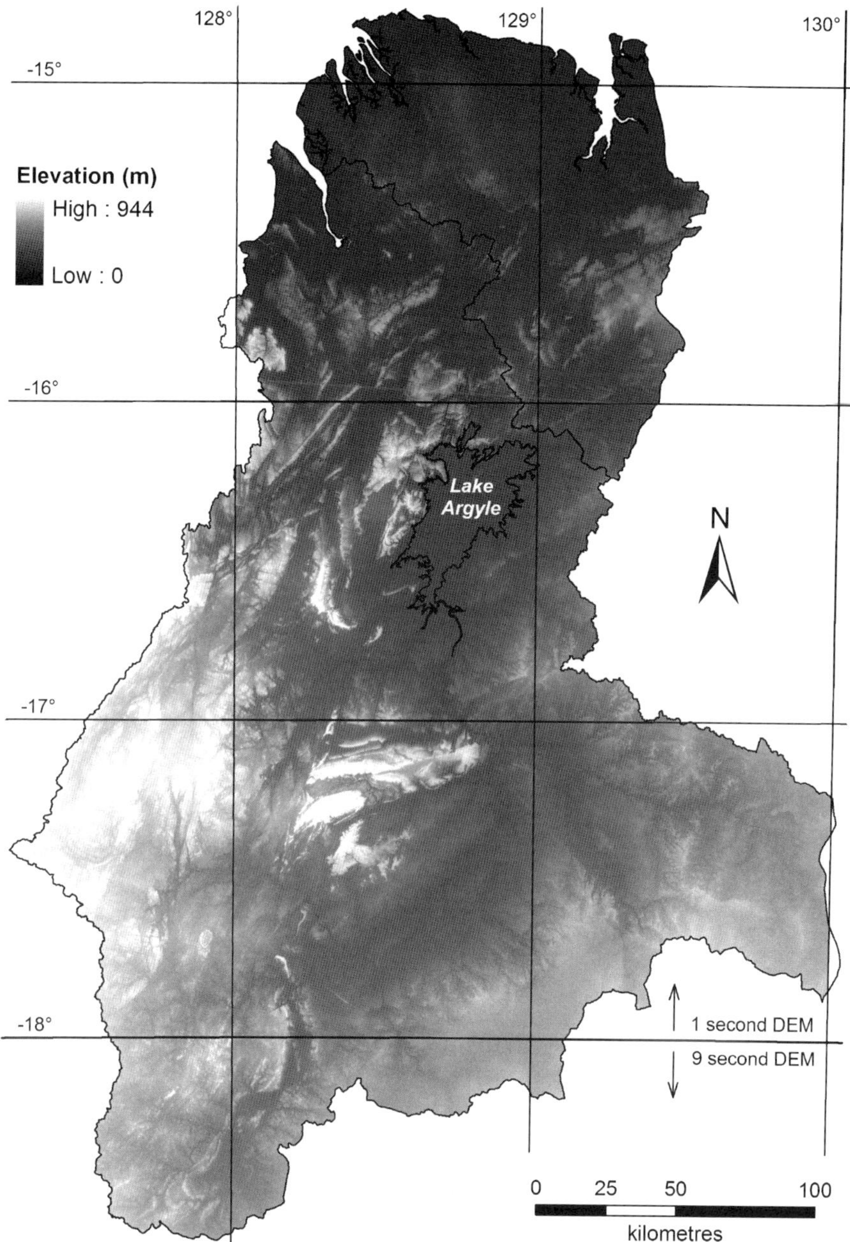

Figure 15.3. *Visualization of the 1-s DEM derived from the Army topographic data.*

5. Transport power (incoming)
6. Transport power (outgoing)
7. Change in transport power
8. Sediment coarseness index (incoming)
9. Sediment coarseness index (outgoing)
10. Deposition path length
11. Erosion path length

15.2.2 Biophysical dataset 2: Geology

At the time of modelling, digital geology was available only for the Dixon and Lissadell 1:250,000 map sheets (Fig. 15.1), and the Dixon, Mount Remarkable, Bow river, McIntosh and Turkey Creek 1:100,000 map sheets (subsets of the Lissadell and Dixon 1:250,000 geology). For the remainder of the area (Gordon Downs, Lansdowne, Cambridge Gulf and Medusa Banks 1:250,000 map sheets), the mosaic of national geology coverage at compiled for the national land and water resources audit was used (Johnston et al., 2003).

The legend was simplified into 20 lithology categories as described in Table 15.1.

Table 15.1. *Lithology codes.*

Land-unit lithology code	Geology
A	Basalt, dolerite, gabbro
B	Chert, jasperlite, banded ironstone
C	Colluvial
d	Limestone, dolomite, calcrete
e	Granite, gneiss
f	Lateritic
g	Sandstone, quartzite, conglomerate
h	Siltstone, shale, mudstone
j	Aeolian
k	Porphyry
l	Alluvial
m	Metamorphics (slate, schist, migmatites)
n	Acid igneous lavas (rhyolite, dacite)
o	Ultramafic
p	Other duricrusts (silcrete)
q	Lacustrine
z	Black soil
r	Metamorphosed volcanics
s	Coastal silt and evaporites
x	Water/disturbed land

15.2.3 Biophysical dataset 3: climate

Nine predictor datasets were generated from the digital climatic surfaces (ANUCLIM climate surfaces (250-m grid cells) and QDNR SILO climate surfaces (5-km grid cells) (Jeffrey et al., 2001).

1. Isothermality (mean diurnal temperature range as a proportion of mean)
2. Annual temperature range
3. Temperature seasonality (coefficient of variation)
4. Maximum temperature of warmest month
5. Temperature annual range
6. Precipitation of wettest month
7. Precipitation seasonality (coefficient of variation)
8. Highest month soil moisture index
9. Soil moisture index seasonality (coefficient of variation)

15.2.4 Biophysical dataset 4: Landsat 7 TM

Twelve predictor datasets were derived from four bands of Landsat 7 TM imagery acquired in September 1994.

1. TM band 1 (1 s of a degree)
2. TM band 1 mean value over window with 10-s radius
3. TM band 1 range of values over window with 10-s radius
4. TM band 2
5. TM band 2 mean
6. TM band 2 range
7. TM band 4
8. TM band 4 mean
9. TM band 4 range
10. TM band 5
11. TM band 5 mean
12. TM band 5 range

15.2.5 Predictive modelling

The modelling approach is explained below step by step and shown schematically in in Figure 15.4.

Step 1 Randomly sample training data and create an environmental cross-attribute matrix from the predictor datasets. Sampling density is set as a function of map scale, estimated reliability and extent.

Step 2 Randomly sample 50% of the environmental cross-attribute matrix and use the output to create a classification tree model (Table 15.2) using C5.0

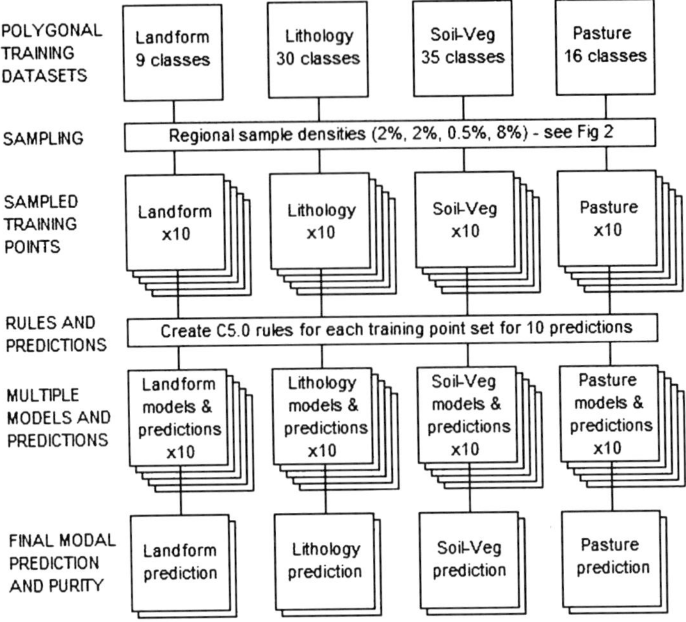

Figure 15.4. *Schematic diagram of modelling process.*

Table 15.2. *A portion of a C5.0 classification tree.*

Classification tree example for assigning a value to an individual pixel

```
slopepercent <= 1.809347:
:...intdem <= 116.8682:
:   ...riv_d <= 14.2:
:   :   :...intdem <= 29.10445:
:   :   :   :   ...riv_d>2.8:
:   :   :   :   :...slopepercent <= 0.578: 7 (3633/148)
:   :   :   :   :   slopepercent>0.578:
:   :   :   :   :   :...relief <= 47.49: 7 (432/29)
:   :   :   :   :       relief>47.49:
:   :   :   :   :       :...relief>85.55: 7 (376/26)
:   :   :   :   :           relief <= 85.55:
:   :   :   :   :           :...riv_d <= 8.2: 8 (146/80)
:   :   :   :   :               riv_d>8.2: 7 (63/11)
:   :   :   :   riv_d <= 2.8:
etc....
Result: pixel assigned to class "1aktCTHP"
```

software (http://www.rulequest.com.au). Pruning is set so that the minimum number of cases going into a leaf or a split is 50.

Step 3 Randomly sample 50% of the environmental cross-attribute matrix and use the output to create another tree that will be used as a test of the classification tree generated in Step 2.

Step 4 Compare the result from the tree model in Step 2 with output from the test of the decision tree in Step 3.

- If the error rates in the trees from Steps 2 and 3 are similar (i.e. the prediction is fairly consistent), apply the classification tree model to unmapped pixels.
- If the error rates in the trees from Steps 2 and 3 are not similar (i.e. the prediction is fairly inconsistent), go back to Step 1 and change model constraints (e.g. sampling rate or pruning) and repeat process.

Step 5 Repeat the above steps 10 times.

Step 6 From the 10 predictions for each pixel across the whole study area determine the modal value of prediction and the frequency of the modal value.

Step 7 Use the modal value for each pixel to create a land unit map.

Step 8 Use the frequency of the modal value for each pixel to determine the reliability of the land unit map. For example, if most of the 10 predictions predict the same value for a given pixel then the reliability of prediction for that pixel is assessed as high. Conversely, if most of the 10 predictions predict a different value for a given pixel then the reliability of prediction for that pixel is assessed as low.

Note that a reliable prediction does not mean that the correct value has been assigned; it just means that the model has a high level of confidence that the predicted value will be correct.

15.3 Results and Discussion

The new WA land unit mapping and 1:50,000 NT soil mapping as training data, together with the 30-m DEM, Landsat TM and more detailed digital lithology produced what appeared to be a better prediction. Nevertheless, there were several problems. Discrepancies were apparent between the land unit landform classification as compared to the continuous terrain attributes derived from the DEM, and there were differences in land unit coding and detail across the NT and WA state boundary.

Rather than predicting land units directly, predictions were made for each land unit attribute as individual themes (in other words, landform, geology, soil type and vegetation type were predicted separately) and the predictions reaggregated into land unit polygons. Differential sampling density over the different training areas (Fig. 15.2), as a function of map scale and average polygon size, was implemented but the resulting map was still found to have shortcomings. One problem that could not be overcome for the soil and vegetation attributes was the lack of training data in the westernmost part of the Ord basin where unique land units had not been mapped. For example, it is clear from the DEM (Fig. 15.4) that the prediction of 'riverine system' in the western Ord basin (Plate 15, see Colour Plate Section) is wrong. Although there were 20 lithology classes in the land unit attribute list, only 11 were represented in the training areas (Fig. 15.2); therefore, only 11 classes could be predicted by C5.0 trees. A nonsensical default value of 'alluvial' was allocated over much of the central and western Ord basins (Plate 15, see Colour Plate Section) even though the trees were very consistent in their (erroneous) predictions (Fig. 15.5) and the predictions of landforms and geology (Plate 1, see Colour Plate Section) were consistent with each other. Note that this problem probably would have been averted if the new field work had covered the recommended, randomly chosen training regions in the central and western Ord basins (Fig. 15.2).

We used the DEM and terrain analysis to implement expert-generated rules to adjust the landform classification. We spliced other digital geology data from Johnston et al. (2003) into the 'alluvial' class in the western region but had no other basis for adjusting the soil and vegetation predictions outside the training areas, although it may have been possible to ascertain soil type in the problematic areas by disaggregation (e.g. Bui and Moran, 2001) of land systems map units from Aldrick et al. (1978) and Stewart (1970).

In August 2003, training data areas were revisited and, in addition, new sites were examined and described in areas outside of the training-data collection areas to assist validation of the model. What became immediately apparent from the field trip was that the modelling made a fair-to-good prediction within the training data areas – consistent with the κ-coefficients obtained from contingency tables (Bonham-Carter, 1994), which were >0.75, made a fair-to-poor prediction close to the training data but was a very poor predictor at distance from the training data.

An assessment of the accuracy of model output was made by examining all site data and comparing how well the model predicted the land unit attribute values compared with the actual land unit attribute values assigned to the site in the field. A total of 103 sites were within the training data areas and 107 were outside. The success of prediction is summarised in Table 15.3.

There are a number of reasons for the apparent failure of the modelling approach, the main ones being:

- lack of pure training data (the existing training data was heterogeneous in many areas and would have required preliminary disaggregation);
- absence of any representative training data in some areas (Fig. 15.2);
- lack of high-resolution datasets over the entire area.

Table 15.3. *Results of field assessment of prediction success for final model.*

Land unit attribute	Prediction success (%)	
	Sites within training data polygons	Sites outside training data polygons
Landform	53	24
Geology	73	34
Soil	45	16
Vegetation	39	15
Land unit (above 4 combined)	26	3

Some very valuable lessons were learnt in the mapping process even though a satisfactory final product was not achieved.

On the basis of the experience gained from this project, the essential requirements for a successful land unit mapping project are:

- representative field data for all land unit attribute values,
- high-resolution digital datasets of predictor variables (especially terrain, geology), all stored in the same datum and projection,
- aerial photography and
- incorporation of expert knowledge at all stages of the mapping process.

These requirements have been built into a revised land unit mapping methodology that is a hybrid between traditional techniques and the use of digital data. An interactive technique is being used that involves onscreen mapping using all digital datasets, with expert knowledge incorporated as the lines are drawn. For example, the hillshaded DEM and a 10-m contour map generated from the DEM show up slopes and relief very effectively and are used to delineate hilly land units. Low-relief plains are better interpreted through various Landsat scenes, aerial photography and field information. By carefully selecting and combining the various datasets that highlight different landscapes, land units can be effectively and efficiently identified by experts.

15.4 Conclusion

Although computer-aided modelling can reproduce soil maps used as training data with a high level of accuracy and consistency, it cannot overcome a sampling program that does not comprehensively represent the region to be mapped. This is the major lesson learnt from the sampling of Ord river basin. To remedy this problem, a sampling campaign that ensures every combination of environmental covariates is represented, for example, one designed using a

Latin-hypercube (Minasny and McBratney, Chapter 12) is necessary. The field surveyors must then make every effort to cover those sampling regions.

References

Aldrick, J.M., Howe, D.F. and Dunlop, C.R., 1978. Report on the lands of the Ord River catchment, Northern Territory. Technical Bulletin 24. Animal Industry and Agriculture Branch, Department of the Northern Territory, Darwin, NT.

Bonham-Carter, G.F., 1994. Geographic information systems for geoscientists: modelling with GIS. Computer Methods in the Geosciences. Vol13. Pergamon Press, Oxford.

Bui, E.N., Moran, C.J., 2001. Disaggregation of polygons of surficial geology and soil maps using spatial modeling and legacy data. Geoderma 103, 79–94.

Bui, E.N., Moran and C.J., 2003. A strategy to fill gaps in soil survey over large spatial extents: an example from the Murray-Darling basin of Australia. Geoderma 111, 21–44.

Jeffrey, S.J., Carter, J.O., Moodie, K.B., Beswick, A.R., 2001. Using spatial interpolation to construct a comprehensive archive of Australian climate data. Environ. Modell. Software 16, 309–330.

Johnston, R.M., Barry, S.J., Bleys, E., Bui, E.N., Moran, C.J., Simon, D.A.P., Carlile, P., McKenzie, N.J., Henderson, B.L., Chapman, G., Imhoff, M., Maschmedt, D., Howe, D., Grose, C., Schoknecht, N., Powell, B., Grundy, M., 2003. ASRIS: The database. Australian J. Soil Res. 41, 1021–1036.

Moran, C.J., Prosser, I. and Cannon, G., 2001. Specification of the SEDUM model for modelling patterns of sediment transport based on unit stream power. CSIRO Land and Water Technical Report 25/01. http://www.clw.csiro.au/publications/technical/

Moran, C.J., Bui, E.N., 2002. Spatial data mining for enhanced soil map modelling. Int. J. Geogr. Inf. Systems 16, 533–550.

Salis, J. de., 1993. Resource inventory and condition survey of the Ord River Regeneration Reserve. Department of Agriculture, Western Australia. Miscellaneous publication 14/93.

Stewart, G.A., 1970. Lands of the Ord-Victoria area, Western Australia and Northern Territory. Land Research Series no. 28. CSIRO, Melbourne.

Van-Cuylenburg, H.R.M., McLeod, P.J., Fett, D.E.R. and Lang, M.P., (in prep.) Land Resources of Victoria River District. Northern Territory Department of Lands, Planning and Environment, Darwin, NT.

E. New environmental covariates for digital soil mapping

Environmental covariates representing soil-forming factors are a key approach to soil spatial prediction (Chapter 1). Chapters 16–21 overview the newer aspects of environmental covariates. These include remotely-sensed data at various portions of the electromagnetic spectrum and digital elevation models and terrain attributes and ways of combining these attributes.

Chapters 16–18 deal with covariates obtained from remote sensing. Chapter 16 discusses airborne gamma radiometrics which can be considered as either intrinsic soil property or a surrogate for parent material. These data seem to have been most widely used in Australia, although there are reports from the UK, Brazil and Finland. The widespread adoption of gamma radiometrics data will enhance soil predictability. In Chapter 17, we can see that there is much information to be gained from detailed spectra in the visible and near-infra-red parts of the EM spectrum. This is being developed for rapid soil assessment and its field application through airborne hyperspectral sensors is starting to emerge as a competitive technology. In Chapter 18, classical multispectral Landsat data are processed to give landcover classes which act as a conveyor of the o factor in the scorpan approach. The repetitive nature of such imagery allows dynamic aspects of the soil cover to be predicted.

Chapters 19 and 20 deal with the notions of landscape and soilscape which yield more sophisticated covariates which can be derived either numerically or using expert knowledge from digital elevation models and other ancillary variables. These concepts need to be better integrated into the digital soil mapping methodology.

In Chapter 21, the appropriate spatial scale or resolution is investigated through wavelet analysis. Interestingly, it was found that coarser resolutions can give better predictions. Once again, this kind of analysis seems to be appropriate for pre-processing covariates.

This section gives a general flavour of the possibilities of new environmental covariates. The challenge of obtaining useful information on the distribution of soils over a large area is being made more cost-effective by the use of increasingly ubiquitous, newer and accurate covariates. There is a need to develop new tools and spatial prediction techniques to take advantage of the information-rich world. Much more is expected from these techniques, especially in our quest to improve the spatial prediction quality and to meet the immediate and future requirements for dynamic modelling for the monitoring of earth's natural resources.

Developments in Soil Science, volume 31
P. Lagacherie, A.B. McBratney and M. Voltz (Editors)

Chapter 16

THE USE OF AIRBORNE GAMMA-RAY IMAGERY FOR MAPPING SOILS AND UNDERSTANDING LANDSCAPE PROCESSES

J. Wilford and B. Minty

Abstract

Airborne gamma-ray spectrometry is a passive remote sensing technique that measures the natural emission of gamma-ray radiation from the upper 30 cm of the earth's surface. The principle gamma-ray emitting isotopes used in airborne geophysical surveys are ^{40}K, and the ^{232}Th and ^{238}U decay series. These are used to estimate potassium, thorium and uranium abundances, respectively.

Gamma rays emitted from the earth's surface mainly relate to the mineralogy and geochemistry of the bedrock and weathered materials (e.g. soils, saprolite, alluvial and colluvial sediments). Gamma-ray imagery can thus be regarded as a surface geochemical map showing the distribution of the radionuclides in rocks, regolith and soil. Where the bedrock contains K-bearing minerals, the loss of K in the soil can often be used as a surrogate for mapping the degree of surface weathering and leaching. Potassium is also associated with potassic clays such as illite, and can be found in smaller amounts where it is absorbed onto clays such as montmorillonite and kaolinite. In contrast, U and Th are often associated with more stable weathered constitutes in the soil profile. Uranium and Th released during weathering are readily absorbed onto clay minerals, oxides (Fe and Al) and organic matter in soils. Elevated U and Th can also be associated with resistate minerals. Therefore, Th and U concentrations often increase as K decreases during bedrock weathering and soil formation.

In erosional landscapes the gamma-ray response will relate mainly to mineralogy and geochemistry of the bedrock, weathering characteristics and both past and present landscape processes. Responses of soils in depositional landscapes will reflect the geochemistry and mineralogy of the source rocks, or regolith, from which the sediments are derived, and the sorting of the sediments and weathering of the sediments after deposition. In both erosional and depositional landscapes, when the radioelement characteristic of the sources are well understood, gamma-ray data can be used to predict specific soil characteristics and provide information about erosional, depositional and weathering processes

16.1 Introduction

The airborne gamma-ray spectrometric method was originally developed as a uranium exploration tool. However, the method soon found additional uses, including geological mapping, the direct detection of mineral deposits and the estimation of snow water equivalents. The method is widely used for assessing the radiation dose to human populations from the natural environment, and identifying areas of natural radiation hazard. More recently, the method has been used for mapping soils and understanding landscape processes. The method is now seen as providing one of the key biophysical datasets required for catchment planning (George et al., 1997). This is because gamma-ray imagery can often provide information on quite specific soil/regolith properties, and help in understanding landscape processes at a range of scales (Bierwirth, 1996; Biggs and Philip, 1995; Cook et al., 1996; Dauth, 1997; Dickson and Scott, 1997; Foote, 1964; Gessler et al., 1995; Martz and de Jong, 1990: McKenzie and Ryan, 1999; Roberts et al., 2002; Ryan et al., 2000; Pickup and Marks, 2000; Pracilio et al., 2003; Taylor et al., 2002; Thomas et al., 2003, Wilford, 1995, 2004; Wilford et al., 1997). Applications of gamma-ray spectrometry to various aspects of digital soil mapping (DSM) can be found in Chapters 21, 24, 26 and 33. It appears that the use of this covariate data source has been most widespread in Australia.

This chapter describes the fundamentals of gamma-ray spectrometry, survey design for use in soil survey, the radioelement responses of bedrock, regolith and soil materials and how the concentrations of these radioelements change during pedogenesis. Examples from different geomorphic regions are used to illustrate the application of gamma-ray imagery for mapping soil properties and in understanding landscape processes.

16.2 Fundamentals of gamma-ray spectrometry

Some isotopes are unstable and change to a more stable state by the emission of ionising radiation. These are the so-called radioactive isotopes or "radioisotopes". All rocks and soil contain radioactive isotopes, and their decay gives rise to a natural gamma-ray flux at the earth's surface. Almost all gamma radiation detected near or at the earth's surface is derived from the natural radioactive decay of just three elements–potassium (K), uranium (U) and thorium (Th). Gamma-ray surveys map the distribution of these elements at the earth's surface. Most gamma rays pass through moderately dense vegetation, allowing direct measurement of soil geochemistry without the masking effect of crops or trees that can limit the use of other remotely sensed datasets such as optical sensors (e.g. Landsat TM). Where the vegetation is dense, the partial attenuation of gamma-ray signal needs to be considered when interpreting the imagery.

Gamma rays emitted from the natural decay of K, U and Th can penetrate up to 35 cm of rock and several hundred metres of air. As a result, they can be measured by detectors mounted on low-flying aircraft. Most airborne gamma-ray detectors used today consist of crystals of thallium-activated sodium iodide (NaI). Gamma rays absorbed in the crystals result in a scintillation of light being emitted, the intensity of which is proportional to the energy of the absorbed photon. Airborne detectors measure a gamma-ray spectrum (Fig. 16.1), i.e. both the number of gamma rays recorded during a specific sample period and the energy of each photon. The number of gamma-rays recorded is proportional to the concentration of the radioelements in the source, and the energies of the gamma-rays can be used to determine the composition of the source isotopes. Potassium abundance is measured using the 1.46 MeV gamma-ray photons emitted when ^{40}K decays to Ar. U and Th abundances are measured from daughter nuclides in their respective decay chains. Distinct emission peaks associated with ^{214}Bi (a daughter product in the ^{238}U decay series) and ^{208}Tl (a daughter product in the ^{232}Th decay series) at 1.76 and 2.61 MeV (Fig. 16.1) are used to estimate the concentrations of U and Th, respectively.

The estimation of U and Th using daughter isotopes in their respective decay series is based on the assumption that their respective radioactive decay series are in equilibrium. However, disequilibrium is common in the ^{238}U decay series, and this should be taken into consideration when interpreting estimated U abundances. For example, U anomalies can be caused by the accumulation of radium (^{226}Ra) in ground waters (Giblin and Dickson, 1984). U and Th concentrations derived from gamma-ray spectrometry are normally expressed in units of "equivalent" parts per million (eU and eTh), as a reminder that these estimates are based on the assumption of equilibrium in their respective decay series.

16.2.1 Essential corrections

Gamma-ray spectra are typically recorded at a frequency of 1 Hz.Ancillary data would include positional information (GPS navigation), temperature, pressure

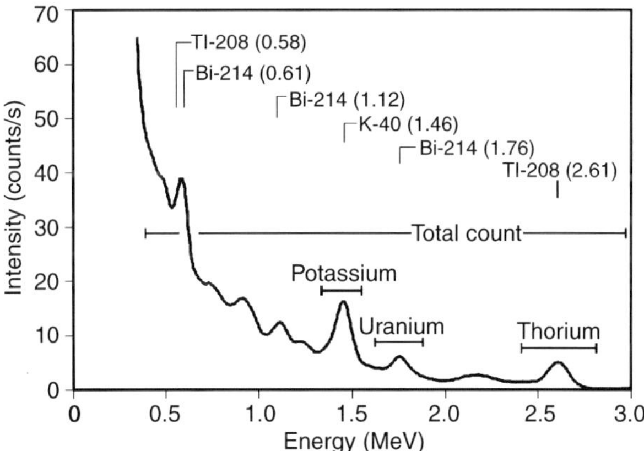

Figure 16.1. *An airborne gamma-ray spectrum (averaged over a long period of time) showing the diagnostic photopeaks and the positions of the K, U and Th windows used in gamma-ray spectrometry.*

and the height of the detector above ground level in the case of airborne surveys. The recorded data require substantial processing before accurate estimates of the ground concentrations of K, U and Th can be made. Statistical noise is usually first removed from the raw multichannel gamma-ray spectra using a statistical noise-reduction technique such as noise adjusted singular value decomposition (NASVD, Hovgaard, 1997; Hovgaard and Grasty, 1997) or maximum noise fraction (MNF, Green et al., 1988; Dickson and Taylor, 1998). These methods use a principal component type analysis to extract the dominant spectral shapes from the survey data. These "principal components" are used to reconstruct spectra that have most of the original signal, but little of the noise. The main corrections that are applied to gamma-ray data are for equipment lifetime, energy drift, background radiation, channel interactions (stripping correction), the height of the aircraft above the ground and the sensitivity of the detector. These corrections are explained in detail by Minty (1997).

16.2.2 Airborne survey design

Survey design for airborne gamma-ray spectrometry is governed mainly by the precision required for the estimates of K, U and Th concentrations, and by the spatial resolution required for these estimates. Spatial resolution is governed by the flight-line spacing, and by the sample interval and speed of the aircraft. The line spacing is a trade-off between spatial resolution and the cost of the survey. Typically, airborne surveys have line spacings between 50 and 400 m. Sample spacing along the lines is governed by the speed of the aircraft, and is typically 50–60 m for a fixed-wing aircraft and a sample interval of 1 s.

Gamma-ray surveys are usually flown at less than 100 m above ground level on regular parallel flight lines. Gamma-rays recorded by the aircraft originate from an area on the earth's surface that is several hundred metres in diameter. The size of this "circle of investigation", or footprint, depends on the survey height. At 100 m height, about 80% of recorded photons would originate from a circle below the aircraft with diameter of about 600 m (Fig. 16.2). Thus, a single airborne estimate of radioelement concentrations is representative of the average concentrations over a large area. Lower flying heights reduce the size of the footprint and increase the spatial resolution and measurement precision of the radioelement estimates. Selection of the optimum survey specifications (flight line spacing, flying height and detector size) should be based on the scales at which soils change and processes occur, within the landscape being investigated.

Portable, hand-held gamma-ray spectrometers are widely used in mineral exploration and environmental studies. However, for mapping applications, vehicle-borne surveys are more commonly employed. The gamma-ray detectors

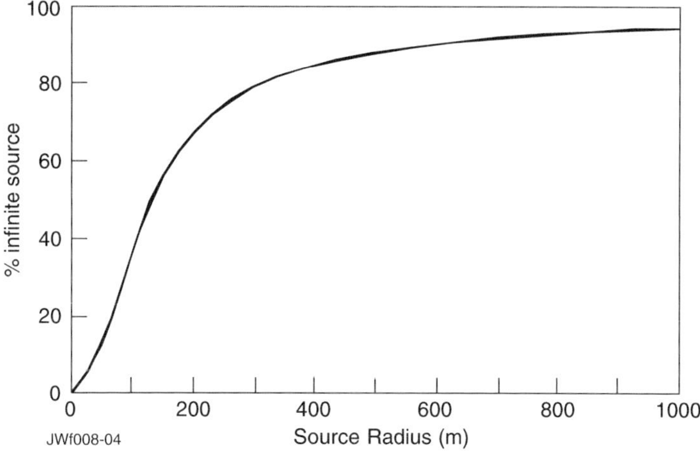

Figure 16.2. *Percentage of the total signal originating from a circle with specified radius below the detector for Th gamma-rays at 2.61 MeV and a detector height of 100 m (from Minty, 1997).*

are mounted on a motor vehicle or quad-bike for continuous recording as the survey area is traversed. The gamma-ray detectors are usually airborne detectors modified for use from a vehicle, and the acquisition, navigation and processing of the data is essentially the same. A big advantage of ground-based surveys is that the background component of radiation is much smaller (as a fraction of the signal) than for airborne surveys. Consequently, a much simpler background correction can be applied to ground-based data. A significant limitation of ground-based surveys is that source-detector geometry effects must be accommodated in the interpretation of the data.

16.3 The radioactivity of rocks, regolith and soil

16.3.1 Rocks

Potassium has an average crustal abundance of 2.3% and is much more common than U and Th, with estimated crustal averages of 3 and 12 ppm, respectively. K, Th and U contents in rocks generally increase with increasing silica content (Fig. 16.3). This is reflected in a general trend of increasing K, U and Th from basic to acid igneous lithologies. Potassium occurs in alkali feldspars, muscovite, alunite and sylvite. High K is typically associated with acid igneous rocks including granite, rhyolite and pegmatite. Potassium is absent or at very low concentrations in mafic minerals and associated mafic to ultramafic rocks such as basalts, dunites, serpentinite and peridotites. Uranium occurs as two main

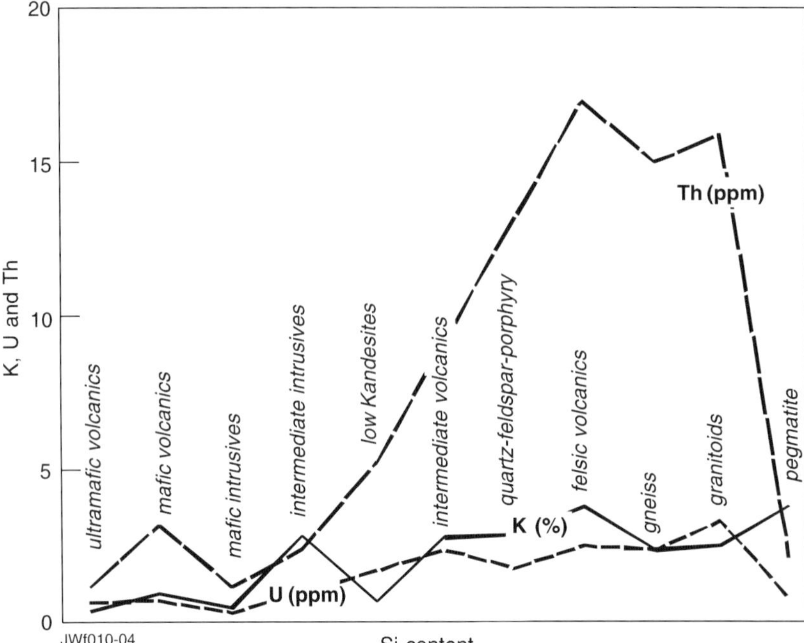

Figure 16.3. *Concentrations of K, Th and U in common rocks generally increase in accordance with increasing silica content (from Dickson and Scott, 1997).*

valency states: U^{+4} and U^{+6}. The oxidised U^{+6} forms complexes with oxygen as the uranyl ion (UO_2^{+2}). Uranyl ions are mobile and typically form chemical complexes with carbonate, sulphate and chloride ions. The mobility of U^{+6} is modified by absorption to hydrous iron oxides, clay minerals and colloids. Under reducing conditions, the more reduced U^{+4} form is contained in insoluble minerals. High uranium is associated with pegmatites, syenites, radioactive granites and some black shales. Thorium occurs in a single valency state (+4), and therefore its mobility does not alter under changing redox conditions. Thorium solubility is generally low, although it can be soluble in acid solutions or at neutral pH when it is associated with organic complexes. Thorium is found in minerals such as thorianite and thorite.

In rocks, it is associated with granite, pegmatite and gneiss. U and Th are found in accessory and resistate minerals such as zircon, titanite (sphene), apatite, allanite, xenotime, monazite and epidote.

16.3.2 Regolith and soils

Weathering modifies the distribution and concentration of radioelements from the initial bedrock concentrations. Fortunately, from a soil science perspective,

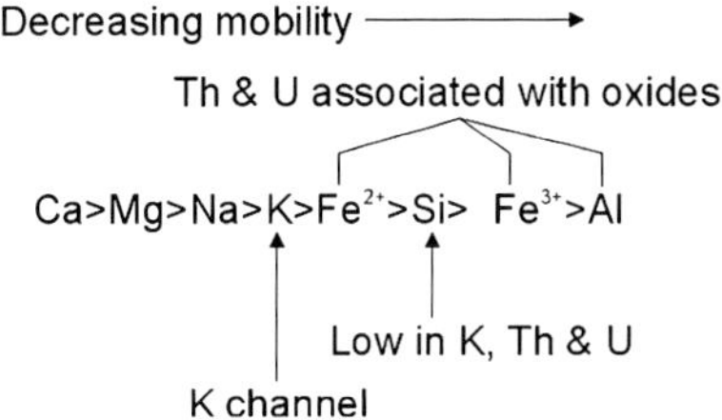

Figure 16.4. *Element mobility index and associated radioelement response.*

K, U and Th behave quite differently from one another during bedrock weathering and pedogenesis. As a general rule, K concentration decreases with increasing weathering. This is because K is soluble under most weathering environments and tends to be leached from a soil/regolith profile. Exceptions to this are where K is incorporated into potassic clays such as illite, where it is absorbed onto clays such as montmorillonite and kaolinite, or where K is associated with either large K-feldspar phenocrysts or mica that take time to weather. In contrast, U and Th are associated with more stable weathering products in soil profiles. Uranium and Th released during weathering are readily absorbed onto clay minerals, Fe and Al oxyhydroxides and organic matter in soils. In addition, U and Th also reside in resistate minerals that persist for a long time in the soil. It is therefore not uncommon for relative concentrations of U and Th to increase in highly weathered soils, as other more soluble minerals are lost in solution. These relationships are summarised in Figure 16.4.

An example of the application of gamma-ray imagery to the mapping of highly weathered soil and bedrock materials is given in Plate 16a(A) (see Colour Plate Section). The image shows highly leached soils and deeply weathered bedrock in blue hues, and thin soils over fresh bedrock in red hues. Separating highly weathered soils and bedrock from areas of relatively shallow weathering was facilitated using a residual modelling technique that combined geological map units with the gamma-ray imagery (Wilford, 2004). The residual modelling technique effectively separated the bedrock and soil geochemical responses based on the preferential loss of K-bearing minerals as the rock weathered. Identifying zones of highly weathered bedrock in the region is of environmental interest because they can contain relatively large stores of salt that can lead to land and river salinisation.

16.3.3 Interpreting regolith and soil responses

Digital elevation models (DEMs) can greatly enhance the interpretation and separation of bedrock and soil radioelement responses. A ternary gamma-ray image draped over a DEM shows the complex radioelement patterns associated with erosional and depositional landscapes (Plate 16a(B)). Perspective drapes are useful in interpreting gamma-ray imagery because they provide a dynamic link between the surface geochemistry and landscape processes. In many cases, the gamma-ray variation within the major bedrock/geomorphic units relate to changing soil and regolith characteristics (Plate 16a(B)).

Factors that need to be considered before coming to an understanding of the distribution of radioelements in a soil over a given bedrock type are summarised by the main processes that affect soil formation, including additions, removals, transformations and translocations (Fig. 16.5). Additions might include aeolian dust, removals; surface erosion, transformations; weathering of primary minerals and translocations; clay illuviation. All these processes need to be considered when interpreting gamma-ray responses of soils from different parent materials. Individual soil types are unlikely to have unique radioelement signatures because similar radioelement responses can reflect different bedrock types, and weathering and geomorphic processes. For this reason, it is best to consider each of the major geomorphic units (including erosional and depositional landscapes and lithological–geochemical groups) separately when interpreting the gamma-ray response in terms of soil-forming processes.

The benefits of integrating the gamma-ray imagery with other ancillary datasets is illustrated by a study of the airborne gamma-ray responses over the Jamestown region in South Australia (Plate 16b(A) (see Colour Plate Section)).

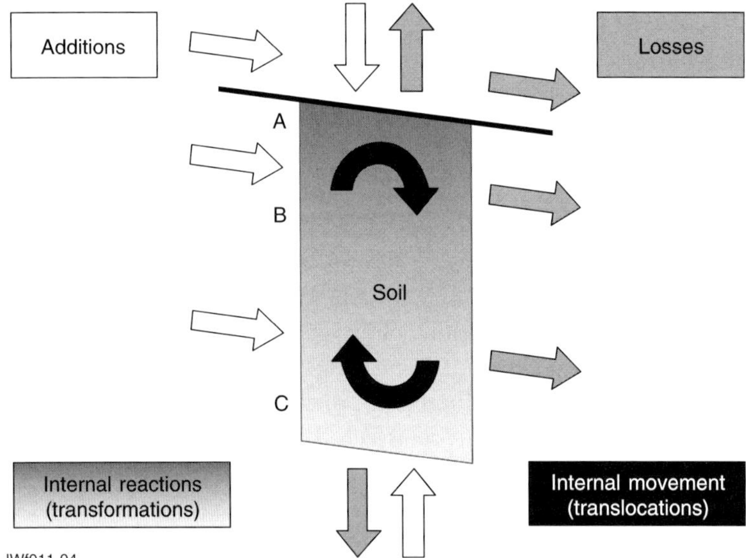

Figure 16.5. *Radioelement concentrations are influenced by the removal, addition, transformation and translocation processes operating in soils (from McKenzie et al. 2004).*

The responses are varied, reflecting complex basement geology, including Proterozoic siltstone, dolomite, sandstone, quartzite, limestone, tillite and more recent Quaternary valley fill sediment. However, useful soil property information was obtained from the airborne imagery by first stratifying the region into erosional and depositional landscapes prior to interpreting the gamma-ray imagery. Surface soil texture and mineralogy of colluvial and alluvial sediments was predicted using K concentration estimates derived from the gamma-ray data. Relatively high K values are associated with silt size mica (muscovite) and illite grains, whereas, low K values reflect a higher proportion of coarser quartz size grains (Wilford, 2004). These relationships were used to produce thematic maps of soil texture within the more productive alluvial and colluvial soils in the study area (Plate 16b(A)).

16.3.4 Mapping catenas

Mapping catenary sequences using gamma-ray imagery illustrates the close relationship between geomorphic process and gamma-ray response. Plate 16b(B) shows an example from an area near Cowra, New South Wales, where moderate relief landforms have developed on Ordovician volcaniclastic sandstone and siltstone. A K/Th ratio image combined with a topographic wetness index (TWI) are used to delineate discrete soil sequences along a slope. Skeletal soils over the hill crest and upper slopes correspond to high K/Th values compared with relatively low ratio values associated with clay rich colluvial soils on mid to lower slopes. The gamma-ray imagery in combination with a TWI is used to identify solodic soils associated with poor drainage (Plate 16b(B)).

16.3.5 Gamma-ray responses in erosional landscapes

The gamma-ray responses from erosional landscapes broadly correlate with bedrock geology. However, variations within major lithological groups relate to soils and regolith materials. The gamma-ray response for a given bedrock type largely depends on the landscape history (e.g. preservation of old weathering profiles) and denudation balance, or the relative rates of erosion verses weathering (Fig. 16.6). Areas in the landscape with relatively high geomorphic process rates are likely to have thin soils and a gamma-ray response that reflects the bedrock geochemistry and mineralogy. Landscapes with stable surfaces that are less geomorphically active will preserve weathered materials and associated soils (Wilford, 1992, 1995). The gamma-ray response of these soils is likely to mainly reflect secondary materials such as clays and oxides.

16.3.6 Responses of transported materials

Responses of alluvial and colluvial sediments will reflect the geochemistry and mineralogy of the source rocks or regolith from which they are derived, and

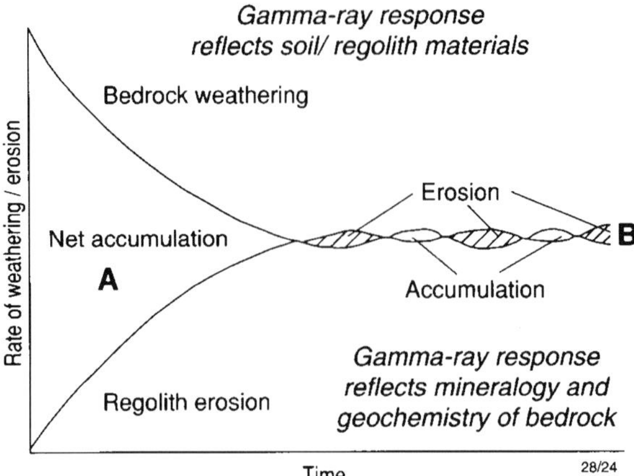

Figure 16.6. *Denudation balance processes and the corresponding gamma-ray response. (A) Where weathering rates are higher than erosional rates the gamma-ray responses will reflect soil composition. (B) Where weathering and erosional rates are similar or where erosional rates exceed weathering rates the gamma-ray response reflects bedrock geochemistry/mineralogy (modified from M.J. Crozier, 1986).*

modifications due to weathering and sorting of the sediments. Where erosion, transportation and deposition of the sediments are relatively rapid, the gamma-ray response is likely to largely reflect the source rock geochemistry. Where these processes are relatively sluggish, weathering of the sediments will modify the gamma-ray response. Therefore, in many cases gamma-ray imagery can trace the provenance of floodplain sediments and recognise changes due to sorting and weathering or age of the sediments. Pickup and Marks (2001), using airborne gamma-ray imagery and DEMs, identified and developed an improved understanding of large-scale erosional and depositional processes from contrasting catchments.

Once connections are made between gamma-ray response, geomorphic/weathering process and surface materials, then variations in gamma-ray radiation have the potential to predict specific soil properties (e.g. soil mineralogy, texture, pH and degree/depth of weathering). However, relationships determined for one region may not apply to another due to differing geological and landscape histories.

16.4 Discussion

A gamma-ray image is a geochemical map showing the distribution of the radioelements K, Th and U in rocks and soils. Gamma-ray data can be used in conjunction with other data in soil landscape mapping or in environmental correlation modelling for DSM. Interpretation of the data can provide an improved understanding of both past and present weathering and geomorphic processes that, in turn, can lead to more robust models of pedogenesis. Geological and geomorphic stratification of the landscape greatly assist in the interpretation of gamma-ray imagery for mapping soil and regolith materials. However, as with all remotely sensed datasets, gamma-ray imagery requires ground validation, and its application should be predicated on an understanding of the strengths and limitations of the technology.

The application of the gamma-ray spectrometric method to soil mapping does have several limitations. Soils do not have unique radioelement signatures. This means that similar responses can relate to different soils. Also, airborne gamma-ray acquisition systems have relatively large "footprints". This results in a poor spatial resolution for farm-scale applications. However, ground-based surveys using quad-bike mounted sensors can be used to obtain better spatial resolutions over small survey areas. For ground-based surveys, source-detector geometry effects need to be considered in the interpretation of these data. Gamma-rays are also partially attenuated by vegetation. In areas with dense vegetation cover (e.g. rainforest), the attenuation of gamma-rays due to attenuating cover can be significant.

References

Bierwirth, P., ss96. Investigation of airborne gamma-ray images as a rapid mapping tool for soil and land degradation: Wagga Wagga, NSW. Australian Geological Survey Organisation, Record 1996/22.

Biggs, A.J.W., Philip, S.R., 1995. Soils of Cape York Peninsula. Queensland Department of Primary Industries, Mareeba, Land Resources Bulletin QV95001.

Cook, S.E., Corner, R.J., Groves, P.R., Grealish, G.J., 1996. Use of airborne gamma-radiometric data for soil mapping. Aust. J. Soil Res. 34, 183–194.

Crozier, M.J., 1986. Landslides: Causes, consequences and environment. Croom Helm, London and Dover.

Dauth, C., 1997. Airborne magnetic, radiometric and satellite imagery for regolith mapping in the Yilgarn Craton of Western Australia. Explor. Geophy. 28, 199–203.

Dickson, B.L., Scott, K.M., 1997. Interpretation of aerial gamma-ray surveys–adding the geochemical factors. AGSO J. Aust. Geol. Geophy. 17, 187–200.

Dickson, B., Taylor, G., 1998. Noise reduction on aerial gamma-ray surveys. Explor. Geophy. 29, 324–329.

Foote, R.S., 1964. Application of airborne gamma-radiation measurements to pedologic mapping. In: D.C. Parker (Ed.), Proceedings of the 5th Symposium of Remote Sensing of the Environment, University of Michigan, Ann Arbor, MI, USA.

George, R.J., McFarlane, D.J., Nulsen, R.A., 1997. Salinity threatens the viability of agriculture and ecosystems in Western Australia. Hydrogeol. J 5, 6–21.

Gessler, P.E., Moore, I.D., McKenzie, N.J., Ryan, P.J., 1995. Soil-landscape models and the spatial prediction of soil attributes. Int. J. Geogr. Inf. Sci. 9, 421–432.

Giblin, A.M., Dickson, B.L., 1984. Hydrogeochemical interpretations of apparent anomalies in base metals and radium in groundwaters near Lake Maurice in the Great Victorian Desert (abstract). J. Geochem. Explor. 22, 361–362.

Green, A.A., Berman, M., Switzer, P., Craig, M.D., 1988. A transformation for ordering multispectral data in terms of image quality with implications for noise removal. IEEE Trans. Geosci. Remote Sens. 26, 65–74.

Hovgaard, J., 1997. A new processing technique for airborne gamma-ray spectrometer data (Noise adjusted singular value decomposition): American Nuclear Society. Sixth Topical Meeting on Emergency Preparedness and Response, pp. 123–127, San Francisco, April 22–25,1997.

Hovgaard, J., Grasty, R.L., 1997. Reducing statistical noise in airborne gamma-ray data through spectral component analysis. In: A.G. Gubins (Ed.) Proceedings of Exploration 97: Fourth Decennial Conference on Mineral Exploration, pp. 753–764.

Martz, L.W., de Jong, E., 1990. Natural radionuclides in the soils of a small agricultural basin in the Canadian Prairies and their association with topography, soil properties and erosion. Catena 17, 85–96.

McKenzie, N., Jacquier, D., Isbell, R., Brown, K., 2004. Australian Soils and Landscapes An Illustrated Compendium. CSIRO Publishing, Victoria, Australia.

McKenzie, N.J., Ryan, P.J., 1999. Spatial prediction of soil properties using environmental correlation. Geoderma 89, 67–94.

Minty, B.R.S., 1997. Fundamentals of airborne gamma-ray spectrometry. AGSO J. Aust. Geol. Geophys. 17, 39–50.

Pickup, G., Marks, A., 2000. Identifying large scale erosion and deposition processes from airborne gamma radiometrics and digital elevation models in a weathered landscape. Earth Surf. Process. Landf. 25, 535–557.

Pracilio, G., Asseng, S., Cook, S.E., Hodgson, G., Wong, M.T.F., Adams, M.L., Hatton, T.J., 2003. Estimating spatially variable deep drainage across a central eastern wheat belt catchment, Western Australia. Aust. J. Agric. Res. 54, 789–802.

Roberts, L., Wilford, J., Field, J., Greene, R., 2002. High resolution ground based gamma ray spectrometry and electromagnetics to assess regolith properties, Boorowa, NSW. In: I.C. Roach (Ed.), Regolith and Landscapes in Eastern Australia. CRC LEME, Adelaide, Australia, p. 136.

Ryan, P.J., McKenzie, N.J., O'Connell, D., Loughhead, A.N., Leppert, P.M., Jacquier, D., Ashton, L.U., 2000. Integrating forest soils information across scales: spatial prediction of soil properties under Australian forests. Forest Ecol. Manag. 138, 139–157.

Taylor, M.J., Smettem, K., Pracilio, G., Verboom, W.H., 2002. Investigation of the relationships between soil properties and high resolution radiometrics, central eastern Wheat belt, Western Australia. Explor. Geophys. 33, 95–102.

Thomas, M., Fitzpatrick, R.W., Heinson, G.S., 2003. Mapping complex soil-landscape patterns using radiometric K%: a dry saline land farming area case study near Jamestown, SA. In: C. Ri (Ed.), Advances in Regolith. CRC LEME, Adelaide, Australia, pp. 411–416.

Wilford, J.R., 1992. Regolith mapping using integrated landsat TM imagery and high resolution gamma-ray spectrometric imagery – Cape York Peninsula. Bureau of Mineral Resources, Australia, Record 1992/78.

Wilford, J.R., 1995. Airborne gamma-ray spectrometry as a tool for assessing relative landscape activity and weathering development of regolith, including soils. AGSO Research News 22, 12–14.

Wilford, J.R., 2004. Regolith-landforms and salt stores in the Angus-Bremer Hills, CRC LEME Report 206, Geoscience Australia, Canberra.

Wilford, J.R., Bierwirth, P.N., Craig, M.A., 1997. Application of airborne gamma-ray spectrometry in soil/regolith mapping and applied geomorphology. AGSO J. Aust. Geol. Geophys. 17, 201–216.

Developments in Soil Science, volume 31
P. Lagacherie, A.B. McBratney and M. Voltz (Editors)

Chapter 17

VISIBLE–NIR HYPERSPECTRAL IMAGERY FOR DISCRIMINATING SOIL TYPES IN THE LA PEYNE WATERSHED (FRANCE)

J.S. Madeira Netto, J.-M. Robbez-Masson[†] and E. Martins

Abstract

Remotely sensed data might be an important tool for acquiring geographical information about soil attributes especially in areas where the soil surface is permanently or temporarily exposed, as in the extensive vineyard region of the Languedoc Roussillon region of France. By providing detailed spectral signature for every pixel, imaging spectrometry can potentially be used to identify the nature and abundance of some soil surface components. This chapter reports results obtained in the analysis of soil spectral data of the Low Languedoc Plains natural region of southern France acquired by an image spectrometer (Hyperspectral Mapper – HyMap) after laboratory calibration to make quantitative estimations of $CaCO_3$ and clay content of surface soil horizons.

Two different approaches for extracting information from soil reflectance spectra were used: (a) band depth analysis for estimating calcium carbonate and clay abundances and (b) a redness index to identify soil colour variations. Validation data are available on surface soil samples as visible–NIR and X-ray spectra, and chemical characterisations.

For areas with soils developed from sedimentary materials of variable calcium carbonate and particle-size composition, we have demonstrated the possibility of obtaining quantitative information for $CaCO_3$ content from the intensity of absorption of the feature centred at 2341 nm. $CaCO_3$ estimates were highly satisfactory and at least as good as field effervescence test usually used in fieldwork. Clay particle-size content classes can also be estimated using the absorption depth of the clay minerals absorption feature at 2206 nm. Redness index along with clay and $CaCO_3$ estimates allows characterisation and explanation of the intra-unit variability of a pre-existing 1:100,000 soil map, and could be further used as new covariates in order to refine limits or segment existing units.

17.1 Introduction

Generic frameworks as well as specific models are gaining importance in the soil science community for predicting soil variability in the landscape. Reviewing recent approaches to make digital soil maps, McBratney et al. (2003) indicate that soils can be predicted from other soils attributes at the same location or from soil and environmental attributes at neighbouring locations. Remotely sensed data are quoted to be especially relevant in providing direct estimations of useful end-user soil covariates and in aiding the delineation of soil units for areas where soil surface is permanently or temporarily exposed.

Hyperspectral imagery provides detailed spectral signature for every pixel and can be potentially used to identify the nature of some soil surface components (e.g. organic or mineral matter), or soil properties (soil colour and texture) as well as their abundance. But the complexity of soil composition and the spectral reflectance signature specificity of each soil component, in addition to the spatial and spectral resolution of images, require some research effort to define what can be obtained from the images and to establish relationships between the signal contained in the image (or derived from it) and valuable soil information.

At this point, two complementary approaches can be distinguished: data-driven or knowledge-driven. In the following, we only detail the second one, arguing that (1) soils under agriculture are not entirely to be discovered and (2) methods are available for coupling existing knowledge, new methods and new data (see Chapter 23).

In this sense, some general hyperspectral techniques have been developed and good reviews about them are available: see Kruse (1998) and Van der Meer and De Jong (2002) for example. However, the clue to making efficient use of spectral data is to relate the desirable soil characteristics to a spectral parameter derived from it. Ryan and Lewis (2000) have used band rationing and matched filtering of Hyperspectral Mapper (HyMap) images to discriminate soils at sub-paddock scale in southern Australia, and related the differences in spectral brightness to the texture of the surface horizon. HyMap images have also proved to be useful in determining the spatial distribution of the "desert-like" soil surface features like desert pavements, saline soils, calcareous soils and exposed gypsum, which could be discriminated using the spectral angle mapper algorithm (Margate and Shrestha, 2001). Raggatt et al. (2004) report successful identification of kaolinite and iron oxides through their absorption features in the 600–800 nm and 1950–2300 nm on HyMap images of South Australia by using the spectral feature fitting algorithm. Moreover, quantitative estimates of soil mineral components have also been possible by using hyperspectral images. Hauff et al. (1990) and Kruse et al. (1991) estimated kaolinite content in the interstratified kaolinite/smectite clays of the Paris basin using dissymmetry measures in the band centred at 2200 nm; Baptista and Madeira Netto (2001) were able to model the Ki index ($[Al_2O_3]/[SiO_2]$) using AVIRIS through the absorption bands of kaolinite and gibbsite at about 2200 and 2265 nm for Brazilian latosols. Van der Meer (2004) combined three absorption band parameters (depth, position and asymmetry) for clay minerals and carbonate absorption bands aiming at identifying six mineral phases in the hydrothermal alteration zone of Cuprite, USA. In these studies, the band depth is assumed to be proportional to the abundance of the minerals.

The present chapter attempts to gauge the potential of an airborne HyMap imaging spectrometer for spatial estimation of calcium carbonate and clay/sand abundance in a small cultivated area located in Low Languedoc Plains natural region of southern France. Calibration makes use of laboratory work already conducted to establish quantitative relationships between soil composition and parameters extracted from spectra. Independent field observations allow verification of the extrapolation potential of laboratory results to the field scale.

17.2 Materials and methods

17.2.1 Study area and main soils

The study area is the lower La Peyne watershed (43°30′N, 3°21′E), located in the Languedoc-Roussillon region, Southern France, about 60 km west of Montpellier. This 70 km^2 catchment area is dominated by vine cultivation and represents fairly well the French Mediterranean coastal plain with respect to climate, relief, geology, geomorphology, agricultural practices, vineyard management and vine species. Soil classes cover a broad range from calcareous, calcic and fersialitic soils to andic brown soils, and occur in various geomorphic positions derived mainly from (1) marine and continental sediments of late Tertiary – early Quaternary, (2) alluvial terraces and (3) volcanic material (Bonfils, 1993). Carbonate and clay contents, stoniness and colour are the main observable, distinctive factors between soil units, whose short-range variability remains unknown but is suspected to be high, making soil survey a headache in some cases. Organic matters contents are low, and do not colour the soil.

17.2.2 Remotely sensed data

The HyMap airborne imaging spectrometer collects data in the visible–near-infrared range (VIS–SWIR: 400–2500 nm) in 126 channels with bandwidths of 15–20 nm and average sampling intervals of 13–17 nm.The instantaneous field of view – IFOV – is 2.5 mrad along track and 2.0 mrad across track for a 60° field of view, and covers a width of 512 pixels with a signal-to-noise ratio of more than 500:1 (Table 17.1). The scene used is centred over La Peyne watershed (Alignan-du-Vent community) at a spatial resolution of 5 m by 5 m, and is part of an image acquired on 13 July 2003 under very dry conditions. Previously, this image had been geometrically, atmospherically and topographically corrected by the German Aerospace Center using PARGE and ATCOR4 procedures described in Schläfer and Richter (2002) and Richter and Schläfer (2002), and a high-resolution IGN digital elevation model and DGPS reference points furnished by the Faculty of Geographical Sciences – Utrecht University (personal communication).

Table 17.1. *Bandwidth and sampling interval for the HyMap sensor.*

Spectral range (nm)	Band numbers	Bandwidth (nm)	Average sampling interval (nm)
438–895	1–32	15–16	15
895–1340	32–62	15–16	15
1340–1807	62–94	15–16	13
1807–2483	94–126	18–20	17

17.2.3 Fieldwork and laboratory analyses

Seventy-five soil samples have been collected from the upper soil horizon in the main morpho-pedological units in La Peyne watershed, and divided into two subsets named CAL and VAL. The CAL subset (37 samples) is intended to cover the regional value ranges concerning carbonate and clay contents, stoniness and colour, and was used for acquiring an adequate comprehension of samples composition and quantifying their effects over the spectra characteristics. Extensive laboratory work included spectrophotometry and spectroradiometry (visible and SWIR spectra in the 1100–2500 nm range), X-ray diffractometry (mineralogical composition) and chemical determinations of $CaCO_3$ content and granulometry, which were used for calibration. The VAL subset (38 samples) comes from bare soil sites extracted from the HyMap image (cf. further) and will be used for validation purposes. A GPS allowed for locating all samples with 5–8 m accuracy. In addition, we used 5-level HCl effervescence grades for more than 500 auger pit descriptions, realised and accurately located by an experienced soil scientist.

17.2.4 Creation of new covariates

Emphasis was placed in the creation of three new, combined bands from the ones already available, namely redness rating, $CaCO_3$ indicator and clay indicator.

Redness rating is expected to help in discriminating soils of different colours. We used the formula suggested by Madeira Netto et al. (1997):

$$\mathrm{RR} = R_{646}^2/R_{450} \times R_{555}^3 \tag{17.1}$$

where R_{646}, R_{555} and R_{450} state for the reflectance at $\lambda = 646$ nm (HyMap band 15), 555 nm (band 9) and 450 nm (band 2), respectively.

Combination tones of the OH stretch and OH-Al bending modes are known to express a strong absorption at 2208 nm. In the study area, previous work strongly suggested that montmorillonite, kaolinite and illite minerals are the main silicate minerals present in the studied soils that may produce this spectral feature. Concerning $CaCO_3$, a sharp feature centred at 2346 nm can be attributed to the CO_3^{2-} overtone vibrations according to Gaffey (1987). Figure 17.1(a) illustrate laboratory reflectance spectra obtained for a selected set of CAL samples, where these features are clearly visible. Figure 17.1(b) shows the

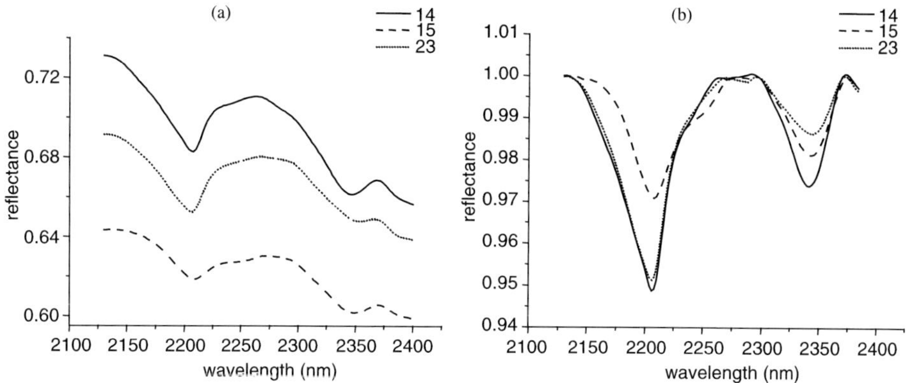

Figure 17.1. *(a) Reflectance spectra of three selected samples: 14, 15 and 23 with $CaCO_3$ contents of 513, 196 and 90 $g\,kg^{-1}$ respectively. (b) Shows the continuum removed spectra for the same samples.*

continuum removed (CR) spectra for the same samples calculated as suggested by Clark and Roush (1984):

$$\mathrm{CR}_{\lambda} = R_{\lambda}/\mathrm{BL}_{\lambda} \qquad (17.2)$$

where:

R_{λ} = Reflectance at wavelength λ

BL_{λ} = Reflectance at wavelength λ for the baseline.

With the HyMap image, the bands centred at $\lambda = 2206$ nm (band 109) and $\lambda = 2341$ nm (band 117) were considered when calculating CR spectra for clay minerals (CR_{2206}) and for $CaCO_3$ (CR_{2341}), respectively.

17.2.5 Statistics

Simple statistical correlation and linear regression analysis were used to relate reflectance parameters (CR_{2206} and CR_{2341}) to the abundance of clay and $CaCO_3$ in the samples. The CAL subset was used for establishing the regression equation, while the validation step used the VAL subset. According to the previous work on these data, we can assume that these two variables display a linear relationship.

To take into account the 5–8 m geometric accuracy of sample points, we decided to replace CR_{λ} by $\overline{\mathrm{CR}}_{\lambda}$, where $\overline{\mathrm{CR}}_{\lambda}$ is the mean of CR_{λ} values of the pixels comprised in an 8-m radius around each point.

17.3 Results and discussion

17.3.1 The laboratory study

To clarify the principles used in the image analysis procedures we summarise some relevant results obtained on CAL samples.

Composition of the samples and main spectral characteristics

According to the X-ray diffraction data, quartz is present in all samples and is the dominant mineral. Calcite, illite and feldspars (albite, microcline and orthoclase), kaolinite and smectites are also present in the fine earth of the samples in general. However, their relative abundance among the samples is quite variable. In general calcite, illite and feldspars are more abundant in molassic or alluvial soils than in older materials from the Pliocene or Plio-quaternary. For the soils derived from Pliocene sediment materials, calcite shows a broad variation in content. For soils from the Plio-quaternary materials calcite is virtually absent or present in very small amounts. These results are in agreement with the presumed ages and weathering stages of the main soils in the region. For the clay fraction ($<2\,\mu m$), illite, kaolinite and smectites are the dominant components, even though quartz and calcite may also be present in some samples.

Chemical analysis showed that calcium carbonate content of the samples varied from 20 to $530\,g\,kg^{-1}$ with an average of $164\,g\,kg^{-1}$. Colour varies from pale yellow to yellowish red according to the Munsell notation, and texture varied from clay to sandy loam ($100–317\,g\,kg^{-1}$). VAL samples exhibit comparable values. Thus, the sample sets cover a wide range of compositions, making them useful for a further calibration of regression relationships.

Regression results for carbonate and clay contents

Table 17.2 shows the statistical relationship obtained between the two CR_λ band depths, calcium carbonate and clay abundances. Both regression equations are highly significant ($p<0.001\%$) and explain most of the stated variability. For $CaCO_3$, the intercept departure from 1 may be related to the presence of illite in

Table 17.2. *Regression parameters for clay and $CaCo_3$ contents against defined continuum removed band depths.*

Regression parameters	Clay content ($g\,kg^{-1}$)		$CaCO_3$ content ($g\,kg^{-1}$)
	Before $CaCO_3$ removal	After $CaCO_3$ removal	
Explanatory variable	CR_{2206}	CR_{2206}	CR_{2341}
R^2_{VAL}	0.55[a]	0.74[b]	0.92
p	<0.001	<0.001	<0.001
Slope	−0.0003	−0.0003	−0.0009
Intercept	0.9754	0.9826	0.9687
n	38	38	38

Note: Regression parameters, with: R^2_{CAL}: coefficient of determination for validation set of samples; p: significance level, slope and intercept of the regression equation; n: number of samples.

[a]Improved to 0.74 by removing three samples developed on basaltic rocks.

[b]Improved to 0.83 by removing three samples developed on basaltic rocks.

most of the samples, which expresses an absorption near 2345 nm (Chabrillat et al., 2001). Clay content was determined before and after $CaCO_3$ removal, and regression analysis results show a significant improvement in the regression coefficient after $CaCO_3$ removal. This is coherent with the fact that the band depth at 2206 nm is affected by the clay minerals and not by $CaCO_3$.

R^2_{VAL} values show that by inverting these regression equations, CaCO3 and clay contents could be estimated by simple reflectance measurements in the 2000–2400 nm range. For $CaCO_3$, standard error of prediction – as formulated in Ben-Dor and Banin (1990) – reaches 65 g kg^{-1}, which (i) is similar to the values obtained by this author and (ii) appears to be adequate for pedological survey.

Usefulness of laboratory results for subsequent airborne remote sensing

Samples have not been subjected to any destructive or preparation process (neither by heating nor by addition of chemicals to the samples), thus, the pre-treatment conditions seem to be roughly similar for laboratory and field conditions.

Soil samples exhibit a linear relationship between spectral characteristics and useful variables. By inverting the model, $CaCO_3$ and clay contents can be predicted from the variables derived from spectral reflectance in laboratory conditions. These results point out the possibility of using imaging spectrometer data for quantitative estimation of carbonates and clay contents for surface soil samples. This is the objective of the next part of the work.

Despite these satisfactory results, some supplementary hypotheses must be assumed before going further:

- Clay minerals are contained in the clay particle-size class ($<2\,\mu m$), and are the prevailing type of material in this size class. Other components such as carbonates, amorphous quartz, iron and aluminium oxides may be present in this class, but under our conditions, they occur in very limited amounts (statistical results reported above support this hypothesis).
- HyMap bands are wider and slightly shifted, compared with the ones used in the laboratory: we assume that peaks will still be recognisable and their intensities directly related to $CaCO_3$ and clay abundance.

It is also important to keep in mind that:

- Sample sizes are very different: laboratory samples represent around 0.01 m^2 on the ground, while each HyMap pixel will embrace a 25 m^2 area; we do not have yet any information about spatial structures of the considered variables.
- Laboratory analyses were performed on the fine earth of the surface soil horizon samples, while HyMap spectral data considers the overall materials present on the soil surface, which includes gravel and stones partially or totally covering the fine earth.
- Although geometrically corrected, the frequent occurrence of ditches, terraces and reaches (not represented in the DEM) may produce discrepancies between field and image location.

- Similarity of radiometric characteristics between soil surface features and mixed soil samples remains to be proved.

The possible deviations from the hypotheses and the points mentioned above are source of errors, if we want to apply the regression equations determined with laboratory data directly to the HyMap data. Thus, we will further assume that a linear equation does represent the nature of relationship between the variables under study, but we will carefully reconstruct its parameters (slope and intercept) from VAL subset while using image data.

17.3.2 The HyMap study

Extraction of bare soils on the HyMap image

At the time the image was acquired (July 2003) most of the soil surface was covered by green vegetation (mainly vineyards, native forests) or crop residues. Only a part was ploughed or recently planted into vineyards, and had its soil surface fully exposed, so that the spectral information collected by the sensor really registered soil properties. Those bare soil areas are the only ones we were interested in for further analysis.

To isolate the bare soil areas, we masked the image with NDVI values over an expert-calibrated threshold (a value of 0.26 was used after considering several visited parcels and "salt-and-pepper" patchiness of the resulting mask). For discarding the areas with an important dry vegetation cover, we used the absorption band centred at 2100 nm (band 103), this feature being related to O-H and C-O bonds of cellulose (Kokaly and Clark, 1999). The same band has been used by other authors to estimate crop residue cover (Daughtry et al., 2003; Nagler et al., 2000). It was present in almost all pixels with recognised mineral absorption features, so we applied a filter to the CR image, limiting the values of the band depth of this feature to a second expert-based threshold (one dolomite and one bauxite quarry presented very crisp boundaries and were known to have no vegetation at all in some specific parts). A morphological erosion with a 3×3 filter eliminated the urban features and road networks (whose widths are smaller than 15 m in this area). We were finally left with 370 polygons of sufficient area (area $> 400\,m^2$) covering a total of $1.22\,km^2$ in the La Peyne watershed.

The relationship between HyMap $\overline{CR}_{2341}$ and $CaCO_3$ contents

Figure 17.2 shows the regression analysis between the $CaCO_3$ contents in the fine earth of the VAL subset with HyMap $\overline{CR}_{2341}$. The regression is significant at a probability level >99.99%, and more than 64% of the $\overline{CR}_{2341}$ variability is explained by the variations of the $CaCO_3$ contents in the fine earth of the samples taken at the soil surface. It is noticeable that for samples 36 and 38, the

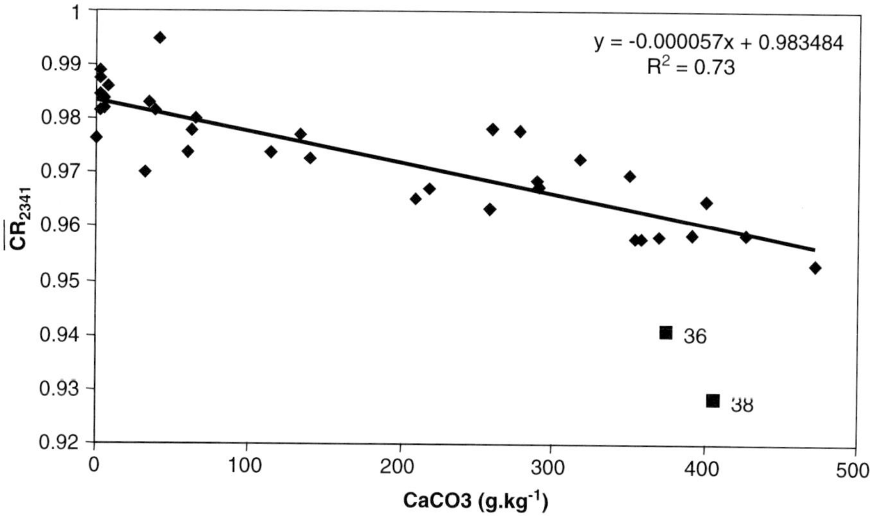

Figure 17.2. *Correlation between $CaCO_3$ contents in the air-dried and 2-mm sieved soil samples and the $\overline{CR}_{2341}$ obtained from HyMap spectra from the sites of sample collection. The two samples (36 and 38) taken from sites with a strong stoniness are represented by square dots.*

$\overline{CR}_{2341}$ is exacerbated as compared with the other values ($\overline{CR}_{2341}$ much higher than would be expected by the general trend defined by the other points if the two points were not introduced in the regression analysis). The zones where these samples were acquired are gravelly to very gravelly at the surface, and the coarse materials are composed essentially of limestone, which is very rich in $CaCO_3$. $\overline{CR}_{2341}$ reproduces this situation.

A comparison between HyMap $\overline{CR}_{2341}$ and field-measured HCl effervescence classes

In order to estimate if HCl effervescence measurements made in the field during soil survey work could be completed by reflectance data, we compared effervescence field tests obtained independently at the same locations by a trained surveyor against $\overline{CR}_{2341}$ values. HCl effervescence is graded in five levels: no effervescence (0), very slightly effervescent (1), slightly effervescent (2), effervescent (3) and very effervescent (4), according to French normalised soil field description methods Baize and Jabiol (1995). Effervescence grades are attributed following visual and audible criteria and are somehow subjective, so that the relationship between actual $CaCO_3$ contents and effervescence grades can be slightly surveyor-dependent.

Table 17.3. *Ranges of $CaCO_3$ content found and $\overline{CR}_{2341}$ for each effervescence class.*

Effervescence	Code	$CaCO_3$ (g kg^{-1}) usual range (for info)	$CaCO_3$ (g kg^{-1}) found	$\overline{CR}_{2341}$
None	0	<20	<41	1–0.9799
Very slight	1	20–100	41–134	0.9799–0.9735
Slight	2	100–250	134–228	0.9735–0.9672
Medium	3	250–500	228–321	0.9672–0.9605
Strong	4	>500	>321	<0.9605

Note: Usual ranges are given here for additional information, according to Lozet and Mathieu (2002).

To build a common scale from known $CaCO_3$ contents and the 0–4 effervescence scale, the same surveyor was asked to grade each sample for the CAL subset, allowing for estimating by regression the $CaCO_3$ thresholds for the five effervescence classes (Table 17.3).

The $CaCO_3$ content ranges found experimentally are not very much different from usual ones (excluding the very effervescent class), so that they have been retained here for taking into account the actual accuracy and bias of the surveyor.

These thresholds were then applied to the $\overline{CR}_{2341}$ image band, resulting in a 5-class image shown in Plate 17(upper) (see Colour Plate Section). Twenty-eight supplementary auger pits descriptions for surface horizons falling over the bare soil plots were used for comparison with the image class values: 64% of the points (17 points) belonged to the same effervescence class, 14% (4 points) differed from one or two units and 4% (1 point) differed from more. We can therefore consider that the results obtained are satisfying, and that $\overline{CR}_{2341}$ provides a good estimate of $CaCO_3$ abundance in the soil surface horizon, while being very likely more accurate than if produced by using HCl effervescence tests.

The relationship between HyMap $\overline{CR}_{2206}$ and clay content

A correlation analysis was also done with clay size contents and $\overline{CR}_{2206}$ values obtained from the HyMap image. The correlation coefficient *R* equals 0.8, and the regression is highly significant ($p<0.001$). So, we can be 99.9% confident that there is a negative correlation between the clay content in the fine earth of the soils in the sampled area and the depth of absorption at the 2206 nm wavelength, and that 64% of the $\overline{CR}_{2206}$ variation is explained by the clay content in the fine earth. It is important to remind that $\overline{CR}_{2206}$ is theoretically an estimation of the clay minerals content in the fine earth. As discussed previously in the clay fraction there are other minerals other than the clay minerals (calcite, quartz, ...). Also, in the surface of some soil units are present variable amounts of rock fragments (quartz, schists, ...), which can affect the precision of the estimates.

The sample points considered comprise the conditions mentioned above, and the regression results are quite good.

The linear model was inverted to estimate the clay content of the soils based on the $\overline{CR}_{2206}$ band. The resulting image was density sliced in four clay content classes (resulting image and legend in Plate 17(lower) see Colour Plate Section).

At the time of writing this chapter, additional measurements of clay size abundance at the soil surface were not available to verify the quality of the results. However, field inspection of the results indicates good correspondence with reality. For example, the occurrence of molasses fits very closely to the zones mapped as low clay content (blue class). The areas mapped at the highest clay content class correspond to the zones that have clay sediments as soil parent material. At the agricultural plot level, several instances were verified of concordance in the clay abundance mapped and detected by field tests.

Additional evidence of the quality of the results can be illustrated by variations in the clay content and the CR_{2206} for a regular-spaced sampling sequence taken at the interior of a plot shown in Figure 17.3. The maximum $\overline{CR}_{2206}$ possible value ($\overline{CR}_{2206} = 1$) was subtracted from the actual measured values to make the representation easier to interpret. The actual clay content determined by analytical procedures and the HyMap $\overline{CR}_{2206}$ follows the same trend, and the procedure proposed is very sensitive to the small variations in clay size abundance.

Redness rating: a distinctive factor for the basalt-derived soils

Laboratory studies of clay size estimation showed that the model does not apply to the basalt-derived soils. A robust identification is needed in order to process them separately. Since those soils have a particular colour characteristic (dark brown to brown), the redness rating RR (eq. 17.1) – available from the image – was tested for their discrimination.

RR values ranged from 8×10^{-8} to 3.2×10^{-5}. The lowest values were found to be related to the molassic soils that present whitish to pale yellow colours. Soils derived from more recent, continental sediments from the Villafranchian presented redness ratings of the same magnitude even though the soil matrix presents a reddish colour. The highly reflecting quartz gravels partially covering their surface are responsible for this behaviour. The highest RR values were observed in soils developed from basalt with values 15 times greater than soils derived from molasses. This is mainly due to the presence of dark materials from the fresh basalt rocks and/or from the opaque magnetite that are abundant in the surface of those soils, resulting in low values for the green band (R_{555}). Intermediate values were observed for yellowish to reddish colours common to the continental sediments (Pliocene and Villafranchian). RR values could therefore be used for moving the basaltic soils apart from the analyses and getting better correlation.

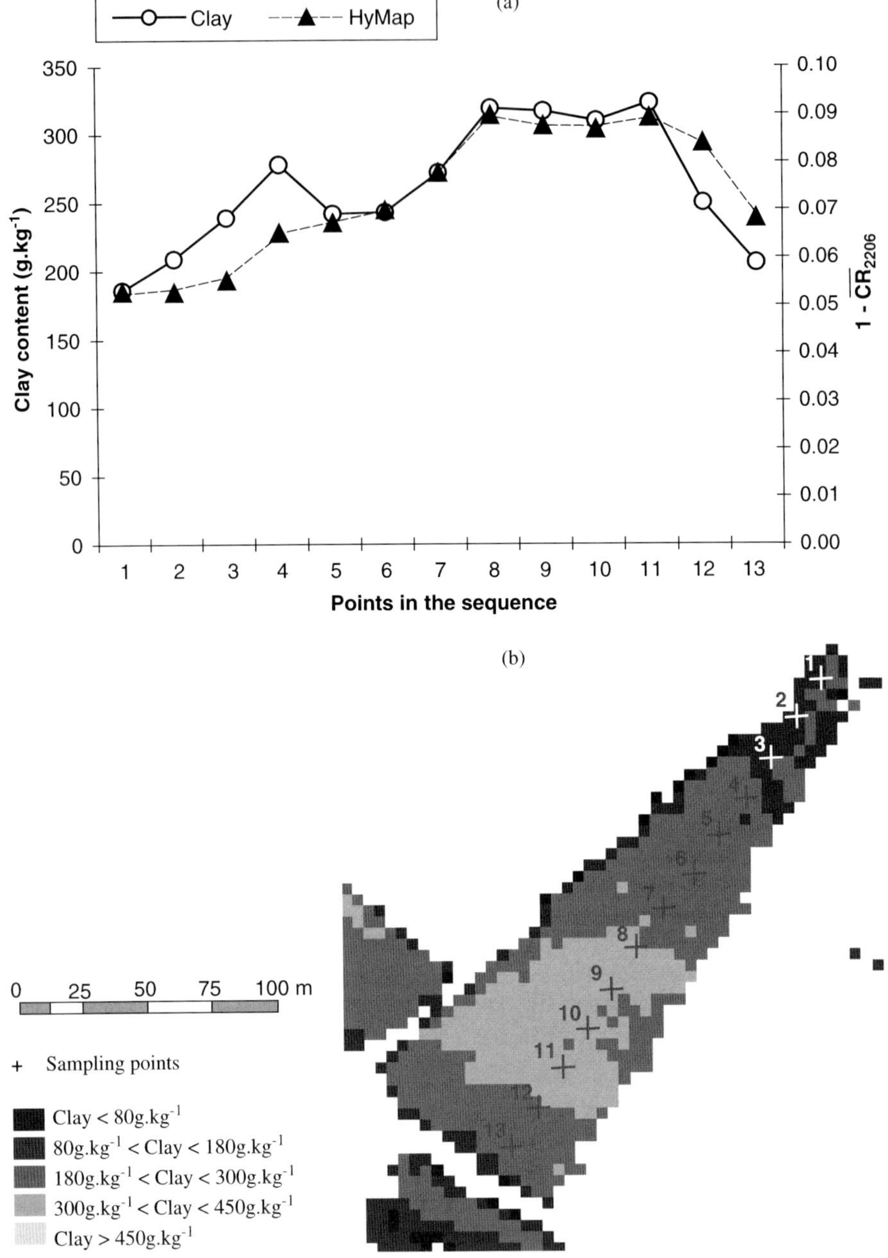

Figure 17.3. *(a) Clay content ($g\,kg^{-1}$) compared with depth of absorption at 2206 nm of the HyMap acquired spectra ($\overline{CR}_{2206}$) for a regularly sampled sequence of points in an agricultural plot. $1-\overline{CR}_{2206}$ was considered instead of $\overline{CR}_{2206}$ for clarity of the representation. (b) "Clay" image and location of the points in the sequence.*

17.4 Conclusions

This research exploited two different approaches for information extraction from soil reflectance spectral data: (1) band depth analysis for estimating calcium carbonate and clay abundances; and (2) redness rating index for identifying soil colour variations.

For soils developed from sedimentary materials of variable calcium carbonate and particle size composition, we have demonstrated the possibility of obtaining:

- Quantitative information for calcium carbonate content from the intensity of absorption of the feature centred at 2341 nm.For spectra acquired in the laboratory, precision of estimates was close to 70 g kg^{-1}, which makes the estimates highly satisfactory and the method reliable for most of soil survey work.
- Clay particle size content estimation, by using the absorption depth of the clay minerals at 2206 nm.Soils having significant amounts of highly absorbent materials in the near infrared (e.g. magnetite) as is the case of basalt-derived soils, need to be treated differently. A redness index allowed their discrimination.

These estimations are at least as accurate as the ones made by experienced surveyors using field tests, with the advantage of providing at one go a bi-dimensional view of the variability of the estimated parameters. The method is simple and requires very little sample and data manipulation. Thus, CR spectra band depth analysis may be a powerful tool for getting quantitative information about soil composition.

This preliminary work shows a great potential for the use of HyMap images to make semi-quantitative estimations of soils factors that are important in the discrimination of soil classes. It makes possible to (1) build more accurate sampling strategies in order to collect data on soil mineralogy and chemical properties, and (2) create a segmentation scheme from these relevant variables for such soils whose spatial patterns are difficult to delineate on the ground by traditional means.

Further work is planned for refining the procedure and to understand the factors influencing the lack of precision when HyMap images are used (as compared with the laboratory-acquired data). It is necessary to clarify how other objects (as well as their intrinsic properties and their spatio-temporal dynamics) normally present at the soil surface, for example plant residues, photosynthetically active vegetation, gravels, crusts, cracks, etc., affect the prediction potential of the parameters suggested in this work. Knowledge of the relationship between soil surface features dynamics and soil properties is likely among the main issues to be overcome.

It is important also to identify the extension of the geographical zones for which it is possible to acquire images with a reasonable exposure of the soil surface. For the vineyard area of the Low Languedoc Plains natural region, we estimate that the potential of acquiring images with a significant soil surface exposure is relatively high. Before the start of vegetative growth in the spring, when plants are pruned and most of the vineyards inter-rows have the soils cultivated, may be a good opportunity. In this case however, the potential issue will be to get moisture-independent measurements.

Once those problems have been solved, the results produced by similar studies can act as an aid to soil scientists conducting conventional survey work, as well as in numeric predictions of soil variability in the landscape as proposed by McBratney et al. (2003).

The fast development in the spectroscopic industry allows to suppose that in the near feature spectroradiometers, or equipment using spectral reflectance principles, will be available at a reasonable cost, covering broader ranges of wavelengths (e.g. gamma-ray, see Chapter 16), thus making possible their use in fieldwork. The repeatability of these measurements *over time* may be used to capture soil dynamics and processes (see example in Chapter 18), while the extensive nature of these measurements *over space* allows for dealing with the significance of spatial patterns (see Chapter 19).

Acknowledgements

We are indebted to Professors Raymond Sluiter and Steven De Jong, Utrecht University, for making available the HyMap images used in this work. We acknowledge Mr. Didier Brunet from the Matière Organique des Sols Tropicaux (IRD) laboratory for helping in the acquisition of reflectance spectra at the laboratory, Mr. Guillaume Coulouma from INRA-Montpellier/LISAH for his advice in the fieldwork and for making available his current soil database, Dr. Noële Guix (INRA-Montpellier/LISAH) for making available some of the samples and to Dr. Ali Bedidi (Université Marne-la-Vallée) for his comments.

References

Baize, D., Jabiol, B., 1995. Guide pour la description des sols. Techniques et Pratiques. INRA éditions.

Baptista, G.M.d.M., Madeira Netto, J.S.d., 2001. RCGb index: a tool for mapping the weathering degree of the tropical soils in Brazil. The AVIRIS Earth Science and Applications Workshop. Pasadena, CA, USA.

Ben-Dor, E., Banin, A., 1990. Near infrared reflectance analysis of carbonate concentration in soils. Appl. Spectrosc. 44, 1064–1069.

Bonfils, P., 1993. Carte pédologique de la France au 1/100.000, Feuille de Lodève (notice+carte). INRA SESCPF, Orléans.

Chabrillat, S., Goetz, A.F.H., Olsen, H.W., Krosley, L., 2001. Use of hyperspectral imagery in identification and mapping of expansive clays in Colorado. Remote Sens. Environ. 82, 431–435.

Clark, R.N., Roush, T.L., 1984. Reflectance spectroscopy: quantitative analysis techniques for remote sensing applications. J. Geophys. Res. 89, 6329–6340.

Daughtry, C.S.T., Hunt, E.R., Jr., Doraiswamy, P.C., McMurtrey, J.E., III, Russ, A.L., 2003. Remote sensing of crop residue cover and soil tillage intensity. Geoscience and Remote Sensing Symposium, IGARSS '03 IEEE International.

Gaffey, S.J., 1987. Spectral reflectance of carbonate minerals in the visible and near infrared (0.35–2.55 μm): anhydrous carbonate minerals. J. Geophys. Res. 92, 1429–1440.

Hauff, P.L., Kruse, F.A., Thiry, M., 1990. Spectral identification and characterization of kaolinite/smectite clays in weathering environments. The 5th Australasian Remote Sensing Conference. Perth, Australia, pp. 898–905.

Kokaly, R.F., Clark, R.N., 1999. Spectroscopic determination of leaf biochemistry using band-depth analysis of absorption features and stepwise linear regression. Remote Sens. Environ. 67, 267–287.

Kruse, F.A., 1998. Advances in Hyperspectral Remote Sensing for Geologic Mapping and Exploration. Proceedings 9th Australasian Remote Sensing Conference. Sydney, Australia, pp. July 1998.

Kruse, F.A., Thiry, M., Hauff, P.L., 1991. Spectral identification (1.2–2.5 μm) and characterization of Paris basin kaolinite/smectite clays using a field spectrometer. 5th International Colloquium – Physical Measurements and signatures in Remote Sensing. Courchevel, France, pp. 181–184.

Lozet, J., Mathieu, C., 2002. Dictionnaire de Science du Sol. coll. Tec et Doc. Lavoisier, Paris.

Madeira Netto, J.S.d., Bedidi, A., Pouget, M., Cervelle, B., 1997. Visible spectrometric indices of hematite (Hm) and goethite (Gt) content in lateritic soils: the application of a Thematic Mapper (TM) image for soil-mapping in Brasilia, Brazil. Int. J. Remote Sens. 18, 2835–2852.

Margate, D.E., Shrestha, D.P., 2001. The use of hyperspectral data in identifying desert-like soil surface features in Tabernas area, Southeast Spain. In: S. Crisp, AARS (Ed.), 22nd Asian Conference on Remote Sensing, Singapore, pp. 736–741.

McBratney, A.B., Mendonca Santos, M.L., Minasny, B., 2003. On digital soil mapping. Geoderma 117, 3–52.

Nagler, P.L., Daughtry, C.S.T., Goward, S.N., 2000. Plant litter and soil reflectance. Remote Sens. Environ. 71, 207–215.

Raggatt, T.J., Lewis, M., Fitzpatrick, R.W., 2004. Spectral discrimination of soil and regolith attributes within Herrmanns Catchment, Mount Lofty Ranges, SA. In: I.C. Roach (Ed.), Regolith 2004, CRC LEME. Adelaide, Australia, pp. 287–291.

Richter, R., Schläfer, D., 2002. Geo-atmospheric processing of airborne imaging spectrometry data. Part 2: atmospheric/topographic correction. Int. J. Remote Sens. 23, 2631–2649.

Ryan, S., Lewis, M., 2000. Discriminating and mapping soils using HyMap hyperspectral imagery, Barossa Valley, SA, Tenth Australasian Remote Sensing & Photogrammetry Conference (ARSPC), Adelaide.

Schläfer, D., Richter, R., 2002. Geo-atmospheric processing of airborne imaging spectrometry data. Part 1: parametric orthorectification. Int. J. Remote Sens. 23, 2609–2630.

Van der Meer, F.D., 2004. Analysis of spectral absorption features in hyperspectral imagery. Int. J. Appl. Earth Obs. Geoinf. 5, 55–68.

Van der Meer, F.D., De Jong, S., 2002. Imaging Spectrometry: Basic Principles and Prospective Applications. Kluwer Academic Publishers, Dordrecht, the Netherlands.

Developments in Soil Science, volume 31
P. Lagacherie, A.B. McBratney and M. Voltz (Editors)

Chapter 18

LAND-COVER CLASSIFICATION FROM LANDSAT IMAGERY FOR MAPPING DYNAMIC WET AND SALINE SOILS

S. Kienast-Brown and J.L. Boettinger

Abstract
Wet and saline soils have been recognized as an important and complex component of wetland ecosystems in arid environments. Analysis and classification of remotely sensed spectral data is an effective method for discerning the spatial and temporal variability of soils. The East Shore Area (ESA) of the Great Salt Lake soil survey update is focused on updating soil map units containing wet and saline soils. The ESA provides a unique environment for the use of remotely sensed spectral data for map unit refinement because of low relief and a large extent of soils that are wet and saline to various degrees. Map units in the ESA containing wet and saline soils were updated and refined using Landsat 7 imagery. Five land-cover classes are related to dominant soil types that vary in soil wetness, salinity, calcium carbonate concentration and vegetation cover type. Supervised classification of the imagery was performed using the five land cover classes. The final classification resulted in 14 land cover classes, including nine additional classes that help describe the variability in the original five classes. The classification results were validated using visual inspection in the field, a priori knowledge of the area and an error matrix. The results of the classification were used to enhance original soil map units and calculate map unit composition in the final soil mapping process. This information was then incorporated into the updated soil map. Temporal variation in land cover classes has the potential to be considered in map-unit refinement to reflect the dynamic nature of the margins of the Great Salt Lake, Utah.

18.1 Introduction

Satellite data have been used to examine the Earth's surface for several decades. Remote sensing techniques have provided scientists with a more efficient method of conducting large-scale studies of the Earth's surface. Landsat thematic mapper (TM) and Landsat multispectral scanner (MSS) spectral data have been the most widely used satellite data for remote sensing research. Much information about the Earth's surface, relating to vegetation, rocks, minerals and soils can be extracted from spectral data.

Remote sensing imagery has also facilitated the study and mapping of salt- and sodium-affected soils. Metternicht and Zinck, (1997) used image classification of Landsat TM data, combined with field observations and laboratory data, to discriminate salt-affected versus sodium-affected soils in Bolivia. Transformed divergence analysis was used to determine the best band combination for maximum separability between salt- and sodium-affected soil classes, and resulted in accuracies from 64 to 100% in the classification. The spectral behaviour of salt-affected soils was examined using Landsat TM false color composite (FCC) images and in situ radiometric measurements (Rao et al., 1995). The middle infrared region (MIR) bands were added to the FCC image and greatly improved the ability to identify salt-affected soils. Singh (1994) used Landsat TM FCC imagery to successfully monitor the increase or decrease in the extent of salt-affected soils over time, and to delineate moderately versus severely salt-affected soils.

The ability to distinguish soil properties from spectral data has led to the use of remote sensing for soil mapping. Landsat MSS digital number data was used to delineate soil associations in Nebraska (Lewis et al., 1975). Soil associations mapped by conventional field techniques were compared to soil associations mapped on the satellite image by photo-interpretation techniques. Soil associations were successfully delineated in areas where topography and vegetation could be discriminated on the satellite image, and related to field data. In an Arizona soil survey, Landsat MSS digital number data was clustered according to soil parent material and landform (Roudabush et al., 1985). This information was used to refine map unit boundaries and composition, and increase the understanding of map unit variability. They also concluded the creation of a soil map from spectral data to be a valuable prefield mapping tool.

An update soil survey of the East Shore Area (ESA) of the Great Salt Lake, Utah, USA, headed by the USDA Natural Resources Conservation Service (NRCS), began in 1999. A soil survey of the ESA was published in 1975 as part of the East Box Elder County soil survey (Chadwick et al., 1975). The ESA update is focused on refining soil map units containing wet and saline soils. The ESA is heavily influenced by fluctuations in precipitation and lake level, and anthropogenic controls on fresh water flow from diked areas. The interaction between these factors affects local groundwater level, soil moisture and soil salinity. Therefore, the ESA is a dynamic and complex system, resulting in a challenge for map unit refinement.

The ESA provides a unique environment for the use of remotely sensed spectral data for map unit refinement because of low relief and a large extent of soils that are wet and saline to various degrees. Interpretation of remotely sensed spectral data is an effective method for mapping wet and saline surfaces at large scales. Five major land cover types exist in the ESA and are related to soil moisture, salinity, calcium carbonate concentration and vegetation cover type. These land cover types are salt flat, pickleweed flat, upland, saltgrass, and sedges-rushes. A supervised classification of Landsat 7 imagery was conducted to determine spatial extent and variability of the five major land cover types so that mapping of the dominant associated soil types can be refined.

18.2 Methods and materials

18.2.1 Study area

The ESA is located in northern Utah, USA, between 41° 41′18″ N, 112° 14′ 46″ W and 40° 44′ 46″ N, 112° 5′ 14″ W. The ESA encompasses 166,905 ha on the northeastern shore of the Great Salt Lake (Fig. 18.1). There is very little elevation change over the ESA, with elevations ranging from 1280 m in the southern part to 1368 m in the northern part. The mean annual temperature in the ESA is 8–10°C, and the mean annual precipitation is 300–380 mm (Chadwick et al., 1975). The soil climate is mesic and xeric or xeric/aquic.

The ESA contains natural and managed waters. Freshwater inputs include Blue Creek, West Canal and Sulfur Creek, which flow into the ESA from the northwest, north and northeast, respectively. Runoff from the Promontory Mountains flows into the ESA from the west. The majority of these freshwater sources are managed through dikes by private land owners in the area. The Bear River drains into the ESA from the east, and is heavily managed by the Bear River Migratory Bird Refuge.

All the freshwater that enters the ESA eventually flows south into the Great Salt Lake. Lake levels fluctuate radically as a function of precipitation and evaporation. As the lake level rises, highly saline water from the Great Salt Lake moves north and mixes with the freshwater. As the lake level lowers, the lake water retreats and evaporation concentrates salts in the soil and on the soil surface. In both scenarios, the salinity of the freshwater increases dramatically as it flows through the ESA to the Great Salt Lake.

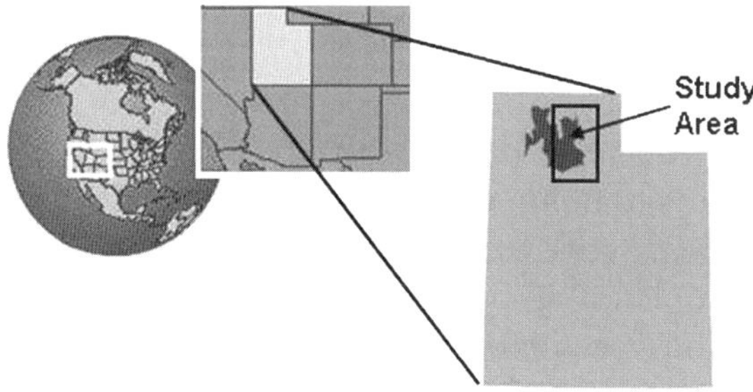

Figure 18.1. *Location of the ESA study area in northern Utah, USA.*

The ESA is composed entirely of lacustrine sediments from Pleistocene pluvial Lake Bonneville (13,000–25,000 years ago) and from the present-day Great Salt Lake. The Great Salt Lake is the existing remnant of Lake Bonneville. The lake sediments in the ESA are mainly fine-grained, stratified silts and clays (Stokes, 1988). The entire ESA study area is located on a lake plain. The area is essentially flat, with an elevation gradient of about 90 m over a 16-km distance, resulting in an average slope of 0.5%, dipping to the south. The only topographic features occurring in the ESA are upland "islands" located throughout the lake plain. These islands are approximately 9 m above the lake plain and are composed of stratified coarse silt lake sediments.

18.2.2 Major land cover over types and associated soil properties

The ESA contains five major land cover types that are related to dominant soil types that vary with respect to moisture, salinity (reported as saturated paste electrical conductivity (EC) to a depth of 150 cm), calcium carbonate concentration (reported to a depth of 150 cm) and vegetation cover type. Table 18.1 shows selected characteristics of the land cover types. The land cover types in the order of increasing wetness are upland, pickleweed flat, salt flat, saltgrass and sedges-rushes. The land cover types in the order of increasing calcium carbonate concentration are upland, sedges-rushes, saltgrass, pickleweed flat and salt flat. The land cover types in the order of increasing salinity are upland, saltgrass, sedges-rushes, pickleweed flat and salt flat. Figure 18.2 shows the representative landscape, calcium carbonate profile and EC profile for the salt flat and upland land cover types as examples of the two extremes in the area.

18.2.3 Imagery and analysis

Landsat 7 imagery from May 21, 2003, path 38 rows 31 and 32, was used for all analyses. The two images were mosaiced and then a subset of the ESA study area was made. *ERDAS Imagine 8.6* was used for all image processing, analysis and classification.

Training sites for the supervised classification were identified using GPS points from field data collection, and from digital ortho-photo quads (DOQs) and Landsat imagery using a priori knowledge of the area. A minimum area equivalent to a 3×3 Landsat pixel area (8100 m^2) was used for training site selection. Originally, 90 training sites were selected in the following 10 classes: water, shallow water, very shallow water, sedges-rushes, sedges-rushes-saltgrass, saltgrass, upland, agriculture, salt flat and pickleweed flat.

A spectral signature set containing a spectral signature for each of the 90 training sites was created in *Imagine.* Optimum index factor (OIF) was calculated to determine the optimum three-band combination for the subset Landsat 7

Table 18.1. *Selected characteristics of major land cover types and associated soils.*

Land cover type	Percent of total area	Associated soil classification[a]	Max EC (dS/m)	Min EC (dS/m)	Ave EC (dS/m)	Max $CaCO_3$ concentration (%)	Min $CaCO_3$ concentration (%)	Ave $CaCO_3$ concentration (%)
Sedges/ rushes	9	Typic Endoaquepts	38	29	30	26	21	23
Saltgrass	4	Typic Halaquepts	52	11	29	25	22	24
Salt flat	36	Typic or Calcic Aquisalids	107	64	87	30	14	26
Pickleweed flat	5	Typic or Calcic Aquisalids	34	12	30	33	17	25
Upland	19	Typic or Aquic Natrixeralfs Typic or Aquic Natrixerolls Sodic or Typic Calcixerepts	15	9	12	15	10	13
Agriculture[b]	<1	Typic Haploxeroll						
Water[b]	27							

[a]Soil Survey Staff, 2003.
[b]Not considered a major land cover type for the purposes of this study, but did occur in the study area.

image (Jensen, 1996). Transformed divergence analysis was used to evaluate the separability of the spectral signatures and determine the band combination that would result in maximum separability of the signatures (Metternicht and Zinck, 1997). The OIF-calculated three-band combination was evaluated against other combinations of the six Landsat 7 bands in the transformed divergence analysis. Mean spectral signature plots for the 90 classes were generated, and like signatures were merged to achieve maximum separability with the minimum number of classes. Spectral signature histograms were evaluated for each of the resulting 14 classes to determine appropriate classification algorithm based on data distribution, normal versus multimodal.

18.2.4 Classification

Supervised classification was performed using fuzzy classification with two best classes assigned to each pixel (Jensen, 1996). The fuzzy classification method was chosen because of the high variability over small distances in the ESA and the high likelihood of mixed pixels, especially in the salt flat and pickleweed flat classes. The fuzzy classification image and associated distance image were processed using a fuzzy convolution filter to reduce speckle, or "salt and pepper," in the classification. The fuzzy convolution filter uses the fuzzy classification image and the associated distance file to calculate a total weighted distance for all classes within the filter window. Then, the class with the largest total inverse distance, summed over the entire set of fuzzy classification layers

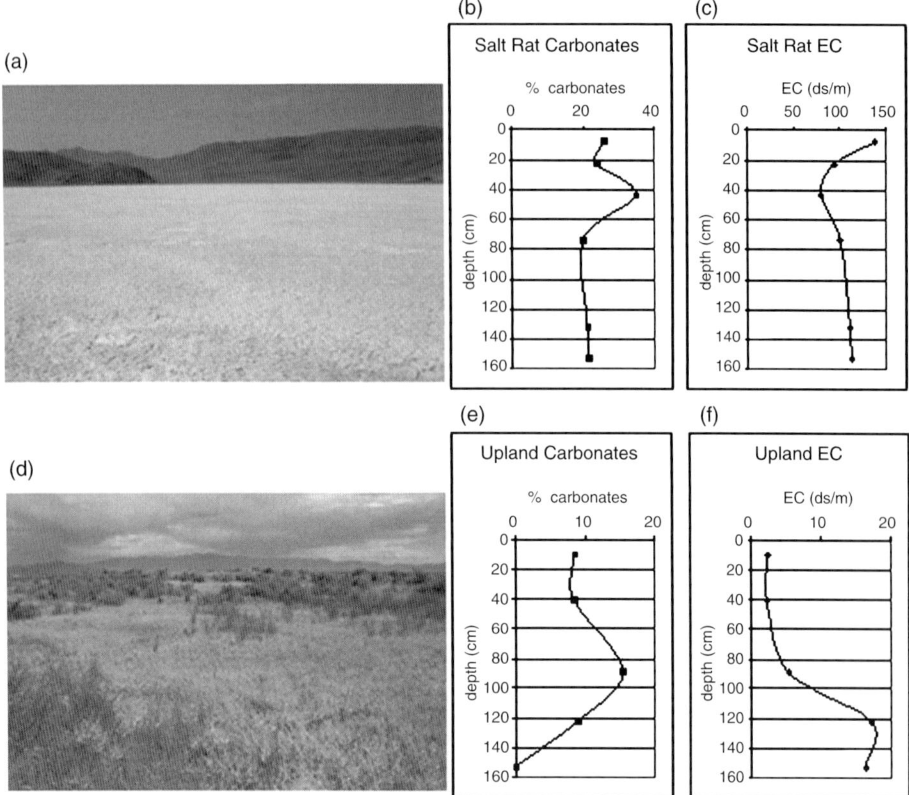

Figure 18.2. *Representative landscape, calcium carbonate profile and EC profile for the salt flat (a), (b), (c) and upland (d), (e), (f) land cover types.*

(in this case, two), is assigned to the central pixel of the filter window. A 3×3 pixel neighbourhood was used for the filter window. The fuzzy classification was processed using both the minimum distance to means and the maximum likelihood algorithms. The resulting images from the two different classification algorithms were compared to determine which algorithm best represented the variation in the classes across the ESA. The classified image was validated by visual field inspection and a priori knowledge of the field area. An accuracy assessment was conducted using 135 random points and an error matrix (Congalton and Green, 1999). The land cover type class composition for the ESA study area was also calculated (Table 18.1). These data will be used in the final survey updating process to refine soil map-unit composition throughout the ESA survey area. Spatial distribution of the map units was also updated based on the results of the classification.

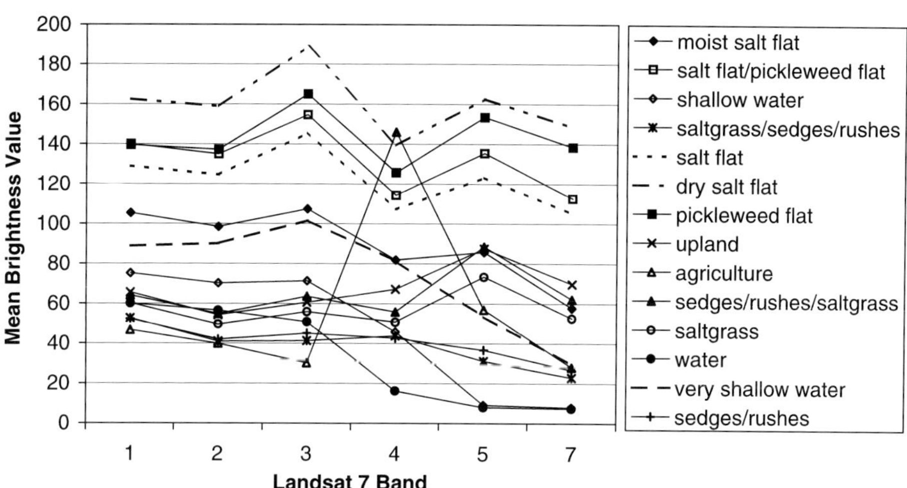

Figure 18.3. *Mean spectral signatures of the final 14 classes resulting from original 90 training sites.*

18.3 Results and discussion

18.3.1 Spectral signature analysis

The OIF calculation suggested that the combination of Landsat 7 bands 3, 5 and 7 would provide the most information for the ESA image. When the spectral signature separability was evaluated for the OIF-band combination, a transformed divergence value of 1860 resulted. When compared to the transformed divergence value calculated for all six bands, 1982, the OIF combination value was lower. When compared to other 2-, 3-, 4- and 5-band combinations with transformed divergence values ranging from 1860 to 1890, the OIF band combination value was very similar. The combination of all six bands resulted in the greatest separability, as indicated by the highest value of transformed divergence. Therefore, all six Landsat bands were used for the supervised classification.

After evaluating class statistics and mean reflectance in each band for the original 90 training sites, like signatures were merged. This process resulted in a final signature set containing 14 classes (Fig. 18.3) that would be used for the image classification. The final 14 classes were salt flat, moist salt flat, dry salt flat, salt flat-pickleweed flat, pickleweed flat, saltgrass, sedges-rushes-saltgrass, saltgrass-sedges-rushes, sedges-rushes, upland, agriculture, water, shallow water and very shallow water. Compound classes, such as saltgrass-sedges-rushes, indicate that the majority of the cover in the class is the first cover type listed (i.e. saltgrass).

The evaluation of the histograms for each of the resulting 14 classes showed a mixture of normal and multimodal distributions. Within a single class, variation in data distribution existed between the six Landsat 7 bands. In some bands, the data distribution was normal, and in other bands, it was multimodal.

18.3.2 Classification

Given the mixture of normal and multimodal data distributions among the classes, the fuzzy classification was processed using both the minimum distance to means and the maximum likelihood classification algorithms, to determine which would yield the best results. Both classification algorithms produced reasonable results; however, the results of the minimum distance to means classifier seemed to represent the variation on the landscape better than the maximum likelihood classifier. This assessment is based purely on the field experience and knowledge of the soil scientist performing the classification, and a bona fide accuracy assessment would be the preferred method to determine the optimal classification algorithm.

Plate 18 (see Colour Plate Section) shows the thematic map resulting from the supervised fuzzy classification and the application of the fuzzy convolution filter. Identifying the extent and variability of the salt-flat and pickleweed-flat classes is a top priority, as together they cover a significant portion of the ESA, are highly variable, and least accessible (Table 18.1). Validation of the final classified image was completed by visual field inspection, a priori knowledge of the ESA and an error matrix. The error matrix used 135 points randomly stratified over the 14 final classes. Each point was visited in the field, and land cover type recorded. The error matrix determined the overall map accuracy to be 88%, which indicates strong agreement between predicted and observed classes.

18.4 Discussion

Supervised classification of Landsat 7 imagery for the ESA produced a reasonable thematic representation of the 14 land cover type classes across the landscape. This classification also provided much information regarding the spatial distribution of the classes and the variability within the major classes of salt flat, pickleweed flat, sedges-rushes and saltgrass. The salt flat and pickleweed flat classes were of particular interest because together they cover a significant proportion of the area and are least accessible; also, their spatial variability was not well understood. The classification gave new information about the spatial variability of these land cover types, and the information was presented in a context that can be interpreted with greater ease than aerial photos or DOQs. This information enhanced the understanding of soil-landscape relationships in the ESA and allowed for refinement of existing map units and polygon lines (Plate 18, see Colour Plate Section). Supervised classification also provided a method for refining wet and saline map units based on the spectral characteristics of the area. Examining the spectral characteristics and relating them to existing knowledge of the area produced quantitative, physical data on which to base map unit refinement decisions. The quantitative data can then be archived and enhanced as other data become available.

Another interesting application of remotely sensed data is the ability to examine the temporal variability of an area. This is of particular interest in the ESA due to the dynamic response of soil moisture and salinity to fluctuations in local water table caused by natural and anthropogenic controls. Figure 18.4 shows four Landsat 7 images that highlight changes in the landscape over one season (April and June, 2000) and over several years of drought (2000–2003). The dynamic nature of the ESA landscape presents an interesting challenge for map unit refinement. However, temporal variation in the landscape has the potential to be considered in map unit refinement by comparing Landsat 7 images and classifications of spectral characteristics. Changes in the proportional area for each class can be calculated and presented as documentation for map unit

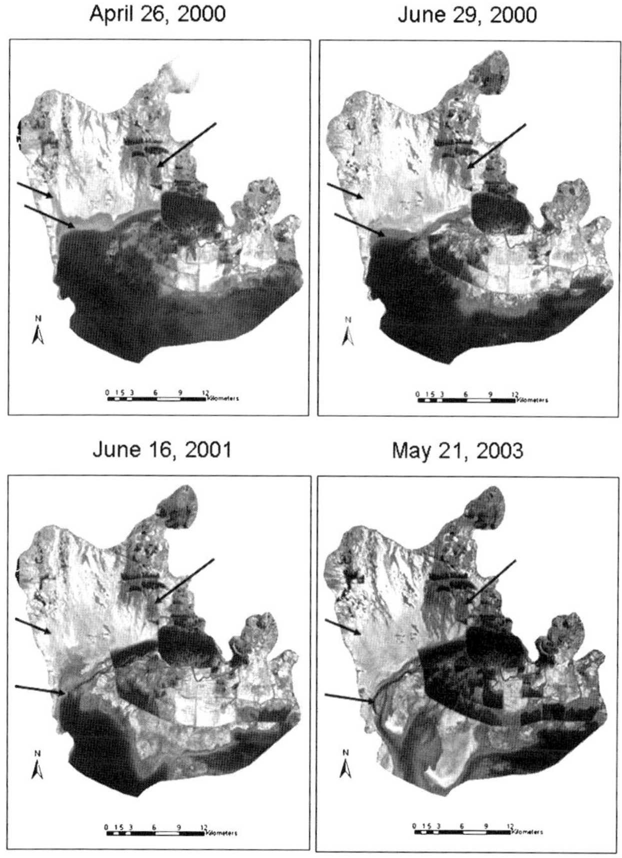

Figure 18.4. *Temporal variation of the ESA due to natural and anthropogenic controls.*

refinement. Ultimately, a single pixel and its classification changes related to soil moisture and salinity can be tracked through time.

18.5 Conclusions

Supervised classification of Landsat 7 imagery for the ESA provided insight into the spatial variability and extent of major land cover types across the landscape. Information gained from the classification was used to enhance and refine the original soil map units by establishing the relationship between the new set of land cover classes and soil properties. The final soil map resulting from this process will provide users with more complete information regarding the wet and saline resources in the ESA than did the original soil survey. The use of remotely sensed data and techniques, such as supervised classification, provides soil scientists with tools to create robust soil maps that present soil-landscape relationships based on quantitative, physical data.

Acknowledgments

1. Utah Agriculture Experiment Station, Journal paper Number 7739.
2. USDA Natural Resources Conservation Service.

References

Chadwick, R.S., Barney, M.L., Beckstrand, D., Campbell, L., Carley, J.A., Jensen, E.H., McKinlay, C.R., Stock, S.S., Stokes, H.A., 1975. Soil Survey of Box Elder County, Utah, Eastern Part. U.S. Government Printing Office, Washington D.C.

Congalton, R.G., Green, K., 1999. Assessing the Accuracy of Remotely Sensed Data: Principles and Practices. CRC Press, Boca Raton, Florida.

Jensen, J.R., 1996. Introductory to Digital Image Processing; A Remote Sensing Perspective, 2nd edition. Prentice-Hall, Englewood Cliffs, New Jersey.

Lewis, D.T., Seevers, P.M., Drew, J.V., 1975. Use of satellite imagery to delineate soil associations in the Sand Hills region of Nebraska. Soil Sci. Soc. Amer. Proc. 39, 330–335.

Metternicht, G., Zinck, J.A., 1997. Spatial discrimination of salt- and sodium-affected soil surfaces. Int. J. Remote Sens. 18, 2571–2586.

Rao, B.R.M., Ravi Sankar, T., Dwivedi, R.S., Thammappa, S.S., Venkataratnam, L., Sharma, R.C., Das, S.N., 1995. Spectral behaviour of salt-affected soils. Int. J. Remote Sens. 16, 2125–2136.

Roudabush, R.D., Herriman, R.C., Barmore, R.L., Schellentrager, G.W., 1985. Use of Landsat multispectral scanning data for soil surveys on Arizona rangeland. J. Soil and Water Conser. 40, 242–245.

Singh, A.H., 1994. Monitoring change in the extent of salt-affected soils in northern India. Int. J. Remote Sens. 15, 3173–3182.

Stokes, W.L., 1988. Geology of Utah. Utah Museum of Natural History and Utah Geological Survey. Salt Lake City, Utah.

Soil Survey Staff, 2003. Keys to Soil Taxonomy, 9th edition. U.S. Government Printing Office, Washington, D.C.

Developments in Soil Science, volume 31
P. Lagacherie, A.B. McBratney and M. Voltz (Editors)

Chapter 19

PRODUCING DYNAMIC CARTOGRAPHIC SKETCHES OF SOILSCAPES BY CONTEXTUAL IMAGE PROCESSING IN ORDER TO IMPROVE EFFICIENCY OF PEDOLOGICAL SURVEY

J.-M. Robbez-Masson†

Abstract

Producing a preliminary cartographic sketch of soil-forming factors is required by pedologists while preparing their ground survey. With this aim, we define "soilscapes" (*pédopaysages*) as complex spatial objects observable at a determined spatial resolution and potentially recognisable by an original composition of landscape elements, the co-occurrence and topological relationships of these elements or the size of their spatial pattern.

The proposed approach describes each point to be mapped by characteristics of its spatial neighbourhood and allocates it to pre-defined reference classes using a mathematical distance. The result is a numeric dashboard with continuous and crisp maps, graphs and cartographic and statistical quality indicators.

An application example presents a small-scale cartographic sketch, as required by a French inventory program (Inventaire, Gestion et Conservation des Sols de France, IGCS), delineating soilscape units on a 50,000-ha area of southern France from 50-m resolution lithological map and topographical derivatives. A satisfactory agreement is found with an existing soilscape map made by trained surveyors.

Such an approach intends to be a quantitative, iterative and interactive tool in order to (1) model the pattern of soil-forming factors in a generic, explicit and reproducible way, and (2) enrich and hasten the field work.

19.1 Introduction

Soil survey of countries like France is still incomplete. This statement gave birth to the "Inventaire, Gestion et Conservation des Sols de France" (IGCS) program, co-led since 1990 by the French Ministry of Agriculture and INRA. IGCS aims to provide, for the whole French nation, an operational knowledge of the soil heritage for protection and durable use. It mainly relies on the generation of the Référentiel Régional Pédologique map delineated at the 1:250,000 scale and representing the spatial organisation of soils in the form of soilscapes named "pédopaysages" (Favrot et al., 1994).

†Deceased 2005.

Making such an inventory requires costly field sampling of accurate but local observations (auger holes, profiles and reference sectors). The cost and duration of elaborating such knowledge lead to the idea of increasing their utility by transposing them to broader areas, and to optimise their location by means of a sampling scheme that takes existing field knowledge explicitly into account.

With this aim, the conventional surveyor usually compiles a set of available maps related to soil-forming factors in the form of a cartographic sketch, which will then be put to test in the field.

Nowadays, the advent of Geographical Information Systems (GIS) and image processing systems, linked to the availability of accurate data sources, change our capabilities and promise possible assistance to these tasks by computers. A GIS procedure allowing a prior segmentation of the area to be mapped and an assessment of its quality seems to be an important and realistic step towards a first cartographic sketch to be generated and further refined in the field. Moreover, geomatics should add a missing dynamic dimension to this sketch.

Prior work has modelled and implemented this segmentation by using the concept of landscape.

(1) Thus, we shall briefly review the modelling principle that we have been using.
(2) Then, we will show, by a concrete example of segmentation in a rather complex area, how the procedure can be used by a pedologist for *quantitatively predicting soilscape classes*.
(3) Finally, we will argue that such a task is aimed to improve the efficiency of a conventional survey, but can also be understood as the production of new quantitative covariates describing spatial patterns of soil-forming factors.

19.2 A principle for segmentation of landscapes

Making a cartographic sketch firstly consists in choosing the variables that will be used: the major ones are the main soil-forming factors that can be supposed as time-stable at the scale of the soil forming process. Regional monographs, helped by local experts, allow us to determine the easily available data playing this role: relief, geology, land cover and geomorphology (Brabant, 1989). The joint presence of given values for these variables at each point defines areas named *landscape elements* with a generic meaning (Girard, 1983), or landscape facets, landform units or landform elements when dealing only with geomorphology (MacMillan et al., 2000; Schmidt and Hewitt, 2004).

The simple point-by-point overlay of cartographic layers in a GIS is not sufficient for a compliant physiographic synthesis of landscape elements, for several reasons:

- The more variables that are overlaid, the larger will be the number of possible combinations, which will make it difficult to correctly display the obtained synthesis; this will yield an unreadable map having a salt-and-pepper effect, unusable by a surveyor.

- The relationship between each point and the spatial context in which it takes place are not taken into consideration; it is impossible to identify complex objects such as soilscapes with the composition of typological units that constitute it, for example "10% fersialsols and 90% lithosols", or according to the functional links between these units "valley soils covered by sloping colluvial soils";
- The point-by-point overlay is very sensitive to the respective georeferencing errors between the layers to be overlaid.

There have been several approaches developed in the recent years aimed at defining spatial segmentation models that are compliant with a contextual approach. They are based on a landscape paradigm as defined by (Hénin, 1993), *"recording of the synchronous presence and of the relative disposition of some objects in a given space, having specific properties, delineated by a visual field covered from a given point: the viewing point"*. The definition of Hole (1978) will be retained here for dealing with the pedological view on the landscape, named *soilscape:*

"A soilscape is the pedologic portion of a discrete stretch of terrain, that is at least partially covered with soil, as seen from a particular vantage point, but also as interpreted by the observer, who carries a mental image of a 'view from the aircraft' and a fund of knowledge about the data. To understand a soilscape, topographic, geologic, hydrologic, biotic, and pedologic studies are needed, as well as those of human impact on the environment". This object, which can be described and segmented, has been used by conventional pedologists for many years for making "reconnaissance" maps.

19.2.1 Define and segment the landscapes: a framework

Some authors consider that landscape (and thus – *soilscape*) can validly be characterised by taking into account the distribution of a descriptor in the spatial neighbourhood of the current point (Robbez-Masson et al., 1995; Wharton, 1982). This proposed approach, named Clapas, requires several steps: (1) choose the relevant soil-forming factors; (2) define unambiguously the neighbourhood in which the relevant variables will be acquired (*"landscape is what's lying around you"*) and (3) determine which aggregation criterion of this distribution will be used. Then one needs to (4) compare these measurements obtained at each point in order to (5) decide to which pre-defined soilscape category to assign the point, and finally (6) validate the obtained result.

19.2.2 Choosing relevant soil-forming factors (1)

Let $v(x)$ be a variable describing the elementary soilscape at each site x, that is the point-to-point combination of soil-forming factors. $v(x)$ is a categorical variable taking its values in $\{v_1, v_2, \ldots, vi, \ldots, v_p\}$, the set of the p elementary soilscape classes of the region. Each elementary soilscape class v_i is defined by a unique combination of soil-forming factor classes on a point-to-point basis. Soil-forming factor classes are either mapping units (e.g. geological units, landform facets or land-use classes) or derived from a pre-classification of quantitative variables (e.g. elevation, slope gradient of a DEM and NDVI index). The pedologist is responsible for choosing such adequate variables and classes. p classes of multivariate objects will then be identified, p being less or equal to the product of the number of factors and the number of categories of each soil-forming factor.

19.2.3 Describing the neighbourhood of each point (2 and 3)

The soilscape of site x is then defined by a "cover-frequency vector" (Wharton, 1982) $\mathbf{l}(x) = \{f_1(x), f_2(x), \ldots, f_i(x), \ldots, f_p(x)\}$, where $f_i(x)$ is the relative frequency of class v_i within an area delineated around x to include the set of neighbouring sites that must be taken into account for describing the soilscape at x. The size and the shape of this area are user-defined. In our approach, the neighbourhood area is elliptic, which requires the setting of the two main radii of the ellipse and its orientation. We are now able to describe the multivariate neighbourhood of each site as an information vector (Fig. 19.1).

19.2.4 Calculating distances between soilscapes (4 and 5)

The description of soilscape at site x can then be compared quantitatively with the one of a reference site x_0 by computing the distance $d(x, x_0)$ between the

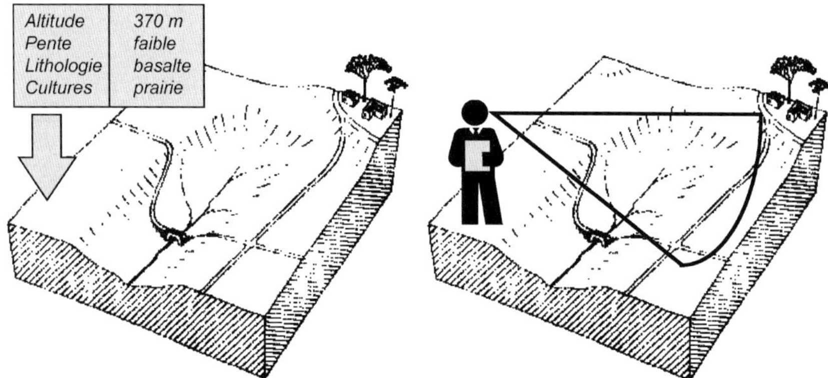

Figure 19.1. *A definition of landscape. A landscape can be defined through the observation of a set of characteristics (left) in a spatial, explicitly defined neighbourhood (right). The synchronous occurrence of these characteristics allows the definition of landscape elements, whose synthesis inside the spatial neighbourhood characterises the landscape of the point.*

vectors $\mathbf{l}(x)$ and $\mathbf{l}(x_0)$. The following Manhattan distance was preferred among distances dealing with qualitative variables because of its robustness (Borne, 1990; Gong and Howarth, 1992):

$$d(x, x_0) = \sum_{i=1}^{p} \left| f_i(x) - f_i(x_0) \right| \tag{19.1}$$

The distances calculated with Eq. 19.1 range between 0, that is same composition of classes within the explored areas, and 2, that is disjoined cover-frequency vectors with no classes in common. These distances can then be used for allocating individuals to pre-defined reference soilscapes. In Clapas, allocation is performed on a nearest neighbour basis.

Let X_0 be the set of the q selected reference soilscapes $x_{01}, x_{02}, \ldots, x_{0i}, \ldots x_{0q}$ obtained at various sites. The distance between soilscapes of these reference sites and the one of a given site x is defined as follows:

$$d'(X_0, x) = \min_{j=1,q} \left(d(x_{0j}, x) \right) \tag{19.2}$$

Plate 19 (upper, see Colour Plate Section) provides a simple example of a Clapas application. First, $p = 4$ elementary landscape classes are defined by combinations of soil-forming factor maps (Plate 19(upper, a)). Then, the soilscape of the black-dotted site of Plate 19(upper, a) is quantitatively described by a cover frequency vector histogram (Plate 19(upper, b)), which is computed from an elliptic neighbourhood of the site. Finally, the soilscape description of the black-dotted site is compared with both soilscape descriptions of reference sites 1 and 2 (Plate 19(upper, c)). In this example, the computed Manhattan distances reveal that the studied soilscape is closer to reference site 1 (distance = 0.311) than to reference site 2 (distance = 1.870). Then the soilscape at this location will be assigned to the first category. Reproducing this at each location will yield a segmented image containing both categories 1 and 2, depending on which class the point has been assigned to, according to the composition of its spatial context.

Let us give a simple example of the whole approach:

- Each point will be described by the composition of its neighbourhood, for example *"I'm at point M5, there are around me 80% basaltic rocks on weak slopes and 20% argilites on steep slopes"*.
- This composition will be then compared with each of a set of reference compositions, for instance computed from training sites created by an expert: *"the P12 soilscape is made of 85% basaltic rocks on weak slopes and 15% argilites on steep slopes"*.
- The comparison will yield a vector of soilscape distances allowing to assign each point to the soilscape whose composition is the nearest – *"point M5 belongs to the P12 soilscape"*.

19.2.5 Validation of the segmentation by an iterative approach (5)

Statistical quality of the segmentation can be assessed at each site x by the vector of distances $\mathbf{d}(x) = \{d(x_{01}, x), d(x_{02}, x), \ldots, d(x_{0j}, x), \ldots, d(x_{0q}, x)\}$ between the composition of the site neighbourhood and of the neighbourhood of each reference site. A simple way for pooling these distances is to keep the minimal distance $d'(X_0,x)$ (cf. Eq. 19.2) encountered in the distance vector, that is corresponding to the category at which the point has been assigned – *"point M5 got a rather satisfactory assignment"*

The minimal distance permits definition of the areas with well-assigned points, that is these where examination of existing documented information would allow a match of the observed case to a known one, and, in contrast, the poorly assigned ones. Thus, the density of the field-sampling scheme could be adjusted according to the value of this distance.

From the cartographic point of view, examination of sizes and shapes of the polygons obtained is another indicator of the sketch quality (Robbez-Masson, 1994).

The obtained knowledge can be integrated into an iterative and interactive framework (for instance, one can add a new reference into an area that has a bad statistical score at the previous iteration, get a new sketch and so on).

19.3 A case study: the cartographic sketch of the Lodévois region (France)

We tried to yield a sketch map of soilscapes, as required by a pedologist going into the field in the framework of the IGCS program. Such a program requires the graphical rendering of the map to be legible at a 1 to 250,000 scale. The Clapas procedure has been used to achieve this by synthesising a set of images characterising main soil-forming factors in a rather complex region, and generalising in one go the attribute, topological and spatial dimensions of the data.

In the Lodévois region (Hérault, France), researchers agree in considering slope (r component) and lithology (p component) as the most relevant and discriminant pedogenetic factors, while recognising the complexity of their spatial patterns. The sketch has been made from both accurate, $\rho = 50$ m resolution raster lithological and slope class maps available for ~50,000 ha. The point-by-point overlay of the available data would have produced 231,000 polygons (mean area of 0.21 ha) assigned to 84 cross-categories. Preparation and pre-processing used the capabilities of ArcView™ and Idrisi™ GIS.

Classification was used on 11 reference sites chosen in the 11 major lithological materials. The neighbourhood chosen was circular with a radius of 750 m. The distance image of the first segmentation shows the areas where a new reference site is needed, and allows the user to decide on splitting or merging of the existing ones (Plate 19 (lower, see Colour Plate Section)). Then a second iteration yielded 16 soilscapes and only 465 polygons, along with some

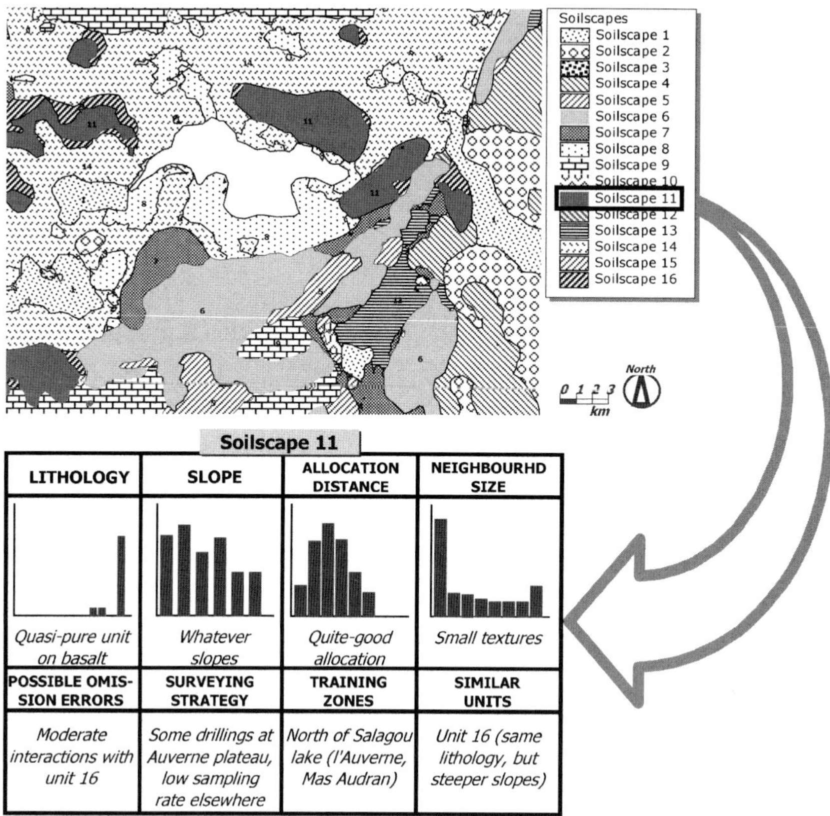

Figure 19.2. *The litho-morphological sketch of Lodévois (Hérault, France). This sketch was realised by assignment of cover-frequency vectors informed in terms of slope and lithology. A first iteration is done according to a set of reference areas. Examination of the results allows to give advice on the location for new reference areas, and to yield a second iteration, whose sketch is presented above, accompanied by a legend for each class under the form of an indicator panel. From left to right and from top to bottom, the panel indicates, respectively, the class definition according to lithology and slope, the global statistical performance of the assignment and the preferred size of the pattern to be retained. Moreover, it can indicate the cartographic and thematic interactions between classes, and some advice for a ground sampling (for validation of the landscape-soil relationship, and for pedological survey ss). Other indicators can easily enrich the legend, hence constituting a genuine interface for interactive modelling.*

charts and ancillary images. These allow the production of a preliminary legend, to show those areas that should be prioritised for field sampling (Fig. 19.2), those areas that are most representative of the soilscape units and those areas where confusions may occur with other soilscapes, and also highlight which calibration parameters seem to be the most predictive, and the cartographic synthesis level (Robbez-Masson, 1994).

The sketch thus produced was compared with the available soilscape map, the survey of which had required several working years by experienced surveyors (Bornand et al., 1992). Results proved to be encouraging since it was possible to find,with a 0.6 probability, the whole area of a known soilscape unit by looking at only 12.5% landscapes of the sketch. Furthermore, this approach improves the reproducibility of the segmentation by proposing a panel of statistical and cartographic indicators. Therefore, the results could be improved in three ways:

- Quality of original data (e.g. the lithological map does not mention the whole set of superficial rocks, like infra-metre-thick aeolian and colluvial deposits).
- Choice of an optimal reference set (e.g. how to decide when optimisation of statistical indicators is worthy).
- Modelling of surveyor's experience (e.g. fix weights between themes and define neighbourhood geometry).

19.4 Discussion and perspectives

Although the above case is virtual – the soilscape map was available at the time when the sketch was made – such an approach is actually being used for a subset of the Burgundy region of France, where a test case was done on a 200 km^2 area before being applied in the IGCS production process on the western part of the Nièvre Department (Gourmelon, 2003).

Further, this methodology and these tools have been successfully used in other pedological applications:

- delineation of complex soil map units under forest cover with aerial photographs and remote sensing imagery (Bornand et al., 1997);
- recognition of soil horizons on profile photographs (Wassenaar and Robbez-Masson, 1998) and
- use of soilscape distance by means of a fuzzy-set approach, in order to delineate representative areas around reference areas, as in Lagacherie et al. (2001).

Some issues have been encountered that need to be overcome for the easier use of such digital cartographic sketches.

(1) Currently, delineation of units (the shape of the soilscape polygons produced) is a consequence of individual classification of the pixels (image cells): taking jointly into account shape and content constraints while building the map should be fruitful, producing more eloquent maps.
(2) Pedological processes do not play a direct part in our model. A significant effort must be made in this direction.
(3) A more complete linking with the entire survey production process, including step-by-step refinement with field information, as in Ledreux et al. (1994) has to be done.

Although the Clapas software was originally designed for describing quantitatively, comparing and classifying soilscapes in small-scale soil surveys (Robbez-Masson, 1994), it was successfully used in other topics than soil science, particularly land cover and land use from satellite and aerial images: see for instance Malenovský (2001) and Meunier-Caldairou (1999)[1]. The software allows the choice of other neighbourhood sizes and shapes, other descriptors and metrics and other expressions of the results. Since its creation, it has been widely used as a more generic approach for other purposes (Robbez-Masson et al., 2001). A similar approach, more dedicated to remotely sensed data, was developed by Girard et al. (1991). This broader experience could be useful for further exploration of soil survey issues.

19.5 Conclusion

We attempted to quantitatively predict soilscape classes S_c at (x, y) location as a function f of its neighbourhood $(x + u, y + v)$ taking into account such *scorpan* variables as r and p: $\{S_c(x, y)\} = f(\{r, p\}(x + u, y + v))$ using the terminology of (McBratney et al., 2003). But, whereas most authors firstly overlay quantitative variables and finally get new objects (polygons, facets, ...) – for example (Burrough et al., 2000; Schmidt and Hewitt, 2004) – we developed an original and complementary approach by firstly defining elementary objects (or using existing ones) and then measuring their spatial pattern. This approach produces fuzzy surface ("taxonomic" distance images) from which deriving crisp but organised objects (soilscapes) is made easy.

Using this prior modelling of the soilscape, this procedure allowed us to: (1) propose locations of reference areas; (2) yield a segmentation into spatially complex units of which these reference areas are representative, in the form of a cartographic sketch; (3) validate a posteriori the choice of given reference areas (use of the soilscape distance image); (4) propose a "smart" field-sampling strategy, combining prior knowledge and modelling results and (5) merge the acquired knowledge into an iterative and interactive decision-helping system for soil surveyors.

Our approach produces a prior compilation of relevant cartographic data by quantitative modelling, in order to assist the soil surveyor into map delineation, legend creation and the field-sampling scheme. It is aimed to improve its efficiency, being a preliminary step to the creation of units into which soil attributes can further be measured or predicted. According to situations, other indicators derived from elevation (Burrough et al., 2000) or remote sensing (see examples in Chapters 17 and 25) can be overlaid for making landscape elements.

Soilscape distances to references can also be used in order to show boundary fuzziness or class overlap as proposed in Chapter 41. In fact, they should be considered as new quantitative covariates expressing knowledge about the soil-landscape model.

Indeed such tool approaches question our working habits, but they have to be considered as complementary to our field experience and knowledge by proposing a reactive indicator panels able to objectivise the delineation of soilscapes that are spatially complex) This new vision involves objective procedures and field experience interacting in a collaborative way (see a review in Chapter 22). Indeed there are still numerous issues concerning (1) the availability of relevant data sources and (2) the supply by experts of calibration parameters for such modelling. Nevertheless, these skills have to be overcome while soil scientists have more and more to give efficient, quick and rational responses concerning landscape planning and environmental management.

Notes

1. Complete list and links on http://sol.ensam.inra.fr/Produits/Clapas/Default.asp.

Acknowledgement

The author gratefully acknowledge Dr. Philippe Lagacherie for his fair and efficient reading of the first version of this paper.

References

Bornand, M., Barthès, J.P., Bonfils, P., 1992. Carte des pédopaysages du Languedoc Roussillon à l' échelle du 1/250000. Laboratoire de Science du Sol, Montpellier.

Bornand, M., Robbez-Masson, J.M., Donnet, A., Lacaze, B., 1997. Caractérisation des sols et paysages des garrigues méditerranéennes. Typologie et extrapolation spatiales par traitement d'images satellitaires. Étude Gestion Sols 4, 27–42.

Borne, F., 1990. Méthodes numériques de reconnaissance de paysages, application à la région du lac Alaotra, Madagascar. Ph.D. Thesis, Université Paris VII, Paris, 213pp.

Brabant, P., 1989. La cartographie des sols dans les régions tropicales: une procédure à cinq niveaux coordonnés. Sci. Sol. 27, 369–395.

Burrough, P.A., van Gaans, P.F.M., MacMillan, R.A., 2000. High-resolution landform classification using fuzzy k-means. Fuzzy Sets Systems 113, 37–52.

Favrot, J.C., Arrouays, D., Bornand, M., Girard, M.C., Hardy, R., 1994. Informatisation et spatialisation de la ressource sol: le programme "Inventaire, gestion et conservation des sols". Cah. Agric. 3, 237–246.

Girard, M.C., 1983. Recherche d'une modélisation en vue d'une représentation spatiale de la couverture pédologique. Application à une région des plateaux jurassiques de Bourgogne. Sols 12, 295.

Girard, M.C., Mougenot, B., Ranaivoson, A., 1991. Présentation d'un modèle d'organisation et d'analyse de la structure des informations spatialisées: OASIS, "Caractérisation et suivi des milieux terrestres en régions arides et tropicales". ORSTOM, Bondy, pp. 341–350.

Gong, P., Howarth, P.J., 1992. Land-use classification of SPOT HRV data using a cover-frequency method. Int. J. Remote Sens. 13, 1449–1471.

Gourmelon, J., 2003. Faisabilité de l'utilisation des méthodes de segmentation sous SIG pour une stratification assistée des paysages. Application à la réalisation d'un plan d'échantillonnage pertinent dans le cadre de la constitutiondes bases de données géographiques "Sols et Territoires de Bourgogne". D.E.S.S. Espace Rural et Environnement Thesis, Université de Bourgogne, Dijon, 54 (+15) pp.

Hénin, S., 1993. Le paysage, une entité pour l'appréciation du milieu? C.R. Acad. Agric. Fr. 79, 39–43.

Hole, F.D., 1978. An approach to landscape analysis with emphasis on soils. Geoderma 21, 1–23.

Lagacherie, P., Robbez-Masson, J.M., Nguyen The, N., Barthès, J.P., 2001. Mapping of reference area representativity using a mathematical soilscape distance. Geoderma 101, 105–118.

Ledreux, C., Jeansoulin, R., King, D., Lagacherie, P., 1994. Un raisonnement spatial symbolique-numérique pour la reconnaissance d'unités de sol dans SAPRISTI, Journées de la Recherche CASSINI, Lyon, pp. 35–43.

McBratney, A.B., Mendonca Santos, M.L., Minasny, B., 2003. On digital soil mapping. Geoderma 117, 3–52.

MacMillan, R.A., Pettapiece, W.W., Nolan, S.C., Goddard, T.W., 2000. A generic procedure for automatically segmenting landforms into landform elements using DEMs, heuristic rules and fuzzy logic. Fuzzy Sets Systems 113, 81–109.

Malenovský, Z., 2001. Possibilities of using satellite data for mapping the vegetation formation types in the forested area of Mediterranean region. J. Forest Sci. 47, 114–123.

Meunier-Caldairou, V., 1999. Analyse des transformations de l'information dans des images satellitales classées au cours de procédures de généralisation de leur contenu : influence de différents niveaux de précision cartographique et de nomenclatures de type "occupation du sol", "utilisation du sol" et "motif paysager". Ph.D. Thesis, Un. Paul Sabatier, Toulouse III, Toulouse, 135pp.

Robbez-Masson, J.M., 1994. Reconnaissance et délimitation de motifs d'organisation spatiale. Application à la cartographie des pédopaysages (recognition and segmentation of spatial patterns. Application to survey of soilscapes). Ph.D. Thesis, Ecole Nationale Supérieure Agronomique de Montpellier, Montpellier, 161pp.

Robbez-Masson, J.M., Borne, F., Girard, M.C., 1995. Description et Segmentation de Motifs d'Organisation Spatiale. Application à l'obtention d'esquisses paysagères. In: INRA (Ed.), Actes du Colloque INRA "Phénomènes Spatiaux en Agriculture", La Rochelle (France), pp. 65–79.

Robbez-Masson, J.M., Foltête, J.C., Cabello, L., Flitti, M., 2001. Prise en compte du contexte spatial dans l'instrumentation de la notion de paysage. Application à une segmentation géographique assistée. Rev. Int. Géomatique 9, 173–195.

Schmidt, J., Hewitt, A., 2004. Fuzzy land element classification from DTMs based on geometry and terrain position. Geoderma 12, 243–256.

Wassenaar, T., Robbez-Masson, J.M., 1998. Application of remote sensing techniques on soil profile photographs : a new area in soil profile description? IUSS World Congress of Soil Science, Montpellier, p. 12.

Wharton, S.W., 1982. A contextual classification method for recognizing land use patterns in high resolution remotely sensed data. Pattern Recog 15, 317–324.

Developments in Soil Science, volume 31
P. Lagacherie, A.B. McBratney and M. Voltz (Editors)

Chapter 20

CONCEPTUAL AND DIGITAL SOIL-LANDSCAPE MAPPING USING REGOLITH-CATENARY UNITS

R.N. Thwaites

Abstract

This chapter emphasises the geomorphological component to landscape modelling and simulation for soil resource assessment. This concept views the soil resource as a dynamic 3-dimensional geomorphological landscape and presents information as regolith–terrain (R–T) data in the context of Regolith-Catenary Units (RCUs) in an explicit and repeatable process. RCUs are 3-dimensional R–T systems and are viewed as composite R–T entities. They are described through soil geomorphic techniques. Observations of the regolith are limited, so conceptual models of the R–T as RCUs are necessary. Spatial expression of RCUs is achieved through the predictive capabilities of digital terrain analysis using derivative functions from digital terrain models. Landform attributes are combined as RCU components through a set of fuzzy rules to form simulated RCUs. These are closer to the conceptual and linguistic definition of hillslope components and their geomorphic processes. The concept is applied to predicting soil and regolith attributes derived through fuzzy classification to produce continuous data surfaces for a forested terrain in southeast Queensland. These then serve as decision support 'maps' of the R–T attributes that relate directly to site-specific management of the forest land resource

20.1 Introduction

An understanding of how soil landscapes function is essential to developing solutions for an increasing range of environmental problems. We require models that describe the complex relationship of the land surface and the regolith (which includes the soil material layers) and include the geomorphic processes acting within them. We also require models that connect with the functionality of the landscape for land use management and planning purposes in a practicable manner. The *regolith–terrain* (R–T) relationship, as a more holistic concept of the soil-landscape system, can be expressed through the spatial observation of the landscape from remote sensing and field survey, and the infrequent field observations of the regolith itself. The following describes a modelling framework that captures the important 3-dimensional processes operating both on the land surface and in the regolith, which are expressed in terms of fuzzy classification using the semantic rules to define them. This allows for more realistic expression of the R–T in spatial digital models as Regolith-Catenary Units (RCUs), whilst also focussing on the relevant attributes as the essential environmental covariates. These are then included within a predictive spatial model of R–T qualities relevant to site-specific management in a forestry environment.

The notion builds on existing geomorphological concepts that categorise the landscape as readily recognisable and repeatable landform units or components, for example hillslope summits, valley backslopes, footslopes and free faces. These are somewhat vague and semantic descriptions based on geomorphic systems processes as well as morphology. These have been explored through digital terrain modelling also by Burrough et al. (2000), MacMillan et al. (2000) and Schmidt and Hewitt (2004), also expressed with fuzzy classification. Thus they are viewed as major covariates for the soil/regolith attributes that are relevant to land resource management. They are therefore functional landform components, which is the focus of this study for their more explicit and precise expression within 3-dimensional catenary context. It forms part of a Ph.D. study (Thwaites, 2001) that was undertaken to develop the inclusion of site-specific management for upland forestry. Devising the expression of functional landform components in the RCU context is the conceptual basis for the prediction of R–T attributes for land resource assessment and planning purposes using an array of further environmental covariates from field survey, remote sensing and soil-geomorphological theory.

20.2 The regolith–terrain framework

The framework required for defining the R–T system must be predictable and consistent. It should be formed from the factors that have influenced the development and the distribution of regolith materials. The resulting landscape would represent sequences or juxtapositions of the R–T system components or spatial units, where changes in one R–T sub-system may impact upon an adjacent or subsequent R–T sub-system.

The 'catena' concept (Milne, 1935) expresses a natural R–T relationship that emphasises the topography and parent material factors as well as geomorphological processes. Adjacent soil types and regolith materials link at different elevations along a hillslope by lateral migration of biogeochemical elements in a largely predictable manner in the soil landscape. Specific attributes of the regolith along this gradient covary with the associated terrain position, both in absolute and relative terms.

The conventional catena concept is expressed 2-dimensionally. Consequently it is perceived as steady state; it only truly represents a straight thalweg or a hillslope cross-section with straight overland- and through-flowlines. If the catena is viewed in broader scales, in three dimensions, the elementary R–T unit that parallels the catena is the geomorphic drainage basin unit, and inter-basin units, of the R–T system. This generic unit is here termed the *'regolith-catenary unit'* or *RCU* (Thwaites, 2000). Like the catena, the RCU consists of a unique composition of soil-geomorphological properties in a recognisable and predictable pattern that makes it distinctive and 3-dimensional.

20.2.1 The regolith-catenary unit

The RCU is adopted as the fundamental organisational unit in the R–T system, which, as it equates with the hydrological drainage basin, neatly agrees with the concept of a fundamental 3-dimensional functional landscape unit at the same scale (Figure 20.1).

The material components of the RCU are the mineral skeletal material, the soil plasma and the soil solution. The processes are those of the pedologic system, as defined by Simonson (1959): addition, removal, transfer (translocation) and transformation of materials, as well as the hydro-geomorphic processes of the hillslope surface and near surface (erosion, deposition, mass movement, surface flow, throughflow and deep drainage).

The lateral boundaries to the conventional catena in the landscape cannot be defined. The boundaries to the RCU can be. They are perceived to be 'fuzzy', as the materials and processes of one RCU component (e.g. hillcrest and backslope) will merge, gradually, to a greater or lesser degree through morphology and processes, with adjacent components.

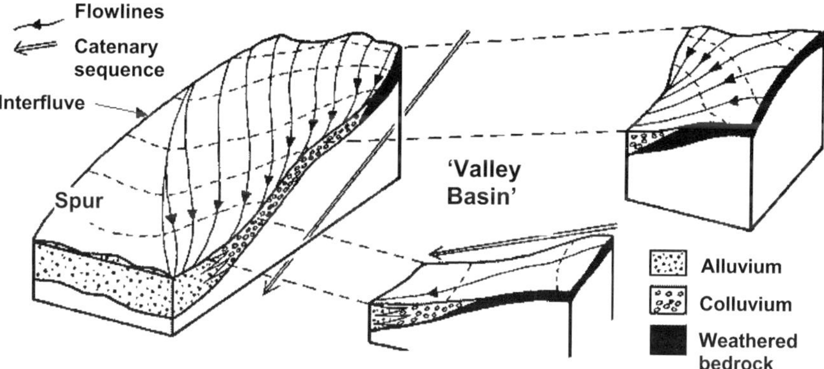

Figure 20.1. *Cross-sections of the Regolith-Catenary Unit (here depicted as the 'valley basin' unit) showing surface process and catenary relationships in a 3-dimensional form (adapted from Huggett, 1975).*

As a functional soil-geomorphic unit, the RCU relates to land use systems. Therefore, it is possible to equate the RCUs to functional land management units for not only traditional land use planning but also towards the more specific goal of site-specific management for any land use system.

The main advantages of developing RCUs are:

- they possess explicitly defined elements and boundary conditions,
- they are functional, 3-dimensional, open geomorphological systems,
- they conform to an open system of energy and matter flux and conservation,
- the relationships between elements and boundaries can be established through empirical hypotheses and conceptual models,
- their dynamics of form and process may be simulated by digital terrain modelling and
- they can therefore be subjected to digital spatial analysis.

The terrain imperative to this model is the third dimensional component of water and solution/suspension movement. If conductivity of surface and sub-surface water is assumed to be isotropic in the x and y directions then the flowline path will be determined by slope gradient and shape. Hence concavity leads to flow concentration and convexity to flow divergence in both profile and plan curvature, as expressed by Huggett (1975) amongst others. This is the basis to digital terrain model analysis (Wilson and Gallant, 2000). In addition, the assumption is that there will be convergent and divergent *infiltration* and *throughflow* in isotropic soil and regolith materials.

The definition of the RCU is:

An area of the earth's surface encompassing a volume of earth surface materials which is delineated by surface drainage features and represents a characteristic system of soil-geomorphological processes within a hierarchy of scales (Thwaites, 2000).

The RCU concept requires:

- boundary constraints: the land surface; watersheds; the weathering front or fresh bedrock, or 'non-regolith',
- that the RCU forms part of a more extensive (hierarchical) R–T system,
- that it functions as a natural open system (endorheic drainage basins can also be accommodated),
- that it can be quantified.

20.3 The regolith-catenary unit components

The RCU components (Plate 20a (see Colour Plate Section)) can be:

- open drainage features broadly termed 'valley basins' (RCU_{vb}),
- closed drainage features, or 'closed basins' (RCU_{cb}),
- crestal inter-basinal units, or 'summit surfaces'(RCU_{ss}), for example hill crests/ridge crests (RCU_{ssc}), plateaux (RCU_{ssp}),

- other inter-basinal units, or 'inter-basins' (RCU_{ib}), divided into two subsets: erosional (RCU_{ibe}), for example spur-ends, cliffs; and depositional (RCU_{ibd}), for example floodplains, terraces.

Delineation of any of these units is scale-dependent, particularly upon the observational or sampling grain of the investigation and data resolution.

The most important and dominant of the RCUs is the valley basin (RCU_{vb}), this represents a 3-dimensional catena (see Figure 20.1). It is the fundamental soil-geomorphic unit of the landscape, dominating in erosional landscapes, but commonly occurring in predominantly depositional and residual landscapes.

Acknowledging scale-dependency the RCU generically possesses some or all of the following (Plate 20a):

- a drainage outlet to a subsequent drainage system,
- a dominant *profile* (long) axis which has a *proximal* end (the crest of the headslope) of higher altitude than the *distal* end (the drainage outlet at either the drainage confluence or at the higher order drainage floodplain margin), and a shorter, sub-dominant *cross* (short) axis that is generally normal to the profile axis,
- relative relief between the proximal and distal ends greater than 5% of its profile axis (the relative relief of the valley basin maybe more than that of its profile axis),
- its boundary defined by incipient water-shedding slopes (ssc) which encircle the drainage basin,
- a *core* (vbc) defined by the channel(s) of the surface drainage features,
- a predominantly water-concentrating concave plan-profile (basinal) landform,
- surface and sub-surface water-concentrating drainage processes,
- simple, convexo–concave, or coarse-scale complex sideslopes (vbs),
- predominantly 'transitional' (erosional–depositional) slope processes, with erosional processes dominating the peripheral regions and depositional processes dominating the 'core' (drainage line) regions and
- an overall size that has resource management functionality, which can be variably defined.

The closed (surface) drainage system (RCU_{cb}) has a similar definition to that of the RCU_{vb}, except that it does not have a drainage outlet to a subsequent drainage system, and that depositional processes are more prevalent (particularly around the *sink* region where the drainage concentrates).

The 'summit surface' (RCU_{ss}) is either a level or convex landform with predominantly 'residual' (vertical drainage processes, minimal erosion or deposition) or 'erosional' geomorphic processes. It possesses commonly curvilinear plan-form as an inter-fluve, or drainage divide (RCU_{ssc}), although broad plateaux (RCU_{ssp}) may take a variety of plan-forms.

The RCU_{ib} may have several forms but is conveniently divided into two types:

- *Erosional* (RCU_{ibe}). Dominated by the spur-end (or nose) landform between valley basins and terminating at the subsequent drainage line (or its floodplain). Other erosional forms are cliffs.
- *Depositional* (RCU_{ibd}). Floodplains with complex micro-topography, terraces and ephemeral features as well as large, irregularly formed debris slopes, for example talus and scree.

20.4 The regolith-catenary unit and regolith–terrain analysis method

Regolith-Catenary Units can be defined digitally through terrain analysis. The satisfactory application of this method is scale-dependent and the resolution and accuracy of the digital elevation model (DEM) and the original elevation data are ultimately limiting.

Regolith-Catenary Units can be delineated intuitively from aerial photographs and topographic maps, with field checking. The finer the scale of investigation (i.e. the finer the grain and the smaller the extent) the more difficult it is to delineate RCUs precisely with line boundaries. Boundaries will always be arbitrary to some degree, depending upon the interpretation of landform shape and dominant surface process. This problem is exacerbated at finer scale to a point where delineation becomes redundant because of the 'crisp' categorisation, or dichotomy, in the concept (the boundary defines absolutely *either* one component *or* the other). The components of RCUs, like other R–T data, are not naturally discrete entities: imprecise definition of RCUs (or any terrain features) can be handled effectively by employing fuzzy sets, and their digital expression can be enhanced in the process.

A 'fuzzy' classification of RCUs is generally more acceptable (although it can be argued to be unnecessary for the broadest-scale purposes). Fuzzy RCUs have been developed and tested as part of a R–T study for the Benarkin Key Area (BKA) in southeast Queensland to predict relevant regolith attributes for site-specific management purposes at 1:20,000 mapping scale for the Queensland Department of Natural Resources Forestry Division.

Regolith–terrain patterns can be predicted from surrogates for regolith attributes that relate to the factors determining R–T variation in the landscape (i.e. hillslope processes and parent material, as well as climatic and biological influences over time–again dependent on scale). Representativeness was achieved through exploratory data analysis of relevant landscape components such as lithology, soil, catenas and landforms, remotely sensed data (airborne gamma-ray spectrometry data in this case), as well as primary and secondary derivatives from the DEM using regression tree and partitioning analysis (Thwaites, 2001). Rules are then constructed from the knowledge-based conceptual models, using the landscape component data and surrogate data, to produce digital models of the RCUs.

20.4.1 A case study in a forested terrain

The BKA study site lies within Benarkin State Forest in southeast Queensland, Australia, some 125 km northwest of Brisbane (Figure 20.2). This undulating and dissected plateau serves as the northwest headwaters to the Brisbane river, the major drainage basin in southeast Queensland.

The BKA represents around 10 km^2 of a partially dissected, and actively eroding, exhumed tertiary geomorphic surface, modified by a remnant veneer of deeply weathered basalt in the western and northern parts.

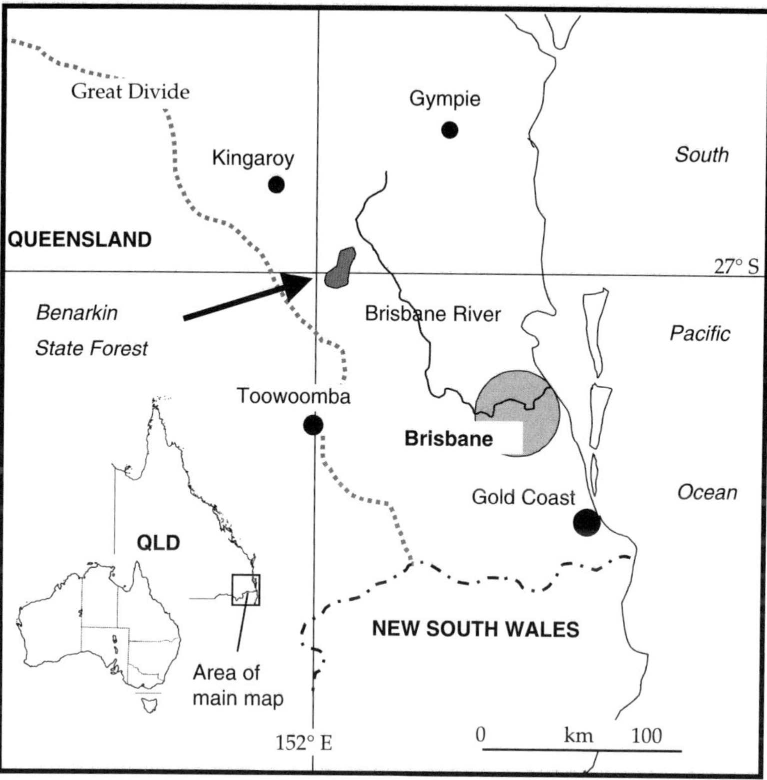

Figure 20.2. *Location of the Benarkin State Forest in southeast Queensland, The Benarkin Key Area is situated on the eastern margin of the State Forest.*

The drainage pattern in the BKA is convergent dendritic, and represents the 4th order, headward component of a convergent tributary network to the Brisbane river.

The methodological model for the R–T analysis and regolith attribute prediction comprises a synthesis of four major analytical components applied to the R–T system: remote sensing analysis, soil-geomorphic analysis, GIS analysis, and RCU analysis with soil–regolith attributes as the geomorphic covariates.

The first three components were not performed in a linear sequence but were synthesised to allow the RCU analysis and soil–regolith attribute prediction to be undertaken.

This process led to the conceptualisation of the covarying relationships to the independent variables from the topographic data, remotely sensed data and field data, for the RCUs. The models are based on the conceptual understanding of the soil-geomorphic processes in the landscape and on the R–T model. These conceptual models were expressed through simple, schematic catenary profile sketches of representative units, and the use of notional relationship (or sufficiency) diagrams–expressing the perceived covariate relationships, to develop rules for the fuzzy set analysis.

A 20-m DEM for the BKA was created with ANUDEM v4.6.1 (Hutchinson, 1989) using elevation and stream input data from published 1:25,000 topographic mapping in ArcInfo GIS. Drainage enforcement and 'sink' removal was used to render the elevation surface as faithfully as possible as hydrologically correct. The DEM was then iteratively corrected and refined against the original contour and drainage data. A series of DEM terrain derivatives were generated using TAPES-G (Gallant and Wilson, 1996), UPSUM-G (a program within the TAPES-G set) and 'ERA tools' (R. Searle, Department of NR&M, Queensland, personal communication, 1998), a set of Avenue script macros for ArcView GIS.

All the themes derived from the DEM and GIS analysis were classified for subsequent input into the fuzzy analysis. The class thresholds were iteratively manipulated to: either (a) provide an equal frequency distribution of classes (e.g. elevation classes), or (b) provide a normal distribution of classes, or (c) reflect the trends of the specific landscape. The method of classification depended on the derivative type and the trend of its original frequency distribution.

These classes were combined with remotely sensed data, field survey and geomorphological data to generate digital models of RCUs, both by 'crisp' (discrete) classification and 'fuzzy' classification means. It was also done to:

- provide grid cell-based data for terrain attributes to correlate with field regolith attributes,
- provide a fuzzy classification of selected regolith attributes and
- present a spatial distribution of fuzzily classified soil–regolith attributes as continuous data surfaces.

This was achieved through both an explicit modelling with the DEM primary derivatives directly, and by fuzzy classification, using rules to incorporate the DEM primary and secondary derivatives, and the remotely sensed and field data. This required the expression of a series of separate semantic rules consistent with fuzzy logic for each of the RCU components.

The semantic rules were developed from charting notional relationships between the RCU component response-variable and the terrain derivative and environmental covariates (explanatory variables). The resultant digital RCUs were then used for spatial prediction of regolith attribute variables and to provide digital R–T maps for forest management.

20.5 Summary and analysis

The discretised depiction of the major RCU components is closer to the reality of their topographical definition than the simple linear boundaries and areal polygons drawn from aerial photography. Nevertheless, class boundaries in this classification are still crisp and the classes are discrete.

Definition of the RCU components is particularly suited to fuzzy classification because of the strong linguistic nature of describing the landform variables that constitute RCUs (e.g. "concave at base", "broadly convex and low gradients", "steepening near the upper part", etc.). The fuzzy depiction of RCU components (e.g. RCU_{ssc}, Plate 20b (see Colour Plate Section)) resulted from a substantial rule-base which not only improved upon the discrete classification, but also reproduced the dynamic nature of the model by introducing parameters relating to hillslope processes, for example compound topographic index, dispersive area, upslope curvature and distance from stream (Gallant and Wilson, 1996). This is now a model of the R–T *system*. Crests are no longer simple watershed lines or uniform areas of one class, but summit regions influenced by hillcrest processes. Similarly, plateaux are presented as summit surfaces defined by surface water and material movement processes rather than delineation of uniform areas.

The boundaries between units are fuzzy where the rules depicting each component unit necessarily weaken. The central concept of each component unit is shown by the strength of colour, which reflects the goodness-of-fit to the rules defining the component. In the graphical sense a landform unit (RCU component) fades in definition at its margins to merge with neighbouring units. The classification in this GIS analysis produced the best depictions of summit surfaces (RCU_{ss}), crests (ssc) and backslopes (vbs) of the valley basins and valley cores (vbc) but difficulties arose with consistent depiction of plateaus, the steepest, concave to straight slopes and inter-basin (ib) units (Plate 20c (see Colour Plate Section)).

The main reasons for these difficulties were various: unsatisfactory explicit semantic rules defining these units, an inappropriate grain of the DEM to represent sometimes small and complex land surface forms, and the inadequacy of the DEM derivatives to portray the processes and landform sufficiently (insufficient relationship to the landform components in the conceptual models).

Erosional inter-basin units (RCU_{ibe}) were the hardest to define explicitly. Many attempts were made to confine these component units to the vicinity of creek junctions and distributive, non-basinal areas. The result is acceptable in some instances, but it does not reflect the manual interpretation of these units sufficiently overall. Clearly the DEM grain is not precise enough to render these units distinguishable from valley basin slopes (vbs). A higher grain DEM is necessary and further explicit rules, distinguishing non-valley basins, are required to enhance the RCU_{ibe} classification.

The main outcomes of this modelling exercise are twofold:

- The digital RCU model is repeatable (it was iterated and improved upon several times), and it has the potential to be transferred to other landscapes as it is rule-based and thus generic, although some rules are locality-specific. It can be used dynamically as a digital model (to recreate and model regolith-catenary processes, as well as for more general pedogenesis and landscape development purposes).
- The accuracy and precision of the DEM is not enough to recreate the real landscape at this scale of investigation (1:20,000 presentation scale), and the variable hierarchical scales of R–T processes therein. The digital model is faithful to the RCU concept.

20.6 Conclusions

Defining RCU components is a convenient and a geomorphologically sympathetic way of classifying the R–T (or the soil-geomorphological landscape). They can be defined through field survey and aerial photographic interpretation as well as digitally through defining appropriate rules for DEM derivatives and environmental covariates. Their expression has been further enhanced by fuzzy classification means. The successful definition of fuzzy RCUs and components substantially aids the process of predicting soil/regolith attributes, for example regolith depth, top layer depth, top layer stoniness, texture, permeability, infiltration capacity, drainage, which was successfully achieved as the ultimate purpose of the overall study (Plate 20d (see Colour Plate Section)). The generally unsuccessful fuzzy definition of some RCUs (e.g. inter-basinal units) is potentially confounding for soil–regolith attribute prediction. Although Schmidt and Hewitt (2004) showed that hill spurs can be expressed as 'form elements' they still do not equate to the RCU_{ibe} units in this case but form some component of it. Their spurs elements can also be confused with ridge and ridge crest components of RCUs and they do not adequately encompass the convex plan and profile curvatures of nose slopes. This part of component expression is the subject of further investigation but it is possible that unequivocal definition of erosional inter-basins (RCU_{ibe}) may elude the best attempts with the suite of DEM derivative algorithms currently available.

For this study, the fuzzy definition of the summit surfaces (RCU_{ssc}; RCU_{ssp}) and the valley basin slopes (RCU_{vbs}) was achieved successfully at this scale (grain and extent). The distinction between valley basin 'cores' (RCU_{vbc}) and depositional inter-basins (RCU_{ibd}) needs further refinement. It is noted that Schmidt and Hewitt (2004) have defined flood plain forms well in their 15 × 15 km upland landscape, which could possibly equate with RCU_{ibd} in this case, however they do not express an equivalent of the valley cores (RCU_{vbc}). The sampling scheme used by Burrough et al. (2000) at their 68 ha Hussar site (at 5 m grid resolution) appears to have merit when used with the topographic wetness index and their Class 1 from the 5 Class system depicting valley bottoms is a possible equivalent to RCU_{vbc}, but a similar procedure used on the BKA has not improved the digital distinction between the two RCU components as yet.

The valley basins and summit surface crests were the only RCU components to be investigated and sampled in the field for R–T relationships and to be validated correctly to within 5% significance (Thwaites, 2001). The environmental covariates relevant to inter-basin units and plateaux for predictive R–T analysis has yet to be fully explored. The R–T processes for depositional inter-basins and plateau summit surfaces are not well defined as yet. This will be necessary to validate them as regolith-catenary components.

When fully definable by digital means, RCUs have the potential to become the basis for any soil-geomorphological stratification of the landscape for predicting the spatial variation of soil-landscape or R–T attributes as a means towards defining functional sites for land management.

References

Burrough, P.A., van Gaans, P.F.M., MacMillan, R.A., 2000. High resolution landform classification using fuzzy k-means. Fuzzy Sets Systems 113, 37–52.

Gallant, J.C., Wilson, J.P., 1996. TAPES-G: a grid-based terrain analysis program for the environmental sciences. Comput. Geosci. 22, 713–722.

Huggett, R.J., 1975. Soil landscape systems: a model of soil genesis. Geoderma 13, 1–22.

Hutchinson, M.F., 1989. A new method for gridding elevation and streamline data with automatic removal of pits. J. Hydro. 106, 211–232.

MacMillan, R.A., Pettapiece, W.W., Nolan, S.C., Goddard, T.W., 2000. A generic procedure for automatically segmenting landforms into landform elements using DEM, heuristic rules and fuzzy logic. Fuzzy Sets Systems 113, 81–109.

Milne, G., 1935. Some suggested units of classification and mapping, particularly for East African soils. Soil Res 4, 183–198.

Schmidt, J.S., Hewitt, A., 2004. Fuzzy land element classification from DTMs based on geometry and terrain position. Geoderma 121, 243–256.

Simonson, R.W., 1959. Outline of a generalised theory of soil genesis. Soil Sci. Soc. Am. Proc. 23, 152–156.

Thwaites, R.N., 2000. Pedogeomorphic terrain analysis using Regolith Catenary Units for land resource survey in forest uplands. In: J.A. Adams and A.K. Metherell (Eds.), Soil 2000: New Horizons for a New Century. Proceedings of the Australian and New Zealand Second Joint Soils Conference 3–8 December 2000, Lincoln University, New Zealand. Vol. 3, pp. 205–206.

Thwaites, R.N., 2001. Pedogeomorphic Terrain Analysis for Forestland Resource Assessment, Vol. 1 & 2. Ph.D. Thesis, Department of Geographical Sciences, The University of Queensland. Australia. Digital Thesis Program, UQ, Brisbane.

Wilson, J.P., Gallant, J.C., 2000. Digital terrain analysis. In: J.P. Wilson and J.C. Gallant (Eds.), Terrain Analysis. John Wiley & Sons, New York.

Developments in Soil Science, volume 31
P. Lagacherie, A.B. McBratney and M. Voltz (Editors)

Chapter 21

SOIL PREDICTION WITH SPATIALLY DECOMPOSED ENVIRONMENTAL FACTORS

M.L. Mendonça-Santos, A.B. McBratney and B. Minasny

Abstract

Prediction of soil attributes and soil classes in digital soil mapping relies on finding relationships between soil and the predictor variables of soil-forming factors and processes. The predictor variables can be remotely or proximally sensed images of soil, landscape, parent material or climatic factors. Till date, most prediction methods are based on performing regression on the predictor variables directly to predict soil attributes or classes. There are problems using data layers from different sources, particularly, multicollinearity, and the fact that the relationships between soil and environmental variables can change with spatial scale. To overcome the problem of correlation between variables, principal component analysis can be performed on the predictor variables. With respect to the spatial dependency, each of these variables can be decomposed into separate spatial components and mapped separately. One of the methods of achieving this is wavelet analysis, which decomposes the variables into separate hierarchical spatial components of decreasing spatial resolution. These components could all be derived and subsequently used as separate layers in predicting soil classes or soil attributes. In this chapter, data are decomposed using the wavelet method and examples of predictions of soil classes and surface-clay content are shown, in order to evaluate the effect of using the decomposed layers in comparison with the original data.

21.1 Introduction

Digital information from remote and proximal sensing, as well as environmental correlates data are currently used in digital soil mapping (McBratney et al., 2003). Nevertheless, their interpretation as well as the relationships between variables can be very complex, principally when considered with regard to other information as is the case for digital soil mapping. The studied phenomena may be scale dependent, and short-range variation in two variables, which are uncorrelated between them, might obscure underlying relationships.

Most soil prediction methods for digital soil mapping directly relate the environmental variables obtained from different sources to soil attributes or classes. There are problems using data layers from different sources;

multicollinearity and the relationships between soil and environmental variables may change with spatial scale. To overcome the problem of correlation among variables, principal-component analysis can be performed on the predictor variables. And for the problem of spatial dependency, each of these variables can be decomposed into separate spatial components and mapped separately. Spatial decomposition and its application to digital soil mapping have been reviewed briefly by McBratney et al. (2003). They viewed it as a possible approach to improve predictions in soil and environmental variables, through the detection of scale-dependent relationships and appropriate spatial associations. This can be achieved either by geostatistical methods dealing with scale-dependent correlation (e.g. Oliver et al. (2000) applied to Système Pour l'Observation de la Terre (French Earth Observation Satellite) (SPOT) images) or by wavelet decomposition (e.g. Epinat et al. (2001) applied to airborne normalized difference vegetation index (NDVI) imagery).

The geostatistical approach uses the so-called factorial kriging method. It involves modelling the correlation structure in the imagery by decomposing the variogram into independent spatial components, and then taking each component in turn and kriging it, thereby separating it from the others. Nevertheless, geostatistical methods assume stationarity of the variation (Goovaerts and Webster, 1994).

Wavelets are mathematical functions that cut up data into different scale components, and then study each component with a resolution matched to its scale (Graps, 1995). Wavelet represents signals (information) using a linear combination of wavelet functions. Wavelet allows us to analyse spatial data according to scale or resolution, which is very useful to analyse data with sharp discontinuities. A detailed explanation of the theory and application of the wavelets in soil science can be found in Lark and McBratney (2002), Lark and Webster (2004) as well as in Chapter 23 of this book.

This method offers additional insight in signal and image processing where it is possible to describe the local phenomena and tell where a particular event took place. If the signal is stationary, this isn't very important. However, soil data contain numerous nonstationary or transitory characteristics: drift, trends, abrupt changes and beginnings and ends of events. These characteristics are often the most important part of the image.

Wavelet analysis permits a signal to be represented in terms of a set of basis function, including both, short-high and long-low frequencies. These basic functions allow the signal to be dilated by a scale parameter and translated to a location to provide local and scale-specific analysis of the data, providing access to information that can be obscured by other time-frequency methods, like Fourier analysis. Another difference between wavelets and Fourier analysis is that the wavelets have an infinite set of possible basis functions, not only sine and cosine.

Wavelet analysis has been developed since the early 1980s with a wide range of applications, and recently it has attracted analysis of soil data (McBratney, 1998; Lark and Webster, 1999). In this chapter, we will explore the spatial decomposition of the environmental data layers using the wavelet method and evaluate the effect of using the decomposed layers in predicting soil classes and soil attributes.

21.2 Material and methods

21.2.1 The study site

The study site is located at Edgeroi area, near Narrabri, New South Wales (NSW), Australia (Fig. 21.1). It is a typical part of the northwestern slopes and plains of NSW described by Ward (1999) in his study to provide detailed information on the regional soil pattern.

The soil dataset consists of 317 sampling units, from which 210 sampling units are arranged on a systematic, equilateral triangular grid with approximately 2.8-km spacing between sites, and an additional 106 sampling units are distributed more irregularly or on transects (Fig. 21.1). Soil attributes (both morphological and chemical data), vegetation and landform information were recorded by McGarry et al. (1989). Under the scope of this work, soil profiles were classified using the Australian soil classification system (ASC) (Isbell, 1996) at the family level. Ancillary data related to the scorpan model (McBratney et al., 2003) were also recorded for the soil dataset and the ancillary dataset at a grid spacing of 25 m.

The environmental factors used to predict soil attributes are Landsat 7 ETM$^+$ images from year 2003 (bands 1, 2, 3, 4, 5 and 7). The Landsat bands were used to calculate the following indices: NDVI (band 4–band 3)/(band 4+band 3) and clay index (band 5/band 7). Data from a gamma-radiometric survey (see Chapter 16 for details of such surveys) (K, Th, U and KThU) as well as digital elevation model (DEM) and its derivatives were interpolated onto a common 25-m grid using kriging with local variograms. Data from these environmental layers were used to predict soil classes and soil properties for the purpose of digital soil mapping, as shown in Table 21.1.

21.2.2 The wavelet decomposition

The data layers were spatially decomposed using a two-dimensional discrete wavelet transform. The theory and application to soil data is given by Lark and Webster (2004). The Wavelet toolbox of MATLAB software (MathWorks, 2004) was utilised to perform the analysis employing Daubechies wavelet with two vanishing moments as the wavelet function. The data layers (Table 21.1) were decomposed into four levels (L_1, L_2, L_3 and L_4), which correspond to pixel

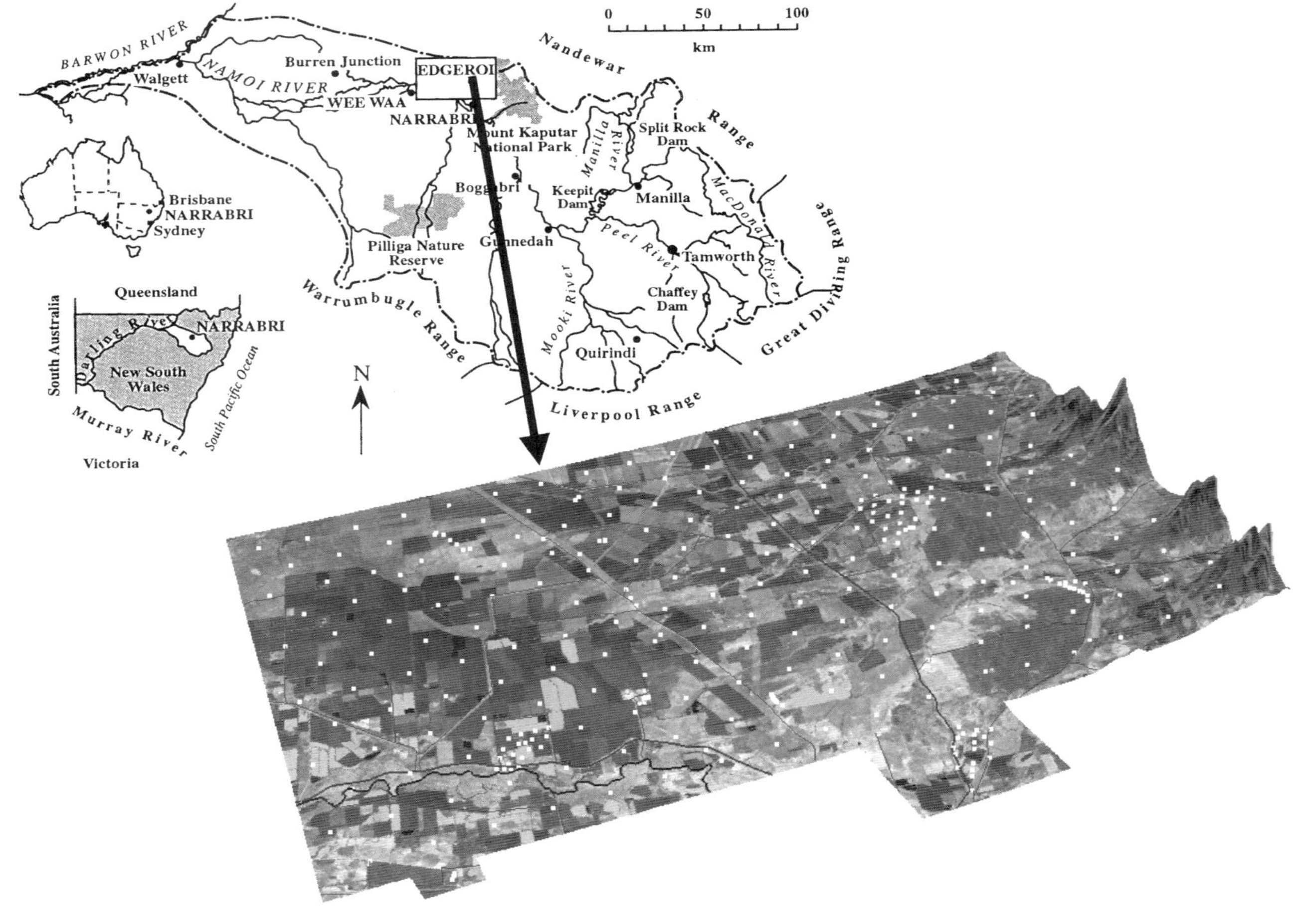

Figure 21.1. *Study site and soil profile locations.*

Table 21.1. *Predicted variables, prediction models and predictor variables.*

Predicted variable	Prediction model	Predictor variables
Topsoil clay content (0–10 cm)	Multiple linear regression	Landsat 7 ETM$^+$ bands 1, 2, 3, 4, 5 and 7, NDVI, clay index,
Soil suborder (ASC)	Classification tree	Gamma radiometric K, Th, U, KThU, DEM derivates

resolution of 50, 100, 200, and 400 m, as illustrated in Figure 21.2. These decomposed data and their corresponding components (a_1, a_2, a_3 and a_4) formed part of the dataset, which will be used as predictors of soil classes and attributes. Shown in Figure 21.3 is the effect of decomposition on the semivariogram. The spatial decomposition takes away a small part of the variance while still retaining the shape of the spatial variation.

21.2.3 Prediction of soil properties and soil classes

In order to illustrate this method, clay content from a depth of 0–10 cm and soil suborders according to the ASC system (Isbell, 1996) were predicted using the original data and the decomposed components (Table 21.1). For the continuous data (i.e. clay content) stepwise linear regression was first used to identify the useful predictors, then a multiple linear regression was formed using the selected predictors. To predict soil classes, a classification tree program called C5.0, (RuleQuest Research, 2003) was used. A tree structure is generated by partitioning the data recursively into a number of groups, each division being chosen so as to maximise some measure of difference in the response variable in the resulting groups.

We compare the prediction using original variable, approximated variables where different spatial components have been filtered: L_1, L_2, L_3 and L_4, and approximate variables with its spatial components (L4 with a_1, a_2, a_3 and a_4). To compare the performance of the model and accounting for the number of parameters used, Akaike's information criterion (AIC) (Akaike, 1973) is used as a quality index:

$$\mathrm{AIC} = -2l + 2m \tag{21.1}$$

where l is the log-likelihood of the prediction, and m, the number of parameters used in the model. This index is a compromise between the goodness of fit and the parsimony of the model. The best model is the one that has the smallest AIC.

The log-likelihood for class prediction ($k = 1, .., K$) is calculated as follows (Hastie et al., 2001):

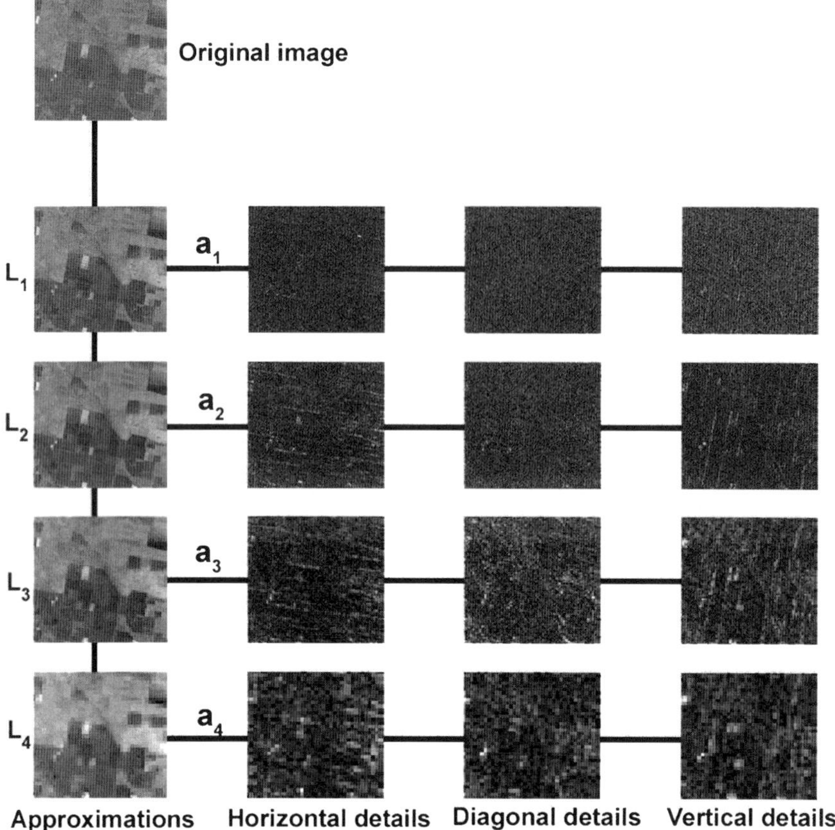

Figure 21.2. *Decomposition of Landsat 7 ETM*$^{+}$ *Band 1 image at four levels (L_1, L_2, L_3 and L_4), showing the horizontal, diagonal and vertical components. Note: Original* $= L_1+a_1$, $L_1 = L_2+a_2$, *and so on.*

$$l = \sum_{k=1}^{K} \log \hat{p}_k \tag{21.2}$$

where $\hat{p}_k$ is the proportion correctly allocated to class k:

$$\hat{p}_k = \frac{1}{N_k} \sum I(y_i = k) \tag{21.3}$$

For continuous variables AIC is approximated by

$$\text{AIC} = N \ln \left(\sum_{i=1}^{N} (\hat{y}_i - y_i)^2 \right) + 2m \tag{21.4}$$

where N = total number of data (soil sampling units).

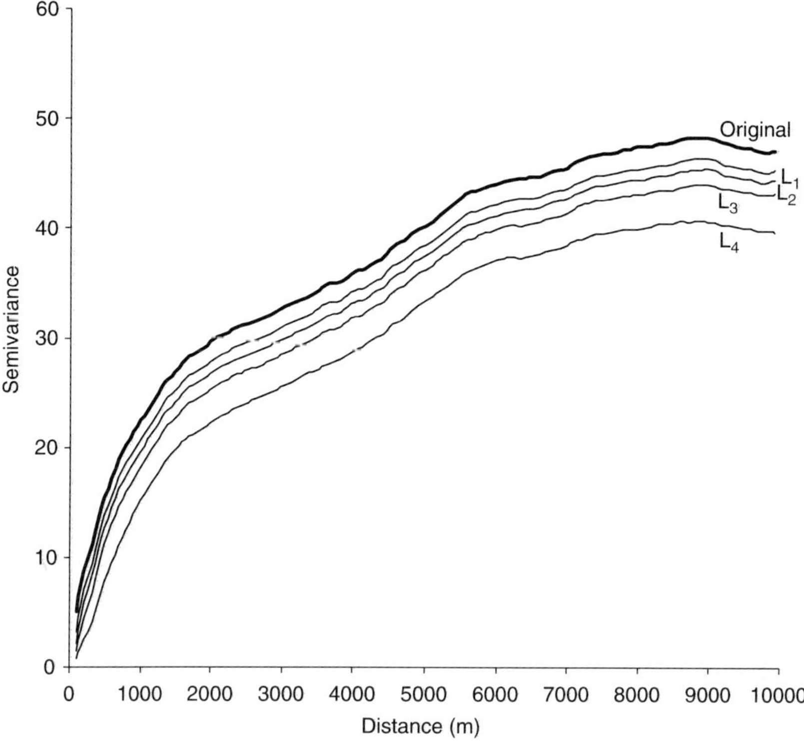

Figure 21.3. *Semivariograms of the Landsat Band 1 image and its decomposed layers.*

21.3 Results and discussion

21.3.1 Prediction of clay content

Results for the prediction of the 0–10-cm clay content is given in Table 21.2. Stepwise linear regression was applied to the original data, and the analysis identified the following variables as useful to fit the model: Landsat 7 ETM$^+$ bands 1, 2, 3, 4 and 5, NDVI, elevation, slope and K. Using these variables as predictors, linear regression was formed to predict the surface clay content using original data, L_1 (data with pixel scale 50 m), L_2 (scale 100 m), L_3 (scale 200 m), L_4 (scale 400 m) and decomposed data with its spatial components (L_4 with a_1, a_2, a_3 and a_4). For the wavelet decomposed data, stepwise regression found the following variables as useful predictors: Landsat band 1 (L_4), band 2 (L_4), B5 (L_4), elevation (L_4), slope (L_4, and a_2, a_3), NDVI (L_4, and a_2, a_3) and K (L_4 and a_4).

The results in Table 21.2 showed that data with different pixel resolutions can give different prediction results. The decomposed layers (L_1, L_2, L_3 and L_4) can be viewed as data with "noise" removed or data with different scale components. It is worth noting that L_4 (pixel scale 400 m) has the largest prediction capability

Table 21.2. *Results of the prediction for clay content (0–10 cm), using multiple linear regression, for the original variables, each of the decomposed levels, and all the decomposed variables with their spatial components.*

	R^2	RMSE	m	AIC
Original	0.48	12.41	10	3103
L_1	0.47	12.53	10	3110
L_2	0.49	12.24	10	3095
L_3	0.49	12.27	10	3097
L_4	0.51	12.09	10	3087
Decomposed (L_4,a_1,a_2,a_3 and a_4)	0.57	11.32	14	3054

compared to the original data and data at other scales. Using the spatial components in addition to the smoothed data (which essentially sum up to the original variables) enhanced the prediction. It is interesting to note that the spatial components of the Landsat bands were not found to be significant predictors, implying much noise in these. Slope and NDVI with components a_2 and a_3 were found to be useful predictors. The results suggest that the first component (a_1) might be composed of short-range noise and the a_2, a_3 and a_4, which represents components with spatial scales of 100, 200 and 400 m, respectively, contains useful spatial information that can represent the soil attributes in the landscape. The prediction of clay content improved slightly (average of 1%) using the spatial components. The fact that the decomposed layers with the components use more predictors (14 as opposed to 10 parameters) is justified by the AIC.

21.3.2 *Prediction of soil class*

To predict the soil suborders based on the ASC system (17 suborders), a classification tree was used with all variables (Table 21.1) as predictors. The algorithm in C5.0 then selects the useful predictors to build a tree. The results of the prediction are presented in Table 21.3, where the 'tree size' represents the number of 'nodes' in the tree and is used as the number of parameters m for calculation of AIC. Similar to the prediction of clay content, data representing different scales possess different predictive capabilities. In the case of soil class prediction decomposed data layers (L_1, L_2 and L_4) made a better model compared to the original data as revealed by the AIC (less parameters and a still adequate fit to the data). Using the spatial components only improved the performance slightly (2% misclassification improvement).

It is most interesting to note that data layers at L_4 (scaled 400 m) gives the best model to predict soil suborders and also clay content. This implies that there are a lot of redundant data present in the data layers, which is not related to the soil properties. Clearly filtering the data can enhance the prediction of soil attributes.

Table 21.3. *Results of the prediction for ASC soil suborders using a regression tree.*

Predictors	Tree size	% misclassification	AIC
Original variable	61	18.0	97.1
L_1	47	22.1	72.4
L_2	49	21.5	76.9
L_3	62	15.5	99.8
L_4	43	22.1	65.8
Decomposed (L_4,a_1,a_2,a_3 and a_4)	58	15.8	91.3

21.4 Conclusion

The spatial decomposition of the environmental variables shows that data layers at different scales can be used to enhance the prediction of soil attributes for digital soil mapping. Decomposing the layers remove some noise that may be redundant or may be artifactual. In the examples described above the smoothed version of the data layers at a resolution of 400 m gives a better prediction model compared with using the original variable at a finer resolution.

Wavelet analysis provides a powerful technique as more and more rapid-sensing methods for soil become available. Compared with factorial kriging, the wavelet method is more robust as it does not assume stationarity and does not need the variogram to be modelled. Using a nested model for the variogram can be quite arbitrary and the scale of the image that can be decomposed will depend on the 'range' parameter of the variogram. Wavelets also provide more efficient computation compared with factorial kriging. We recommend wavelet decomposition of environmental covariates to be performed as part of the digital soil mapping process. The only problem with this approach is the computational one of proliferation of potential covariate data layers.

Acknowledgement

This work is funded by an Australian Research Council Discovery project on Digital Soil Mapping.

References

Akaike, H., 1973. Information theory and an extension of maximum likelihood principle. In: B.N. Petrov and F. Csaki (Eds.), Second International Symposium on Information Theory. Akademia Kiado, Budapest, pp. 267–281.

Epinat, V., Stein, A., de Jong, S.M., Bouma, J., 2001. A wavelet characterization of high-resolution NDVI patterns for precision agriculture. ITC J 2, 121–132.

Goovaerts, P., Webster, R., 1994. Scale-dependent correlation between topsoil copper and cobalt concentrations in Scotland. Eur. J. Soil Sci. 45, 79–95.

Graps, A., 1995. An introduction to wavelets. IEEE Comput. Sci. Eng. 2, 50–61.
Hastie, T., Tibshirani, R., Friedman, J., 2001. The Elements of Statistical Learning: Data Mining, Inference and Prediction. Springer Series in Statistics. Springer-Verlag, New York.
Isbell, R.F., 1996. The Australian Soil Classification. CSIRO Publishing, Melbourne.
Lark, R.M., McBratney, A.B., 2002. Wavelet analysis. Chapter 1–Soil sampling and statistical procedures. In: J.H. Dane and G.C. Topp (Eds.), Methods of Soil Analysis. Part 4–Physical methods. SSSA Book Series 5. Soil Science Society of America, Madison, Wisconsin, pp. 184–195.
Lark, R.M., Webster, R., 1999. Analysis and elucidation of soil variation using wavelets. Eur. J. Soil Sci. 50, 185–206.
Lark, R.M., Webster, R., 2004. Analysing soil variation in two dimensions with the discrete wavelet transform. Eur. J. Soil Sci. 55, 777–797.
Mathworks, 2004. Matlab version 7.0. The Mathworks Inc., Natick, MA.
McBratney, A.B., 1998. Some considerations on methods for spatially aggregating and disgaggregating soil information. Nutrient Cycl. Agroecosystems 50, 51–62.
McBratney, A.B., Mendonca-Santos, M.L., Minasny, B., 2003. On digital soil mapping. Geoderma 117, 3–52.
McGarry, D., Ward, W.T. and McBratney, A.B., 1989. Soil studies in the lower Namoi valley – Methods and data. Vol. 1 and 2. The Edgeroi data set. CSIRO Australia, Division of Soils, Glen Osmond, SA.
Oliver, M.A., Webster, R., Slocum, K., 2000. Filtering SPOT imagery by kriging analysis. Int. J. Remote Sens. 21, 735–752.
RuleQuest Research, 2003. See5/C5.0 version 1.20. RuleQuest Research Pty Ltd., Sydney, Australia. http://www.rulequest.com
Ward, W.T., 1999. Soils and landscape near Narrabri and Edgeroi, NSW, with data analysis using fuzzy k-means. CSIRO Land and Water Technical Report 22/99, July 1999.

F. Quantitative modelling for digital soil mapping

Chapters 22–24 overview quantitative modelling approaches in some detail. Chapters 22 and 23 present modelling tools that currently seem useful for digital soil mapping. Chapter 22 is an overview of the different strategies that have been tested in the recent past to embed soil survey knowledge in numerical algorithms of spatial soil prediction. Chapter 23 deals with multiscale issues, one of the burning questions of the new discipline of digital soil mapping (see Chapter 1). Chapter 24 describes a practical and complete exercise in digital soil mapping based on a few pedogenetic indicators derived from DTMs and gamma radiometrics to help predict and map user-requested soil variables.

Developments in Soil Science, volume 31
P. Lagacherie, A.B. McBratney and M. Voltz (Editors)

Chapter 22

INTEGRATING PEDOLOGICAL KNOWLEDGE INTO DIGITAL SOIL MAPPING

C. Walter, P. Lagacherie and S. Follain

Abstract
Classical soil survey usually integrated existing pedological knowledge to enhance its efficiency and compensate for very low standard sampling densities. This approach has been criticised because the information taken into account was not explicitly specified and validation procedures were not developed. We advocate that pedological knowledge of soil landscape distribution, soil-forming factors and soil processes is essential to modern soil spatial analysis and may be rigorously integrated into soil mapping. Basic reasons for such integration are an increase in prediction efficiency and also the necessity to link soil maps to dynamic modelling, enabling risk assessment and impact studies. Several approaches are reviewed including spatial prediction techniques using existing soil maps and spatial modelling based on soil-forming factors. Combination in the near future of space and time modelling demands additional integration of the dynamics of physical and biochemical soil processes.

22.1 Introduction

Soil spatial variability appears to have been recognised for millennia. Boulaine (1989) quotes extracts of ancient texts indicating that soil diversity has been appreciated since antiquity. Its scientific study is much more recent and dates only from the end of the nineteenth century, initiated by the precursors of pedology, in particular Dokuchaev (1893). The approach developed by these pioneers was from the beginning of a deterministic one that is based on the idea that the main factors of soil variability could be recognised in the landscape and therefore soil mapping could be derived from a limited number of direct soil observations. Jenny (1941, 1961) published a soil-forming factor equation integrated in an open system analysis, which was the basis of most soil surveys undertaken in the subsequent decades (Bockheim et al., 2005). Soil spatial analysis was therefore strongly linked to the study of soil formation.

Despite its success, this approach has always encountered difficulties from several sources:

- Soil-forming factors are numerous, interact and are likely to vary in time (Yaalon, 1971; Webster, 2000). Moreover, human activity (tillage, fertilization, amendment and contamination) often superimposes in a prevalent way and greatly modifies the soil processes and properties. Soil is therefore a very complex entity, which can hardly be characterised by a small set of diagnostic properties (Butler, 1980; Ruellan et al., 1989). Many authors have highlighted large variability among soil samples initially chosen from within an apparently identical soil type or mapping unit (Beckett and Webster, 1971; Banfield and Bascomb, 1976; Wilding and Drees, 1983);
- sampling density applied to soil spatial studies is in general very low, even for detailed studies (Table 22.1): the sampled fraction is much lower than a millionth of the volume of interest (Legros, 1996). Direct measurements of soil cover are therefore scarce and hence soil-forming factor recognition or soil mapping is difficult in most cases.

Since the 1970s, statistical tools, mainly based on geostatistical interpolation procedures, have been developed to predict soil properties in geographic space without underlying assumptions about soil formation. Indeed, research on spatial statistics has been very productive over the last 30 years and new approaches have very often been tested with soil property examples (Cressie, 1991; Goovaerts, 1997). Large panoply of statistical methods is now available to predict soil properties or classes at unvisited locations from neighbouring punctual measurements, with some indication of the associated prediction error.

Compared with classical soil maps, maps derived from interpolation procedures appear generally very smooth: Figure 22.1 gives an example of soil depth estimation over a 2000 ha region in Brittany (France) derived from a classical soil map or obtained by ordinary kriging (Walter, 1990). Owing to short-distance variability, kriging tends to smooth the soil depth variability over this region, as

Table 22.1. *Soil cover sampling densities adopted in France for spatial studies of decreasing resolution.*

	Structural analysis (Ruellan et al., 1989)	French soil survey standards (Jamagne, 1967)			French national soil tests database (Walter et al., 1997)
		1/10,000	1/25,000	1/100,000	
1 auger pit for ...	0.005–0.1 ha	1–5 ha	5–10 ha	20–100 ha	20–100 ha
1 profile for...	0.05–0.5 ha	10–50 ha	50–100 ha	500–10,000 ha	-
Sampled/total volume	1/3000–1/700,000	1/6,000,000–1/3,000,000	1/3,000,000–1/6,000,000	1/12,000,000–1/60,000,000	1/12,000,000–1/60,000,000

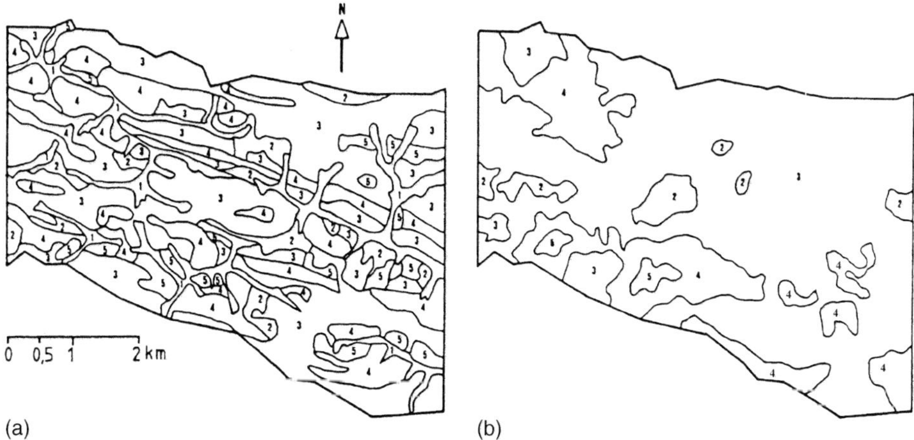

Figure 22.1. *Comparison of soil depth prediction over a 2000 ha area in Brittany, France, derived from (a) a classical soil map and (b) estimated by ordinary kriging. Both techniques use the same sampling dataset (after Walter, 1990).*

the distance between sampling points appears too large to reveal by interpolating the soil depth pattern within the area. By opposite, the finer details of the classical soil map cannot been justified by the existing soil observations and they reveal the auxiliary information taken into account by the surveyor. This information has not been integrated in an explicit modelling approach and no precision criteria can therefore directly be associated to the map.

More generally, in the context of operational soil science applications, the sampling effort suggested by statistical approaches appears too high with respect to the cost of fieldwork. Except in situations where a precise knowledge of prediction error is crucial, such as for polluted soil remediation, qualitative or expert approaches are therefore commonly preferred to quantitative ones, particularly for decision-making studies.

Alternative approaches therefore have to be developed gathering additional information useful for soil mapping; this information should be of lower cost or more efficient than the classical soil sampling procedures. The application of new techniques enabling higher sampling densities, such as geophysics (Tabbagh et al., 2000) or direct field measurements (Viscarra Rossel and Walter, 2003) appears to be promising ways to achieve such objectives.

Another approach considered here is to include pre-existing knowledge of soil variation in space and to link it to more widespread information than direct soil measurements. The concept of soil digital mapping (McBratney et al., 2003; Chapter 1), an approach that merges the classical soil mapping approach with advanced numerical techniques, gives a framework for such improvement.

The aim of this chapter is to review several approaches integrating pedological knowledge into soil spatial prediction procedures. First, we analyse new issues for soil spatial predictions before illustrating current approaches to combining pedological knowledge and quantitative mapping techniques. Finally, we consider further developments necessary to predict soil properties or classes jointly in space and time.

22.2 New issues for soil spatial prediction

22.2.1 Evolution of concepts

Soil spatial analysis classically focused on two types of soil properties: first, properties considered to be stable in time and intrinsic to soils (texture, mineralogy and acidity); second, properties likely to influence the vegetation growth (soil depth, waterlogging and nutrient content). This choice was based on a double logic: (i) only properties stable in time seemed to be worth spatial analysis (Legros, 1996); (ii) soil was primarily perceived through its function of supporting crops, with agronomists aiming to evaluate or enhance its production potential (Sebillotte, 1989).

This perception of the soil, mainly considered through its fertility, remained dominant within the community of soil surveyors until the end of the 1980s, than changed quickly following the new environmental issue of water quality in agricultural areas. The role of soil in pollutant transfer to the water resource emerged first, then its function as a regulator of fluxes to other portions of the ecosystem (air, fauna and flora) was recognised. The idea that the soil by itself may be considered as non-renewable resource requiring preservation also progressed, leading to the concept of soil quality (Doran et al., 1994; Hoosbeek and Bouma, 1998).

22.2.2 Implications for soil spatial analysis

This evolution modified the procedures and aims of soil surveys. First of all, soil properties subject to a spatial analysis are no more systematically "selected" by a surveyor for their accessibility or their diagnostic role: they are often imposed by other specialists to be integrated in their own models; they can also derive from new regulations establishing threshold values not to be exceeded, for example heavy metal contents in soil.

Integration of soil data into geographical information systems for their combination in models with other environmental information (digital elevation model (DEM), databases, ...) also generates strong constraints on the integration support which can be a pixel, an agricultural field, a watershed or even an administrative entity (Burrough and McDonnell, 1998). Soil spatial information is therefore requested at resolutions and for entities unusual for classical soil surveys.

Finally, the temporal dimension can no longer be neglected, one of the major issues being the evaluation of the modification of the soil and its processes under human activity, for example changes in farming practices, direct or indirect contamination by pollutants, etc. As suggested by Boiffin and Stengel (2000), *"soil knowledge needs to be quantified and integrated in complex decision-making systems, which enable to simulate the impacts of production strategies, land planning and land use in general"*.

22.3 Spatial predictions using pedological knowledge

22.3.1 What renders pedological knowledge?

When mapping soils, surveyors generally not only rely on soil direct observations, but also use collective knowledge on soil properties and processes. We distinguish five broad domains with existing knowledge on soils that are possibly useful for spatial predictions:

(i) Relative distribution of soil entities (profiles, horizons) within the landscape.

A wealth of soil surveys have been made in the past few decades; we may consider that pedological systems have been extensively recognised, and in many regions of the world we can find high-resolution soil maps describing the local soil distribution.

(ii) Identification of soil-forming factors.

Climate, organisms including human activity, topography, parental material and time have early being recognised as the main factors of soil formation (Dokuchaev, 1893; Jenny, 1961), but the first attempts at dynamic modelling of soil formation appeared only recently (Heimsath et al., 1997; Hoosbeek et al., 1999).

(iii) Correlation between soil properties.

Soil properties are generally strongly correlated, albeit not always linearly, as they undergo common processes. These correlations can be used directly in the spatial prediction, to estimate laborious or expensive-to-determine soil properties from more accessible ones (McBratney et al., 2002); they could also be introduced simply as prediction constraints, for example to avoid predicting at the same place a low pH and a high carbonate content.

(iv) Spatial structures of soil properties.

Spatial structure information derives from the numerous studies involving variography of soil properties, giving access to main characteristics of spatial structures (range, short-distance variability, spatial variance). McBratney and Pringle (1999) first compared variograms of identical soil properties established in various contexts and found that, to a given extent, they were comparable and could be estimated from a sparse sample assuming a given variogram model.

(v) Temporal dynamics of physical and biochemical processes.

Soils represent a complex reactor governed by physical laws, for example conservation of mass and energy, by biochemical processes, for example organic matter turnover, and transfer functions, like water and solute movements. They vary therefore in time at different time scales, from the millennium (mineral alteration) to the second (microbial processes, adsorption–desorption) and with a strong influence of the seasonal cycles (Horn and Baumgartl, 2000).

22.3.2 Soil properties prediction based on existing soil maps

Existing soil maps describe variations of major soil characteristics in the landscape and may therefore be the support for spatial prediction of a given soil property. Nevertheless, many studies have showed that variability within mapping units may be large (Beckett and Webster, 1971; Marsman and De Gruijter, 1986) and that special attention should be given to evaluate the residual variability. The concept of a 'representative' profile classically used by soil surveyors to characterise the mapping units (Soil Survey Staff, 1951; Boulaine, 1980; Bouma et al., 1980) with a soil profile purposively positioned, should therefore be used with care, as the underlying hypothesis of invariability within the same stratum appears too strong in most landscapes and no indication of prediction error is available.

Two different approaches have been developed to integrate soil maps into quantitative prediction procedures: (i) the discrete approach: soil mapping units are considered as entities with sharp limits which may be described by the classical sampling theory (Cochran, 1977); (ii) the mixed approach: the boundaries on the soil map delineate abrupt changes in the landscape and geostatistics are used to infer the spatial variation between these limits or taking them into account (Voltz and Webster, 1990; Gascuel et al., 1993; Heuvelink and Webster, 2001).

The discrete approach, or design-based estimation (Brus and De Gruijter, 1997), generally supposes to implement an additional sampling scheme, which is independent from the soil map construction phase. Several sampling strategies have been proposed to estimate position and dispersion parameters of soil properties within mapping units: simple random sampling (Wright and Wilson, 1979), random transects with fixed spacing between points (Steers and Hajek, 1979; Leenhardt, 1991; Wang, 1984), stratified sampling to distinguish short distance variability from differences between polygons (De Gruijter and Marsman, 1984; Walter, 1990). Several authors developed nested sampling strategies

(Oliver and Webster, 1986; Walter, 1990) to evaluate roughly the spatial structures within a given mapping unit.

Plate 22a (see Colour Plate Section) illustrates another technique to implement the discrete approach, by the introduction of existing soil information (Schvartz et al., 1998). In this example, soil samples sent by farmers to laboratories for chemical fertility assessment are considered: the analytical results from more than 800,000 topsoil samples were collected in France over the period 1990–2000 (Walter et al., 1997); methods and procedures appeared sufficiently standardised between laboratories to enable the integration of the data in a unique national database. Only an imprecise location (with an approximate accuracy of 3 km) of the samples could be stored in the database because of confidentiality regulations; nevertheless, this geographical reference enabled an intersection of the data with the 1:1,000,000 soil map of France (CEC, 1985). For instance, statistical parameters of soil texture within the mapping units could be derived from the 30,000 samples on which particle size analysis had been measured. The resulting map of modal textural class (Plate 22a) could therefore be compared with the one established by using the pre-existing database established by expert knowledge (King et al., 1994).

Cazemier et al. (1998, 2001) developed another way of implementing the discrete approach to estimate soil–water retention from small-scale maps. As soil physical measurements are very time-consuming, the implementation of independent sampling schemes to estimate statistical parameters within mapping units cannot be done routinely. Therefore, Cazemier et al. (2001) infer properties like soil–water retention from more accessible soil data, for example descriptions of basic properties available in soil databases (King et al., 1994) and pedotransfer functions (McBratney et al., 2002). They used a fuzzy logic approach to integrate imprecision sources into the water-retention estimation procedure, namely the variability of soil depth within a mapping unit, the fuzziness of the boundary between two units and the estimation error of the pedotransfer function. Instead of the classical prediction error, they proposed representing the possibility that a given water-retention threshold may be exceeded (Fig. 22.2); in some areas, a lack of information may lead to the impossibility of reaching a conclusion.

The mixed approach attempts to merge the discrete approach of the soil maps to the continuous model-based approach developed by geostatistics. A basic reason for this is the inability of the classical geostatistical approach to deal with sharp boundaries, for example the abrupt transition from well drained soils on hillslopes to waterlogged soils in valleys.

A classical solution takes into account the soil map limits during the variogram computation and to limit the kriging neighbourhood to observations included in a given mapping unit (Stein et al., 1988; McBratney et al., 1991).

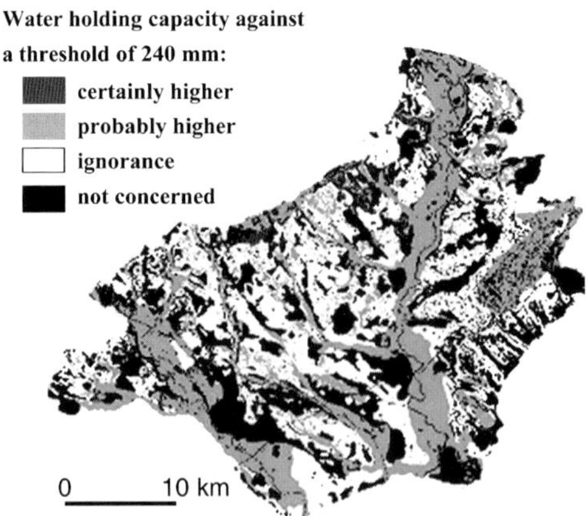

Figure 22.2. *An example of the fuzzy logic approach to integrate sources of uncertainties when predicting from soil maps: probability that the soil water holding capacity, derived from the 1:250,000 soil map of the Herault valley (France), is greater than 240 mm (after Cazemier et al., 1998).*

Nevertheless, this solution requires a very large number of samples to infer a variogram for each mapping unit. Voltz and Webster (1990) defined therefore a mean intra-strata variogram, by selecting only pairs where both points are included into the same mapping unit. This describes a mean variogram internal to a mapping unit, which is considered to be representative of all mapping units. Van Meirvenne et al. (1992) suggest an alternative way by computing relative intra-strata variogram where each semi-variance is divided by the general variance of the mapping unit; this approach weakens the previous assumption of an identical variance between mapping units. Spatial prediction is performed using intra-strata kriging (Voltz and Webster, 1990; Van Meirvenne et al., 1992) using the intra-strata variogram and neighbours selected only within the same mapping unit. Voltz and Webster (1990) compared this approach to ordinary kriging and soil map estimations to map soil physical properties in a region with crisp soil boundaries and found that intra-strata kriging performed best due to a better representation of the gradual variations within mapping units.

Several improvements have been brought to this basic idea including the use of fuzzy set theory to distinguish several types of boundaries, more or less gradual (Burrough, 1989; Lagacherie et al., 1994). Kriging with an external drift (Heuvelink, 1996), which may be qualitative (Monestiez et al., 1999), is also an interesting solution as it performs well over a large range of situations, from strong discrete soil entities to gradual transitions between soils (Heuvelink and Webster, 2001).

Finally, the association of existing soil maps with classical statistics, geostatistics or fuzzy logic appeared successful from a methodological point of view:

the complex information included in soil maps, based on direct or indirect soil observations as well as on soil surveyor expertise, may be quantitatively validated. Spatial predictions derived from soil maps may therefore be integrated into quantitative procedures where predictions errors are rigorously assessed. Nevertheless, there are structural limits to this approach, the first being that large areas are not covered by detailed enough soil maps (e.g. half of France) or existing soil maps have still to be digitised. A second limit is that most of the existing soil maps were not conceived to be linked with statistical procedures: the number of mapping units defined by the surveyor is generally too large to be separately statistically assessed. Moreover the boundaries between polygons are not described, so that their sharp or gradual nature is difficult to evaluate afterwards. Merging of similar mapping units or boundary description can hardly be undertaken from the soil map legend and associated report and should be undertaken by the soil surveyors themselves.

22.3.3 Soil prediction based on pedological knowledge of forming factors

Classical soil mapping or geostatistics do not explicitly identify the underlying causes of soil spatial variability. Historically, statistically approaches were precisely developed to disconnect the soil spatial prediction from the identification of the causes, source of abusive or erroneous interpretations (Webster, 2000). As suggested by R. Thom (1993), prediction does not presume comprehension and experience gathered from statistical studies over the past 40 or so years clearly confirms that prediction may be inferred without understanding of the whole process having led to the current soil organisation.

Nevertheless, explicitly reinstating major determinants in the soil prediction process seems essential for both epistemological and practical reasons. First, scientific knowledge on the factors producing a given soil pattern within a region has to be improved. Predictive efficiency should not be the unique criterion for method comparison and one may prefer a method with a larger uncertainty, but explicitly founded on factors and processes, likely to be validated, refined progressively and extrapolated away from the site where the model was constructed. The second reason is linked to the increasing availability of exhaustive databases of soil-forming factors, for example topography described by DEMs with a resolution of 50 m or better, and geological databases increasingly digitised. These new information layers open perspectives for producing maps for whole regions or countries, based on duly established relations between soil and extrinsic factors (Lagacherie, 1992; Robbez-Masson, 1994; Dobos et al., 2000; Bui and Moran, 2003).

A full review of recent approaches to make digital soil maps from existing environmental data, including from soil-forming factors, has been proposed by

McBratney et al. (2003) and Lagacherie and McBratney (Chapter 1). In a more focused presentation, we distinguish three main groups of methodologies that explicitly integrate pedological knowledge in the procedures linking soil properties or soil classes to their forming factors, according to the embedded level of explanation of their relation.

(i) Deterministic approach.

In few cases, simplified physical laws may be related under given assumptions to the spatial distribution of soil properties (Heimsath et al., 1997). The distribution of hydric soils in a landscape is a good example of such an approach as their genesis is related to the waterlogging duration, which may be derived from hydrological modelling. Beven and Kirkby (1979) developed a physically based hydrological model to predict the extent and distribution of saturated areas in a watershed. Under the assumptions that the hydraulic gradient is under the control of the surface topography and that the soil hydraulic conductivity decreases exponentially with depth, and also assuming a uniform discharge and a succession of steady states, the authors show that the spatial distribution of the saturated areas in a catchment can be described by an index, so-called soil-topographic index, $\log(a/T \tan\beta)$, where a is the drainage area per unit contour length and T is the transmissivity. Given that, in most cases, the variations of transmissivity can be neglected as compared with the variations of slope and drainage area; the resulting index, becomes $\log(a/\tan\beta)$, is called the topographic index. The potential saturation of soil increases with the value of this index.

Many studies have applied this topographic index, easily derived from DEMs, to predict hydric soil distribution in space (Moore et al., 1993; Mérot et al., 1995; McKenzie and Ryan, 1999; Chaplot et al., 2003; Chaplot and Walter, Chapter 38). In the context of the Armorican Massif, where the underlying assumptions generally apply, prediction of hydric soils based on topographically controlled waterlogging appeared efficient, equivalent to detailed soil survey and better than interpolation approaches based exclusively on punctual observations (Mérot et al., 1995; Curmi et al., 1998). Prediction errors were mainly explained by uncertainties in the topographic index computation due to the use of DEMs with too coarse a resolution (Chaplot et al., 2000). The prediction could also be enhanced by the use of the upper boundary of saprolite rather than the present surface topography (Chaplot et al., 2004) or the additional consideration of the elevation above stream which may be linked to the water table associated with the stream (Crave and Gascuel, 1996; Chaplot et al., 2000). Finally, Mérot et al. (2003) applied this approach to contrasted situations in Europe: they took into account the effective rainfall, in addition to topographic factors, in order to achieve prediction in areas with different rainfall conditions. They found that the approach was able to predict the structure and general extent of wetlands without any local calibration of the model, except where the permeability of the soil surface is important and very different from the other sites (Plate 22b (see Colour Plate Section)).

Another example of a deterministic approach is the one proposed by Park et al. (2001) on the basis of a "nine-unit soil landscape model" (Conacher and Darlympe, 1977). According to this model, each point on a slope can be allocated to one of nine soil units, each having distinct pedogenic characteristics reflecting the influence of soil water–gravity interrelationships governed by surface forms. Park et al. (2001) developed a numerical formulation of this model that uses two topographic indices derived from DEMs, namely upslope contributing area and surface curvature. The model is considered easily transferable to another slope in a predictive manner as it is based on a deterministic soil-formation model. However, it is restricted to situations of marked relief, which predominantly influences soil variability.

(ii) Conditional probability approaches using maps of reference areas.

In many situations, the distribution of soil properties or classes cannot be easily derived from generating processes as they may be too complex or subject to feedback, so that a unique spatial prediction may not be derived from a given set of conditions. An alternative is to use a more local kind of knowledge that can be derived from the soil survey of a reference area (Favrot, 1981) i.e. an area selected to be representative of the region of interest. As argued by Bui (2004), the soil map issued from a soil survey can be considered as a structured knowledge of the soil spatial distribution. The mental model used to make soil maps can therefore be captured by some kind of expert rules and used for mapping the region. This supposes obviously that much attention is paid to selecting and mapping the reference area (Lagacherie et al., 2001; Bui and Moran, 2003). Two types of expert rules have been derived from a soil map, that is soil landscape rules and soil patterns rules, which are briefly described.

Soil landscape rules use as predictors the scorpan factors (Lagacherie and McBratney, Chapter 1)), for example DEM, remote sensing, and predict soil class at unvisited places. They have been derived from soil maps using decision trees (Lagacherie, 1992; Lagacherie and Holmes, 1997; Bui et al., 1999) that provide a set of probabilistic rules expressed as follows:

if	*the elevation at x_0 is greater than 20.5 m*
and if	x0 is on the left bank of the river
and if	*x_0 is included in geological unit no 1, 3 or 4*
then	probability estimates of the presence of soil classes 11, 12, 13, ..., 3 at x_0 *are respectively 37, 33, 23, ..., 1%*

The application of the soil landscape rules for predicting soils in the Herault valley (Lagacherie, 1992) gave fairly good results: precision of the resulting map was intermediate between the existing 1:100,000 and 1:250,000 soil maps with a purity (74%) similar to the one measured for the conventional 1:250,000 soil map (69%). Soil-landscape rules were applied more extensively to map the Murray-Darling Basin of Australia at 1:250,000 scale from a set of existing soil surveys (Bui and Moran, 2003).

Soil pattern rules use a set of sampling points and DEMs to predict soil classes at unvisited places (Lagacherie et al., 1995). They translate the knowledge of the soil pattern derived from the reference area map. For example, if in the reference area soil classes A and B are systematically separated by soil class C, unit C will be predicted to be somewhere between two observation points where A and B have been identified. Soil patterns rules may be expressed as follows:

If *the soil mapping unit* s_5 *is identified at observation point* $\mathbf{x}_i$,
if point $\mathbf{x}_0$ *is located between* d_k *metres and* d_k*+50 metres from* $\mathbf{x}_i$,
and if point $\mathbf{x}_0$ *is higher than* $\mathbf{x}_i$,
then probability estimates of the presence of soil classes $s_1, s_2, \ldots, s_j, \ldots, s_p$ *are respectively 3, 69, 23, ..., 1%*

The procedure using soil pattern rules for predicting soil classes was tested in the Valley of Herault in south of France (Lagacherie et al., 1995). The levels of purity obtained by the procedure overall ranged between 55 and 90% depending on the density of observed sites and the complexity of the soil patterns, therefore equivalent to conventional soil maps based on a similar sampling density. Soil pattern rules were further used for predicting soil properties from a limited set of measurements located at the representative profile of the reference area (Lagacherie and Voltz, 2000).

(iii) Expert system approaches.

The knowledge on soil–environment relationship can also be provided directly by soil surveyors having substantial experience in a given region. Several numerical mapping procedures have been developed to combine this knowledge with spatial datasets of environment variables. They differ both in the theoretical frameworks for representing the soil knowledge and in the technique used to capture the knowledge (McBratney et al., 2003).

Bayesian inference systems like Prospector (Duda et al., 1978) were adapted to soil mapping by Skidmore et al. (1991) and Cook et al. (1996) to predict soil classes. In Skidmore et al. (1991), expert knowledge is conveyed by two sets of probabilities: (i) the conditional probabilities of an environmental situation given an observed soil class (e.g. the probability that slope is less than 5% when soil class C was observed, is equal to 0.6) and (ii) the probability of occurrence of each soil class over the whole study area. These two sets of probabilities are provided by an experienced soil surveyor. To simulate the intuitive weighting approach of a soil surveyor, Cook et al. (1996) used the odds format of Bayes theorem, which accepts as input two likelihood ratios. The first evaluates how strong is the belief that a given soil class may occur in a given environmental situation whereas the second evaluates how damaging it is to a given soil class occurrence if this environmental situation does not occur. For both approaches, a pooled estimate of probability of each soil classes is provided that takes into account all the considered environmental situations and their respective uncertainty of occurrence. The results are fairly good, for example 33% of misclassified soil classes (Skidmore et al., 1991). However, to avoid inconsistencies and large prediction errors, Bayesian inferences require a precision of inputs that is often higher than the precision expected from soil expertise. Cook et al. (1996) suggested therefore the use of a set of observation points to estimate the likelihood ratios.

Fuzzy logic is an alternative that seems more adapted to the imprecise knowledge conveyed by soil surveyors. The Soil Land Inference model (Zhu et al., 1996) is a good example of fuzzy inference scheme for mapping soil classes. A prediction is provided under the form of a vector of membership values (the Soil Similarity Vector) describing the similarity of the soil at a given point to predefined soil classes. This prediction is inferred by fuzzy-logic algorithms from inputs provided by soil surveyors. To obtain these inputs, a structured interview of soil surveyor is undertaken. Two types of knowledge are formulated: the central concepts of each soil class under the form of ranges of values for the environmental variables (e.g. soil class A is found in areas with less than 5% slope and on granite) and the tolerance for a soil series to the variations of environmental conditions (to which extent am I still in soil class A if slope differs from the reference slope?). A very similar approach is applied by Martin-Clouaire et al. (2000) for disaggregating compound soil landscape units into soil classes using spatial datasets of environmental variables. The difference is that the fuzzy inputs are provided by information stored in classical soil databases, which greatly extends the possibility of using such techniques. The main interest of such techniques is that the knowledge of soil surveyor is accurately represented, that is without additional hypothesis and parameters that artificially reduces its imprecision. It may however lead to soil predictions that are often too imprecise for decision-making. In such situations, it may reveal and locate the necessity of additional data to solve a given problem.

Finally, the approaches mentioned above (deterministic, based on reference maps, expert system) differ strongly with respect to the existing pedological knowledge taken into account: the deterministic approach applies detailed processes occurring in the soil and landscape, whereas the conditional probability approaches rely on a list of factors presumably correlated to soil variation as shown by a reference area soil map. Despite these strong differences, steps to apply these approaches are essentially the same: (i) calibration/training on comprehensive datasets or reference areas; (ii) extrapolation over larger areas where only the environmental factors are available; (iii) validation based on independent soil observations or maps. The last step, namely validation, is essential, as prediction errors need to be quantified so that departures from assumptions or reference conditions may be detected.

22.4 Perspectives: integration of process dynamics to predict soils in space and time

Purely state-space approaches are not sufficient to integrate soil variation in decision-making procedures, which may be the main issue of soil science in near future. Soil management practices should be assessed before their effective implementation through their potential impact on both local and regional scales integrating evolutions over time.

Relatively few studies integrate both spatial and temporal dimensions. Heuvelink and Webster (2001) review different approaches to combine these dimensions and also remind that physical laws of continuity and conservation of mass and energy must be obeyed in time, which relies on knowledge of processes occurring in the soil. In comparison with purely space approaches, development of spatio-temporal models of soil variation requires therefore additional integration of pedological knowledge namely the dynamics of physical and biogeochemical processes.

Soil organic matter, as a major soil quality indicator, was first considered for its joined spatial and time variations (Jolivet et al., 1997; Nieder and Richter , 2000). Arrouays (1995) studied at a regional level the carbon-stock evolution following deforestation using a C dynamic model derived from δC^{13} measurements and aerial photographs over a 30-year period. Walter et al. (2003) developed a method for field-scale simulations of the spatio-temporal evolution of topsoil organic carbon at the landscape scale over a few decades and under different management strategies. A virtual landscape with characteristics matching part of Brittany (France) was considered for the study. Stochastic simulations (Goovaerts, 2000) and regression analysis were used to create a virtual landscape with characteristics matching a given region. Spatial fields with known spatial structures were simulated: short- and long-range variability was extracted from experimental variograms; medium-range variability derived from lower mineralisation rates in waterlogged soils. Agricultural influence was simulated by considering different land uses and their evolution overtime was simulated using transition matrices. Evolution of soil organic matter was estimated each year for each pixel through a rudimentary balance model that accounts for land use and the influence of soil waterlogging on mineralisation rates (Fig. 22.3). This spatio-temporal simulation approach at the landscape level allowed the simulation of several scales of soil variability including within-field variability.

Simulations over longer time periods have also been recently developed at landscape scale (Heimsath et al., 1997; Minasny and McBratney, 1999, 2001) to infer soil distribution in space from underlying processes of soil production through alteration and soil transport (erosion–sedimentation). At such time scales (thousands of years), small variations in the initial state of the soil may lead to final large differences and the perturbations occurring over time are difficult to document. Such simulation approach appears therefore more useful to study what kinds of soil patterns may be generated by given soil processes rather than to predict soil properties in a real landscape.

Finally, the main limitation of spatio-temporal approaches lies in the lack of models to predict the impact on soil of external constraints. To a first approximation, such models exist to predict evolution of stocks and fluxes for water and chemicals, but not for the evolution of the soil physical and particularly biological states. We should therefore be able to develop predictive models involving mass conservation (organic matter, nutrients and trace elements), but this appears not foreseeable for soil physical and biological qualities. Monitoring networks able to detect evolution of these properties are therefore critically needed to supply the missing modelling approaches.

22.5 Conclusion

Much effort has been devoted over the last few decades to the development of new tools for soil spatial analysis: we have taken advantage from methodological progress (statistical inference, deterministic modelling, stochastic simulation, etc.) to combine soil information with environmental factors, for spatial, and progressively temporal, prediction. We should therefore be able to move to soil-landscape modelling that integrates major soil processes and therefore develop a dynamic representation where risk or impact assessment may be evaluated.

'Field' application of these new tools will become a crucial issue. Soil scientists with field experience appear to become rare in most countries and therefore the main challenge of soil survey consists in the training of a new generation of specialists capable of good fieldwork and also mastering modern spatial analysis techniques.

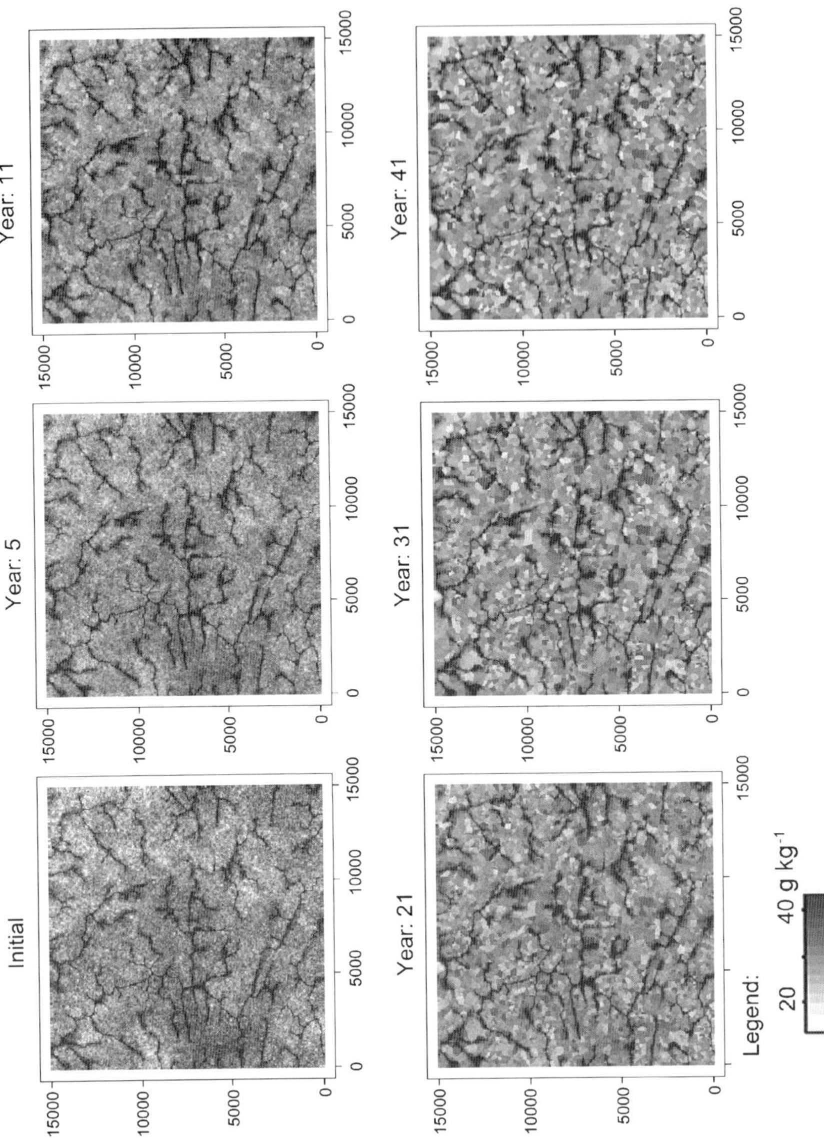

Figure 22.3. *Example of soil spatio-temporal simulation using a framework of stochastic and deterministic approaches: simulated organic carbon content evolution in a virtual agricultural landscape over a 50-year period given a scenario of strong land use change (after Walter et al., 2003).*

References

Arrouays, D., 1995. Analyse et modélisation spatiales de l'évolution des stocks de carbone organique des sols à l'échelle d'un paysage pédologique. Ph.D. Thesis ENSA Montpellier, INRA Orléans, France (in French).

Banfield, C.F., Bascomb, C.L., 1976. Variability in three areas of the Denchworth soil map unit. II. Relationships between soil properties and similarities between profiles using laboratory measurements and field observations. J. Soil Sci. 27, 438–450.

Beckett, P.H.T., Webster, R., 1971. Soil variability: a review. Soils Fertilizers 34, 1–15.

Beven, K.J., Kirkby, M.J., 1979. A physically based, variable contributing area model of basin hydrology. Hydrol. Sci. Bull. 23, 419–438.

Bockheim, J.G., Gennadiyev, A.N., Hammer, R.D., Tandarich, J.P., 2005. Historical development of key concepts in pedology. Geoderma 124, 23–36.

Boiffin, J., Stengel, P., 2000. Réapprendre le sol: nouvel enjeu pour l'agriculture et l'espace rural. In: Demeter 2000, Economie et Stratégies agricoles, Armand Colin, pp. 152–194.

Boulaine, J., 1980. Pédologie Appliquée. Masson, Paris.

Boulaine, J., 1989. Histoire des Pédologues et de la Science des Sols. INRA Editions, Paris.

Bouma, J., de Laat, P.J.M., Awater, R.H.C.M., van Heesen, H.C., van Holst, A.F., van de Nes, T.J.M., 1980. Use of soil survey data in a model for simulating regional soil moisture regimes. Soil Sci. Soc. Am. J. 44, 808–814.

Brus, D.J., De Gruijter, J.J., 1997. Random sampling or geostatistical modelling. Choosing between design-based and model-based sampling strategies for soil (with Discussion). Geoderma 80, 1–59.

Bui, E.N., 2004. Soil survey as a knowledge system. Geoderma 120, 17–26.

Bui, E.N., Moran, C.J., 2003. A strategy to fill gaps in soil survey over large spatial extents: an example from the Murray-Darling basin of Australia. Geoderma 111, 21–44.

Bui, E.N., Loughhead, A., Corner, R., 1999. Extracting soil-landscape rules from previous soil surveys. Aust. J. Soil Res. 37, 495–508.

Burrough, P.A., 1989. Fuzzy mathematical methods for soil survey and land evaluation. J. Soil Sci. 40, 477–492.

Burrough, P.A., McDonnell, R., 1998. Principles of Geographical Information Systems, 2nd edn. Oxford University Press, Oxford.

Butler, B.E., 1980. Soil Classification for Soil Survey. Oxford University Press, Oxford.

Cazemier, D., Lagacherie, P., Martin-Clouaire, R., Bornand, M., 1998. Mapping soil hydraulic properties from imprecise information contained in a small scale pedological map. XVIth World Congress of Soil Science, Montpellier.

Cazemier, D.R., Lagacherie, P., Martin-Clouaire, R., 2001. A possibility theory approach for estimating available water capacity from imprecise information contained in soil databases. Geoderma 103, 113–132.

Chaplot, V., Van Vliet, B., Walter, C., Curmi, P., Cooper, M., 2003. Soil spatial distribution in the Armorican Massif, Western France: effect of soil-forming factors. Soil Sci 168, 445–454.

Chaplot, V., Walter, C., Curmi, P., 2000. Improving soil hydromorphy prediction according to DEM resolution and available pedological data. Geoderma 97, 405–422.

Chaplot, V., Walter, C., Curmi, P., Lagacherie, P., King, D., 2004. Using the topography of the saprolite upper boundary to improve the spatial prediction of the soil hydromorphic index. Geoderma 123, 343–354.

Cochran, W.G., 1977. Sampling Techniques. Wiley & Sons, New York.

Conacher, A.J., Darlympe, J.B., 1977. The nine unit land surface model: an approach to pedogeomorphic research. Geoderma 18, 1–153.

Cook, S.E., Corner, R., Grealish, G.J., Gessler, P.E., Chartres, C.J., 1996. A rule-based system to map soil properties. Sci. Soc. Am. J. 60, 1893–1900.

Crave, A., Gascuel, C., 1996. The influence of topography on time and space distribution of soil surface water content. Hydrol. Process. 11, 203–210.

Cressie, N., 1991. Part I. Geostatistical data, In: Statistics of Spatial Data, John Wiley & Sons, New York, pp. 29–382.

Curmi, P., Durand, P., Gascuel-Odoux, C., Mérot, P., Walter, C., Taha, M., 1998. Hydromorphic soils hydrology and water quality: spatial distribution and functional modelling at different scales. Nutr. Cycling Agroecosyst. 50, 127–142.

De Gruijter, J.J., Marsman, B.A., 1984. Transect sampling for reliable information on mapping units. Proceedings of Workshop of ISSS and SSSA, Las Vegas, USA, Pudoc Wageningen. pp. 150–165.

Dobos, E., Micheli, E., Baumgardner, M.F., Biehl, L., Helt, T., 2000. Use of combined digital elevation model and satellite radiometric data for regional soil mapping. Geoderma 97, 367–391.

Dokuchaev, V.V., 1893. Russian Chern ozems. Israel Prog. Sci. Trans., Jerusalem, Israel, 1967, translated by N. Kaner.

Doran, J.W., Coleman, D.C., Bezdicek, D.F., Stewart, B.A., 1994. Defining Soil Quality for a Sustainable Environment. SSSA Spec. Publication No. 35, SSSA/ASA, Madison, WI.

Duda, R.O., Hart, P., Barrett, J.G., Gashnig, K., Reboh, R., Slocum, J., 1978. Development of PROSPECTOR consultation system for mineral exploration. SRI Projects 5821 and 6415, SRI International Artificial Intelligence Center, Menlo Park, CA.

Favrot, J.C., 1981. Pour une approche raisonnée du drainage agricole en France : la méthode des secteurs de référence. C.R. Académie d'agriculture de France, séance du 6 mai 1981, 716–723.

Gascuel, C., Walter, C., Voltz, M., 1993. Intérêt du couplage des méthodes géostatistiques et de cartographie des sols pour la prédiction spatiale. Sci. du Sol 31, 193–213.

Goovaerts, P., 1997. Geostatistics for Natural Resources Evaluation. Oxford University Press, New York.

Goovaerts, P., 2000. Estimation or simulation of soil properties? An optimisation problem with conflicting criteria. Geoderma 97, 165–186.

Heimsath, A.M., Dietrich, W.E., Nishiizumi, K., Finkel, R.C., 1997. The soil production function and landscape equilibrium. Nature 388, 358–388.

Heuvelink, G.B.M., 1996. Identification of field attribute error under different models of spatial variation. Int. J. GIS 10, 921–935.

Heuvelink, G.B.M., Webster, R., 2001. Modelling soil variation: past, present and future. Geoderma 100, 269–301.

Horn, R., Baumgartl, T., 2000. Dynamic properties of soils. In: M.E. Sumner (Ed.), Handbook of Soil Science. CRC Press, Boca Rotan, Florida, pp. E77–E116.

Hoosbeek, M.R., Bouma, J., 1998. Obtaining soil and land quality indicators using research chains and geostatistical methods. Nutr. Cycling Agroecosyst. 50, 35–50.

Hoosbeek, M.R., Amundson, R.G., Bryant, R.B., 1999. Pedological modeling. In: M.E. Sumner (Ed.), Handbook of Soil Science. CRC Press, Boca Rotan, Florida, pp. E77–E116.

Jamagne, M., 1967. Bases et techniques d'une cartographie des sols. Ann. agron. 18 (No h.s.), 142.

Jenny, H., 1941. Factors of Soil Formation: A System of Quantitative Pedology. McGraw Hill Book Company, New York.

Jenny, H., 1961. Derivation of state factor equation of soils and ecosystems. Soil Sci. Soc. Am. Proc. 25, 385–388.

Jolivet, C., Arrouays, D., Andreux, F., Leveque, J., 1997. Soil organic carbon dynamics in cleared temperate forest spodosols converted to maize cropping. Plant Soil 191, 225–231.

King, D., Daroussin, J., Tavernier, R., 1994. Development of a soil geographic database from the soil map of the European Communities. Catena 21, 37–56.

Lagacherie, P., 1992. Formalisation des lois de distribution des sols pour automatiser la cartographie pédologique à partir d'un secteur pris comme référence. Cas de la Petite Région Naturelle de la moyenne vallée de l'Hérault. Ph.D. Thesis, ENSA-INRA Montpellier, France (in French).

Lagacherie, P., Andrieux, P., Bouzigues, R., 1994. Fuzziness and uncertainty of soil boundaries: from reality to coding in GIS. In: P.A. Burrough and A. Frank (Eds.), Spatial Conceptual Models for Geographic Objects with Undetermined Boundaries. GISDATA 2. Taylor & Francis, London, pp. 275–287.

Lagacherie, P., Legros, J.P., Burrough, P., 1995. A soil survey procedure using the knowledge of soil pattern established on a previously mapped reference area. Geoderma 65, 283–301.

Lagacherie, P., Holmes, S., 1997. Addressing geographical data errors in a classification tree for soil unit predictions. Int. J. Geogr. Inf. Sci. 11, 183–198.

Lagacherie, P., Robbez-Masson, J.M., NguyenThe, N., Barthès, J.P., 2001. Mapping of reference area representativity using a mathematical soilscape distance. Geoderma 101, 105–118.

Lagacherie, P., Voltz, M., 2000. Predicting soil properties over a region using sample information from a mapped reference area and digital elevation data: a conditional probability approach. Geoderma 92, 141–165.

Legros, J.P., 1996. Cartographie des Sols : de l'Analyse Spatiale à la Gestion des Territoires. Presses Polytechniques et Universitaires Romandes, Lausanne.

Leenhardt, D., 1991. Spatialisation du bilan hydrique. Propagation des erreurs d'estimation des caractéristiques du sol au travers des modèles de bilan hydrique. Cas du blé dur d'hiver. Ph.D. Thesis, Montpellier, France (in French).

Marsman, B.A., De Gruijter, J.J., 1986. Quality of soil maps: a comparison of survey methods in a sandy area. Soil Survey Papers. Netherlands Soil Survey Institute, Wageningen.

Martin-Clouaire, R., Cazemier, D., Lagacherie, P., 2000. Representing and processing uncertain soil information for mapping hydrological soil properties. Comput. Electron. Agric. 29, 41–57.

McBratney, A.B., Hart, G.A., McGarry, D., 1991. The use of region partitioning to improve the representation of geostatistically mapped soil attributes. J. Soil Sci. 42, 513–532.

McBratney, A.B., Mendonça-Santos, M.L., Minasny, B., 2003. On digital soil mapping. Geoderma 117, 3–52.

McBratney, A.B., Minasny, B., Cattle, S.R., Verwoort, R.W., 2002. From pedotransfer functions to soil inference systems. Geoderma 109, 41–73.

McBratney, A.B., Pringle, M.J., 1999. Estimating average and proportional variograms of soil properties and their potential use in Precision Agriculture. Precision Agric 1, 219–236.

McKenzie, N.J., Ryan, P.J., 1999. Spatial prediction of soil properties using environmental correlation. Geoderma 89, 67–94.

Mérot, P., Ezzahar, B., Walter, C., Aurousseau, P., 1995. Mapping waterlogging of soils using digital terrain models. Hydrol. Process. 9, 27–34.

Mérot, P., Squividant, H., Aurousseau, P., Hefting, P., Burt, T., Maitre, V., Kruk, M., Butturini, A., Thenail, C., Viaud, V., 2003. Testing a climato-topographic index for predicting wetlands distribution along an European climate gradient. Ecol. Model. 163, 51–71.

Minasny, B., McBratney, A.B., 1999. A rudimentary mechanistic model for soil production and landscape development. Geoderma 90, 3–21.

Minasny, B., McBratney, A.B., 2001. A rudimentary mechanistic model for soil production and landscape development: II. A two-dimensional model. Geoderma 103, 161–179.

Monestiez, P., Allard, D., Navarro Sanchez, I., Courault, D., 1999. Kriging with categorical external drift: use of thematic maps in spatial prediction and application to local climate interpolation for agriculture. In: J. Gomez-Hernandez, A. Soares, and R. Froidevaux (Eds.), GeoENV II. Geostatistics for environmental applications 2. European Conference, Valencia, pp. 163–174.

Moore, I.D., Gessler, P.E., Nielsen, G.A., Peterson, G.A., 1993. Soil attribute prediction using terrain analysis. Soil Sci. Soc. Am. J. 57, 443–452.

Nieder, R., Richter, J., 2000. C and N accumulation in arable soils of West Germany and its influence on the environment – Developments 1970 to 1998. J. Plant Nutr. Soil Sci. 163, 65–72.

Oliver, M.A., Webster, R., 1986. Combining nested and linear sampling for determining the scale and form of spatial variation of regionalized variables. Geogr. Anal. 18, 227–242.

Park, S.J., McSweeney, K., Lowery, B., 2001. Identification of the spatial distribution of soils using a process-based terrain characterization. Geoderma 103, 249–272.

Robbez-Masson, J.M., 1994. Reconnaissance et délimitation de motifs d'organisation spatiale. Application à la cartographie des pédopaysages. Ph.D. Thesis, ENSA Montpellier, France (in French).

Ruellan, A., Dosso, M., Fritsch, E., 1989. L'analyse structurale de la couverture pédologique. Sci. du Sol 27, 319–334.

Schvartz, C., Walter, C., Daroussin, J., King, D., 1998. Statistical review of soil tests made in France and comparison with the 1:1,000,000 soil map. Proceedings of the 16th World Congress of Soil Science Montpellier. Symposium 17 "New tools and methods in soil survey".

Sebillotte, M., 1989. Fertilité et systèmes de production. Essai de problématique générale. In: M. Sébillotte (Ed.), Fertilité et Systèmes de production. Editions INRA, Paris, pp. 13–57.

Skidmore, A.K., Ryan, P.J., Dawes, W., Short, D., O'Loughlin, E., 1991. Use of an expert system to map forest soils from a geographical information system. Int. J. Geogr. Inf. Sci. 5, 431–445.

Soil Survey Staff, 1951. Soil Survey Manual. US Department of Agriculture Handbook 18. Government Printing Office, Washington, DC.

Steers, C.A., Hajek, B.F., 1979. Determination of map unit composition by a random selection of transects. Soil Sci. Soc. Am. J. 43, 156–160.

Stein, A., Hoogerwerf, M., Bouma, J., 1988. Use of soil map delineations to improve (co-) kriging of point data on moisture deficits. Geoderma 43, 163–177.

Tabbagh, A., Dabas, M., Hesse, A., Panissod, C., 2000. Soil resistivity: a non-invasive tool to map soil structure horizonation. Geoderma 97, 393–404.

Thom, R., 1993. Prédire n'est pas expliquer. Entretiens avec Emile Noël. Collection champs. Flammarion, Paris.

Van Meirvenne, M., Scheldeman, K., Baert, G., Hofman, G., 1992. Quantification of soil textural fractions of the Bas-Zaïre using soil map polygons and/or point observations. Proceedings of Pedometrics-92: Developments in Spatial Statistics for Soil Science, Wageningen, pp. 222–238.

Viscarra Rossel, R.A., Walter, C., 2003. Rapid, quantitative and spatial field measurements of soil pH using an ion sensitive field effect transistor. Geoderma 119, 9–20.

Voltz, M., Webster, R., 1990. A comparison of kriging, cubic splines and classification for predicting soil properties from sample information. J. Soil Sci. 41, 473–490.

Walter, C., 1990. Estimation de propriétés du sol et quantification de leur variabilité à moyenne échelle. Cartographie pédologique et géostatistique dans le sud de l'Ille-et-Vilaine (France). Ph.D. Thesis, Université Paris VI, France (in French).

Walter, C., Schvartz, C., Claudot, B., Bouedo, T., Aurousseau, P., 1997. Synthèse nationale des analyses de terre. II. Descriptions statistique et cartographique de la variabilité des horizons de surface des sols cultivés. Etude et Gestion des Sols 3, 205–219.

Walter, C., Viscarra Rossel, R.A., McBratney, A.B., 2003. Spatio-temporal simulation of the field-scale evolution of organic carbon over the landscape. Soil Sci. Soc. Am. J. 67, 1477–1486.

Wang, C., 1984. La méthode du transect et son application aux problèmes de la prospection pédologique. Bul. Tech., 1984-4F, Dir. Gale de la Rech. du Canada, 35pp.

Webster, R., 2000. Is soil variation random? Geoderma 97, 149–163.

Wilding, L.P., Drees, L.R., 1983. Spatial variability and pedology. In: L.P. Wilding, N.E. Smeck, G.F. Hall (Eds.), Pedogenesis and Soil Taxonomy I. Concepts and Interactions. pp. 83–116.

Wright, R.L., Wilson, S.R., 1979. On the analysis of soil variability, with an example from Spain. Geoderma 22, 297–313.

Yaalon, D.H., 1971. Soil-forming processes in time and space. In: D.H. Yaalon (Ed.), Paleopedology – Origin, nature and dating of paleosols. Int. Soc. Soil. Sci. and Israel Universities Press, Jerusalem, Israel, pp. 29–39.

Zhu, A.X., Band, L.E., Dutton, B., Nimlos, T.J., 1996. Automated soil inference under fuzzy logic. Ecol. Model. 90, 123–145.

Developments in Soil Science, volume 31
P. Lagacherie, A.B. McBratney and M. Voltz (Editors)

Chapter 23

DECOMPOSING DIGITAL SOIL INFORMATION BY SPATIAL SCALE

R.M. Lark

Abstract
Digital soil mapping (DSM) depends on the identification of relationships between soil variables and informative covariates. These have components of variation that correspond to different spatial scales, so an analysis of variation by spatial scale may be both informative and of practical value for improving predictions of soil properties. In this chapter multiresolution analysis on wavelet basis functions is explained and illustrated, using both continuous variables and indicator variables defined on soil classes. A case study is presented in which a decomposition of sensor data on soil electrical conductivity resulted in better predictive models for soil properties. The use of wavelet methods for denoising data is illustrated. Because of the local properties of wavelet coefficients it is possible to smooth noise while retaining local variation at high spatial frequencies that appears to be significant. Finally, it is shown how wavelet packet transforms and best-basis identification can be used to develop a wavelet basis that is adapted to the properties of a particular data set.

23.1 Introduction

The spatial variability of the soil is caused by many different factors, and these factors operate over a range of spatial scales. The goal of digital soil mapping (DSM) is more effectively to map the actual variations of soil using information on covariates (such as digital elevation data or data from remote sensors). Covariates will also vary at different spatial scales, and the variation at some scales may be more closely correlated with soil variation than at others. A consequence of this is that the spatial variations of particular covariates may be more useful for predicting the variation of the soil at some spatial scales than at others. That spatial variables may be correlated in a scale-dependent way has been widely recognised (e.g. Rodriguéz-Arias and Rodó, 2004). Quantitative modelling of variation in the data that we use in DSM must account for this scale dependence. For example, when we explore the relationships between soil properties and covariates in a preamble to DSM it would be useful to decompose those variables for which we have appropriate data into components of different spatial scale.

The contention of this chapter is that the identification of predictively useful relationships between covariates used in DSM and key soil variables of interest will be facilitated by an analysis of their variation into components associated with different spatial scales. It is also argued that the most appropriate way to do this is with the wavelet transform, since this entails a minimum of assumptions about the variability of the variables in question. Most particularly, it requires no assumption that the data are a realisation of a spatially stationary process.

This chapter is primarily an introduction to multiresolution analysis of data by spatial scale using the wavelet transform. The aim is to help the reader understand fully the output of software for wavelet transforms, which is becoming widely available, and to encourage the use of the method in DSM. A case study is presented in which such a multiresolution analysis, on data from a sensor measuring the electrical conductivity of the soil, substantially improved predictions of soil properties from the data. Then two further aspects of the wavelet transform are addressed that are relevant in the context of DSM. The first exploits properties of the wavelet transform in order to 'denoise' data.In the second, I consider the valid criticism of the ordinary discrete wavelet transform (DWT) that, unlike factorial kriging, it imposes a set of spatial scales into which our data are decomposed. It is shown how it is possible to find a basis, in a library of wavelet packet bases, that is the best in some sense for representing spatial variation in a particular data set.

23.2 Multiresolution analysis on wavelets

23.2.1 Some mathematical background

To understand the wavelet transform we require some results from vector algebra. In particular we need to develop the idea of a vector space. A vector space, V, is a set of vectors such that:

$$(\mathbf{a} + \mathbf{b}) \in V, \quad \text{if } \mathbf{a} \in V \text{ and } \mathbf{b} \in V, \tag{23.1}$$

and

$$k\mathbf{a} \in V \quad \forall\, \mathbf{a} \in V \quad \text{for any scalar } k, \tag{23.2}$$

and such that certain rules of associativity and commutativity govern scalar products and sums of vectors in the space. An inner product space is a vector space in which the inner products of vectors obey certain rules. Two vectors, **a** and **b** both in V, are considered orthogonal if and only if

$$\langle \mathbf{a}, \mathbf{b} \rangle = 0 \tag{23.3}$$

where $\langle , \rangle$ denotes the inner product of the terms in the angle brackets.

An inner product space in which any convergent sequence of vectors converge to a vector that belongs to the space is a Hilbert space. This includes the set of twice differentiable real-valued functions, L^2 ($\Re^1$), such as the values of any variable of finite variance on a linear transect.

If the maximum number of linearly independent vectors that can be chosen from V is n, then n is the dimension of v. One of the properties of a Hilbert space is that such a set of vectors constitutes what is called an orthogonal basis for the set $\mathbf{a}_1, \mathbf{a}_2, \ldots, \mathbf{a}_n$ such that, for any $\mathbf{v} \in V$

$$\mathbf{v} = \sum_{i=1}^{n} \langle \mathbf{v}, \mathbf{a}_i \rangle \mathbf{a}_i. \tag{23.4}$$

In short, we can represent any vector in V as a linear combination of all the vectors in the basis, the coefficients being the inner product of the vector and each basis vector. This is essentially what a wavelet analysis does, the useful properties of the analysis arising from the form of the basis vectors.

Another property of Hilbert spaces will be required later. Any such space, V, can be represented as the direct sum of two complementary subspaces, Y and Z, if every $\mathbf{x} \in X$ can be represented as

$$\mathbf{x} = \mathbf{y} + \mathbf{z}, \quad \mathbf{y} \in Y \quad \mathbf{z} \in Z. \tag{23.5}$$

If

$$\langle \mathbf{y}, \mathbf{z} \rangle = 0, \quad \forall \mathbf{y} \in Y, \quad \mathbf{z} \in Z, \tag{23.6}$$

then Y and Z are orthogonal complementary subspaces, and we may denote Z by $Y^{\perp}$. The projection theorem states that, if Y is any complete subspace of a Hilbert space H, then $H = Y + Y^{\perp}$, such that for any $\mathbf{x} \in H$ there exists some $\mathbf{y} \in Y$ such that $\mathbf{x} = \mathbf{y} + \mathbf{z}$, $\mathbf{z} \in Y^{\perp}$. The vector $\mathbf{y}$ is the *orthogonal projection* of $\mathbf{x}$ onto Y and $\mathbf{z}$ is the orthogonal projection of $\mathbf{x}$ onto $Y^{\perp}$.

23.2.2 Wavelet bases

We denote by f(x) the values of a soil property as a function of location, x, on a one-dimensional (1-D) transect. If we assume that soil variation is, at the limit, continuous, then this transect belongs to the vector space $L^2(\Re^1)$. There are many basis functions on which we could decompose these data as in Eq. (23.4). One such is the Fourier basis in which $\mathbf{a}_i$ is a complex function of sines and cosines with a particular spatial frequency, ω_i. The resulting coefficient is a complex number that contains the amplitude and phase of a sinusoidal component of the data at the frequency ω_i. We may therefore decompose our data into components of different spatial frequency. Such an analysis is rarely useful in the study of soil variation, however. This is because the analysis makes an assumption of stationarity, that is to say it is assumed that the amplitude and phase of f(x) for some frequency are the same for any x. If this does not hold then any interpretation of the coefficients is impossible, and the components into which f(x) is decomposed are unlikely to be informative.

This is the motivation for considering wavelet bases for multiresolution analysis of soil and other environmental variables. A wavelet oscillates locally like a wave, but it also damps rapidly to zero either side of its centre, see some wavelets in Figure 23.1. This property is called compact support in the wavelet literature. The necessary properties of wavelets are described in detail elsewhere (e.g. Lark and Webster, 1999), but the property of compact support is critical for an intuitive understanding of what a wavelet transform does. It is clear

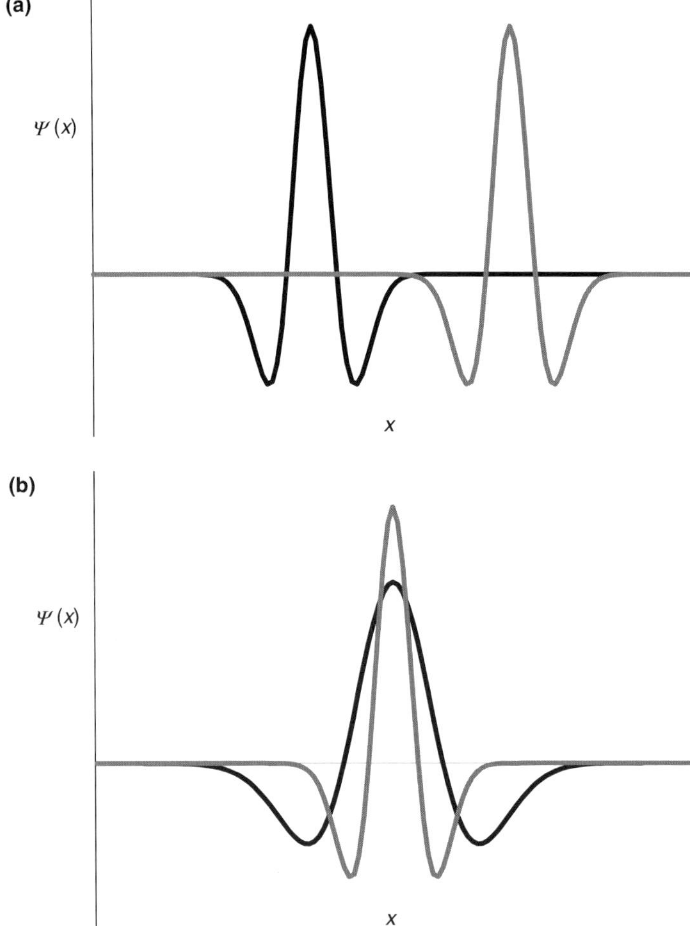

Figure 23.1. *Two translations (a) and two dilations (b) of the Mexican hat wavelet.*

that if vector $\mathbf{a}_i$ contains a basis function that is a wavelet then many of its elements will be zero and the coefficient obtained from its inner product with the data in **v** will represent variation in **v** only over that interval where the wavelet takes non-zero values. This has three implications. First, the resulting wavelet coefficient describes local variation, and so its interpretation relies on no assumptions about the stationarity of the process under investigation. Second, in order to form a complete basis for any data it will be necessary to use wavelets that will respond to variation at other locations; this is achieved by translation of the wavelets (Fig. 23.1a). Third, it is clear that a set of translations of a wavelet will represent variation over a delimited range of spatial frequencies (determined approximately by the length of its compact support and the number of oscillations it makes). We therefore require dilations of the wavelet to provide a complete basis (Fig. 23.1b). If the wavelet is dilated then its coefficients describe local variation at a coarser spatial scale over a wider support.

I now describe the DWT in more detail. A basic or mother wavelet function, $\psi(x)$, is dilated and translated in discrete steps. The dilation is controlled by a scale parameter, λ. In the DWT, $\lambda = \lambda_0^m$ where λ_0 is a dilation step larger than one, conventionally $\lambda_0 = 2$, and $m = 1, 2, \ldots$. The translations are done in scale-dependent steps to location $x = nx_0\lambda_0^m$ where $x_0 > 0$ and $n = 1, 2, \ldots$. By convention $x_0 = 1$. A particular dilation and translation of the wavelet is denoted as $\psi_{m,n}(x)$ where

$$\psi_{m,n}(x) = \lambda_0^{-m/2}\,\psi\left(\lambda_0^{-m}x - nx_0\right). \tag{23.7}$$

The inner product of the data and the wavelet gives a DWT coefficient

$$D_{m,n} = \langle f, \psi_{m,n} \rangle. \tag{23.8}$$

If the mother wavelet has certain properties (Daubechies, 1988) then its dilates and translates, computed in the discrete steps described above, constitute an orthonormal basis for any data set of finite variance. This means that f(x) can be approximated to some arbitrary level of precision as a sum of independent components each corresponding to a particular value of the scale parameter

$$f(x) = \sum_{m=-\infty}^{\infty} \sum_{n=-\infty}^{\infty} \langle f, \psi_{m,n} \rangle \psi_{m,n}. \tag{23.9}$$

Eq. (23.9) contains sums over infinite dilations because f(x) is continuous in x. In practice we must work with discretely sampled data. To do this we must introduce a second function into the basis, this is the scaling function that is intimately related to the wavelet.

If we modify Eq. (23.9) so that the approximation includes only components for which $m > k$, where k is an integer, then the approximation is $P_k f(x)$

$$P_k f(x) = \sum_{m=k+1}^{\infty} \sum_{n=-\infty}^{\infty} \langle f, \psi_{m,n} \rangle \, \psi_{m,n}. \tag{23.10}$$

This is a smooth approximation to the data, for scale parameter 2^k, since all components corresponding to this scale or finer (higher frequency) have been discarded. The discarded component for scale parameter 2^k is called the detail, $Q_k f(x)$

$$Q_k f(x) = \sum_{n=-\infty}^{\infty} \langle f, \psi_{k,n} \rangle \, \psi_{k,n}. \tag{23.11}$$

This is the component of the data extracted by the kth dilation of the wavelet. Comparison of Eq. (23.9) with Eq. (23.11) shows that

$$f(x) = P_k f(x) + \sum_{m=-\infty}^{k} Q_m f(x). \tag{23.12}$$

Since Eq. (23.12) can be written for any arbitrary minimum scale parameter 2^k, we can write

$$P_{k-1} f(x) + \sum_{m=-\infty}^{k-1} Q_m f(x) = P_k f(x) + \sum_{m=-\infty}^{k} Q_m f(x), \tag{23.13}$$

and so

$$P_{k-1} f(x) = P_k f(x) + Q_k f(x), \tag{23.14}$$

that is the smooth representation at some scale is equal to the smooth representation at the next coarsest scale plus the corresponding detail component (the component on the wavelet basis).

Consider a set of increasingly smooth representations of f(x),

$$P_m f(x), P_{m+1} f(x), P_{m+2} f(x), \ldots.$$

Each of these belongs to the vector space L^2 ($\Re^1$), as does f(x) since it is of finite variance. We define $V_m{}^1$ as the vector space that contains the smooth representations at scale parameter 2^m of all vectors in L^2 ($\Re^1$). Now, since it is possible that a signal contains no variation at scale parameters smaller than 2^k, it follows that these subspaces must be nested (i.e. a subspace contains all subspaces of smoother sequences), so

$$V_{m+1}^1 \subset V_m^1 \subset V_{m-1}^1 \subset L^2(\Re^1) \tag{23.15}$$

Mallat (1989) shows that the operation from f(x)$\in L^2$ ($\Re^1$)onto P_mf(x)$\in V_m{}^1$is (i) a linear mapping onto the function in $V_m{}^1$ that resembles f(x) more than any other in V_m^1 and (ii) an operation that would map $P_m f(x)$ onto itself, if it is an orthogonal projection of $f(x)$ onto V_m^1. Orthogonal projections have been defined at the end of Section 23.2.1. By the same argument, $P_{m+1}f(x)$ is the orthogonal projection of $P_m f(x)$ onto V_{m+1}^1, so from the projection theorem (Section 23.2.1) we can write

$$V_m^1 = V_{m+1}^1 \oplus (V_{m+1}^1)^{\perp}, \tag{23.16}$$

so

$$P_m f(x) = P_{m+1} f(x) + B_{m+1} f(x), \tag{23.17}$$

where B_{m+1} $f(x)$ is the orthogonal projection of $P_m f(x)$ onto the orthogonal complement of V_{m+1}^1. Comparison of Eq. (23.17) with Eq. (23.14) shows that B_{m+1} $f(x)$ is the detail component, that is the translations of $\psi_{m+1,n}(x)$ are the basis functions for $(V_{m+1}^1)^{\perp}$. The wavelet therefore defines, via this projection, the complementary space V_{m+1}^1which has some other basis functions. These are the scaling functions, $\phi_{m+1,n}$, $n = 1,2, \ldots$ that may be used to compute the smooth approximation with

$$P_m f(x) = \sum_{n=-\infty}^{\infty} \langle f, \phi_{m,n} \rangle \phi_{m,n}(x), \tag{23.18}$$

where $\phi_{m,n}(x)$ is a dilation and translation of a father scaling function, $\phi(x)$, so

$$P_m f(x) = \sum_{n=-\infty}^{\infty} \langle f, \phi_{m+1,n} \rangle \, \phi_{m+1,n}\,(x) + \sum_{n=-\infty}^{\infty} \langle f, \psi_{m+1,n} \rangle \, \psi_{m+1,n}(x). \tag{23.19}$$

By using the scaling function we are able to compute a multiresolution analysis from a finite number of sampled data, since Eq. (23.19) allows us to compute smooth and detail components without summing over infinite dilations. We still have sums over infinite translations, however, which is a problem because real data occur on a finite interval. In practice, we obtain wavelet coefficients by convolving our data with a wavelet function filter **h** (here a column vector) and smooth approximations using a scaling-function filter **g**. The problem of the finite interval emerges near the limits of the data where these filters overlap the ends. In 1-D analysis, there are adapted versions of Daubechies's (1988) wavelets that can be used near the ends of the data (Cohen et al., 1993) and in two dimensions the data have been padded at the ends by symmetrical extension.

The 1-D DWT is most efficiently implemented using the pyramid algorithm. The data, stored in an array, most simply 2^i of them where i is an integer, are convolved with the filters **g** and **h**. Alternate output from the convolution with g are stored in the first 2^{i-1} positions of the array, and alternate output from the convolution with h are stored in the last 2^{i-1} positions. This downsampling procedure means that when the first 2^{i-1} values are convolved a second time,

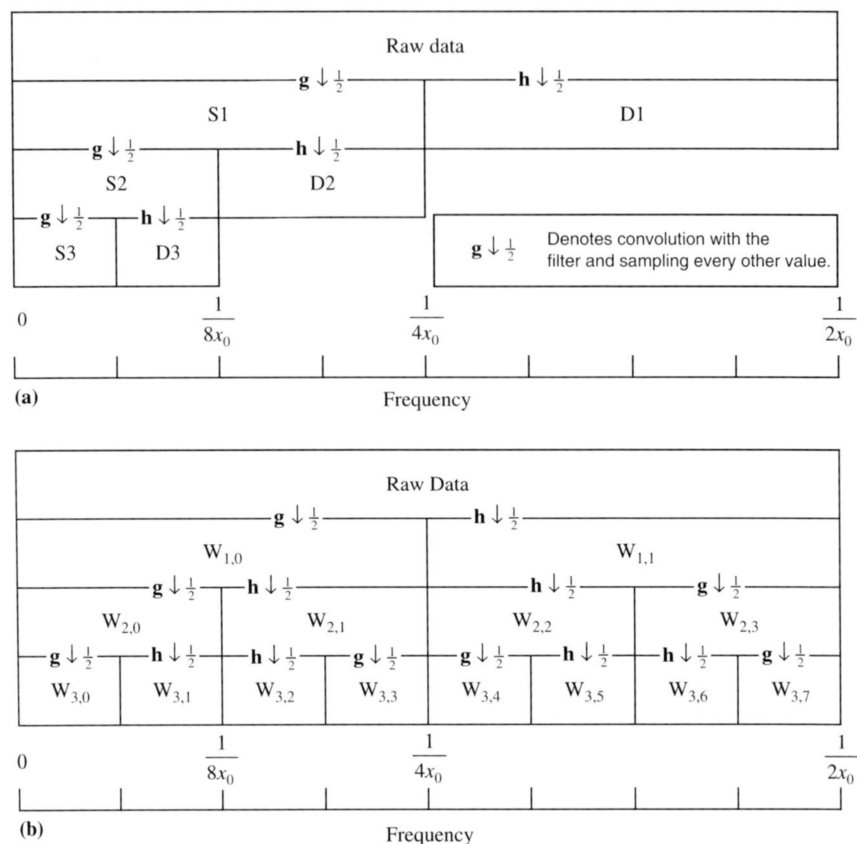

Figure 23.2. *Schematic of top (a) the discrete wavelet transform (DWT); S denotes a scaling-function coefficient and D a wavelet coefficient. Bottom (b) the discrete wavelet packet transform (DWPT).*

with downsampling again, this is equivalent to dilating the wavelet by one step and increasing the translation step by the same amount. Figure 23.2a shows the operation of the pyramid algorithm over three dilations. The result is 2^i coefficients. Of these, 2^{i-1} correspond to the first dilation (scale parameter 2^1, D1 in Fig. 23.2a), 2^{i-2} are wavelet coefficients for the second dilation, 2^{i-3} are wavelet coefficients for the third dilation and the final 2^{i-3} coefficients are the scaling-function coefficients for scale parameter 2^3 (S3 in Fig. 23.2a). The wavelet transform may be inverted, and if we set all but the wavelet coefficients for scale parameter 2^3 to zero before conducting the inverse transform, we would generate the detail component at this scale.

A further development of the basic multiresolution analysis is the shift-averaging procedure of Coifman and Donoho (1995). The multiresolution analysis is done within a moving window, and the results are averaged. This is done to ensure that the output is shift-invariant.

23.2.3 Examples

Figure 23.3 shows a multiresolution analysis of a 1-D data set using the shift-averaging procedure and Daubechies's (1988) wavelet with two vanishing moments. The raw data are measurements of the rate of emission of nitrous oxide from soil cores collected at 4-m intervals on a transect (Lark et al., 2004). Each line represents the detail component at the indicated scale, and at the top the raw data are shown. The scale parameters increase in a dyadic sequence of multiples of the basic sampling interval (4 m). The transect is described in more detail in the reference cited above, but its principal features are mixed drift over Cretaceous sandstone (Lower Greensand) to around position 40, a drainage ditch in alluvium with peaty lenses around position 50, clay drift over Cretaceous clay (Gault Clay) to near position 150, drift of variable texture over the Gault to around position 225 and then clay drift over the Gault to the end of the transect.

The smooth component at scale 128 m shows the broad trend in nitrous oxide emissions with a peak in the alluvium near the ditch and larger emission rates from the drift of mixed texture over the Gault Clay than from the clay drift. The details show complex variation with marked increases in the variability at the finer scales in the vicinity of the ditch and in the drift of mixed texture. In short,

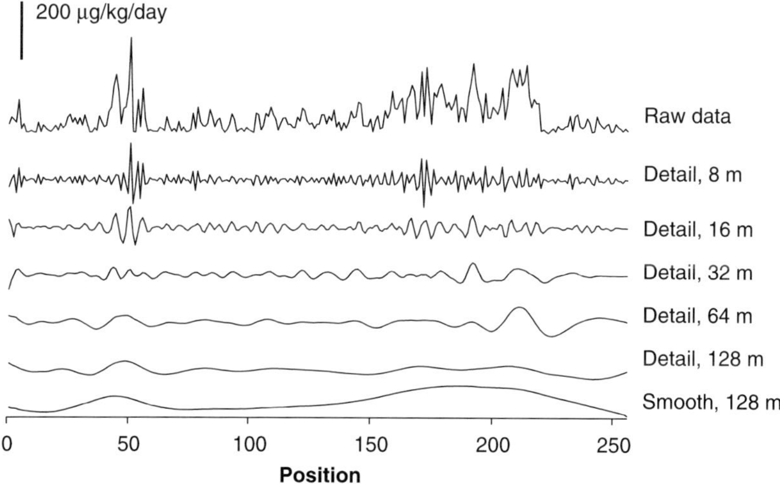

Figure 23.3. *Detail components from a multiresolution analysis of data on nitrous oxide emission from the soil. The top trace shows the raw data. Note that the components are offset for clarity; the scale bar shows the magnitude of an emission rate of 200 μg kg^{-1} per day.*

the multiresolution analysis decomposes the data into components whose features reveal aspects of topography and parent material that dominate the variation of the process at different spatial scales.

The DWT can be extended to two dimensions or more (Mallat, 1989). The mechanism is simple: the 1-D pyramid algorithm is used to transform each column of the image by one step, and the results are stored in the corresponding columns. Each row is then transformed by one step. The result is an image with four sections, effectively obtained by convolving the data with, respectively, the square filters $\mathbf{gg}^{\mathrm{T}}$, $\mathbf{gh}^{\mathrm{T}}$, $\mathbf{hg}^{\mathrm{T}}$ and $\mathbf{hh}^{\mathrm{T}}$, and then downsampling. The first section is a smooth version of the image, and the other three sections contain wavelet coefficients corresponding to variation along the columns, down the rows and along the diagonals of the image, respectively. Lark and Webster (2004) discussed the process in more detail. One problem that they identified is that the adapted wavelets of Cohen et al. (1993) are not suited for two-dimensional (2-D) analysis. For this reason, I have solved the boundary problem by padding the data with the values obtained by symmetrical extension of the original data.

Figure 23.4 shows a multiresolution analysis of a 2-D set of data. These are thickness of the soil over underlying parent material from a large survey on a grid of 100-m intervals at Ivybridge in the south-west of England (Webster, et al., 1979). The surveyed region encompasses uplands (top left of the Fig. 23.4) over a granite intrusion, and lowlands — predominantly slates with a strike running from east-north-east to west-south-west. The landscape is dominated by more or less parallel ridges and valleys trending in that direction. The analysis of these data with wavelets was first described by Lark and Webster (2004). Fig. 23.4 shows the smooth representation of the data at each scale filtering out detail components of all orientations. Note how different features of the variation are emphasised when detail components from different scales are removed. In particular note how the broad patterns related to geology are distinguished from short-range variation at the first scale, and how the pattern resulting from the strike of the slates disappears at scales 800 m and coarser.

Another example is shown in Plate 23 (see Colour Plate Section). Each location in the Ivybridge grid was allocated to a soil series in the original survey, and Webster et al. (1979) generalised this classification to provide a legend of nine soil classes. Some sites were not allocated. Soil class is a multistate variable; its states or levels (soil classes) do not constitute a mathematical field, unlike the continuous variables that we have described so far, and so they cannot be analysed directly with wavelets. Lark (2005) presents a wavelet transform for multistate variables, but this does not give rise to a multiresolution analysis. Here I have transformed the multistate variable to a set of ten indicator variables, one corresponding to each class in the generalised legend and one containing unclassified sites. The indicator is set to 1 at sites where the class is

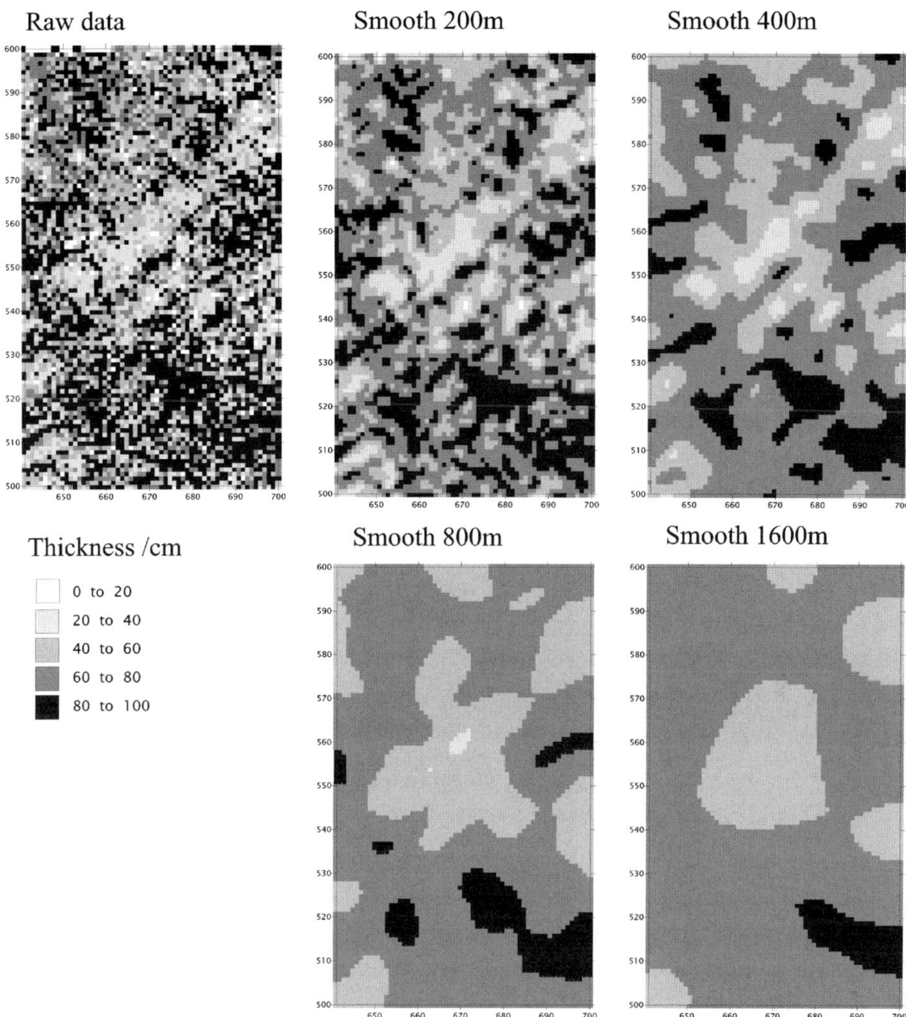

Figure 23.4. *Smooth components from a two-dimensional (2-D) multiresolution analysis of data on soil thickness at Ivybridge, South West England. Figure taken from Lark, R.M., Webster, R., 2004. Analysing soil variation in two dimensions with the discrete wavelet transform. Eur. J. Soil Sci. 55, 777–797, with permission.*

observed and zero elsewhere, so that each site has exactly one indicator with a value 1 and all others zero. A multiresolution analysis was then conducted on all the vectors, using the same method applied to the data on soil thickness. The class for which the corresponding indicator has the largest value in each smooth approximation is shown in Plate 23. Note that nowhere does class 10 (unclassified) appear in any of the smooth approximations. As the scale parameter of

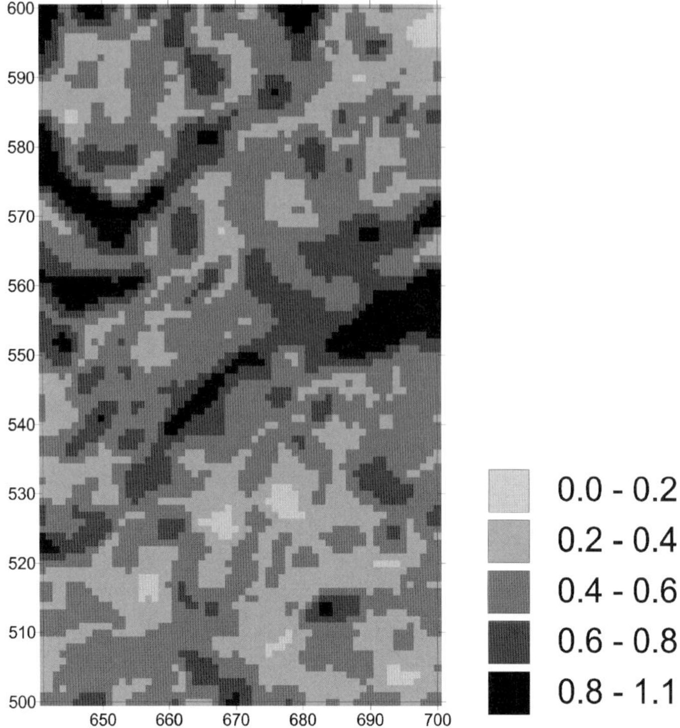

Figure 23.5. *Largest value among the smooth components at scale-parameter 400 m, from a 2-D multiresolution analysis of ten indicator variables defined on soil classes at Ivybridge, South West England.*

the smooth component increases, the pattern of classes appears more generalised and some classes disappear so that only four are represented at the coarsest scale. Such an analysis could be useful for generalising the results of DSM at different scales in a rational way. At any location the dominant local soil class at some scale of generalisation is presented. Figure 23.5 shows the value of the largest indicator (over all classes) at each site for the smooth approximation at scale parameter 400 m. This shows that the soil class pattern at this and finer scales is more complex in some areas (e.g. the south east) than others.

23.2.4 Case study on multiresolution analysis in digital soil mapping

Data and methods

The example that is described here is drawn from a study by Lark et al. (2003), which contains a full account of the analysis. The aim of the study was to predict soil properties from electrical conductivity measurements of the soil.

The study site was a 32-ha field in the Tulare Basin in California's San Joaquin Valley. The soil at the site is a Lethent clay loam (fine, montmorillanitic, thermic and typic Natrargid). These soils are saline-sodic and have little organic matter. An initial cursory survey of the field was conducted using electromagnetic inductance to map the apparent electrical conductivity (EC_a), and 40 sample sites were chosen using the methods of Lesch et al. (1995) to select sites that represented the variation of the EC_a. At each sample site, soil cores were taken at two points roughly 5 cm apart. Soil cores were taken at 0.3-m increments to a depth of 1.2 m. These cores were analysed for a variety of physical and chemical properties related to soil quality, including electrical conductivity of the saturation extract (EC_e).

The EC_a of the field was then surveyed intensively (at over 7000 points) using fixed-array electrodes (Wenner array), which were set to determine EC_a to a depth of 1.2 m. The data were collected on 80 passes east-to-west across the field; EC_a measurements were taken at 4-m intervals.

The problem is to predict the laboratory-determined variable, EC_e, from the field-determined covariate, EC_a. In this study, we test a hypothesis that an initial decomposition of the covariate on a wavelet basis will allow us to make better predictions of the variable than could be made from the raw data. This is because EC_a varies due to many factors, including the salinity of the soil, which is measured by EC_e. It is likely that different sources of variation in EC_a will dominate variability at different spatial scales, so by extracting components of different spatial scale we may be able to form better predictors for particular properties of the soil.

The measurements of EC_a were made at regular (4-m) intervals along the passes, but the passes were not regularly spaced. For this reason, I have not attempted a 2-D multiresolution analysis of the data, but rather analysed each pass as a 1-D data set. The multiresolution analysis was done using Cohen et al.'s (1993) adaptation of Daubechies's (1988) wavelet with two vanishing moments, and the shift-averaging procedure described above. This generated average detail components for scale parameter $x_0 2^m$ where x_0, the basic step, is the 4-m interval along the pass and $m = 1, 2, \ldots, 4$, and a smooth component for the coarsest scale, $x_0 2^4 = 64$ m.

The smooth components of the data at these four scales are shown in Figure 23.6. As Lark et al. (2003) discuss, much of the variation removed in the first two detail components appeared to be non-stationary measurement error due to particular structural conditions of the surface soil in part of the field. Of the four detail components extracted that of coarsest scale (64 m) accounted for most variation. It was therefore decided to evaluate the following four predictors or combinations of predictors for EC_e: raw EC_a data, smooth representation of EC_a at scale 64 m (P_{64}), smooth representation of EC_a at scale 32 m (P_{32}) and P_{64} and the corresponding detail component (Q_{64}).

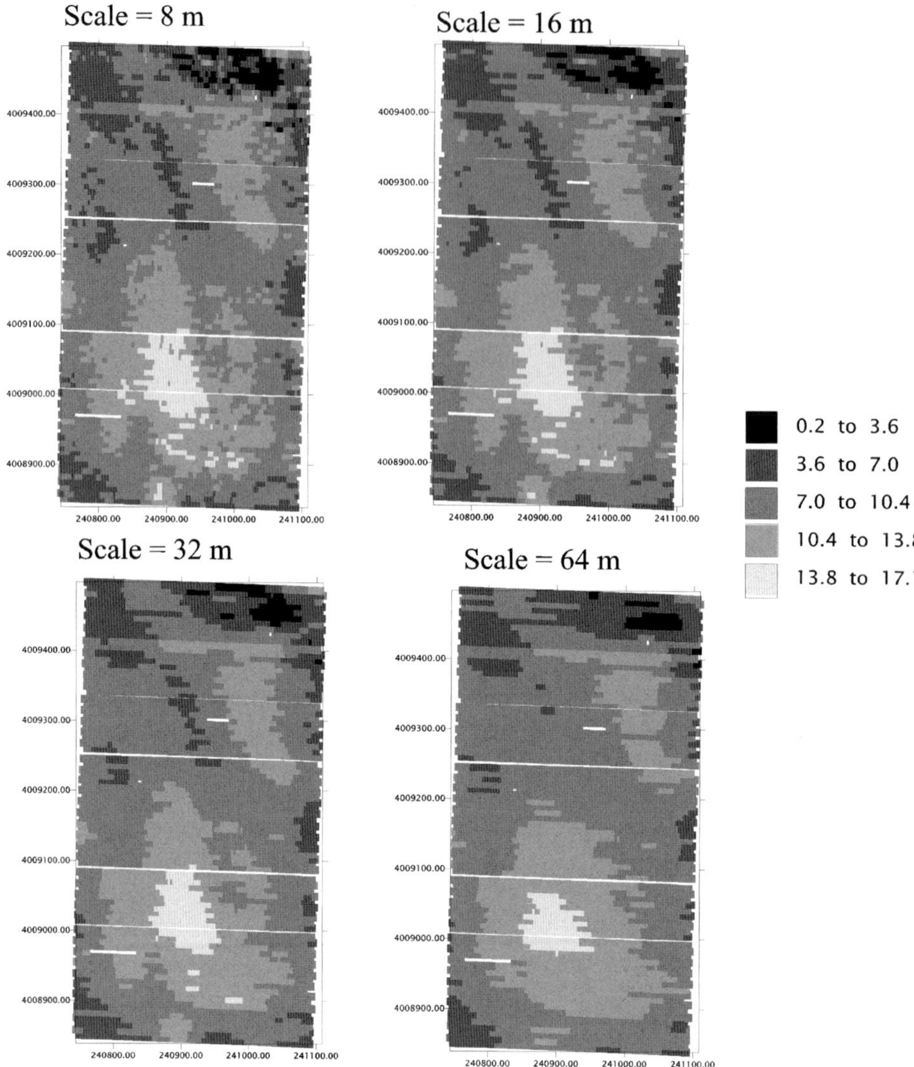

Figure 23.6. *Smooth components of the multiresolution analysis of fixed array measurements of EC_a along passes on the Tulare basin study site.*

These predictors were extracted for each soil sampling site. Multiple linear regressions of EC_e for each of the four depth increments on the four sets of predictors were then computed using maximum likelihood to fit a model in which the residuals are spatially autocorrelated (Lark, 2000). We therefore make no assumptions of independence, which would be compromised by the non-random sampling design. The significance of each regression was tested by computing the Wald statistic, and the Akaike information criteria (AIC) was computed from the likelihood for each model (see Lark et al., 2003 for details). The AIC allows us to compare regressions with different numbers of predictors; that with the smallest value of the AIC is the most parsimonious predictor of EC_e.

Results

Table 23.1 shows results for the prediction of EC_e at each depth from the different sets of predictors. Figure 23.7 shows the predicted value of EC_e for depths 90 and 120 cm using the best regression (smallest AIC). Note that, while there are resemblances between these two maps, they are not simple rescalings of the covariate (EC_a) since both the smooth and the detail components may have different coefficients.

Discussion

These results indicate that the decomposition of a covariate into components of different spatial scale prior to prediction of a soil property using calibration data may be advantageous. In all cases, the best predictor, as judged by the AIC and

Table 23.1. *Results for regression of soil EC_e (electrical conductivity)/$dS\,m^{-1}$ on different predictors.*

Soil property	Predictor(s)	Error variance	Wald statistic	P*	a[a]
EC_e	EC_a	77.6	0.001	0.97	149.5
Depth: 30 cm	P_{64}	78.42	2.02	0.16	147.5
	P_{32}	77.35	0.12	0.73	149.4
	P_{64}, Q_{64}	74.32	7.82	0.02	144.4
EC_e	EC_a	19.69	13.80	<0.001	117.8
Depth: 60 cm	P_{64}	19.01	13.43	<0.001	119.3
	P_{32}	17.80	17.34	<0.001	**115.7**
	P_{64}, Q_{64}	17.84	17.84	<0.001	117.0
EC_e	EC_a	25.4	34.04	<0.001	116.5
Depth: 90 cm	P_{64}	19.3	43.68	<0.001	113.3
	P_{32}	19.1	53.56	<0.001	108.8
	P_{64}, Q_{64}	19.1	54.26	<0.001	110.7
EC_e	EC_a	72.9	30.9	<0.001	117.6
Depth: 120 cm	P_{64}	48.1	15.6	<0.001	122.4
	P_{32}	56.6	40.0	<0.001	112.7
	P_{64}, Q_{64}	49.9	44.8	<0.001	112.3

Note: From Lark et al. (2003). Note that here the scale parameters are quoted in metres.
**p* value for the Wald statistic.
[a]Variable part of the Akaike Information Criterion. The values for any one soil property are comparable and *a* is shown in bold for the regression with the smallest AIC.

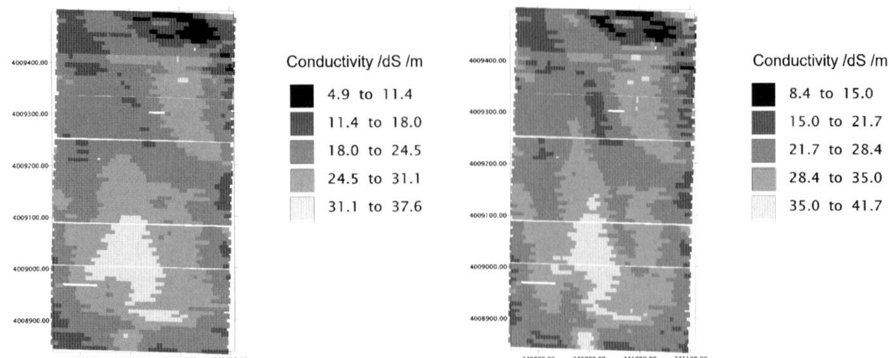

Figure 23.7. *Predicted soil EC_e for (left) 0–30 cm increment and (right) 90–120 cm using best models from Table 23.1.*

the Wald statistic, used components from the analysis of the EC_a data rather than the raw data. Further, because different components may be used for prediction, all maps of properties based on the same covariate may exhibit different patterns of variation, with different spatial structures, since they are not simple linear rescalings of the original covariate.

Limited numbers of data were available in this study, so it was not possible to test the predictive models after fitting, but only to look at the inferential statistics of the fit. I would hope that the growing interest in DSM will provide opportunities to test the hypothesis that decomposing covariates into components of different spatial scale gives the opportunity to create more flexible and effective predictors of soil properties than if we simply use the raw data for prediction. One interesting case study in which a multiresolution analysis of sensor data did lead to improved predictions of soil properties is provided in Chapter 21.

23.3 Denoising data using the wavelet transform

23.3.1 Principles and methods

The discussion above has shown how the wavelet transform provides a basis for a decomposition of digital soil data into components that correspond to different spatial scales. These components can serve the interpretation of data, and, as shown in the case study, might be useful to predict other soil properties. I now discuss how wavelet analysis has been used to denoise data in digital signal processing and image analysis. It is suggested that this could be particularly for the analysis of data layers in DSM, when variables are measured with error, but standard methods of filtering may unduly suppress real variation in some parts of the landscape.

Most variables are measured with error that can reasonably be assumed to have zero mean (unbiased measurements) and some finite variance ($\sigma_n{}^2$) that may be estimated, perhaps by repeated measurements on the same sample. It is assumed that errors are independent random variables, although developments of the method allow this to be relaxed.

The power spectrum of uncorrelated random error is uniform at all spatial frequencies. It is usual practice to assume that the variance of the underlying error-free variable that we measure (the 'signal') is dominated by components of low frequency. If we suppress variation at high frequencies (e.g. by a so-called low-pass filter), while this will not remove all the noise and will have some effect on the signal, the noise will be suppressed more than the signal and the signal to noise ratio will therefore increase. Low-pass filters (such as local averaging filters) are commonly used for this purpose (e.g. Webster et al., 1979). A disadvantage of low-pass filtering to denoise data is that all high-frequency variation is suppressed indiscriminately. If we consider Fig. 23.3, it is clear that some of the wavelet coefficients, describing the variation at the highest frequencies (smallest scale parameters), reflect real variation in the underlying process in some parts of the landscape, and simple high-pass filtering may suppress information on real variations.

The wavelet transform provides another approach to the denoising of data (Donoho and Johnstone, 1994). An uncorrelated random noise with standard deviation σ_n, will give rise to DWT coefficients at all scales with a standard deviation of σ_n. Wavelet denoising proceeds on the assumption that most natural variables will have a more or less sparse representation on a wavelet basis. That is to say, a large proportion of the wavelet coefficients are not significantly different from zero. A large class of functions, by no means all smooth, have sparse wavelet expansions (Silverman, 1999). This contrasts with stationary random noise and suggests that we may denoise a signal efficiently by suppressing those wavelet coefficients at any scale that cannot be regarded as significantly different from zero. The process simply involves the computation of a DWT, replacing with zero any coefficients, $D_{m,n}$, which are smaller than a threshold, τ, and then computing the inverse transform of the thresholded coefficients.

Donoho and Johnstone (1994) showed that if N data are collected with measurement error of standard deviation σ_n then a threshold

$$\tau_{\mathrm{DJ}} = \sigma_n \sqrt{2\log_e N}, \tag{23.20}$$

is optimal in the sense that the risk of incorrectly retaining some coefficient $|D_{m,n}| > \tau_{\mathrm{DJ}}$ is asymptotically close to the risk if we were to use the unknown best or oracle threshold. However, Abramovich and Benjamini (1996) showed that, when we consider the problem of thresholding the whole set of coefficients (rather than on a coefficient by coefficient basis) then Donoho and Johnstone's threshold is equivalent to a Bonferroni threshold and so is lacking in statistical power. They proposed an alternative method based on controlling the false discovery rate, that is the overall probability that a coefficient should have been suppressed given that it has been retained. We set an acceptable false discovery rate, q. We then define $p_{j,n}$ as the two-tailed p value for a null hypothesis that the coefficient $D_{j,n}$ is zero, given the measurement error:

$$p_{j,n} = 2\left(1 - \Phi\left\{\frac{|D_{m,n}|}{\sigma_n}\right\}\right), \tag{23.21}$$

where Φ is the standard normal distribution function. We then order all m of these p values irrespective of scale such that $p_1 \leqslant p_2 \leqslant \ldots \leqslant p_m$ so that any p_i corresponds to $D_{j,n}$ for some j and n. We then find the largest index i_0 such that

$$p_{i_0} \leq \frac{i_0}{m} q \tag{23.22}$$

and the absolute value of the corresponding wavelet coefficient is τ_{FDR}, the false discovery rate threshold. Abramovich et al. (2000) showed that this threshold is an asymptotically minimax solution to the identification of terms when the degree of sparsity of the underlying error-free variable is unknown.

23.3.2 Case study, denoising data on slope from a field survey

Figure 23.8a shows measurements of surface slope made in the field in the grid survey of the Ivybridge region previously cited (Webster et al., 1979). Slope may be a useful data layer in DSM, and Lark and Webster (2004) showed that it was correlated with soil thickness (at coarser spatial scales). It will also be measured with error, but its variability is also likely to be sparse on a wavelet basis. This is because the variation in slope over much of the landscape will be small, but in some areas of marked relief (e.g. around river channels) there is rather more variability.

A root mean square error was assumed for the determination of slope of 5°, and a false discovery rate of 0.05 specified. Coifman and Donoho (1995) was followed in using the 'spin-cycle' method, conducting the thresholding procedure within a moving window on symmetrically extended data, and then incrementing the origin of the window through 2^{j+1} steps in each dimension where j is the number of dilations of the wavelet. Daubechies's (1988) wavelet with six vanishing moments was used. The results of all thresholdings are then averaged. This ensures that the results of the procedure are shift-invariant.

Fig. 23.8b shows the denoised data. Note that, while the slope is much smoother in some parts of the landscape (e.g. on the granite intrusion), the

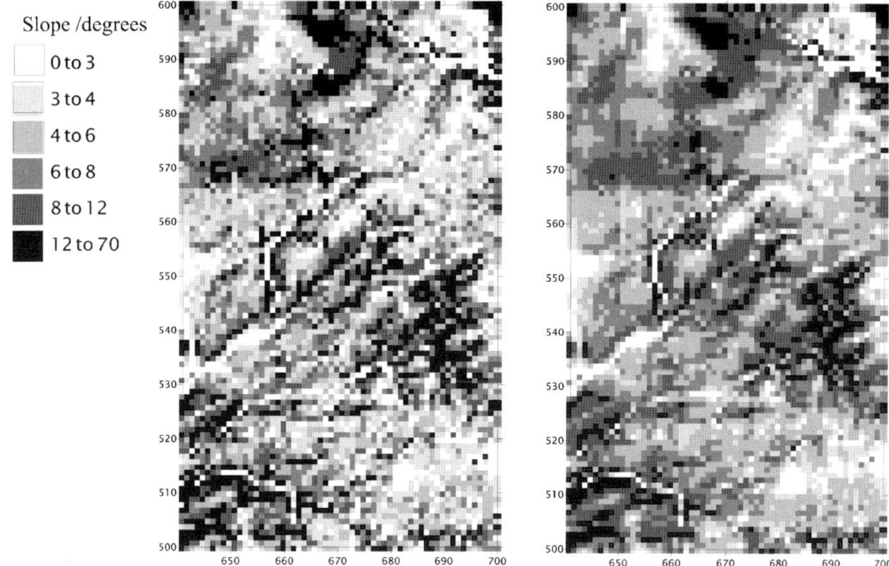

Figure 23.8. *Left (a) raw data on surface slope from Ivybridge survey; and right (b) denoised slope data.*

variability is retained elsewhere, notably in parts of the landscape around drainage channels.

23.4 Adapting the wavelet basis to the data

23.4.1 Principles and methods

The principal advantage of the wavelet transform over other approaches to the decomposition of spatial data into different components is that it requires no assumptions of spatial stationarity. This is in contrast with Fourier analysis, or with kriging analysis (factorial kriging) which requires the assumption of intrinsic stationarity. The principal disadvantage of the DWT is that it provides a predetermined basis for the data set that does not depend in any way on the spatial variation of that data. In kriging analysis, by contrast, the data are decomposed into components identified by modelling the experimental variogram, and so the number of components and the spatial frequencies that they represent depend on the spatial variability of the process under study.

This disadvantage of the DWT is compounded, as Si (2003) points out in the context of soil science, because the DWT does not resolve variation equally at all spatial frequencies. This is seen in Fig. 23.2a. Variation at spatial frequencies

$1/4x_0$ to $1/2x_0$, that is half the frequency range, is associated with a single scale parameter – the first dilation.

These considerations motivate consideration of the discrete wavelet packet transform (DWPT). While widely used in signal analysis, this has yet to be applied to problems in soil science.

A full account of the DWPT is not attempted, and the text of Percival and Walden (2000) is recommended. The principle is illustrated in Fig. 23.2b. In the DWPT, at each dilation of the basic wavelet, the wavelet and scaling-function filters are each applied to both the wavelet and the scaling-function coefficients generated at the previous dilation (by contrast with the DWT where they are only applied to the scaling-function coefficients). As a result, instead of j+1 components from an analysis with j dilations (j detail components and one smooth), we have components from translations of 2^j wavelet packets. Note that the order in which the scaling-function coefficients and wavelet coefficients are passed from one dilation to the next alternates along the array. Because of this, the frequency ordering of the wavelet packets is conserved, and, as the Fig. 23.2b shows, by the jth dilation the frequency interval is divided into 2^j equal bands.

All the 2^j wavelet packets at the final dilation provide an orthonormal basis for data, but it is clear that other bases could be selected. If, in this instance, we selected the set $\{W_{3,0,}\ W_{3,1},\ W_{2,1}$ and $W_{1,1}\}$ from all of the wavelet packets in Fig. 23.2b this corresponds to the ordinary DWT. Other orthonormal bases could be selected from the tree, for example $\{W_{2,0,}\ W_{3,2},\ W_{3,3}$ and $W_{1,1}\}$. This provides another partition of the interval of spatial frequencies $[0,\ 1/2 \times {}_0]$, a finer partition in some parts than in others. This particular set of packets might be more informative about one particular data set than would the DWT if there is particular structure in the spectrum (local or stationary) in those parts of the frequency interval that are better resolved. This suggests that we could select a basis that is in some sense optimal for a particular set of data.

The selection of a best basis from a library of possible packet bases, such as those in Fig. 23.2b, can be done automatically (Coifman and Wickerhauser, 1992). The overall aim in the selection of a best basis is to find a basis with the most inhomogeneous distribution of coefficients, that is, to find the basis on which the sparsest representation is possible. This may be done with an entropy criterion, although others are also used. The entropy of a particular packet, $W_{j,k}$ for the data in the vector $\mathbf{z}$ is computed from the $N_{j,k}$ packet coefficients, $\omega_{j,k,n}$ as

$$-\sum_{n=1}^{N_{j,k}} \left(\frac{w_{j,k,n}}{\|\mathbf{z}\|}\right)^2 \log_e \left(\frac{w_{j,k,n}}{\|\mathbf{z}\|}\right)^2, \tag{23.23}$$

where $\|\mathbf{z}\| = \text{trace}\ (\mathbf{z}\mathbf{z}^{\mathrm{T}})$.

We then consider each wavelet packet at dilation $j-1$, and compare its entropy with that of the two 'child' packets generated at dilation j by it wavelet and scaling function coefficients. If the parent entropy is larger than the sum of entropies of the two child packets then the latter are retained in the basis. This is then iterated until all packets at levels, $j-2$, $j-3$,... have been considered.

23.4.2 Case study

Figure 23.9 shows the wavelet decomposition of data on the concentration of nitrate in the soil on the transect from which nitrous oxide emission rates are shown in Fig. 23.3. Fig. 23.9a shows the decomposition on the full set of wavelet packet generated by three dilations of Daubechies's (1988) wavelet with two vanishing moments. This multiresolution analysis was performed using symmetrical extension of the data. The eight packets are arranged in increasing order of spatial frequency from bottom to top. Fig. 23.9b shows the best basis selected from the corresponding library with the components labelled according to the corresponding packet. There are seven packets in the best basis, again ordered in ascending frequency from bottom to top of the Fig. 23.9b. In this case, the best basis includes the first six packets from the full decomposition, $W_{3,0}$, $W_{3,1}, \ldots, W_{3,5}$, and one packet from the second dilation of the wavelet — $W_{2,3}$.

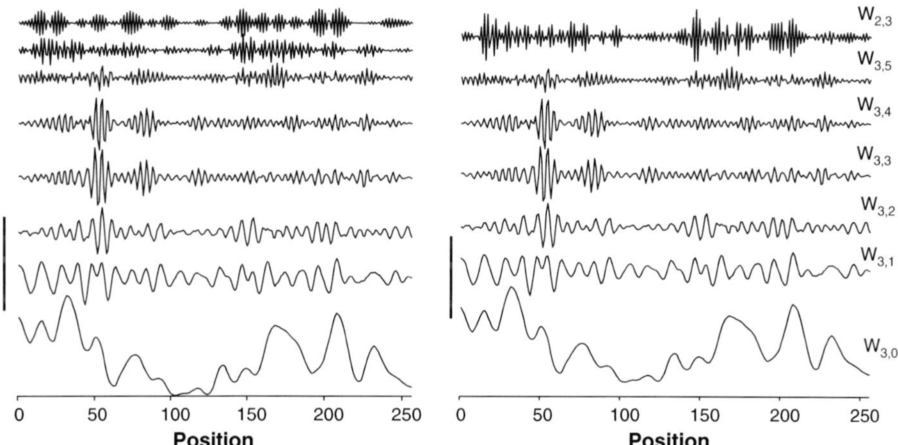

Figure 23.9. *Left (a) decomposition of data on soil nitrate content on all wavelet packets in the third dilation of Daubechies's (1988) wavelet with two vanishing moments and right (b) decomposition on the best basis selected from this library by an entropy criterion; basis packets are labelled. The components are offset; in each case the vertical bar is 50 mg kg*$^{-1}$.

On inspection Fig. 23.9 shows the advantages of the DWPT over the DWT. A DWT to three dilations would represent the three packets of highest frequency in the best basis (Fig. 23.9b) as a single detail component. The DWPT on a best basis requires better resolution over this part of the frequency domain. It is interesting to note that the variation on the mixed drift over Gault Clay (between positions 125 and 200) is predominantly in the packet of highest frequency in the best basis. Packet $W_{3,4}$, which would be part of the component of finest scale in the DWT, is distinguished from others and shows most of its variation in the first part of the transect over the Lower Greensand.

In summary, the DWPT with best basis selection allows us to achieve a decomposition of data on a basis with a partition of the frequency interval that best captures complex and localised spatial variation in our particular data set. This combines the advantages of kriging analysis, an analysis into components identified from the data rather than imposed by the analysis, with the generic advantage of wavelet methods — the possibility of identifying components of the data with markedly local, non-stationary variation as we see in Fig. 23.9b.

23.5 Discussion

The goal of this chapter, as stated in Section 23.1, is to explain multiresolution analysis with the wavelet transform, and to demonstrate some of its properties, with particular reference to the quantitative modelling of spatial variation for DSM. It is hoped that workers concerned with the problems of DSM will use these analyses to map soil properties from covariate data. The case study in Section 23.2.4 should encourage this since there was evidence that the separate components of data from a sensor may be better predictors of a related soil property than are the raw data alone. This is also supported by Mendonça-Santos et al. (Chapter 21), who found similar results for different soil properties and covariates over a larger region. As these authors note, one problem with using multiresolution analysis in this context is that it generates a multiplicity of predictors. This intensifies the urgency with which we need disciplined and statistically respectable methods to select predictive models from the many possible combinations of covariates and their spatial components.

The multiresolution analysis of the indicator variables that represent a multistate variable (in this case, soil classes) may also be useful. Firstly, it provides a procedure for automated generalisation of a soil map at different spatial scales, indicating the local dominant class at some scale of resolution. This could be useful since not all users of a digital soil map will need or want information at the same scale of resolution. Secondly, it is possible that a generalisation of multistate variables could uncover scale dependence in their relation to soil properties, and so increase their predictive power for mapping soil. Such multistate variables might include information on landuse (where broad trends tell us more about soil variation of interest than do variations from field to field), ecological survey data or geological maps. More research is needed in this area.

The demonstration of wavelet denoising may also be of direct practical interest. Most data used in DSM contain significant measurement error, and all contain some. However, we know that the variability of the underlying variable may well be non-stationary; in our example, slope may be more variable in some parts of a landscape than in others. This means that simple approaches to the suppression of error by filtering the data may locally suppress important information. We can exploit this general intuition, without any statement of how intermittent the variation of slope actually is, in order to 'denoise' the data by wavelet methods, suppressing stationary variation which appears to arise from the noise process only and retaining local variation with a sparse wavelet representation.

Finally, it is acknowledged that a drawback of the DWT is that the scales into which the data are decomposed are predetermined. This is by contrast to kriging analysis of variables, where the scales are identified at the initial stages of variogram estimation and modelling. It has been shown how the selection of a best basis from a library determined by a DWPT allows us to adapt the basis to the variability of the data. It remains for further research to discover whether decomposition of covariates on best bases gives substantial advantages in the task of DSM.

The discussion above has considered some of the immediate implications of the work presented in this chapter, but there are other aspects of the wavelet transform as a tool for modelling soil variation — and these may be pertinent to its role in DSM.

Firstly, the case studies have all been applied to relatively small areas, covering from a few tens to a few thousands of hectares. Wavelet techniques can be, and have been, applied just as readily to large digital data sets as to small ones (e.g. Jordan and Schott, 2005). In the context of DSM over large areas, wavelet analysis of covariates may prove useful at the initial stages of planning field sampling. For example, the heterogeneity of the variance of a covariate, showing marked variation in some regions and uniformity in others, can be quantified by the analysis of wavelet coefficients. The implication of such heterogeneity may be that we need to sample the soil in some subregions more intensively than others if we are to use the total sampling effort as efficiently as possible. This could be done by reference to wavelet coefficients of the covariates at critical spatial scales.

Secondly, our covariates are generally available on regular grids (raster images or similar). These are ideally suited to wavelet analysis. Field observations of the soil will generally be somewhat irregularly distributed, which prevents their analysis with standard wavelet methods such as the DWT. This is inevitable since the DWT decomposes observations by spatial scale without assuming stationarity, and so assuming that all spatial scales are represented at any location. This implies uniform sampling. A wavelet transform on irregularly sampled data will therefore not provide us with an analysis of variation by scale, as does the DWT. However, analyses are possible using the lifting algorithm. This recognises that the wavelet transform can be regarded as a partitioning of our data (into odd- and even-indexed components in one dimension or regular sampling in two dimensions). If we then predict the even-indexed data from the odd, the prediction constitutes the smooth approximation and the residual from this prediction constitutes a detail component. This can readily be extended to the irregular case in one dimension, yielding different smooth representations of a variable, although these cannot be associated with one spatial scale at all locations. In two dimensions this is harder, although some progress has been made (Jansen et al., 2001). Whether a useful tool for quantitative mapping of soil variation will emerge remains to be seen.

Finally, this chapter has presented some preliminary results on multiresolution analysis of indicators that represent states of a multistate soil variable, in this case soil classes. These results are promising in that the maps appear to be reasonable generalisations of the distribution of soil classes at different scales. One could envisage, for each of the four scales of the analysis, a different user of soil maps, concerned with management at different spatial scales, for whom the map at this scale of generalisation would be most useful. The properties of these maps require further investigation. Lark (2005) has presented a different wavelet analysis for multistate variables. This does not generate a multiresolution analysis, but does identify non-stationarity of the class pattern. This could be a useful tool for analysis of categorical covariates, such as maps of land cover, over large regions, and may allow us to identify subregions within which it is plausible to make assumptions of statistical stationarity. Since much historical data on the soil consist of categorical variables (profile classes, drainage classes, structural classes, etc.), further work in this area could prove very useful.

23.6 Conclusions

To conclude, quantitative modelling of the variation of soil for DSM must account for complex soil variation over multiple spatial scales. This chapter has demonstrated the value of wavelet transforms for this purpose. Wavelet transforms allow multiresolution analysis of soil variables, so that we can present them at different scales, and extract different, scale-specific, components from covariates which have been shown here and elsewhere to have advantages over the raw data for prediction of soil properties. Wavelet analysis also allows us to denoise data without suppressing short-range variation everywhere, a substantial advantage over simple filtering methods since the variation of natural processes is often non-stationary. In addition, it has been shown how wavelet analyses can be adapted to the specific scale dependence of our variables through the selection of a best wavelet-packet basis.

The scope for wavelet analysis in service of DSM is substantial. Further work is needed on how the advantages of the method can fully be exploited across large and complex landscapes, on the analysis of irregularly sampled data and on the analysis of categorical soil variables such as classes.

Acknowledgments

I am grateful to Dr. Dennis Corwin of the USDA George E. Brown, Jr. Salinity Laboratory, Riverside CA, and to Dr. Stephen Kaffka of University of California, Davis, for permission to present results from our previous study of the Tulare data. I am also grateful to the Lawes Agricultural Trust for permission to use data from the Ivybridge survey and to Professor R. Webster and Blackwell Publishing for permission to reproduce Fig. 23.4 from our paper on two-dimensional wavelet transforms. This research was funded by the UK Biotechnology and Biological Science Research Council through its core grant to Rothamsted Research, and through grant MAF12269.

References

Abramovich, F., Benjamini, Y., 1996. Adaptive thresholding of wavelet coefficients. Comput. Stat. Data Anal. 22, 351–361.

Abramovich, F., Benjamini, Y., Donoho, D., Johnstone, I., 2000. Adapting to unknown sparsity by controlling the false discovery rate. Technical Report, Statistics Department, Stanford University, Stanford, CA. (www-stat.stanford.edu/~imj/Reports/2000/AUSCFDR.pdf)

Cohen, A., Daubechies, I., Vial, P., 1993. Wavelets on the interval and fast wavelet transforms. Appl. Comput. Harmon. Anal. 1, 54–81.

Coifman, R.R., Donoho, D.L., 1995. Translation-invariant denoising. In: A. Antoniadis and G. Oppenheim (Eds.), Lecture Notes in Statistics, Number 103: Wavelets and Statistics. Springer-Verlag, New York, pp. 125–150.

Coifman, R.R., Wickerhauser, M.V., 1992. Entropy-based algorithms for best basis selection. IEEE Trans. Inf. Theory 38, 713–718.

Daubechies, I., 1988. Orthonormal bases of compactly supported wavelets. Commun.Pure Appl. Math. 41, 909–996.

Donoho, D.L., Johnstone, I.M., 1994. Ideal spatial adaptation by wavelet shrinkage. Biometrika 81, 425–455.

Jansen, M., Nason, G., Silverman, B., 2001. Scattered data smoothing by empirical Bayesian shrinkage of second generation wavelet coefficients. In: M. Unser and A. Aldroubi (Eds.), Wavelet Applications in Signal and Image Processing. IX Proceedings of SPIE, Vol. 4478. SPIE, Bellingham, WA, pp. 87–97.

Jordan, G., Schott, B., 2005. Application of wavelet analysis to the study of spatial pattern of morphotectonic lineaments in digital terrain models. A case study. Remote Sens. Environ. 94, 31–38.

Lark, R.M., 2000. Regression analysis with spatially autocorrelated error: examples with simulated data and from mapping of soil organic matter content. Int. J. Geogr. Inf. Sci. 14, 247–264.

Lark, R.M., 2005. Spatial analysis of categorical soil variables with the wavelet transform. Eur. J. Soil Sci. 56, 779–792.

Lark, R.M., Webster, R., 1999. Analysis and elucidation of soil variation using wavelets. Eur. J. Soil Sci. 50, 185–206.

Lark, R.M., Webster, R., 2004. Analysing soil variation in two dimensions with the discrete wavelet transform. Eur. J. Soil Sci. 55, 777–797.

Lark, R.M., Kaffka, S.R., Corwin, D.L., 2003. Multiresolution analysis of data on electrical conductivity of soil using wavelets. J. Hydrol. 272, 276–290.

Lark, R.M., Milne, A.E., Addiscott, T.M., Goulding, K.W.T., Webster, C.P., O'Flaherty, S., 2004. Scale- and location-dependent correlation of nitrous oxide emissions with soil properties: an analysis using wavelets. Eur. J. Soil Sci. 55, 611–627.

Lesch, S.M., Strauss, D.J., Rhoades, J.D., 1995. Spatial prediction of soil-salinity using electromagnetic induction techniques. 2. An efficient spatial sampling algorithm suitable for multiple linear-regression model identification and estimation. Water Resour. Res. 31, 387–398.

Mallat, S.G., 1989. A theory for multiresolution signal decomposition: the wavelet representation. IEEE Trans. Pattern Anal. Machine Intel. 11, 674–693.

Percival, D.B., Walden, A.T., 2000. Wavelet Methods for Time Series Analysis. Cambridge University Press, Cambridge, UK.

Rodriguéz-Arias, M.A., Rodó, X., 2004. A primer on the study of transitory dynamics in ecological series using the scale-dependent correlation analysis. Oecologia 138, 485–504.

Si, B., 2003. Scale and location dependent soil hydraulic properties in a hummocky landscape: a wavelet approach. In: Y. Pachepsky, D. Radcliffe, and H.M. Selim (Eds.), Scaling Methods in Soil Physics. CRC Press LLC, Boca Raton, FL, pp. 169–187.

Silverman, B.W., 1999. Wavelets in statistics: beyond the standard assumptions. Philos. Trans. R. Soc. Lond., A 357, 2459–2473.

Webster, R., Harrod, T.R., Staines, S.J., Hogan, D.V., 1979. Grid sampling and computer mapping of the Ivybridge area, Devon. Soil Survey Technical Monograph No 12. Soil Survey of England and Wales, Rothamsted Experimental Station, Harpenden.

Developments in Soil Science, volume 31
P. Lagacherie, A.B. McBratney and M. Voltz (Editors)

Chapter 24

DIGITAL SOIL MAPPING WITH IMPROVED ENVIRONMENTAL PREDICTORS AND MODELS OF PEDOGENESIS

Neil J. Mckenzie and John C. Gallant

Abstract
Conventional and quantitative methods require synthesis for digital soil mapping to reach its full potential, particularly when survey extents are large (i.e. $>10{,}000\,km^2$), data are sparse and resources are limited. One of the challenges is to express narratives of pedogenesis in an explicit form that can be incorporated into digital soil mapping. We present an applied example where such a narrative is developed – it emphasizes landscape history, provenance of soil parent materials and pedogenic process. The narrative is then expressed as rules that are used to produce digital soil maps. The rules are parsimonious; they encourage testing and enable spatial prediction; and are easy to update. The rules rely on just a few terrain and geophysical variables. We demonstrate how a new terrain variable, the multiresolution valley-bottom flatness (*MrVBF*) index, can be used with the familiar topographic wetness index (*TWI*), to map geomorphic and soil features.

24.1 Introduction

To ensure efficiency, relations between soil properties and more readily observed environmental features are used for soil mapping in surveys with a grain of 20–500 m and extent of 5–50 000 km^2. In conventional survey, these relations have been expressed as narratives. These narratives rarely capture the richness of understanding gained by the field pedologist, and users of surveys find it difficult to separate evidence from interpretation (Austin and McKenzie, 1988; Hudson, 1992; Hewitt, 1993; Scull et al., 2003).

Until recently, there has been a difference between field practitioners and pedometricians. The former preferred conventional methods (e.g. Dent and Young, 1981) and the latter favoured quantitative and statistical methods (e.g. Webster and Oliver, 2001). Pedometrics developed largely in response to deficiencies in conventional practice. In the quest for quantification, limited prominence was given to the qualitative understanding of pedogenesis. There was also a focus on intensive studies of small regions in landscapes already mapped at a more general level using conventional methods.

In our view, synthesis of conventional and quantitative methods is essential for digital soil mapping to reach its full potential, particularly when survey extents are large (i.e. $>10{,}000\,\text{km}^2$). In this chapter, we explore aspects of this synthesis by addressing two general questions.

- Can narratives of pedogenesis and soil distribution be translated into a form that takes advantage of digital technology and remote sensing?
- Can digital soil mapping using environmental correlation be implemented in a cost-effective way to support management of natural resources across large areas?

We outline the logic of environmental correlation. New terrain variables and gamma radiometric remote sensing are then discussed in the context of a practical example of rapid digital soil mapping. Throughout, we draw heavily on the new Australian *Guidelines for surveying soil and land resources* (McKenzie et al., 2006).

24.2 Digital soil mapping using environmental correlation

Many studies have shown that useful predictive relations exist between environmental variables and soil properties (McBratney et al., 2003; Scull et al., 2003). Environmental correlation[1] is a method of digital soil mapping that exploits these relations. The logic of the method is presented in Figure 24.1 and some particular aspects are discussed below. The review by McBratney et al. (2003) addresses most theoretical and technical aspects of the approach. The defining features of environmental correlation are:

- relations between soil properties and environmental variables are explicit,
- environmental variables enable prediction across the complete study region and
- it is easy to update the predictive model and predictions as new information becomes available.

Environmental correlation is now feasible for digital soil mapping across large areas for several reasons.

- Environmental data of good quality are becoming available in countries without a detailed coverage of soil surveys. In particular, terrain analysis has developed to the point where useful climatic, geomorphic, hydrologic and pedologic distinctions can be mapped (Wilson and Gallant, 2000; Gallant and Hutchinson, 2006). Gamma radiometric remote sensing is a complementary technology that detects differences in soil materials, most notably particle size and mineralogy, in the upper part of the profile (Wilford, 2006).
- Geographic information systems and statistical software facilitate analysis.
- Global positioning systems allow field data to be accurately registered with remotely sensed data

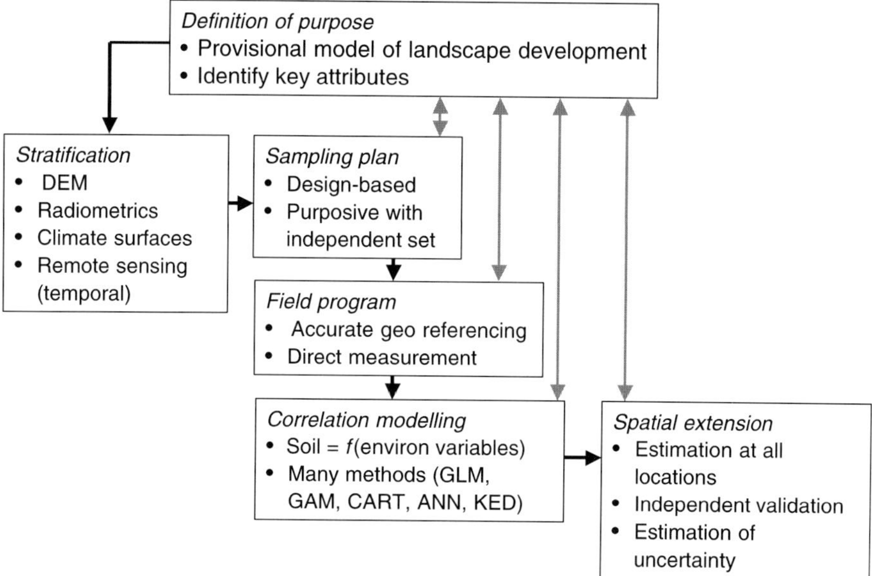

Figure 24.1. *Steps and options involved in environmental correlation.*

Environmental correlation is only feasible when predictors are easy to observe and useful for spatial prediction. Some landscapes have few suitable predictors due to factors such as subtle terrain, lack of reliable vegetative indicators, substantial short-range variation (i.e. at a finer grain than the environmental predictor), heavy imprints of human activity or controls on soil patterns without surface expression (e.g. subsurface structures). Of course, the environmental predictors must be cheaper to measure than the soil variables of interest, otherwise it is more sensible to record the soil directly and use interpolation.

24.2.1 Definition of purpose and provisional models

As with conventional survey, environmental correlation involves the iterative development of predictive models for the study region. The complexity and range of spatial and temporal scales over which processes of soil formation operate, and have operated, makes the development of mechanistic models for spatial prediction almost impossible in routine soil survey (Webster, 2000). Local models of pedogenesis are needed with varying levels of approximation and empiricism, and in most cases these are best expressed using a set of structured rules as discussed by Bui (2004).

The quality of environmental data has a large influence on the success of environmental correlation. In most parts of Australia, prerequisites include a digital elevation model with a fine resolution (i.e. grain and vertical resolution approximately 1% of total relief), and airborne gamma radiometric spectroscopy (flown at a line spacing of 200 m or less). In some environments, these requirements can be waived if suitable data from remote sensing are available (e.g. spectral reflectance, either as time series or hyperspectral).

The survey's purpose determines the target variables for measurement and prediction. Environmental correlation provides predictions of primary soil attributes rather than soil types. This avoids assuming that soil properties are highly correlated, and that general-purpose soil types can be defined and mapped. The goal is to measure and describe the continuum of soil properties and classify later if required. Models of soil and landscape variation are therefore needed for each predicted attribute.

24.2.2 Stratification, sampling and field program

Stratification of the study region provides the framework for sampling, analysis and spatial extension. Maps from earlier surveys can provide a good basis for stratification. If statistical sampling is used, then the inclusion probabilities for field sites must be known – this requires measuring the area of each stratum and the number of sample points. Stratifications that accurately reflect landscape processes (e.g. via delineation of stratigraphic units) will increase efficiency and provide reliable units in which to make predictions. The environmental variables used for stratification will normally be a subset of those with potential as environmental predictors and these are reviewed by Ryan et al. (2000) and McBratney et al. (2003).

There are many options for sampling (see de Gruijter, 2006). In our earlier work, where sampling densities were greater than 1 site per 250 ha (e.g. Gessler et al., 1995; McKenzie and Ryan, 1999), we favoured sampling strategies that were statistical (e.g. stratified random or multistage stratified random). We aimed to collect field data suitable for confirmatory statistical methods (e.g. generalized linear models), and assumed that predictor variables (e.g. terrain variables and gamma radiometric signals) were measured without error – this is rarely the case so the resulting statistical inferences had an unknown level of bias and precision.

However, in practice, when data are sparse and resources are limited, it is logical to use every available line of evidence to build predictive rules and then to assess their success through an independent phase of validation. The rules for mapping can incorporate pedological knowledge gained from other landscapes along with insights from informal observations across the study region.

24.2.3 Correlation modelling and spatial extension

Environmental correlation is so named because it exploits useful correlations with environmental predictor variables. However, predictive models must always be checked to ensure there is a rational interpretation based on physical principles.

The conceptual basis of environmental correlation is similar to the conventional survey – the contrast is in the explicit nature of these relationships and the tools used to derive and express them. The end result of the model building process is a procedure for explicitly producing maps. Our focus here is on simple rules, but a large variety of exploratory and confirmatory statistical methods can be used to formulate predictive models (e.g. Webster and Oliver, 2001; McBratney et al., 2003; Scull et al., 2003). We have found that exploratory and confirmatory methods of analysis only occasionally reveal relations that we have not already detected or suspected during field survey. On the contrary, their main role is to often dispel pet theories on patterns of soil–landscape variation. Correlations between environmental variables and soil properties that cannot be explained by pedologic principles and physical processes should be treated with suspicion. This may seem obvious, but it is easy when analyzing large datasets to overlook results that are spurious or an artefact of the statistical method. For example, overfitting can easily occur when using classification and regression trees. Qualitative methods can also be used to prepare predictive models. The case study below and Cook et al. (1996a) provide examples. Bui (2004) considers methods for expressing rules.

The term *spatial extension* has been used in this chapter to refer to the process of generating spatial predictions, using the correlation model and environmental predictor variables. The term is used in preference to extrapolation and interpolation, neither of which is entirely appropriate. Spatial extension is usually a relatively mechanical task that can be coded using a geographic information system.

24.3 The pedological basis for environmental correlation

It is widely claimed that the state-factor approach provides the only principle of universal value in soil mapping – namely, the hypothesis that "the soil profile is the integrated expression of parent material, climate, topography, living organisms and the age of the landscape" (Dent and Young, 1981). In a similar vein, Scull et al. (2003) state that "all approaches to soil mapping rest on our ability to use knowledge of the process of soil genesis to predict the properties of soils at any point in the landscape."

A challenge in digital soil mapping is to incorporate pedologic knowledge into predictive models. This is difficult because field evidence is often fragmentary and few surveys can afford to develop new or improved models of pedogenesis for each study region. More importantly, the strategies for pedogenic studies differ from those used for mapping (Butler, 1958). While the state-factor model is promoted as a conceptual basis for survey, it has an apparent simplicity that belies the complexity of most landscapes. In our view, it provides limited operational guidance for digital soil mapping besides acting as a check list. The approach has other difficulties as well (see Butler 1963; Chadwick and Chorover, 2001; Johnson and Hole, 1994).

We prefer to develop local and regional models of pedogenesis with a focus on processes and system linkages. General conceptual models of soil evolution by Simonson (1959), Johnson and Hole (1994), Huggett (1998), Paton et al. (1995) and Chadwick and Chorover (2001) provide good starting points. In older landscapes, it is essential to explicitly recognize environmental change over decades, centuries and millennia. Butler's (1982) system for soil studies, with its emphasis on history and stratigraphy, does this but his original account is highly abbreviated to the point of obscurity. Local and regional models of pedogenesis are nearly always qualitative and expressed through narratives, diagrams and sets of rules. In ideal circumstances these models

- explain the nature of materials at or near the earth's surface,
- explain the spatial distribution of these materials,
- provide a basis for sampling and prediction of soil and landscape properties,
- indicate landscape dynamics and likely rates of processes (e.g. erosion, water movement, aeration, solute transport, acidification and nutrient movement) and
- enhance assessment over that provided by simple inventory.

24.4 Case study: digital soil mapping in the Billabong catchment

In this example, we use Butler's (1982) system to develop a local model of pedogenesis, first as a narrative, and then in a form useful for digital soil mapping. Our motivation is to express our knowledge on soil formation in a way that encourages

- testing to determine consistency with biophysical processes,
- translation into a form useful for spatial prediction,
- parsimony and
- adding more complex rules when and where they are needed.

The example comes from the Billabong catchment in southeast Australia (Fig. 24.2). The catchment includes the transition from the southwest slopes of the Dividing Range to the vast riverine plain (Butler and Hubble, 1978). It has productive grazing and cropping lands with a range of environmental problems including soil acidification, dry-land salinity and loss of native plant and animal species. Soil information is needed to help design and implement better systems of land use – of particular importance is the capacity of soils to supply water to plants.

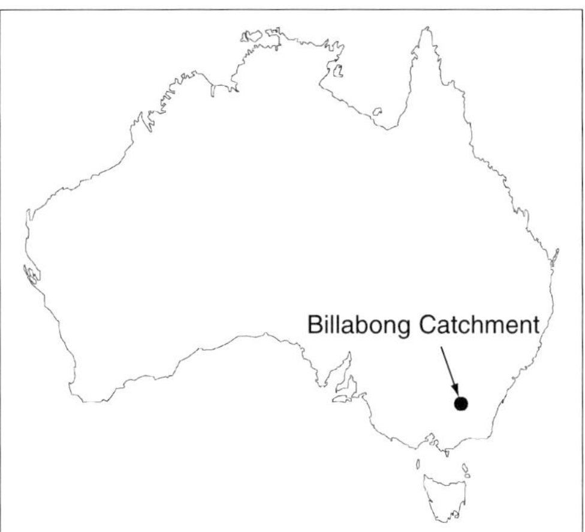

Figure 24.2. *Location of the Billabong Catchment.*

The catchment has an area of approximately 2600 km^2. In this study, limited time and resources were available, so information was drawn from detailed conventional surveys of two small subcatchments (approximately 10% of the region) and some 400 hand-augered or drilled sites across the region. A 2-week reconnaissance of the catchment was also undertaken. Details and data are available from the authors on request – our intention here is to illustrate principles.

24.4.1 Narrative of environmental change

Narratives of environmental change provide context and help explain soil patterns (Richter and Markewitz, 2001; Boyle et al., 2001). A number of events in the evolution of the Billabong catchment are relevant to the contemporary landscape (McKenzie et al., 2004).

The southwest slopes were formed by the uplift of the eastern highlands and subsequent erosion by rivers. The deeply weathered mantle that covered much of the continent during the Paleogene was presumably stripped from most of the areas now occupied by erosional landforms. The relatively steep slopes and narrow drainage lines in the east give way to hills and rises nearer to the riverine plain in the west. Now there are only a few isolated remnants of the deeply weathered profiles. The underlying rocks are predominantly Ordovician metasediments and granites formed during the Silurian and Devonian periods.

By the end of the Neogene, the Billabong catchment had attained its general shape although the climate was still more humid than it is today. As a result, erosion and deposition of sediments were primarily due to water-based erosion and the drainage network, unlike today, was well integrated and dendritic. These alluvial systems are now buried and host good quality aquifers.

The repeating glacial and interglacial periods resulted in changes in wind regimes, and the water balance (Williams et al., 1998; Bowler, 1986; 2002). Exposure of the continental shelf in southern Australia, along with large dry salt-lake beds and arid landscapes, are thought to have provided a significant source for the deflation of clays, silts, carbonates and soluble salts that were subsequently deposited across the Billabong catchment. This resulted in sodic soils in low-lying and poorly drained areas and problematic dry-land salinity.

Large areas, particularly in the more undulating lands to the west, are mantled with a thick layer of aeolian sediment with a characteristic suite of soils – this is the *parna* defined originally by Butler (1956). The mantle is a continuation of the stratigraphic system that has been studied and mapped to the west (Butler and Hutton, 1956), north (Beattie, 1972; Chen et al., 2002), east (Walker and Costin, 1971; Gatehouse et al., 2001) and south (Rowe, 1994) of the Billabong catchment. It would appear from exposures on flat landscapes that some 5–7 m of aeolian material has been deposited across the study area. Subsequent erosion has removed the material from steeper lands. On some flatter areas, erosion by migrating streams and subsequent alluvial deposition has either removed or covered the aeolian material. Inputs of aeolian sediment are still occurring (e.g. Walker and Costin, 1971; McTainsh and Leys, 1993) but rates during the Pleistocene were up to 10 times greater (Chen et al., 2002).

The Holocene saw a transition to present conditions. Reductions in stream flow and bed-load transport in the rivers of the riverine plain resulted in stream incision. Billabong creek attained its current form as a relatively small incised stream within a much larger alluvial plain.

24.4.2 Consequences for soil distribution and spatial prediction

At the broadest level, the Billabong catchment can be stratified into three landscape units on the basis of history, geomorphic process and provenance of soil parent materials. These units are summarized in Figure 24.3 and Table 24.1. Further distinctions and sequences can be identified within each landscape unit, primarily on the basis of topographic position and drainage.

The aeolian unit forms elevated plains and these have been incised to a minor degree. The landforms have a poorly integrated drainage network and the underlying erosional, colluvial and alluvial features have been effectively smothered by sediment. The streams are incompetent and unable to maintain an integrated drainage network. Profiles in flatter areas within the unit have B3 horizons extending for several metres. These horizons have a high degree of pedological organization and morphology is strikingly consistent throughout, although hydromorphic contrasts are common. These features are consistent with upbuilding

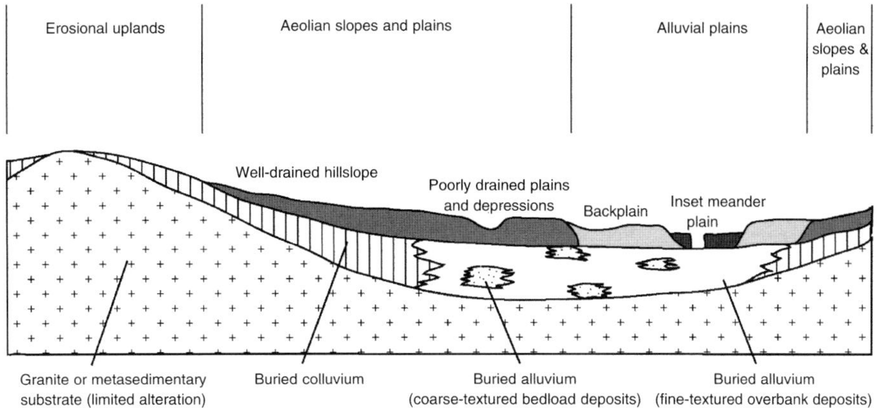

Figure 24.3. *Idealized cross-section of stratigraphic relationships between geomorphic units in the Billabong Catchment (not to scale).*

soil formation (Almond and Tonkin, 1999). There is a predictable soil pattern within the aeolian unit and the major control is local drainage (Table 24.1).

The alluvial landscapes in the catchment have a range of soils. Very young soils occur at point-bar deposits and levees of larger streams. These narrow alluvial landforms and soils are inset within the main alluvial landscape (Fig. 24.3). More extensive alluvial plains are dominated by texture-contrast profiles. Eluvial horizons are often present and water logging is common due to impermeable B horizons. Unlike the texture-contrast soils formed on flat aeolian landscapes, B-horizon development is at a maximum between 0.30 and 1.00 m. Pedological organization diminishes below this depth, suggesting that soil formation has occurred downwards into the alluvium (i.e. top-down soil formation as opposed to upbuilding soil formation in the aeolian unit). Some alluvial areas have more heavily textured soils and these occur where conditions of deposition were less energetic (flatter lands upstream of topographic constrictions).

The erosional uplands support a range of weakly developed shallow soils on upper slopes with texture-contrast soils in lower topographic positions. Minor differences occur due to variations in substrate lithology.

24.4.3 Environmental variables

Here we focus on a few terrain and remotely sensed variables, and, in particular, demonstrate how a new terrain variable, the multiresolution valley-bottom flatness (*MrVBF*) index (Gallant and Dowling, 2003), can be used with the familiar topographic wetness index (*TWI*) to make geomorphic distinctions and predict soil thickness.

Table 24.1. *Descriptions of the geomorphic units, and dominant geomorphic and pedologic processes for the Billabong Catchment. Soil profile class (SPC) is a local classification based on texture, degree of texture-contrast between the A and B horizons, and profile drainage. Approximate equivalents in the Australian Soil Classification (Isbell, 2002) and World Reference Base (Driessen et al., 2001) are provided for general context.*

Geomorphic unit	Dominant processes	Soil variation
Erosional uplands	Hillslope erosion and deposition, active biomantles, weathering of relatively unaltered substrate and colluvium	SPC 1: Shallow Rudosols, Kurosols and Chromosols (Regosols, Lixisols and Luvisols) along with rock outcrop on steep sites. SPC 2: Strong texture-contrast Chromosols (Luvisols) with E horizons and poor drainage in lower landscape positions
Aeolian slopes and plains	Variable rates of dust accumulation from the mid-Pleistocene to present. Mantle thickness ~5–7 m on flat lands. Upbuilding soil formation and redistribution of salts.	SPC 3: Free draining Red Kandosols (Lixisols) on crests and slopes SPC 4: Red/Yellow Chromosols on slopes SPC 5: Yellow/Grey Sodosols in areas with increasingly restricted drainage (Luvisols and Solonetz) SPC 6: Strongly sodic, hydromorphic variants in very poorly drained areas
Alluvial plains	Vertical accretion of fine sediments by overbank flows and lateral accretion of point deposits. Finer sediments deposited on distal backplains and areas upstream of constrictions. Top down soil formation.	SPC 7: Light textured Stratic Rudosols (Fluvisols) on active floodplains and levees of the modern Billabong Creek. SPC 8: Red and Brown Chromosols (Luvisols). SPC 9: Finer-textured variants of SPC8 on the flattest and lowest lying areas.

The digital elevation model was created using the ANUDEM V4.6.3 computer program and input data consisted of elevation contours and drainage lines. These were of several scales (1:25,000 and 1:50,000) depending on the original topographic maps. The drainage layer was edited to ensure downhill flow. Contours were checked for logical consistency and in some places contours were added to define catchment boundaries that could only be discerned in the field. *TWI* (Fig. 24.4) was calculated using

$$TWI = \ln\left(\frac{A_s}{\tan\beta}\right)$$

where A_s is the specific catchment area (m^2/m); and β, the slope gradient in degrees (Wilson and Gallant, 2000).

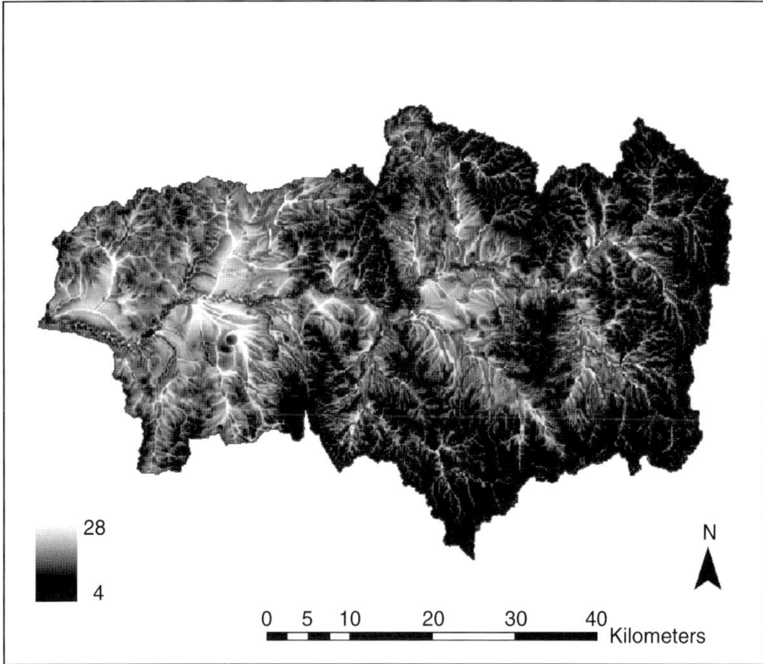

Figure 24.4. *Topographic Wetness Index (TWI) for the Billabong Catchment.*

The *MrVBF* index identifies zones that are flat and low in the landscape at different scales – these are interpreted as valley bottoms (Fig. 24.5). The method is based on three assumptions: valley bottoms are low and flat relative to their surroundings; valley bottoms occur at a range of scales; and large valley bottoms are flatter than smaller ones.

MrVBF is implemented using a slope classification constrained to convergent zones in the landscape. The classification is applied at multiple scales by progressive generalization of the digital elevation model combined with progressive reduction of the threshold for slope. The results at different scales are then combined into a single index representing the degree to which a site is low and flat, with larger values corresponding to flatter and broader low areas.

MrVBF is interpreted as an index of deposition, based on the assumption that flat valley bottoms are flat because they are filled with sediment. The index separates upland terrain dominated by erosional processes from lowland depositional terrain, and further divides the depositional zones into different classes based on slope and areal extent. Values less than 0.5 are considered to be erosional, while values greater than 0.5 are considered to be depositional. Larger values correspond to broader and flatter valley bottoms and presumably to

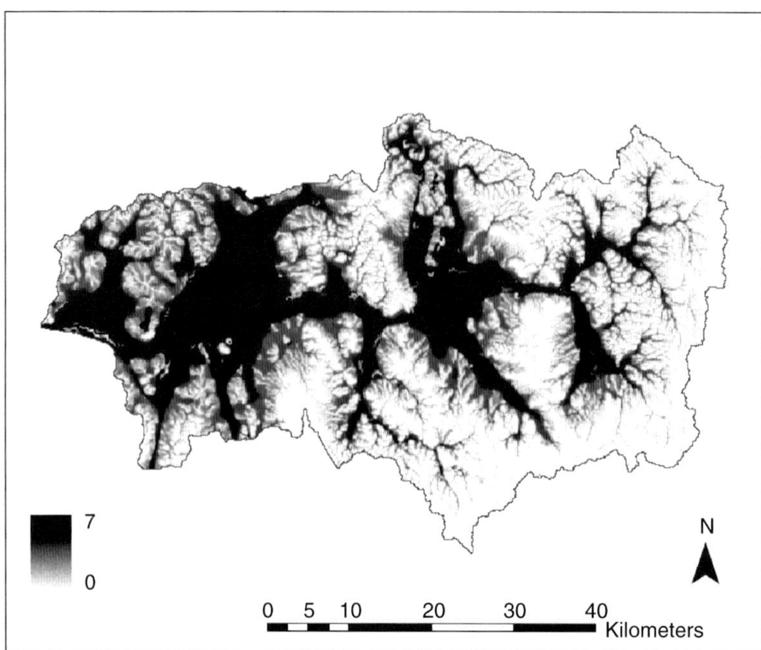

Figure 24.5. *MrVBF for the Billabong Catchment – MrVBF > 0.5 represents zones of deposition – the darkest zones show very flat alluvium upstream of topographic constrictions.*

greater depths of sediment. Each increase of 1 unit in the *MrVBF* index corresponds to a halving of slope and a tripling of length scale.

Airborne gamma radiometric remote sensing uses raw gamma-ray wavelength data to derive abundances of the major gamma-emitting elements – namely, potassium (K), thorium (Th) and uranium (U). Most of the gamma radiation emanates from shallow depths – approximately 90% is from the top 0.30–0.45 m for dry soil (Grasty, 1976). The airborne system used here collected the gamma-ray spectrum at 70-m intervals along profiles that, for most of the catchment, were evenly spaced at 250 m apart. The derived K, Th and U abundances together with a total count are interpolated to create images. Gamma radiometric remote sensing captures soil differences; most notably particle size and mineralogy (see Gessler et al., 1995; Cook et al., 1996b; Bierwirth, 1996; McKenzie and Ryan, 1999; Wilford, 2006). The *K* signal is most useful for predicting soil properties in the highlands, slopes and plains of southern New South Wales. Results for the Billabong catchment are shown in Figure 24.6.

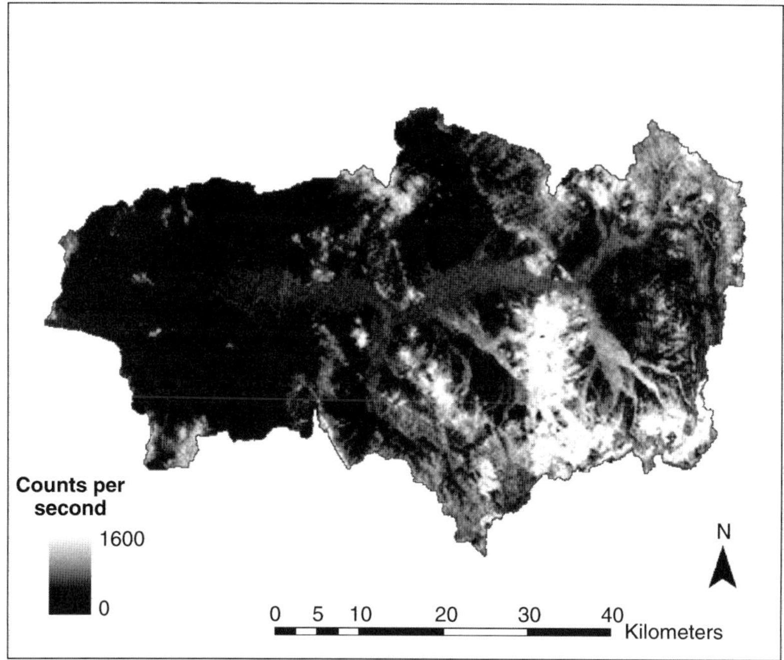

Figure 24.6. *Gamma-radiometric K (cps) for the Billabong Catchment – light zones have shallow soils over unaltered substrate, dark grey zones show alluvial units, and the extensive darkest grey zones correspond to the aeolian mantle.*

24.4.4 Rules for spatial prediction of soil profile classes

We prepared simple rules for spatial prediction that reflect the understanding of soil distribution outlined above. While many refinements are possible, the provisional set of rules, based on four environmental variables, provides plausible predictions.

The 2-week field reconnaissance investigated areas where stratigraphic relationships could be deduced. This involved developing and applying field criteria for aeolian and alluvial sediments, tracing the edges of stratigraphic units, and checking toposequences within these units. Field work involved regular reference to the environmental variables. Sites of an unequivocally aeolian origin were found to have a low gamma radiometric signal ($K<111$ cps) and gentle slopes (<6% – erosion prevents long-term accumulation on steeper lands). These criteria have been used to define the unit. The resulting map is remarkable and many new occurrences were identified across the region. The pedologic insight from a few stratigraphic sections was readily translated into a powerful rule for mapping – a result all the more notable because of the region's

Table 24.2. *Summary of rules for predicting geomorphic units and within unit soil-variation (K – radiometric potassium).*

Geomorphic unit	Definition	Variation within the geomorphic unit
Erosional uplands	$K>111$ cps AND *MrVBF* <2.7	SPC 1: *TWI* <5.5
	OR	SPC 2: *TWI* >5.5
	$K<111$ cps AND *slope* $>6\%$	
Aeolian slopes and plains	$K<111$ cps AND *slope* $<6\%$	SPC 3: *slope* $>2\%$
		SPC 4: *slope* 0.5% – 2%
		SPC 5: *slope* 0.1% – 0.5%
		SPC 6: *slope* $<0.1\%$
Alluvial plains	*MRVBF* >2.7 AND $K>111$ cps	SPC 7 Not mapped[a]
		SPC 8 *slope* $>0.5\%$
		SPC 9 *slope* $<0.5\%$

[a]This narrow unit is always inset and within 500 m of the Billabong creek. The unit could be mapped using a simple measure of local relief if a finer-resolution digital elevation model was available across the complete area.

low relief and subtle landscape. The distribution of soil profile classes within the aeolian unit was based on slope.

The alluvial landscapes have also been defined using terrain and gamma radiometrics. *MrVBF* by itself does not separate alluvial areas from flat lands mantled by aeolian materials. The latter are mostly alluvial landforms covered by aeolian deposits and as a result they retain the general shape of the underlying alluvial plains. The alluvial landscapes were defined by *MRVBF* >2.7 and radiometric potassium >111 cps. The distribution of soil profile classes within the aeolian unit was predicted using slope.

The erosional uplands have been identified quite pragmatically as areas that are of neither aeolian nor alluvial origin. They were identified using radiometric potassium >111 and *MrVBF* index <2.7; or radiometric potassium count <111 and slope $>6\%$. The distribution of soil profile classes within the erosional uplands was predicted using *TWI*.

A summary of the rules for prediction for geomorphic units and soil profile classes is given in Table 24.2 and the results are shown in Figure 24.7.

24.4.5 Prediction of soil thickness

The method for estimating soil thickness was proposed by McKenzie et al. (2003). The approach is empirical and great reliance is placed on landform. *TWI* and *MrVBF* are used to scale soil thicknesses derived from field data. We have assumed that soil profiles will thin towards crests in landscapes where the rate of soil production limits sediment movement. Soil profiles will not thin towards crests in landscapes where the capacity to transport sediment is limited. We

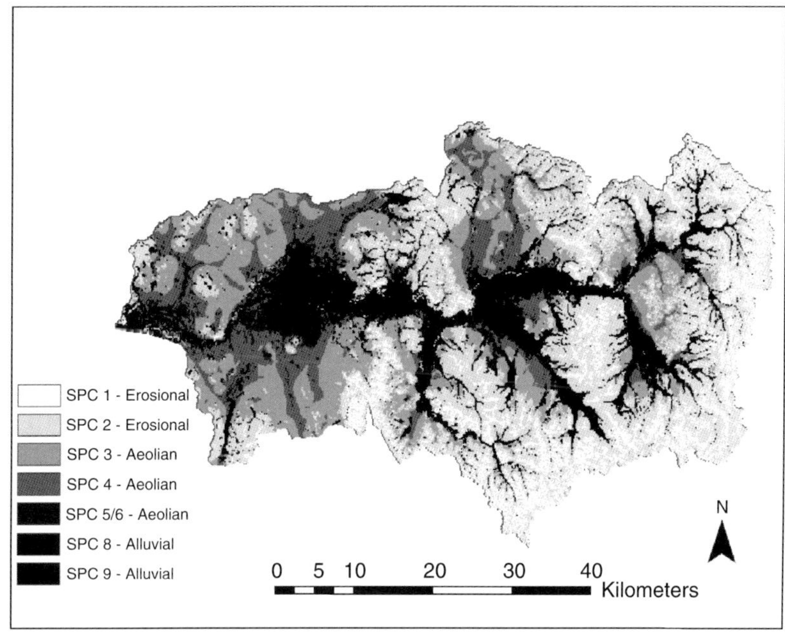

Figure 24.7. *Geomorphic units and soil profile classes for the Billabong Catchment.*

have assumed that supply-limited slopes dominate in the Billabong catchment. We also assumed A-horizon thickness varies with landscape position in erosional areas and is more constant in depositional areas.

We assumed a simple linear relation between *TWI* and soil thickness in upland areas (i.e. zones of net erosion) and empirical support is provided by Gessler et al. (1995) and McKenzie and Ryan (1999). We could not distinguish separate relations for areas with different lithology. On the basis of field observations for the Billabong catchment, we propose

$$\text{Thickness (m)} = -1.8 + 0.95 \times TWI$$

The *TWI* is used to predict soil thickness only in zones where erosion processes dominate. It is less reliable in lower landscape positions because only the information from the point of interest (β) and the upslope area (A_s) is used. However, soil thickness is also affected by factors below the point of interest including landform and the base level of the drainage network.

Soil thickness in depositional zones is estimated with *MrVBF*. Transition zones are estimated with the following scheme for assigning weights. On the basis of preliminary studies in southeastern Australia (Gallant and Dowling, personal communication), we have assumed $MrVBF > 1.5$ delineates areas with more than 5 m of regolith (with a 25-m resolution digital elevation model). For

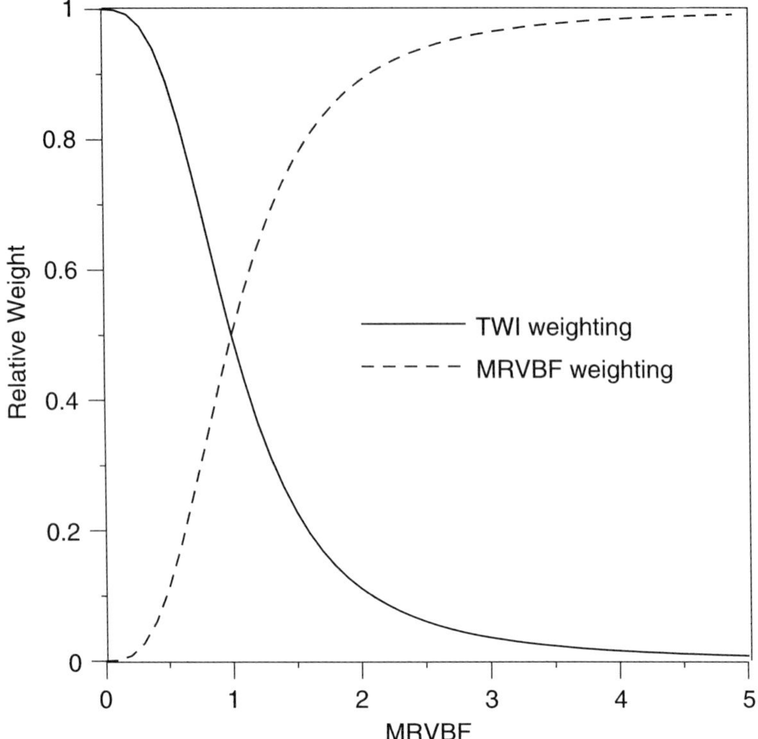

Figure 24.8. *Weighting functions for transitional zones (note $TWI_w + MRVBF_w = 1$).*

the purposes of estimating plant-available water capacity, we have also assumed that soil deeper than 5 m is beyond the reach of most plant roots. Maximum weighting is given to *TWI* in erosional areas (steep, divergent or both) where *MrVBF* is close to zero; the weighting shifts to *MRVBF* at the transition to depositional areas, with equal weighting at $MrVBF = 1$. The functions are shown in Figure 24.8 and apply regardless of the resolution of the digital elevation model. They are defined as follows.

$$TWI_w = \frac{1}{1 + MRVBF^3}$$
$$MRVBF_w = 1 - TWI_w$$

Profile thickness (d_i) is shown in Figure 24.9 and calculated using

$$d_i\ (\mathrm{m}) = TWI_w \times TWI_{pred} + MRVBF_w \times 5$$

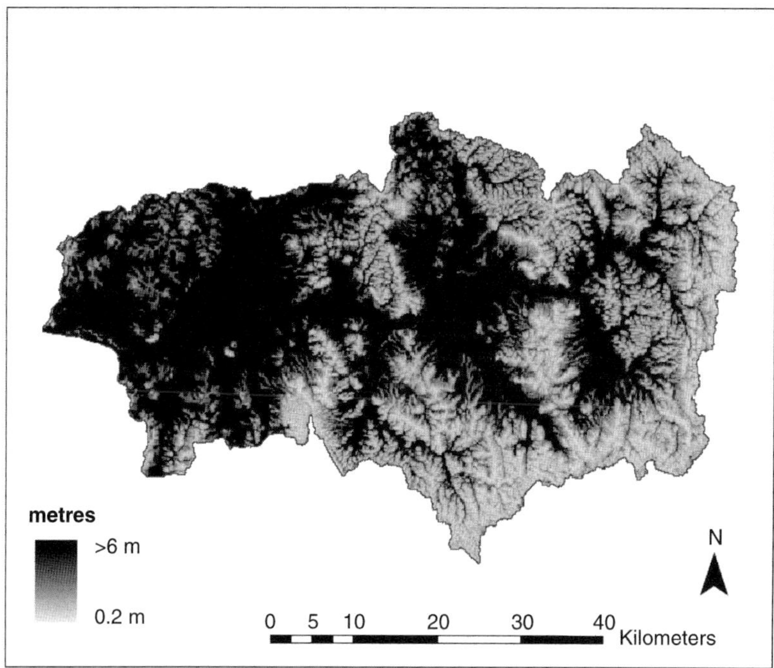

Figure 24.9. *Predicted soil thickness (m).*

24.4.6 Plant-available water capacity

Plant-available water capacity was estimated for perennial vegetation using water retention ($\theta_{-10\,\mathrm{kPa}}$, $\theta_{-1.5\,\mathrm{MPa}}$), horizon thickness and notional patterns of water extraction. Estimation of the available water storage in the A and B horizons required prediction of A-horizon thickness. This was done with the mean A-horizon thickness (d_μ) for each soil profile class (Table 24.3) and profile thickness (d_i).

- The A-horizon thickness ratio (r) was calculated for each soil profile class where $r = d_\mu / d_i$
- A horizon thickness (d_A) was calculated for all locations using

$$d_A = rd_i \text{ for } rd_i < 2d_\mu$$

$$d_A = 2d_\mu \text{ for } rd_i \geq 2d_\mu$$

The approach is pragmatic and further testing is required.

Soil profile class is a nominal variable and it was used to estimate texture, bulk density and grade of structure (Table 24.3). These variables provided inputs to the pedotransfer functions of Williams et al. (1992) for estimating volumetric

Table 24.3. *Estimates of soil attributes for soil profile classes (Texture: L – loam, SL – sandy loam, CL – clay loam, C – clay; Grade of structure: V – massive, W – weak, M – moderate).*

	SPC1	SPC2	SPC3	SPC4	SPC5	SPC6	SPC7	SPC8	SPC9
A thickness (mean)	0.2	0.3	0.3	0.3	0.3	0.3	0.4	0.3	0.3
A texture	L	L	L	L	L	L	SL	L	CL
A bulk density (Mg/m^3)	1.4	1.4	1.4	1.4	1.4	1.5	1.2	1.4	1.4
A structure grade	V	V	V	V	V	V	W	V	V
A K_s (mm/hr)	100	30	100	30	10	10	100	100	30
A −10 kPa θ_v	0.27	0.27	0.27	0.27	0.27	0.26	0.31	0.27	0.30
A −1.5 MPa θ_v	0.12	0.12	0.12	0.12	0.12	0.12	0.14	0.12	0.15
B texture	C	C	C	C	C	C	L	C	C
B bulk density (Mg/m^3)	1.5	1.5	1.5	1.6	1.7	1.75	1.3	1.5	1.6
B structure grade	M	M	M	M	M	M	W	M	M
B K_s (mm/hr)	30	10	30	10	0.3	0.1	100	30	3
B −10 kPa θ_v	0.40	0.40	0.40	0.37	0.33	0.31	0.33	0.40	0.37
B −1.5 Mpa θ_v	0.30	0.30	0.30	0.30	0.30	0.30	0.18	0.30	0.30
X_i (m)	1.4	1.2	2.0	1.5	0.9	0.7	2.0	1.8	1.5

water content at −10 kPa and −1.5 MPa. Soil profile class was also used to estimate average A-horizon thickness and the parameter for scaling water extraction (see below).

The available water capacity was multiplied by the respective thickness of the A and B horizons and then summed. The B-horizon thickness was calculated as the total profile thickness minus the A-horizon thickness. The estimates of available water capacity (i.e. $\theta_{-10\,\mathrm{kpa}} - \theta_{-1.5\,\mathrm{Mpa}}$), when summed over the depth of the profile are known to overestimate the quantity of water extracted in field situations. The main problem is the diminishing ability of plants to extract water from deeper layers. This is due to both the physiology of the plant and subsoil constraints to root growth. Published water extraction patterns across a range of soils and plant types (e.g. Williams, 1983) suggest that a simple correction to the profile available water capacity can be made using a scaling factor that reflects root distribution. Figure 24.10 presents a plausible scaling factor (*f*) for root density used by Verburg et al. (1997).

Individual curves in Figure 24.10 are presented for a range of values for X_i, the depth (m) where 37% of water extraction is from deeper in the profile. This simple physical interpretation provides a basis for estimation because soil profile descriptions provide information on root and macropore distributions with depth. The predicted plant-available water capacity is shown in Figure 24.11 – it is the integral of the profile available-water capacity multiplied by the scaling term for each depth increment. It can be calculated directly for the A and B

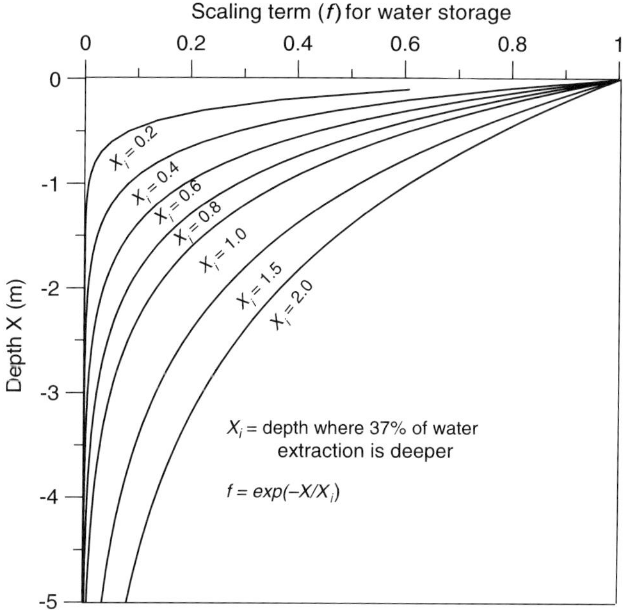

Figure 24.10. *Scaling term for converting profile- to plant-available water capacity.*

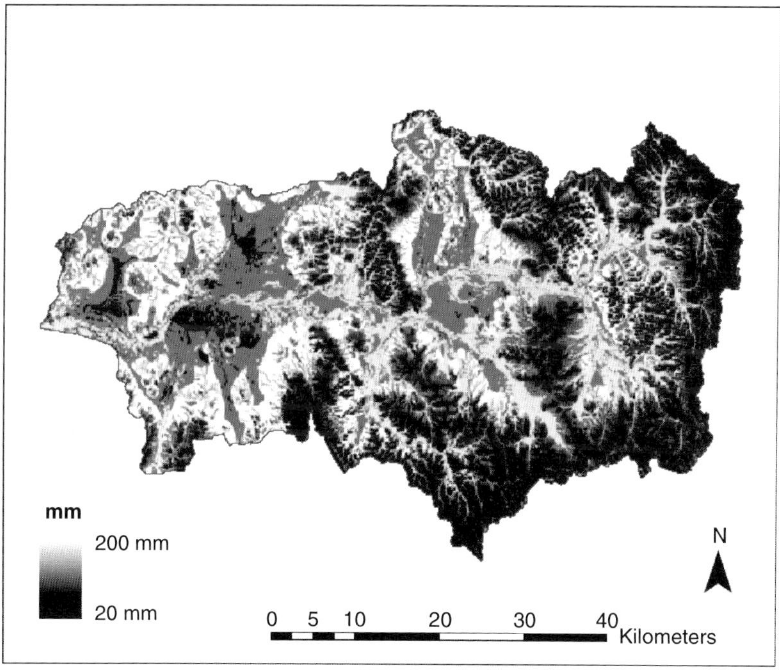

Figure 24.11. *Plant-available water capacity for the Billabong Catchment.*

horizons (A_{Total} and B_{Total}), respectively, both expressed in millimetres (n.b. A_{AWC} and B_{AWC} are in mm/m), by

$$A_{\text{Total}} = A_{\text{AWC}} X_i \left(1 - e^{\frac{-d_A}{X_i}}\right), \quad B_{\text{Total}} = B_{\text{AWC}} X_i \left(e^{\frac{-d_A}{X_i}} - e^{\frac{-d_B}{X_i}}\right)$$

24.5 Discussion

The simple rules for spatial prediction in the Billabong catchment provide a starting point for future soil survey and land evaluation. These rules can be improved. For example, further stratigraphic distinctions might be needed in the alluvial unit but intensive field investigations are required. Similarly, improvements in the resolution of environmental variables would allow discrimination of the Holocene meander plain and levees associated with Billabong creek.

The current rules provide hypotheses for testing. For example, a single relationship between *TWI* and soil thickness was used across a range of lithologies because separate relationships could not be detected from the sparse field observations for the region. Slope profiles differ between granite and metasediment landscapes in the catchment and we expected different relationships between *TWI* and soil thickness. It may well be that broad climatic controls on vegetative cover and similar soil production rates limit sediment supply and ensure a consistent relationship – a statistically based comparison of relationships between *TWI* and soil thickness on different lithologies in erosional areas would test this hypothesis.

The Billabong catchment has a long and complex environmental history. Understanding this history was essential in making the distinction between alluvial and aeolian landscapes – these landscapes appear almost identical in the field. However, soil characteristics and patterns differ within each landscape. Field stratigraphic distinctions combined with access to fine-resolution environmental data provided the basis for formulating the simple rules. It is an open question whether this result would have been possible with a more geographically rather than pedologically focused strategy for sampling. It is worth noting that the aeolian unit is the source of the dry-land salinity problem in the region – this was unknown prior to the current work and it has many practical consequences.

We were initially surprised that such a simple set of rules could provide plausible predictions. Formal statistical validation is the next step and the field program is currently underway. While the results to date are cause for optimism, other soil properties and processes will require a different approach. For example, patterns of soil acidification in the region are related to both natural soil variation and land-use history, so another set of environmental predictors is needed for mapping.

Finally, in applied studies, investment in new information needs to be driven by the degree to which it can reduce uncertainty for decision makers – in many circumstances, the greatest uncertainty comes from other sources (e.g. commodity prices, political uncertainties) and heavy investment in biophysical information is not worthwhile. In our example, statistical validation is worthwhile because large investments are needed to control dry-land salinity and reliable soil information is necessary to determine optimal strategies for revegetation with deep-rooted perennials.

24.6 Conclusions

Digital soil mapping using environmental correlation is feasible when survey extents are large (i.e. $>10{,}000\,km^2$), data are sparse and resources are limited. However, the success of the method will depend on the landscape and the availability of environmental predictors related to pedogenic processes. Unlike conventional survey, the approach is explicit and repeatable. The ease of iteration encourages improvements to predictions when and where they are required.

Notes

1. The name was originally suggested by Alex McBratney.

References

Almond, P.C., Tonkin, P.J., 1999. Pedogenesis by upbuilding in an extreme leaching and weathering environment, and slow loess accretion, south Westland, New Zealand. Geoderma 92, 1–36.

Austin, M.P., McKenzie, N.J., 1988. Data Analysis. In: R.H. Gunn, J.A. Beattie, R.E. Reid, and R.H.M. van de Graaff (Eds.), Australian Soil and Land Survey Handbook. Guidelines for Conducting Surveys. Inkata Press, Melbourne.

Beattie, J.A., 1972. Ground surfaces of the Wagga Wagga region, NSW. CSIRO Australia Soil Publication No. 28.

Bierwirth, P., 1996. Investigation of airborne gamma-ray images as a rapid mapping tool for soil and land degradation – Wagga Wagga, NSW. Canberra, Australian Geological Survey Organization, Record 1996/22.

Bowler, J.M., 1986. Quaternary landform evolution. In: D.N. Jeans (Ed.), Australia, a geography Vol. 1. The natural environment. Sydney University Press, Sydney.

Bowler, J.M., 2002. Lake Mungo: window to Australia's past. University of Melbourne, Melbourne.

Boyle, M., Kay, J.J., Pond, B., 2001. Monitoring in support of policy: an adaptive ecosystem approach. In: T. Munn (Ed.), Encyclopedia of global environmental change Vol. 5. John Wiley and Son.

Bui, E.N., 2004. Soil survey as a knowledge system. Geoderma 120, 17–26.

Butler, B.E., 1956. Parna – an aeolian clay. Australian J. Sci. 18, 145–151.

Butler, B.E., 1958. The diversity of concepts about soils. J. Australian Inst. Agric. Sci. 24, 14–19.

Butler, B.E., 1963. Can pedology be rationalized? Australian Soil Science Society, Publication No. 3, Canberra.

Butler, B.E., 1982. A new system for soil studies. J. Soil Sci. 33, 581–595.

Butler, B.E., Hutton, J.T., 1956. Parna in the Riverine Plain of south-eastern Australia and the soils thereon. Australian J. Agric. Res. 7, 536–553.

Butler, B.E., Hubble, G.D., 1978. The general distribution and character of soils in the Murray-Darling River system. Royal Soc. Victoria Proc. 90, 149–156.

Chadwick, O.A., Chorover, J., 2001. The chemistry of pedogenic thresholds. Geoderma 100, 321–353.

Chen, X.Y., Spooner, N.A., Olley, J.M., Questiaux, D.G., 2002. Addition of aeolian dusts to soils in southeastern Australia: red silty clay trapped in dunes bordering Murrumbidgee River in the Wagga Wagga region. Catena 47, 1–27.

Cook, S.E., Corner, R., Grealish, G.J., Gessler, P.E., Chartres, C.J., 1996a. A rule-based system to map soil properties. Soil Sci. Amer. J. 60, 1893–1900.

Cook, S.E., Corner, R.J., Groves, P.R., Grealish, G.J., 1996b. Use of airborne gamma radiometric data for soil mapping. Australian J. Soil Res. 34, 183–194.

de Gruijter, J.J., Brus, D., Bierkens, M., Knotters, M., 2006. Sampling for natural resource monitoring. Springer, Berlin.

Dent, D., Young, A., 1981. Soil survey and land evaluation. George Allen and Unwin, London.

Driessen, P., Deckers, J., Spaargaren, O. and Nachtergaele, F., 2001. Lecture notes on the major soils of the world. World Soil Resources Reports No. 94, Food and Agriculture Organization of the United Nations, Rome.

Gallant, J.C., Dowling, T.D., 2003. A multi-resolution index of valley bottom flatness for mapping depositional areas. Water Resour. Res. 39, 1347–1360.

Gallant, J.C., Hutchinson, M.F., 2006. Terrain analysis. In: N.J. McKenzie, M.J. Grundy, and A.J. Ringrose-Voase (Eds.), Guidelines for surveying soil and land resources. Australian Soil and Land Survey Handbook Series, Vol. 2. CSIRO Publishing, Melbourne.

Gatehouse, R.D., Williams, I.S., Pillans, B.J., 2001. Fingerprinting windblown dust in south-eastern Australian soils by uranium-lead dating of detrital zircon. Australian J.Soil Res. 39, 7–12.

Gessler, P.E., Moore, I.D., McKenzie, N.J., Ryan, P.J., 1995. Soil-landscape modelling and spatial prediction of soil attributes. Int. J. Geogr. Inf. Systems 4, 421–432.

Grasty, R.L., 1976. Applications of gamma radiation in remote sensing. In: E. Schanda (Ed.) Remote sensing for environmental sciences. Springer-Verlag, New York.

Hewitt, A.E., 1993. Predictive modelling in soil survey. Soils and Fertilizers 3, 305–315.

Hudson, B.D., 1992. The soil survey as a paradigm-based science. Soil Sci. Soc. Amer. J. 56, 836–841.

Huggett, R.J., 1998. Soil chronosequences, soil development and soil evolution: a critical review. Catena 32, 155–172.

Isbell, R.F., 2002. The Australian Soil Classification. CSIRO Publishing, Collingwood, Victoria.

Johnson, D.L. and Hole, F.D., 1994. Soil formation theory: a summary of its principal impacts on geography, geomorphology, soil-geomorphology, Quaternary geology and palaeopedology. In Amundson R., Harden, J., Singer, M. (Eds.). Factors of soil formation: a fiftieth anniversary retrospective. Madison, Soil Science Society of America Special Publication No. 33.

McBratney, A.B., Mendonça Santos, M.L., Minasny, B., 2003. On digital soil mapping. Geoderma 117, 3–52.

McKenzie, N.J., Ryan, P.J., 1999. Spatial prediction of soil properties using environmental correlation. Geoderma 89, 67–94.

McKenzie, N.J., Gallant, J.C. and Gregory, L.J., 2003. Estimating water storage capacities in soil at catchment scales. Technical Report 03/3, Cooperative Research Centre for Catchment Hydrology, Canberra. http://www.catchment.crc.org.au/pdfs/technical200303.pdf

McKenzie, N.J., Jacquier, D.W., Isbell, R.F., Brown, K.L., 2004. Australian soils and landscapes: an illustrated compendium. CSIRO Publishing, Melbourne.

McKenzie, N.J., Grundy, M.J., Ringrose-Voase, A.J., 2006. Guidelines for surveying soil and land resources. Australian Soil and Land Survey Handbook Series, Vol. 2. CSIRO Publishing, Melbourne.

McTainsh, G., Leys, J., 1993. Soil erosion by wind. In: G. McTainsh and W.C. Boughton (Eds.), Land degradation processes in Australia. Longman Cheshire, Melbourne.

Paton, T.R., Humphreys, G.S., Mitchell, P.B., 1995. Soils. A new global view. UCL Press, London.

Richter, D.D., Markewitz, D., 2001. Understanding soil change. Cambridge University Press, Cambridge.

Rowe, R.K., 1994. The influence of Quaternary aridity on soils of non-arid regions: a case-study of soils in the Tallangatta area of southeastern Australia. PhD Thesis, Monash University.

Ryan, P.J., McKenzie, N.J., O'Connell, D., Loughhead, A.N., Leppart, P.M., Jacquier, D.J., Ashton, L., 2000. Integrating forest soils information across scales: spatial prediction of soil properties under Australian forests. Forest Ecology and Management 138, 139–157.

Scull, P., Franklin, J., Chadwick, O.A., McArthur, D., 2003. Predictive soil mapping: a review. Prog. Phys. Geogr. 27, 171–197.

Simonson, R.W., 1959. Outline of a generalized theory of soil genesis. Soil Sci. Soc. Amer. Proc. 23, 152–156.

Verburg, K., Ross, P.J., Bristow, K.L., 1997. SWIM v2.1 user manual. CSIRO Division of Soils Divisional Report 130. CSIRO, Australia.

Walker, P.H., Costin, A.B., 1971. Atmospheric dust accession in south-eastern Australia. Australian J. Soil Res. 9, 1–6.

Webster, R., 2000. Is soil variation random? Geoderma 97, 149–163.

Webster, R., Oliver, M., 2001. Geostatistics for environmental scientists. Wiley, Chichester.

Wilford, J., 2006. Gamma-ray spectrometry. In: N.J. McKenzie, M.J. Grundy, and A.J. Ringrose-Voase (Eds.), Guidelines for surveying soil and land resources. Australian Soil and Land Survey Handbook Series, Vol. 2. CSIRO Publishing, Melbourne.

Williams, J. 1983. Physical properties and water relations. In: Soils: an Australian viewpoint. CSIRO: Melbourne/Academic Press: London.

Williams, J., Ross, P.J. and Bristow, K.L., 1992. Prediction of the Campbell water retention function from texture, structure and organic matter. In M.Th. van Genuchten, F.J. Leij, L.J. Lund (Eds.). Proceedings of the International Workshop on Indirect Methods for Estimating the Hydraulic Properties of Unsaturated Soils. University of California, Riverside.

Williams, M., Dunkerley, D., De Deckker, P., Kershaw, P., Chappell, J., 1998. Quaternary environments, 2nd edition. Arnold, London.

Wilson, J.P., Gallant, J.C., 2000. Terrain analysis: principles and applications. John Wiley, New York.

F.i. Examples of predicting soil classes

Chapters 25–31 give examples of predicting soil classes from Europe (the Czech Republic, England & Wales and Germany), the USA and Australia. Chapters 25–29 deal with the estimation of soil classes from spatial covariates (class scorpan functions) whereas Chapters 30–31 show examples of allocation to pre-existing soil classes from soil attribute (soil allocation functions).

The methods for producing class scorpan functions that are presented in Chapters 25–29 differ mainly in the way with which the soil survey expertise is integrated into the prediction algorithms. In Chapters 25 and 26, existing soil maps are used as training examples for data-mining algorithms currently used in statistic and information theory. The tested algorithms are artificial neural networks, support vector machines, linear regression, learning vector quantisation, classification trees, discriminant analysis and Bayesian belief networks. Chapters 27 and 28 are examples of class scorpan functions mainly derived from simple knowledge-based classifications (trees) that include soil surveyors' prior knowledge and its further revisions as new soil observations are made in the field. Chapter 29 embeds less expertise on soil class distribution by using a classification-tree algorithm applied to a set of site observations.

The most commonly used spatial covariates are by far those derived from Digital Terrain Models. Landsat TM is also largely used to represent parent material and vegetation. Chapter 26 shows the use of the largest range of spatial covariates, including small-scale soil maps, geology maps, climate indices and airborne gamma-ray spectrometry/ magnetometry data.

The mapped areas have generally intermediate sizes–from 400 km^2 to 800 km^2—except in the case of Chapter 29 that deals with a wider area (7 320 km^2). The resolutions of the final maps vary from 10 to 50 m according to the resolution of the input spatial covariates. When point soil observations are used as input, their sample size varies between 120 and 380.

The allocation functions that are presented in Chapter 30–31 are obtained by very different approaches. Chapter 29 presents an inductive-learning approach that consists of applying an Artificial Neural Network algorithm to a set of soil data from a national database. In Chapter 30, there is a comparison of three different strategies, namely fuzzy allocation to a reference profile, a crisp expert rule system and a fuzzy expert rule system. In all cases, basic soil properties currently available from soil profile observation and analysis are required with more ore less tolerance between the strategies with regard to errors, missing data and qualitative descriptions of soil variables.

Further work is needed to orient the choice of a given algorithm to build functions according to the data context that is area size, sample data size, available pedological knowledge and skills, and the nature of soil variation.

Developments in Soil Science, volume 31
P. Lagacherie, A.B. McBratney and M. Voltz (Editors)

Chapter 25

A COMPARISON OF DATA-MINING TECHNIQUES IN PREDICTIVE SOIL MAPPING

T. Behrens and T. Scholten

Abstract

To predict soil maps, data-mining techniques can be utilised. The aim of these techniques is to extract hidden predictive knowledge from large databases. In terms of soil science they are able to learn the relationship between mapped soil classes as well as soil-forming factors, which can be used to predict soil classes in comparable landscape units. Thus, it is possible to automatically build reproducible digital soil maps, helping to speed up field mapping and to reduce costs.

The main objective of this chapter is to compare the ability of different data-mining techniques and algorithms from statistics and information theory, including artificial neural networks (ANNs), support vector machines (SVMs), linear regression, learning vector quantisation and classification trees. The techniques are discussed in terms of prediction accuracy and usability for GIS-based usage.

Altogether 10 data-mining algorithms were tested to predict soil classes on the basis of 65-terrain attributes. Prediction accuracy is tested inside and outside the learning area to compare their generalisation ability

25.1 Introduction

In the context of a growing demand of high-resolution spatial soil information for environmental protection and management, fast and accurate prediction methods are needed. Recent publications indicate that data mining becomes an important and promising pedometrical technique (McBratney et al., 2003; Scull et al., 2003; Behrens et al., 2005). As the term data mining comprises a whole bunch of different methods and algorithms with different heritages and strategies, the question arises which predictor is the best in terms of usability and accuracy concerning digital soil mapping. Thus the aim of this study is to compare different methods and algorithms using the same set of input data, predictor variables and validation data. The results could help decision makers in geological surveys as well as pedometricians who want to use and experiment with these methods.

The comparison comprises 10 different data mining algorithms which are tested in terms of extrapolating mapped soil classes in lower mountain ranges for D3/D4 surveys (McBratney et al., 2003) based on an extended digital terrain analysis (DTA) approach in the feature space (Behrens, 2003).

This study focuses on artificial neural nets (ANNs) (Zell et al., 1995; Bishop, 1995), decision trees (Breiman et al., 1984), support vector machines (SVMs) (Vapnik, 1995), a linear regression model (Rasmussen, 1996) and learning vector quantisation (Kohonen, 1990). All methods are compared in terms of a 2-class problem were each soil class is extrapolated separately.

Concerning ANNs Behrens et al. (2005) show that generally the best prediction results were obtained when relief, geology and land use as soil-forming factors were used together in the modelling approach. Yet, within some regions using relief is sufficient to achieve very good prediction results, indicating the importance of terrain attributes in soil-landscape modelling. Thus, in this study we only use terrain attributes which is more realistic in terms of a "real world" example as in most cases only digital elevation models (DEMs) such as SRTM are available and to expand the range of the results in contrast.

25.2 Material and methods

25.2.1 Study area and existing soil maps

The study area called the 'Palatinate Forest' is located in southern Rhineland-Palatinate between Kaiserslautern in the north and the German–French border in Germany. It is a low mountain range area within the southwest German-Lorraine Triassic escarpment. The highest mountain is the Eschkopf in the central Palatinate Forest, which is 610 m above sea level. The mean annual precipitation varies about 800–900 mm increasing up to 1050 mm at higher elevations (Ministerium für Landwirtschaft, Weinbau und Forsten, 1982).

Since 1996, 12 soil maps within the Palatinate Forest have been surveyed at a scale of 1:50,000 based on a conceptual approach relating to geology and terrain. The dataset contains information about the soil classes and parent material classified according to the current German mapping standard (AG Boden, 1994).

The Palatinate Forest soils have formed from substrates of the Upper Red Bed Sandstone, Bunter and Lower Limestone.

To become an optimal test bed for a detailed validation study we clipped two areas (one for training and one for validation) from the existing soil map, which are completely covered by the same soil units. We found such areas in the north of the Palatinate Forest where six soil classes occur (Table 25.1). Each area comprises about 40 square kilometres based on a resolution of 20 m. Plate 25 (see Colour Plate Section) shows the spatial relationship of the two datasets.

Table 25.1. *Predicted soil classes and their distribution in the training and the validation area.*

ID	Soil class	Parent material	Coverage training area [%]	Coverage validation area [%]
1	Dystric Cambisols	Blocky, sandy slope deposits covering deep-weathered sandstone (Lower Bunter)	11	20
2	Cumulic Anthrosols, Gleyic Anthrosols, Gleysols and Fluvisols	Fluvic soil material in valleys and adjacent to water courses	5	4
3	Dystric Cambisols	Loess-containing sandy Pleistocene periglacial slope deposits covering weathered sandstone (Middle Bunter)	5	2
4	Haplic Podzols	Sandy slope deposits rich in debris covering weathered sandstone (Middle Bunter)	32	32
5	Dystric Cambisols and Cambi-Haplic Podzols	Blocky, sandy slope deposits covering weathered blocky sandstone (Middle Bunter)	12	5
6	Haplic Cambisols	Loess-containing sandy Pleistocene periglacial slope deposits covering weathered sandstone (Middle Bunter)	32	34

Table 25.2. *Features and requirements of the pattern recognition algorithms used in this study.*

Method	Software	Algorithm	Trans.	Link location (URL)
Neural networks	SNNS	Rprop/BPm	Yes	www.ra.informatik.unituebingen.de/SNNS
Decision trees	QUEST, CRUISE	QUEST/CRUISE	No	www.stat.wisc.edu/~loh
Support vector machines	SVM^{Light}	Linear/RBF	Yes	svmlight.joachims.org
Leaning vector quantisation	LVQ_PAK	OLVQ1	Yes	www.cis.hut.fi/research/som_lvq_pak.shtml
Linear model	Lin-1	lin-1	Yes	www.cs.toronto.edu/~delve/methods/lin1/home.html

Note: RProp = resilient backpropagation; BPm = backpropagation with momentum; RBF = radial basis function; Trans. = data transformation required/recommended.

25.2.2 Data-mining software and algorithms

In this study, only freely available data-mining software products were used. The algorithms, as well as important constraints, features and the corresponding link location (URL) are shown in Table 25.2.

The most important preprocessing step is the data transformation, which is recommended for nearly all algorithms. Though theoretically not necessary, it has been proved empirically that a normalisation of all input data to a range of values between 0 and 1 stabilises the performance (Sarle, 2002). This was done by rescaling, that is subtracting the minimum and dividing by the range. Each soil class is presented as a so-called binary dummy variable, where a 1 indicates the occurrence of the respective soil class at this location, and a 0 its absence.

ArcView GIS 3.3 was used to visualise the results. All programs have been compiled for Microsoft DOS (but are available for LINUX as well) and can generally be used via command line interfaces. To work with these software packages more effectively, graphical user interfaces were created for all packages based on ArcView GIS. To speed up data conversion special routines were implemented using C++.

Artificial neural networks

The idea of ANNs has been adopted from data processing in biological nervous systems as there are cells for information reception, others for its forwarding and storage and another group for the outward release of information. The knowledge is described by the weights connecting the cells. The adjustment of these weights, which are randomly chosen at the beginning, is the intrinsic learning process.

The algorithms used in this study are backpropagation with momentum (BPm) and resilient backpropagation (RProp) (Zell et al., 1995). They belong to the most popular group of multilayer feed forward backpropagation networks with information processing from the input cells to the output cell(s) and error propagation in the reverse direction. The difference between the algorithms is the so-called learning rate, describing the amount and the behaviour of how the weights are adjusted. BPm used a momentum term, where a part to the previous weight change is added to the actual weight change to speed up convergence. RProp uses an adaptive learning rate for each weight and has mechanisms to avoid oscillations and to speed up learning. A detailed description can be found in Riedmiller and Braun (1993).

Classification trees

Classification trees are built through binary recursive partitioning. Thereby a dataset is split into two subsets at the "best" split position of a feature. This process is iteratively repeated for each new subset.

Suppose a dataset consists of one continuous input feature x and one 2-class feature to predict y, with N observations, that is, (x, y) for $i = 1, \cdots \cdots, N$. In the binary partition process, the data space is separated into two sub-regions

(He et al., 2003). The split variable x and the split or threshold value λ assign observations to either left or right branch of the node. The observations with $\{x \leqslant \lambda\}$ are assigned to the left branch and those with $\{x > \lambda\}$ to the right respectively.

The algorithms used in this study are QUEST (Loh and Shih, 1997) and CRUISE (Kim and Loh, 2001). QUEST is an acronym for "Quick, Unbiased and Efficient Statistical Tree" and is based on an analysis of variance. CRUISE stands for "Classification Rule with Unbiased Interaction Selection and Estimation". It is a classification tree where multiple subsets can be created at each split point.

Support vector machines

Support vector machines find their roots in statistical learning theory and where invented by Vapnik (1995). SVM correspond to a linear method in a high-dimensional feature space nonlinearly related to input space. The learning algorithm is based on the class of hyperplanes, where the optimal hyperplane is defined as the one with the maximum margin separating two classes. This optimal hyperplane can be constructed by solving a constrained quadratic optimisation problem (Schölkopf, 1997). Classification is based on the distance and the side to the separating hyperplane.

SVM^{light} is an implementation of an SVM learner that addresses the problem of large tasks (Joachims, 1999) and is based on a generalised version of the decomposition strategy proposed by Osuna et al. (1997).

Learning vector quantisation

The method described in this study is based on the learning vector quantisation algorithm OLVQ1 (Kohonen, 1990, 2001) – a combination of a single-layer neural network and a Euclidian distance approach. The goal of the algorithm is to approximate the distribution of a class using a reduced number of vectors where the algorithm seeks to minimise classification errors (Brownlee, 2004).

An LVQ network consists of an array of labelled vectors, so called weight- or codebook-vectors (CV) (Merelo and Prieto, 1994). Each label corresponds to one class. In our (binary) case, a "1" indicates the presence of a soil class, whereas a "0" indicates its absence.

During the learning phase, each CV in the training sample is submitted to the network and the closest CV is determined. If the input CV belongs to the same class it is updated, in such a way that it is moved closer to the class centre, or is moved away from it, if it belongs to a different class (Merelo and Prieto, 1994) resulting in a decision space partitioned by a 'Voronoi net' of hyperplanes (Crone et al., 2004). Thus, the class (decision) boundaries are adjusted during the learning process and the subsequent classification based on a Euclidean distance classifier is more precise.

Linear regression

The linear least squares regression model 'lin-1' is both conceptually and computationally simple. The model is fit to the training data by maximum likelihood, which is found by minimising a cost function (Rasmussen, 1996).

25.2.3 Digital terrain analysis

In this exemplary study only terrain attributes are considered. Whereas most DSM approaches only deal with a few terrain attributes, a "brute force" DTA approach based on 65 continuous terrain attributes and their derivates is carried out, to ensure high prediction accuracy for DSM. The mesh size of the DEM is 20 m by 20 m.

Digital terrain analysis is based on works of Beven and Kirkby (1979), Horn (1981), Zevenbergen and Thorne (1987), Quinn et al. (1991), Kleefisch and Köthe (1993), Huber (1994), Feldwisch (1995), Nogami (1995), Tarboton (1997), Dietrich and Montgomery (1998), Hatfield (1999), Hodgson and Gaile (1999), Shary et al. (2002) and Behrens (2003).

25.2.4 Validation and classification

Validation is based on two independent datasets for each soil class: the training dataset and a validation dataset, which is not used for training.

For each algorithm the prediction results for each binary predicted soil class are evaluated by comparing the mapped and the predicted areas using a cross matrix to derive measures of prediction accuracy (Table 25.3).

Recall (*RC*) is the percentage of correct predictions to all mapped pixels, showing the amount of underestimation. The precision (*PC*) is the percentage of all correct predictions to all predicted pixels, pointing to noise or overestimation (Ishioka, 2003):

$$RC = \frac{tt}{tt + ft} \text{ and } PC = \frac{tt}{tt + tf}$$

Table 25.3. *Cross matrix. On the basis of the matrix the amount of correct and false predicted classes can be compared within the training or the validation area (t: true; f: false).*

		Mapped	
		True	False
Predicted	True	tt	tf
	False	ft	ff

Another commonly used metric is the *F* measure (van Rijsbergen, 1979), representing the harmonic mean of precision and recall:

$$F = 2(PC \times RC)/(PC + RC)$$

Generally larger values indicate a better prediction. To compare the prediction results in this study the *F* measure is used.

Some data-mining algorithms, even if they are implemented as classifiers, return continuous values (ANN, SVM and LM). Hence, an optimal threshold to classify each prediction result must be found. In this study, the break-even point where $RC = PC$ is searched by iteratively classifying the prediction output in small steps from 0 to 1. Each classification is compared with the mapped example. The classification nearest to the break-even point is used for the final prediction. The minimum difference allowed between recall and precision is below 0.005. This methodology ensures that the amount of under- and over-prediction is equal.

25.3 Results and discussion

25.3.1 Parameter settings

The parameters used for predictive soil mapping were generally adopted from the defaults suggested by the authors of the corresponding software. If these parameters returned insufficient results the parameters had been adjusted by experiments according to Table 25.4.

As the datasets we used in this study are large and thin (i.e. ten thousands of samples in terms of pixels in the soil map and compared with that, a small amount of predictor attributes) the prediction is very stable in general. In

Table 25.4. *Algorithm settings used in this study (mindat = minimum number of pattern in an end node of a tree; CV = number of cross-validation samples; SE = square error).*

Algorithm	Settings
RProp	Iterations = 25; 1 hidden layer with 10 units; weight decay = 0.7; min. update value = 0.1; max. update value = 50
BPm	Iterations = 40; 1 hidden layer with 10 units; step width = 0.2; max. error = 0.1; momentum = 0.5; flat spot elimination = 0.25
QUEST	mindat = 10; (CV = 10, SE = 1.0)
CRUISE	mindat = 10; (CV = 10, SE = 1.0)
Linear SVM	Cost factor = 1; hyperplane = unbiased
RBF SVM	Cost factor = 1; hyperplane = unbiased; gamma = 0.1
OLVQ1	Codebook vectors = 300; iterations = 10,000; neighbours = 5; balancing iterations = 3; propinit
lin-1	–

contrast small and fat datasets (sparse sample sizes and many attributes) can return unstable predictions where marginal changes in the training dataset can lead to significantly different prediction results (Breiman, 1996).

25.3.2 *Prediction accuracy and computation speed*

The ability of 10 data-mining algorithms out of 5 different methodological approaches to predict 6 soil classes in lower mountain ranges had been tested on the basis of the *F* measure, which was calculated for the training area as well as an independent validation area. The difference between these two measures can be used to characterise the generalisation ability of each algorithm. For good generalisation ability, the *F* measures should be high in general.

In this study, the highest overall *F* measure is >85 in the training area and >54 in the validation area (Table 25.5). Regarding other studies *F* values >60

Table 25.5. *F measures for the training area (tr) and the validation area (va) for all six soil classes (Table 25.1).*

Soil class		ANN		Classification trees				SVM		LVQ	Linear model
	F1	RProp	BPm	QST (cv)	QST (ds)	CRS (cv)	CRS (ds)	Linear	RBF	OLVQ1	lin-1
1	Tr	49.0	68.2	81.4	84.1	56.1	58.7	16.0	18.1	52.5	59.3
	Va	54.9	53.2	42.8	42.8	39.3	40.1	6.1	6.3	35.2	53.4
2	Tr	67.9	72.7	75.1	78.9	68.9	69.4	60.2	60.1	64.5	64.9
	Va	63.6	62.0	56.4	52.4	57.7	53.2	52.7	48.6	49.9	56.9
3	Tr	64.2	75.1	88.7	92.5	72.1	74.8	7.1	7.4	64.5	58.1
	Va	39.4	24.4	19.3	21.5	22.9	22.0	3	3.1	37.7	22.8
4	Tr	68.7	75.8	85.3	−1	69.6	70.7	53.4	53.7	76.0	73.6
	Va	60.6	64.6	60.4		53.5	53.3	5.2	47.5	58.2	62.5
5	Tr	68.8	71.7	89.1	92.2	77.8	78.4	49.8	46.5	76.8	69.2
	Va	36.3	32.5	42.0	40.3	36.9	36.2	21.9	19.9	43.0	32.8
6	Tr	77.5	81.6	90.5	93.1	82.3	82.8	63.6	71.4	84.5	77.5
	Va	74.5	74.2	74.2	71.5	75.1	75.5	69.7	73.5	73.2	75.3
Mean	ΔF	−11.1	−22.4	−35.8	−42.5	−23.6	−25.8	−15.3	−9.7	−20.2	−16.5
	Tr	66.0	74.2	85.0	88.1	71.1	72.5	41.7	42.9	69.8	67.1
	Va	54.9	51.8	49.2	45.7	47.6	46.7	26.4	33.1	49.5	50.6

Note: To interpret the generalisation ability of the algorithms the difference in $F(\Delta F)$ between the mean F in the training and the validation dataset is given. The higher ΔF the better is the generalisation ability. cv = cross-validation; ds = direct stopping/biggest tree; −1 = computation run out of memory.

Note: To interpret the generalisation ability of the algorithms the difference in $F(\Delta F)$ between the mean F in the training and the validation dataset is given. The higher ΔF the better is the generalisation ability. cv = cross-validation; ds = direct stopping/biggest tree; −1 = computation run out of memory.

are considered to give good prediction results (Sun et al., 2002). In this study the learning example was not trivial, as many soil classes are related to parent material that was not used as a predictor, to expand range of the results in terms of a "real world" example where generally only DEMs are available. Thus overall validation results of about 55 (ranging from 36 to 75) as found by using RProp must be considered remarkably high. Compared with this, the linear model, learning vector quantisation and the classification trees are relatively good competitors, even if the overall difference in accuracy can reach 10%. The only real outlier is the SVM that is not competitive in this study.

In terms of training accuracy the QUEST decision tree outperformed all other approaches about 15%. As the results in the validation area, even of the cross-validated version (cv), are significantly worse compared with the RProp ap proach, the trees seem to overfit the training dataset and thus do not generalise very well.

Concerning computation time SVMs and decision trees were the slowest (Table 25.6). The cvs, returning better results than the biggest not pruned trees (ds), are faster in prediction but slower in training. In general classification trees perform slowly, although there is an advantage in processing time, as no transformation of the input datasets is required.

In terms of computation time the original CART algorithm (Breiman et al., 1984) implemented in QUEST and CRUISE performed slowly. This is why this method was not included in this study. To our surprise the two "simple"

Table 25.6. *Computation time (in minutes) for different steps in predictive soil mapping based on a Pentium 4 (3.4 GHz, RAID 0, 1 GB memory) PC.*

Method	Algorithm	train	pred	class	train data	pred data	sum
Neural networks	RProp						
	Backprop w. momentum	1–2	3	3	1	–	8–9
Classification trees	QUEST (cv)	5–120	2–12	–	1	–	8–133
	QUEST (ds)	3–7	10–60	–	1	–	14–66
	CRUISE (cv)	3–7	2–10	–	1	–	5–17
	CRUISE (ds)	1–2	3–10	–	1	–	4–12
Vector machines	Linear	1–2	2	3	1	7	13–14
	RBF	6	30–200	3	1	7	47–217
Leaning vector quantisation	OLVQ1	1	3	–	1	7	12
Linear model	lin-1	2	3	1	7	12	
kNN	mfs knn	–	~500	–	1	7	507

Note: train = training/learning time; pred = time for prediction; class = time for classification; train data = time to create the training dataset; pred data = time to create the prediction dataset, sum = overall prediction time.

algorithms LVQ and linear regression performed well in terms of prediction ability as well as speed and thus can be recommended for quick overview surveys.

Concerning Dystric Cambisols (3) and Dystric Cambisols and Cambi-Haplic Podzols (5) the recall is very high, but as these soil classes are strongly correlated to a geological unit they are over-predicted and thus have a low precision. In contrast the prediction results for the soil classes Cumulic Anthrosols, Gleyic Anthrosols, Gleysols and Fluvisols (2), Haplic Podzols (4) and Haplic Cambisols (6) is generally satisfying, indicating their relation to relief.

25.4 Conclusions

The comparison of different data-mining methods and algorithms shows that ANNs are the overall best method to predict soil classes. SVMs could not be recommended as the prediction is not accurate and the computation time is slow.

The overall accuracy with an *F* measure of about 55 (36–75) in the validation area for the ANNs could be seen as a promising result for a real-world example without any postprocessing and on the basis of terrain attributes only. Incorporating the geology in the prediction approach would return *F* measures >70 (Behrens et al., 2005).

Another important aspect is the computation time. The results shown indicate a generally fast prediction. As the computation times presented are measured for a single soil class using after all data had been preprocessed, a validated prediction still takes its time if all classes of a soil map should be predicted.

Finally, it is clear that using data-mining techniques for predicting soil classes are seen as a key method in future digital soil mapping.

References

AG Boden, 1994. Bodenkundliche Kartieranleitung. E. Schweizerbart'sche Verlagsbuchhandlung, Stuttgart, 392.

Behrens, T., 2003. Digitale Reliefanalyse als Basis von Boden- Landschaftsmodellen am Beispiel der Verbreitungssystematik periglaziärer Lagen in deutschen Mittelgebirgen. Boden und Landschaft 42, 183 Giessen.

Behrens, T., Förster, H., Scholten, T., Steinrücken, U., Spies, E.-D., Goldschmitt, M., 2005. Digital soil mapping using artificial neural networks. J. Plant Nutr. Soil Sci. 168, 21–33.

Beven, K., Kirkby, M.J., 1979. A physically based, variable contributing area model of basin hydrology. Bull. Hydrol. Sci. 24, 43–69.

Bishop, C.M., 1995. Neural Networks for Pattern Recognition. Oxford University Press, Oxford, England.

Breiman, L., 1996. Bagging predictors. Machine Learn 24, 123–140.

Breiman, L., Friedman, J., Olshen, R., Stone, C., 1984. Classification and Regression Trees. Wadsworth, Belmont, CA.

Brownlee, J., 2004. Learning Vector Quantization (LVQ) Algorithm – WEKA Implementation. Online publication: http://www.users.on.net/~nbcraven/jasonbrownlee/.

Crone, S., Lesmann, S., Stahlbock, R., 2004. Empirical comparison and evaluation of classifier performance for data mining in customer relationship management. In: D. Wunsch, et al. (Eds.), Proceedings of the International Joint Conference on Neural Networks. IJCNN'04, Budapest, Hungary, IEEE, New York.

Dietrich, W.E., Montgomery, D.R., 1998. A digital terrain model for mapping shallow landslide potential. http://socrates.berkeley.edu/~geomorph/shalstab.

Feldwisch, N., 1995. Hangneigung und Bodenerosion. Boden und Landschaft 3, 152 Giessen.

Hatfield, D.C., 1999. TopoTools – A Collection of Topographic Modeling Tools for ArcINFO. http://www.giscafe.com/GISVision/TechPaper/TopoTools.html.

He, P., Fang, K.T., Xu, C.-J., 2003. The classification tree combined with SIR and its applications to classification of mass spectra. J. Data Sci. 1, 425–445.

Hodgson, M.E., Gaile, G., 1999. A cartographic modeling approach for surface orientation-related applications. Photogramm. Eng. Remote Sens. 65, 85–95.

Horn, B.K.P., 1981. Hillshading and the reflectance map. Proc. IEEE 69, 14–47.

Huber, M., 1994. The digital geo-ecological map, concepts, GIS-methods and case studies. Physiogeographica 20 Basel.

Ishioka, T., 2003. Evaluation of criteria for information retrieval. IEEE/WIC International Conference on Web Intelligence (WI 2003), Sponsored by IEEE Computer Society and Web Intelligence Consortium, Oct. 13–17, Halifax, Canada, pp. 425–431.

Joachims, T., 1999. Making large-scale SVM learning practical. In: B. Schölkopf, C. Burges, and A. Smola (Eds.), Advances in Kernel Methods – Support Vector Learning. MIT Press, Cambridge, MA.

Kim, H., Loh, W.-Y., 2001. Classification trees with unbiased multiway splits. J. Am. Stat. Assoc. 96, 589–604.

Kleefisch, B., Köthe, R., 1993. Wege zur rechnergestützten bodenkundlichen Interpretation digitaler Reliefdaten. Geologisches Jahrbuch F, 59–122.

Kohonen, T., 1990. Improved Versions of Learning Vector Quantization. Proceedings of the International Joint Conference on Neural Networks, San Diego, California.

Kohonen, T., 2001. Self-Organizing Maps. Springer, Berlin.

Loh, W.-Y., Shih, Y.-S., 1997. Split selection methods for classification trees. Statistica Sinica 7, 815–840.

McBratney, A.B., Mendonca-Santos, M.L., Minasny, B., 2003. On digital soil mapping. Geoderma 117, 3–52.

Merelo, J.J., Prieto, A., 1994. G-LVQ, a combination of genetic algorithms and LVQ. Online documentation: http://geneura.ugr.es/g-lvq/g-lvq.html.

Ministerium für Landwirtschaft, Weinbau und Forsten, 1982. Wasserwirtschaftlicher Rahmenplan Rheinpfalz: Erläuterungsbericht. Mainz, 42–60.

Nogami, M., 1995. Geomorphometric measures for digital elevation models. Zeitschrift für Geomorphologie Suppl 101, 53–67.

Osuna, E., Freund, R., Girosi, F., 1997. An improved training algorithm for support vector machines. In: J. Principe, L. Gile, N. Morgan and E. Wilson (Eds.), Neural Networks for Signal Processing VII. Proceedings of the 1997 IEEE Workshop, New York, pp. 276–285.

Quinn, P., Beven, K., Chevallier, P., Planchon, O., 1991. The prediction of hillslope flow paths for distributed hydrological modelling using digital terrain models. Hydrol. Process. 5, 59–79.

Rasmussen, C.E., 1996. The linear model lin-1. Delve – Data for Evaluating Learning in Valid Experiments, Learning Methods Documentation. http://www.cs.toronto.edu/~delve/methods/lin-1/home.html.

Riedmiller, M., Braun, H., 1993. A direct adaptive method for faster backpropagation learning: the Rprop algorithm. Proceedings of the ICNN, San Francisco.

Sarle, W., 2002. The IEEE Transactions on Neural Networks (Neural Network FAQ). ftp://ftp.sas.com/pub/neural/FAQ.html.

Schölkopf, B., 1997. Support Vector Learning. R. Oldenbourg Verlag, Munich.
Scull, P., Franklin, J., Chadwick, O.A., McAthur, D., 2003. Predictive soil mapping: a review. Progr. Phys. Geogr. 27, 171–197.
Shary, P.A., Sharaya, L.S., Mitusov, A.V., 2002. Fundamental quantitative methods of land surface analysis. Geoderma 107, 1–35.
Sun, A., Lim, E.-P., Ng, W.-K., 2002. Web classification using support vector machine. Proceedings of the 4th International Workshop on Web Information and Data Management (WIDM 2002), Virginia.
Tarboton, D.G., 1997. A new method for the determination of flow directions and upslope areas in grid digital elevation models. Water Resour. Res. 33, 309–319.
Van Rijsbergen, C.J., 1979. Information Retrieval, 2nd Edn. Butterworths, London.
Vapnik, V.N., 1995. The Nature of Statistical Learning Theory. Number 4, USAF School of Aviation Medicine, Randolf Field, Texas. Springer, Heidelberg.
Zell, A., Mamier, G., Vogt, M., Mache, N., Hübner, R., Döring, S., Herrmann, K.-U., Soyez, T., Schmalzl, M., Sommer, T., Hatzigeorgiou, A., Posselt, D., Schreiner, T., Kett, B., Clemente, G., Wieland, J., 1995. SNNS Stuttgart Neural Network Simulator – User Manual, Version 4.1. University of Stuttgart, Institute for Parallel and Distributed High Performance Systems, Report 6/95.
Zevenbergen, L.W., Thorne, C.R., 1987. Quantitative analysis of land surface topography. Earth Surf. Process. Landf. 12, 47–56.

Developments in Soil Science, volume 31
P. Lagacherie, A.B. McBratney and M. Voltz (Editors)

Chapter 26

DIGITAL SOIL MAPPING: AN ENGLAND AND WALES PERSPECTIVE

Thomas Mayr and Bob Palmer

Abstract
Growing demand for soil information at fine resolutions led to National Soil Resources Institute (NSRI) investigating environmental modelling as an alternative to traditional field surveys. Training data based on the existing records were subjected to data analysis to deduce modelling rules. Techniques were investigated for one nonglaciated and three glaciated terrains. Problems with several predictive data layers were identified. Very limited numbers of data layers were required to model spatial distribution of soils; and these were landscape driven. The success of extrapolation strategies was related to the extent of identified correspondences between trial and test areas. Relationships were more easily defined for nonglaciated areas.

26.1 Introduction

There is a growing demand in England and Wales for soil information at resolutions finer than provided in the existing maps. The drivers for this include the following: the management of water quality under the Water Framework Directive; the integrated management of flood defence through Catchment Flood Management Plans; the control of overgrazing in the uplands, measures for which will shortly be reviewed; and the development and implementation of the Soil Action Plan for England. In addition, detailed soil information would support the implementation of Government policies in the rural sector, by informing and advising the farming community about sustainable management of soil and water and natural and historic heritage. This chapter considers the role of environmental modelling in enhancing the existing soil data to meet these new demands.

Only about 25% of the soils of England and Wales is mapped at scales larger than 1:250,000. Traditional field survey is labour intensive; and furthermore, the necessary mapping skills are disappearing as experienced staff retires. If detailed soil maps are to be produced for areas where coverage currently does not exist, less expensive methods of mapping need to be found.

Research at National Soil Resources Institute (NSRI) to quantify soil–landscape relationships by environmental modelling as a basis for digital soil mapping goes back to 1994. We have recently completed two major projects that allowed us to test various approaches to digital soil mapping in four very diverse test areas (Fig. 26.1) covering a total area of 380 km^2 (Mayr et al., 2001; Mayr, 2002). In this chapter we review this recent work.

26.2 Background

In the NSRI soil classification system, soil profile characteristics are used to define soils at four levels in a hierarchical system; the general characteristics being used at the highest level to give broad separations and more specific ones used at lower levels to give increasingly precise subdivisions. At major group level, divisions are based on the predominant pedogenic characteristics of the soil profile. Soil groups and subgroups are derived based on features that further define the inherent characteristics of the soil material, or modify the basic pedogenic characteristics recognised at major group level. A soil series, the lowest category in the system, is a subdivision of a subgroup based on precisely defined particle size classes, parent material (substrate) types, colour and mineralogical characteristics (Avery, 1980; Clayden and Hollis, 1984).

26.3 Methodology

26.3.1 Modelling approach

In the functional approach to soil systems, soil is considered as a function of five soil forming factors: climate, organisms, relief, parent material and time (Jenny, 1941). The modelling approach adopted in NSRI studies has a form similar to Jenny's functional factorial model but differs in several terms.

$$S_i = f(S_{in}, L_i, T_i, RS_i) \tag{26.1}$$

where for each site, S_i is the soil type identified by the soil surveyor and S_{in}, L_i, T_i and RS_i are explanatory variables taken from spatial coverage of the 1:250,000-scale National soil map (NATMAP) for England and Wales, lithology (as interpreted from standard geology maps), terrain analysis and remote sensing data, respectively. McBratney et al. (2003) define this type of approach as *scorpan* modelling.

26.3.2 Training data

Using the existing detailed soil maps as test areas, we have investigated two approaches. The first is based on auger-bore information; this is the detailed site-specific soil information on texture, depth, colour and soil-water regime

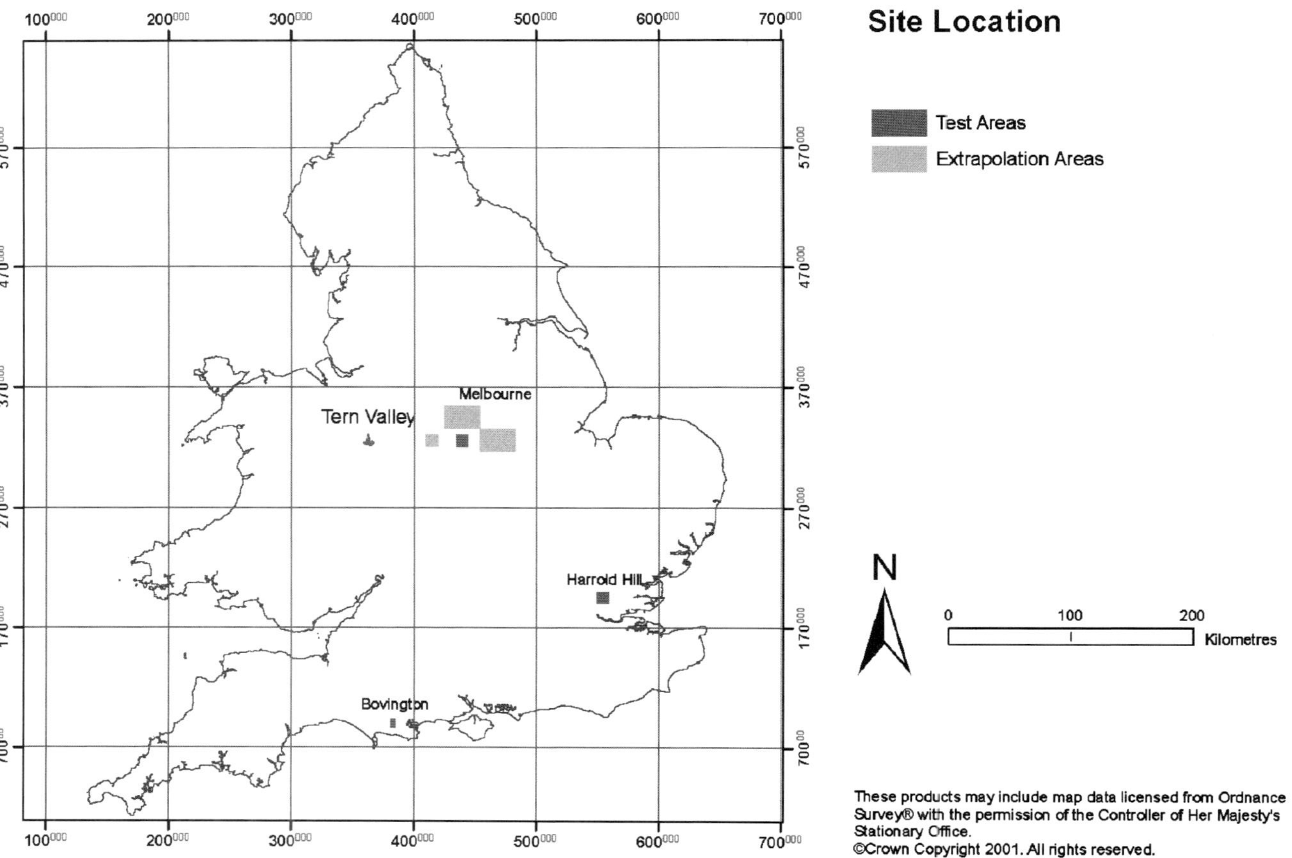

Figure 26.1. *Location of test areas within England and Wales.*

collected by surveyors during the initial survey. The second is based on soil polygons, and assumes that the mental landscape models used by soil surveyors to compile the soil map, particularly the association between soil map units and other environmental spatial data, can be identified.

26.3.3 Rule induction

A range of data analysis methods was applied to develop models for spatial prediction of soil properties or soil types based on their correlation with environmental variables. These include Bayesian belief networks; discriminant function analysis, fuzzy logic, lattice graphs, neural networks and tree-based methods.

26.3.4 Extrapolation strategies

Extrapolation strategies are required when using spatially restricted training areas. These included soilscapes, as defined by the dominant soil types, topography, parent material and geoscapes as well as NATMAP associations.

26.3.5 Accuracy assessment

The classification accuracy was determined by using contingency tables (error matrices). The overall agreement is computed for each matrix based on the difference between the actual agreement of the classification (e.g. agreement between computer classification and reference data as indicated by the diagonal elements) and change agreement that is indicated by the row and column marginals (Congalton and Mead, 1983).

26.4 Study sites

It is well established that patterns of soil distribution in nonglaciated terrain relate more strongly to topography, landscape and, in particular, surface geology, than is the case in landscapes formed in glacial deposits. With this in mind, environmental modelling techniques have been investigated for landscapes located in both glaciated and nonglaciated terrain. The areas were as follows: the Tern Valley that represents an area strongly affected by the Devensian glacial period; Alcester, Harold Hill and Melbourne that represent areas outside the Devensian glacial limits but with remnants of the Wolstonian glacial period; and Bovington Camp that represents an area unaffected by both glacial periods.

26.5 Data layers

Over the past 10 years, a wide range of environmental data has been investigated to identify key variables determining the spatial distribution of soils in the landscape. These include small-scale soil maps and derivatives, lithology derived from geology maps, landscape analysis and classification and remotely sensed information. The selection of appropriate environmental variables is crucial in identifying soil–landscape relationships.

26.5.1 Soil layers

NATMAP was published in 1983 at a scale of 1:250,000 after 5 years of intensive work by the field survey teams of Soil Survey of England and Wales. It covers the whole of England and Wales and represents a synthesis of all previous detailed soil mapping augmented by a reconnaissance survey of the previously unmapped areas, where observations were made at a density of 2–3 per square kilometre. It shows the distribution of 296 soil associations, each identified by a dominant soil subgroup and each containing between one and eight individual soil series (soil types).

In addition, we are in a unique position, in that, most of the dominant soil landscapes found in each county are represented on 1:25,000-scale detailed soil maps. The areas for detailed soil survey were chosen for their pedological, geopmorphological and agricultural interest to form the basis for future county, regional and national soil maps. These detailed surveys correspond to those of the Ordnance Survey 10 km × 10 km Outline Edition 1:25,000 map series.

Soilscapes were derived from the detailed soil maps using expert judgement and information on the lithology of the soil parent material, slope and landscape position.

26.5.2 Geology layers

In the United Kingdom, standard 1:50,000-scale geological maps, produced by the British Geological Survey (BGS), show the distribution of rocks as lithostratigraphical units. These units identify strata principally by their age and order of deposition (the geological succession or stratigraphy) and less rigorously by their rock types (lithology).

In order to generate lithology maps, parent material classes as defined by Clayden and Hollis (1984) were assigned by BGS to each of the lithostratigraphical units shown on published BGS maps. The initial attribution was based on information derived directly from the published 1:50,000-scale paper map sheets and their legends, and from the freely available BGS Lexicon (http://www.bgs.ac.uk/lexicon/lexicon_intro.html). Occasionally, where it was known that local variation was significant, local geological surveyors' experience was incorporated.

Geoscapes were derived by BGS for the Melbourne and Harold Hill test areas using expert judgement to determine the main lithostratigraphical divisions of the geological sequence identified in each area.

26.5.3 Landscape analysis

Landscape analysis characterises the spatial context of a particular soil type by expressing landscape geometry at a particular point and placing that point on the soil catena and within the catchment. This can be achieved through landscape analysis, landscape classification, catchment analysis and catchment classification.

Landscape analysis or quantitative parametric models

Digital terrain analysis allows a suite of variables to be generated that reflects geomorphic, climatic and hydrologic processes. Descriptive statistics of a number of terrain parameters can be used to separate soil classes. Terrain parameters are easily established from gridded digital terrain models (DTMs), but tend to be 'noisy', and the grid size of the DTM does not necessarily reflect the scale at which soil series are distributed in the landscape.

Landscape classification or qualitative morphogenic models

Landscape classification is traditionally based on air photo interpretation although automatic landscape classification based on processing of digital terrain data is currently under investigation. This is a more sophisticated approach than terrain analysis as it organises the land surface according to a formal geomorphological model of landform and interlandform relations.

Catchment analysis or drainage morphometry

Traditional catchment analysis generates a large number of variables describing individual aspects of catchment geometry. For modelling the spatial distribution of soil in the landscape, the critical source area, drainage density and skew as substitute for the hypsometric integral are thought to be most important.

Catchment classification or stream morphometry

Bunting (1965) speculated that soil type and position in a drainage basin are related; this relation is implicit in topofunction analysis but has not been rigorously tested. Arnett and Conacher, (1973) working on soils and drainage basins in the Rocksberg basin, Queensland, also point to a relation between soil types and stream order.

26.5.4 Climate

There are many ways to integrate rainfall, temperature and radiation data to provide indices of local climate. Simple indices of the water balance are amongst the most useful because they integrate biologically important effects of rainfall and evaporation. They give an indication of the potential intensity of leaching and provide a relative measure of potential biological productivity. They are, however, seldom used due to the limited extent of most study areas. As this applied also to our study sites, average incoming short wave and net solar radiation (measured in $\mathrm{MJ\,m^{-2}\,d^{-1}}$) were calculated using the solar radiation modelling program (SRAD) from TAPES-G (Wilson and Gallant, 1998) to provide climatic indices for the test areas. Sunshine fraction, cloudiness, atmospheric transmittance and circumsolar coefficients were calculated according to Wilson and Gallant (1997) and McKenney et al. (1999).

26.5.5 Remote sensing

Airborne gamma-ray spectrometer and magnetometer data were available for the Harold Hill and Melbourne areas. Gamma-ray data are especially appropriate for soil–landscape analysis because they are sensitive to particle size distribution and mineralogy (McKenzie and Ryan, 1999).

The airborne geophysical datasets are in the form of point observations recorded as georeferenced point observations. The gamma-ray data have an inline spacing of about 35 and 400 m, whereas the magnetic data have an inline spacing of approximately 7 and 400 m. A continuous measurement surface was derived by interpolation; whereas the tie lines are used in the correction of the magnetic and very low frequency (VLF) datasets. In the radiometric datasets the values from the tie lines can be used as additional data points for interpolation.

26.6 Results

26.6.1 Data layers

This research has highlighted the difficulties of deriving parent material maps derived from lithostratigraphical units. The extent to which near-surface geology is mapped (quaternary deposits) as well as differences in both terminology and diagnostic descriptions (colour code) currently limit the effectiveness of this approach. Cross-tabulation results using detailed soil maps showed that only 56% accuracy was achieved for Harold Hill and 64% for Melbourne.

The identification of soil-water regime is an extremely important attribute of a soil map. So far, no suitable environmental surrogate layer has been identified that allows an adequate assessment of soil-water regime. In addition, soil surveyor models in order to explain changes in soil-water regime usually rely on subtle variations in landscape position. It is only in very simple landscapes that a 50-m resolution DTM can be expected to identify these relatively small changes. Future work will incorporate a 10-m resolution DTM, which should help improve the prediction of soil-water regime.

26.6.2 Rule induction

The authors investigated a range of data mining techniques including discriminant function analysis, decision graphs, decision trees, neural networks and Bayesian belief networks. Results for the Alcester data (Table 26.1) show differences in accuracy levels among the various data mining techniques; although, a proper comparison cannot be undertaken as the dataset used in the analysis has evolved over time. Although neural networks achieved high accuracy levels, they proved difficult to interpret, which makes them less suitable for extracting mapping rules.

The Bayesian belief networks provide a method to represent relationships between soil and environmental variables, even if the relationships involve uncertainty, unpredictability or imprecision. The software uses the fastest known algorithm for extracting general probabilistic inference in a compiled belief network, which is message passing in a junction tree of cliques. In addition, the rule-based method can provide a numerical estimate of the probability of occurrence for a given attribute, as it varies across the study area, by means of explicit associations with the evidence. Furthermore, it allows for easy incorporation of expert knowledge.

There is an almost infinite number of data layers that can be derived from the above datasets, particularly from the DTM data. However, in all the cases only a very limited number of data layers was required to successfully model the spatial distribution of soils within the landscape, although the combination of particular data layers was very much landscape driven. As the optimal combination of data layers was established by trial and error, a more systematic approach to data selection is required. This, in part, confirms the conclusion of

Table 26.1. *Comparison of accuracy levels for different data analysis methods (Alcester).*

STATISTICA Decision tree	SIPINA Decision graphs	NETICA Bayesian belief network	MATLAB Neural network
Soil series			
40.26	46.31	49.24	–
Soil subgroup			
37.19	53.00	61.06	65.19

Austin et al. (1995) that the type of data mining technique is less important than the type of environmental data used in the analysis.

26.6.3 Stratification strategies

This research has demonstrated the importance of adequate stratification methods. Small-scale soil maps, parent material and landscape classification from fuzzy clustering have been investigated. In all cases, data stratification according to one of the above strategies substantially improved the final predictions (Table 26.2), although the optimum choice of stratification strategy appears to depend on site-specific details and requires further development.

26.6.4 Evaluation of training areas

Various extrapolation strategies were investigated when using locally restricted training areas. For the Harold Hill and Melbourne areas there were marked differences in accuracy of predictions within individual soilscape units. Using contingency tables, overall agreement was calculated 69% (L1), 62.9% (L2) and 54.6% (L2) for Harold Hill and 82% (L2), 79% (L4), 78% (L5), 77% (L1), 69% (L6) and 62% (L3) for Melbourne.

26.6.5 Evaluation of test areas

Extrapolation using soilscapes was investigated when using locally restricted training areas (Plate 26, see Colour Plate Section). The accuracy analysis was based on intersecting the predicted with the published soil map and generating two-way tables for further analysis. Soilscapes, by definition, are broad groupings of soils and landscape and exhibit a defined range of variability. For example, the thick till on hill-top soilscape defined at Melbourne showed subtle but important pedological differences between its occurrence at Melbourne and in the local extrapolation area of Needwood. This soilscape is dominated by Salop series at Melbourne whilst 20% of this unit at Needwood is Crewe series.

Table 26.2. *Stratification results (accuracy levels).*

	Alcester	Harold Hill	Melbourne	Bovington Camp	Tern Valley
	49.2	56.3	61.1	86.6	47.5
NATMAP	76.8	64.6	71.1	99.6	85.6
Geology	80.6	–	–	–	–
Lithology	–	81.3	70.1	98.9	75.0
Landform	64.7	57.6	69.1	100.0	88.5
Soilscape	–	58.1	64.9	–	–
HIRES	–	–	66.4	–	–

Table 26.3. *Extrapolation results using Melbourne training data.*

Derby		Melton Mowbray		Needwood forest	
Landscape unit	Accuracy	Landscape unit	Accuracy	Landscape unit	Accuracy
L2	Poor	L1	Mixed	L1	Mixed
L3	Mixed	L2	Well	L2	Well
L5	Mixed	L4	Poor	L4	very good
L6	Poor				

Note: L1 = thick drift on hill-tops, till remnants; L2 = reddish clay and soft mudstone; L3 = reddish sandstone with soft siltstone; L4 = thick alluvium and Trent terraces in valley bottoms; L5 = clays, soft mudstones and silty shales and subordinate sandstone (Coal measures); L6 = hard sandstone with soft shale and mudstone (mainly Millstone Grit).

For this reason, the 'similar' soil concept was used in assessing the success of the model predictions in extrapolation areas. Soils are defined as similar if they are the same in all but one parameter, and for that parameter differences are represented by adjacent classes in the parameter classification scheme. In the extrapolation, if Salop was predicted rather than Crewe then this would have been treated as an acceptable model output. Consequently, standard two-way contingency tables could not be generated as the row classes (published map) are not the same as the column classes (predicted map). Results of the extrapolation were therefore described in very broad terms such as, very good; worked well; mixed results; and poor (Table 26.3). These terms were intended to be descriptive and no firm definitions were given. In broad terms, >70% success in predicting soils was felt to be very good and <45% was poor. Future work should focus on extrapolating between the soil associations. This should ensure that there is better conformity in environmental conditions between the training and test areas.

26.7 Conclusions

The work has clearly demonstrated the potential for environmental modelling to improve the spatial resolution of soil information for England and Wales. The methodology described in this chapter has been designed to meet the special circumstances encountered in England and Wales, and therefore, reflects the availability of data and especially detailed soil maps. This research has shown that the relationship between the spatial distribution of soils in the landscape and environmental information is more easily defined for nonglaciated than for glaciated areas and the ultimate accuracy will probably be a reflection of the extent of the last ice age. It is anticipated that 1:50,000-scale soil maps will result from this research covering areas where currently the reconnaissance-scale NATMAP is the only spatial representation.

Acknowledgments

Part of this research was carried out from funding from U.K. Department for Environment Food and Rural Affairs (DEFRA) and QinetiQ. In addition, the authors would like to acknowledge the contributions from Russell Lawley from the British Geological Survey, Brian Fletcher from Remote Sensing Applications Consultants, Alec Walker of QinetiQ, V. Sastry from the Royal Military College of Science and Pat Bellamy, Ian Bradley, Rodney Burton and Ian Truckell from the National Soil Resources Institute, as well as Bob MacMillan from Land-Mapper, Canada. We are also grateful to Niklaus Zimmermann for his toposcale AML and Philippe Desmet for his multiple-flow routines.

References

Arnett, R.R., Conacher, A.J., 1973. Drainage basin expansion and the nine unit landsurface model. Australian Geographer 12, 237–249.

Austin, M.P., Meyers, J.A., Belbin, L., Doherty, M.D., 1995. Simulated data case study, Subproject 5, Modelling of landscape patterns and processes using biological data. Division of Wildlife and Ecology, CSIRO, Canberra.

Avery, B.W., 1980. Soil Classification for England and Wales. Soil Survey Monograph No. 14, Harpenden.

Bunting, B.T., 1965. The Geography of Soils. Hutchinson, London.

Clayden, B. and Hollis, J.M., 1984. Criteria for differentiating soil series. Soil Survey Technical Monograph No. 17, Harpenden.

Congalton, R.G., Mead, R.A., 1983. A quantitative method to test for consistency and correctness in photointerpretation. Photogrammetric Engineering and Remote Sensing 49, 69–74.

Jenny, H., 1941. Factors of Soil Formation. A System of Quantitative Pedology. McGraw-Hill, New York.

Mayr, T.R., 2002. Novel Techniques for Soil Mapping. Final Report for Quinetiq project CU009-015109.

Mayr, T.R., Palmer, R., Lawley, R. and Fletcher, P., 2001. New methods of soil mapping. Final Report for DEFRA contract SR0120.

McBratney, A.B., Mendonca Santos, M.L., Minasny, B., 2003. On digital soil mapping. Geoderma 117 (1–2), 3–52.

McKenney, D.W., Mackey, B.G., Zavitz, B.L., 1999. Calibration and sensitivity analysis of a spatially-distributed solar radiation model. Int. J. Geogr. Inf. Sci. 13 (1), 49–65.

McKenzie, N.J., Ryan, P.J., 1999. Spatial prediction of soil properties using environmental correlation. Geoderma 89, 67–94.

Wilson, J.P., Gallant, J.C., 1998. Terrain-based Approaches to Environmental Resource Evaluation. In: S.N. Lane, K.S. Richards, and J.H. Chandler (Eds.), Landform Monitoring, Modelling and Analysis. John Wiley & Sons Ltd., Chichester.

Developments in Soil Science, volume 31
P. Lagacherie, A.B. McBratney and M. Voltz (Editors)

Chapter 27

PEDOGENIC UNDERSTANDING RASTER CLASSIFICATION METHODOLOGY FOR MAPPING SOILS, POWDER RIVER BASIN, WYOMING, USA

N.J. Cole and J.L. Boettinger

Abstract

Vast areas of the earth need new or updated soil survey data, but traditional methods of soil survey are inefficient, expensive and often inaccurate. We developed and tested a methodology that incorporates geographic information systems (GIS), remote sensing and modelling to predict and map soil distribution in the Powder River Basin of Wyoming. Based on conceptual models in which unique soils are the products of unique sets of soil-forming factors, topographic data derived from digital elevation models (DEMs) and Landsat 5 spectral data were selected to represent soil-forming factors. These digital data were analysed using commercially available GIS and image processing software. Unsupervised, supervised and simple knowledge-based classifications were used in the preliminary stage to develop visual representations of soil-landscape patterns and to plan for field data collection. As more was learned about the survey area from data collection, a knowledge-based decision-tree classification model was built and refined. The resulting maps were evaluated qualitatively by local experts and quantitatively using accuracy assessment, and showed good agreement between predicted and observed map units. Continued technological advancements in spatial data and improved GIS and modelling expertise of soil scientists should increase the accuracy and efficiency of the soil survey process.

27.1 Introduction

The traditional process of mapping soils is expensive and time- and labour-intensive. Increased availability and user-friendliness of geographic information systems (GIS) has made it possible to improve the efficiency of soil survey. Now, there is potential for all phases of soil survey – pre-mapping, model development and refinement and map compilation – to be completed in a digital environment.

Jenny (1941) proposed that soils on a landscape are a function of five basic environmental factors: climate (cl), organisms (o), relief ®, parent material (p) and time (t). Knowledge of how these factors affect pedogenesis and soil distribution allows the development of conceptual models showing the spatial extent of soil distribution on a landscape. Zhu (1997, 1999, 2001) developed the Soil Land Inference Model (or, SoLIM) centred on Jenny's (1941) concept that soils are a product of their formative environment and are a continuum on the landscape. The SoLIM approach created maps of continuous soil properties using topographic indices, fuzzy classification and expert knowledge. Although SoLIM was an exciting new approach to mapping soils, the models predicting soil distribution were not transparent, the process was difficult for field soil scientists to grasp and field level personnel and the model developer were separated.

In this chapter, we describe the development of a methodology to quantitatively predict soil distribution based on knowledge of soil-forming factors and their relationships to landscapes in a pilot project in the Powder River Basin of Wyoming. We demonstrate that digital, spatially explicit data representing one or more of the soil-forming factors can be combined to accurately predict the occurrence of unique map units on the landscape. This repeatable methodology employs easy-to-use, commercially available GIS and image processing software to increase the efficiency and accuracy of soil survey while maintaining a close, interactive relationship between field soil scientists and the model developer.

27.2 Study area

The study area is in the Powder River Basin at the eastern base of the Big Horn Mountains, northern Johnson County, north central Wyoming, USA (Fig. 27.1). The 800-km^2 study area is centred at 44° 14′ 29″N, 106° 11′ 27″W (WGS84/NAD83), and is bisected north to south by the Powder River. Near the centre of the study area, mean summer temperature is 19.7°C, mean winter temperature is −4.94°C and mean annual temperature is 7.16°C; average annual precipitation is 28.9 cm, which occurs mostly during the summer months (Western Regional Climate Center, 2000). The soil temperature regime is aridic and the soil moisture regime is mesic.

The Eocene Wasatch Formation, containing variegated sandstones, conglomerates, mudstones, limestone, carbonaceous shale and coal, dominates the soil parent materials in the basin (Flores and Ethridge, 1985). Gently rolling uplands quickly change to dissected badland topography near the Powder River (i.e. the Powder River "breaks"). At the edges of the badlands, young alluvial fans are built upon nearly level stream terraces. Vegetation varies with landscape position and soil texture, but is dominated by various grasses (*Pascopyrum smithii*, *Stipa comata*, etc.) and sagebrush (*Artemisia tridentata*). Moister areas are dominated by stands of Plains Cottonwood (*Populus deltoids*). Soil subgroups include Torrifluvents, Torriorthents, Fluvaquents, Haplocambids, Haplargids and Paleargids (Soil Survey Staff, 2003).

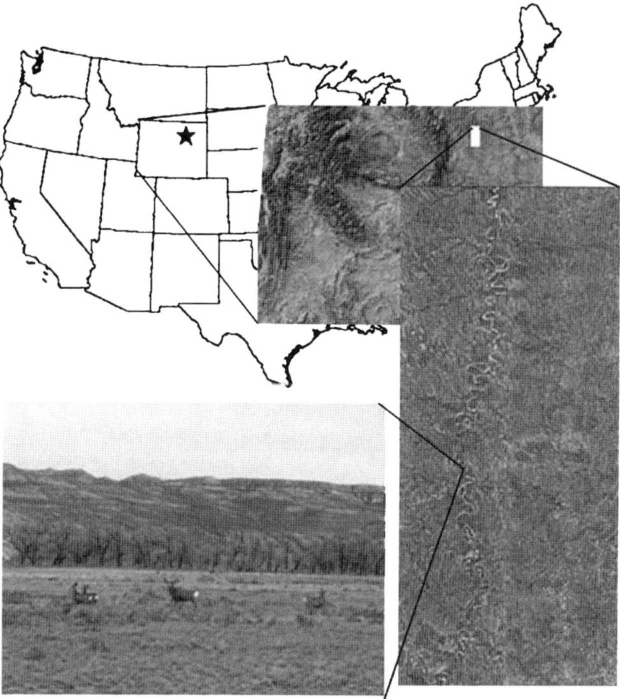

Figure 27.1. *Location of the study area (star) in Wyoming, western USA; digital orthophotos of Wyoming and the Powder River Basin; view looking west over Powder River flood plain with alluvial fans and badlands (breaks) in the background.*

27.3 Developing the pedogenic understanding raster-based classification methodology

The pedogenic understanding raster-based classification (PURC) methodology should be repeatable, incorporating different types of data and software to produce maps predicting soil distribution. For convenience, the methodology is described in three major stages: a preliminary stage, a developmental stage and a finalisation stage (Fig. 27.2).

27.3.1 Preliminary stage

The preliminary stage of the PURC methodology focused on pre-mapping, which is the preparation that must be completed before intensive field data collection for soil survey. This stage consists of data acquisition and review, pre-processing, identification of data layers that proxy for soil-forming factors, stacking of data layers and preliminary classifications of data to recognise patterns on potential map units on the landscape.

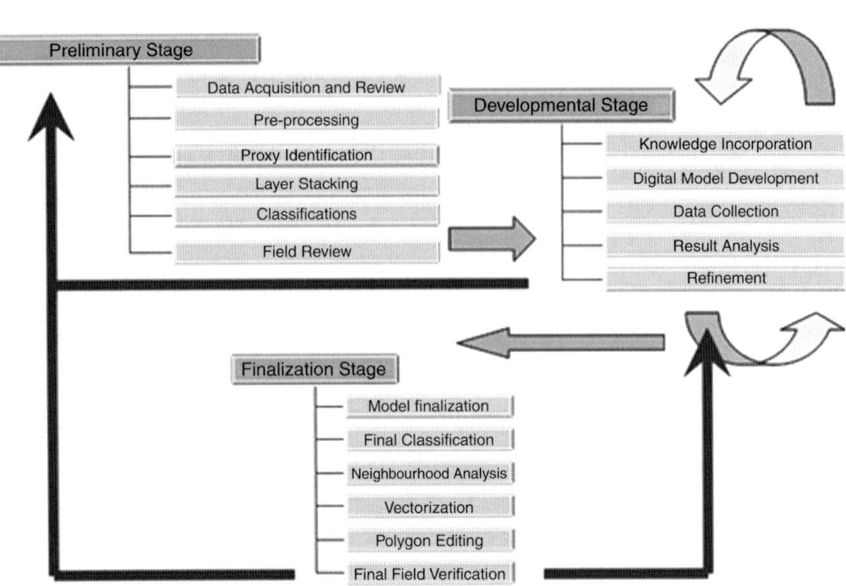

Figure 27.2. *Schematic diagram of the basic pedogenic understanding raster-based classification (PURC) modelling methodology.*

Data acquisition and review

Data acquisition concentrated on obtaining forms of data that could help develop conceptual and quantitative models of soil formation and be used as proxies for soil-forming factors. We acquired hard copy data including topographic maps, bedrock geology maps, soil surveys that either abutted or, in some minor cases, overlapped portions of the study area. We also obtained spatially explicit digital data layers, including vector coverages of precipitation and general geology. United States Geological Survey 10 m digital elevation models (DEMs) of the six 7.5-min quadrangles of the study area were obtained. A Landsat 5 image from August 16, 1993, was chosen as a theme from which to develop all final remotely sensed data layers.

Pre-processing

Digital data were re-projected into Universal Transverse Mercator zone 13 North American Datum (NAD) 1983 using either *ESRI ArcInfo* (DEMs) or *ERDAS Imagine* (Landsat data). Landsat data were atmospherically corrected using the COST method (Chavez, 1996) to yield values of at-sensor reflectance and compensate for selective scattering of light. Landsat data were also re-sampled from 30 to 10 m resolution using cubic convolution to allow for better integration with data derived from 10 m DEMs. The geographic extents of all data sets were clipped to the extent of the study area. Ultimately, all raster (pixel) data layers were converted into Imagine file format (.img) for use in layer stacking and classifications.

Proxy identification

A series of digital data layers that proxy for various components of the soil-forming factor of relief Ⓡ were developed from DEMs. These included elevation, slope, aspect, compound topographic index (CTI) (Beven and Kirby, 1979), relative elevation to the Powder River, slope length, surface roughness (Turner, 1989), profile curvature (Moore et al., 1991) and an ecological landform index (Manis et al., 2002).

Landsat data were used to develop representations of organisms (o), focusing on vegetation. We created a normalised difference vegetation index (NDVI), which is a normalised ratio of the near-infrared (band 4) to the red band (band 5) (Rouse et al., 1973). The NDVI was converted to a fractional vegetation index (FVI), which presents vegetation as an estimate of percent cover (Zeng et al., 2000).

Because the general geology coverages did not capture variation within the Wasatch formation soil enhancement ratios of Landsat data were used to represent parent material (p). Spectral band ratios 3/2, 3/7 and 5/7 have been interpreted to accentuate carbonate radicals, ferrous iron and hydroxyl radicals, respectively, in exposed soil and geologic materials (Amen and Blaszczynski, 2001).

As climate Ⓒ did not vary within the 800-km^2 study area, it was not represented by digital data. Time (t) was also not represented by any specific digital data layers, but may be reflected by other soil-forming factors; for example steep slope and little vegetation cover may represent a geomorphic surface subject to rapid and repeated disturbance.

Layer stacking

Selected raster data layers were combined into multi-band images in Imagine using the Layer Stack function in the Image Interpreter module of *Imagine* (Leica Geosystems GIS & Mapping, 2003). These multi-layer datasets were a combination of data layers selected to represent various soil-forming factors. Stacked data sets included combinations of the three-band soil enhancement image; slope, aspect, relative elevation to the Powder River, surface roughness, profile curvature, CTI, landform index; NDVI and FVI.

Preliminary classification

Unsupervised classification using an ISODATA clustering algorithm was performed on various combinations of data stacks using *Imagine* (Leica Geosystems GIS & Mapping, 2003). Unsupervised classifications are generally viewed as unbiased and data driven. Classifications were performed with a 95% level of convergence. Various data layer stacks and numbers of classes (6–20) were used in a trial and error approach until classifications produced patterns that could represent map units on the landscape interpreted from aerial photography. A 12-class image using the three-band soil enhancement ratios, slope, CTI and FVI was selected as the final unsupervised classification. This map was used as a planning tool for initial field sampling and subsequent classifications.

Supervised classification using the same data stack as the final 12-class unsupervised classification was also performed using *Imagine*. Twenty training sites were identified from soil map units from the adjacent and slightly overlapping Southern Campbell County Soil Survey (United States Department of Agriculture, Natural Resources Conservation Service, 1998). The supervised classification produced a map showing realistic distributions of map units – but only in the same physiographic area as the adjacent survey used for training.

A preliminary knowledge-based classification was developed based on limited field experience and patterns recognised from unsupervised and supervised classifications using the Expert Classifier tool in the Knowledge Engineer of *Imagine*. Input data layers were slope, FVI, CTI and the three-band soil enhancement ratios (3/2, 3/7 and 5/7). Even after several refinements to produce maps that were similar to the adjacent survey area, this initial rule-based classification was largely speculative. However, it was evident that fluvial soils had the highest vegetation and lowest slopes; and badland soils (breaks) had the highest slopes, most exposed parent material (indicated by 3/7 soil enhancement ratio), and lowest vegetation.

Field review of preliminary classifications

Results of preliminary classifications were compared with the actual soils, landscapes and vegetation types on a field trip in May 2003. Local soils experts qualitatively assessed the outputs of the various classification schemes, comparing results with their concepts of soil distribution. After much discussion, it was concluded that (1) unsupervised classification was useful in separating the landscape into classes, but deriving meaning from those classes was difficult; (2) supervised classification closely matched polygons along the survey border but as distance from the survey boundary increased, classification quality decreased significantly; and (3) simple rule-based models stratified the landscape well and showed a great degree of potential because it could be scrutinised by the user and altered as needed.

27.3.2 Developmental stage

Unsupervised, supervised and knowledge-based-classifications developed in the preliminary stage were used to target locations for field data collection.

Because digital data were used to create the preliminary classifications, the initial stratification of the landscape using these classifications allowed the increased rate at which useful information on soil distribution could be acquired through field work.

Knowledge incorporation

To more accurately predict the relationships between soils on the landscape and proxy data layers, it was necessary to incorporate expert knowledge and spatial information on specific soils. We acquired expert knowledge by integrating traditional soil survey and digital modelling expertise. Using a team approach, all information and hypotheses explaining soil occurrence on the landscape were shared daily among all soil survey project members. Climate, relief, organisms, parent material and time were discussed with relation to each developing soil association and map unit. Resulting conceptual models were drawn on paper-lined walls and quantitative models were developed by a soil scientist familiar with modelling techniques and GIS and image processes software. Shortly after field data collection was initiated, a hierarchical classification of each map unit was developed identifying the general physiographic region, temperature regime, soil moisture regime, landscape and landform. This was further developed into a decision tree to aid in the identification of soils in the field.

Digital model development

The conceptual models and decision tree were used to clearly define a set of data layers and quantify specific thresholds and ranges of data that described the environmental conditions for each soil map unit. A digital decision tree model including these specific thresholds and ranges of data was developed using the Knowledge Engineer interface within the Expert Classifier in Imagine. Firstly, specific soil associations were defined (e.g. fluvial soils). Then, specific map units occurring within each soil association were further defined (e.g. Draknab sandy loam, 0–3% slopes, within the fluvial soil association). The decisions and data defining representative soil associations and map units are illustrated in Table 27.1.

Data collection and model refinement

Field data were collected at more than 300 points in 2003, including transects, pedon descriptions and observations (Soil Survey Division Staff, 1993). These data were used to develop and refine the digital decision tree model. Because knowledge acquisition, digital model development and data collection occurred somewhat simultaneously, the digital model evolved and was improved as more data were collected and concepts were revised.

Table 27.1. *Rules used in knowledge-based decision tree classification for generalised soil associations and the eight specific map unit classes evaluated in the accuracy assessment.*

Map unit number	Class name	Classification rules
		Generalised associations
NA	Fluvial soils	Relative elevation to Powder River ≤6 m and slope <2%, or, ≤3 m from Powder River, or ≤5 m in height and ≤50 m distance of small streams
NA	Badland soils	Soil enhancement band 2 (iron) ≥67 and slope ≥8% and not fluvial soils
NA	Uplands	Relative elevation to Powder River ≥60 m and not fluvial and not badland soils
NA	Alluvial fans	Not fluvial and not badland and not upland soils
		Specific map units
938	Water	Ten meter buffer of Powder River line coverage
611	Draknab sandy loam, 0–3% slopes	Fluvial soils = true and soil enhancement band 2 >113 and relative elevation to Powder River ≤5 m, or, fluvial soils = true and relative elevation to Powder River ≤1 and orthophoto value >150 in blue band, does not meet the requirements of any previous decision
613	Haverdad–Kishona loams, 0–3% slopes	Fluvial soils = true and relative elevation to the river ≥10 m, or, fluvial soils = true and slopes >6%, does not meet the requirements of any previous decision
616	Clarkelen–Draknab complex, 0–10% slopes	Fluvial soil = true and near infrared Landsat >60 and fractional vegetation >38, or, fluvial soils with CTI<1, does not meet the requirements of any previous decision
612	Clarkelen fine sandy loam, 0–3% slopes	Fluvial soils = true and fractional vegetation >34, does not meet the requirements of any previous decision
649	Haverdad–Clarkelen complex, 0–3% slopes	Other fluvial soils (dominated by sage and grass community), does not meet the requirements of any previous decision
684	Samday–Shingle–Badland complex, 10–45% slopes	Badland soils with slopes ≥15 and mean slope length factor >1.85, or, badlands having slopes >50%, does not meet the requirements of any previous decision
709	Theedle–Shingle loams, 3–30% slopes	Badland soils = true and mean slope length factor >0.8 and <1.75, does not meet the requirements of any previous decision

Analysis of results

At the end of the first field season, completed mapping was qualitatively assessed by US Department of Agriculture Natural Resources Conservation Service (NRCS) soil data quality specialists. They concluded that the standards expected of a traditional soil survey product were met by the maps produced from the digital decision tree model (Heidt, C. J., personal communication, 2003).

In early spring 2004, a quantitative accuracy assessment was performed on the eight map units that were certified by NRCS data quality specialists; these eight map units covered 28% of the 800-km^2 study area. Ten, semi-random points were located in each map unit, placed in a buffer within 100 m of a road or trail. At each point, the soil map unit classified by observing soil and site characteristics and predicted and observed classes were entered into an error matrix (Congalton and Green, 1999). Overall classification accuracy was 88%, and the Kappa statistic was 0.86, indicating good agreement between predicted and observed classifications.

27.3.3 *Finalisation stage*

The completion of the qualitative and quantitative accuracy assessment initiates either an iterative loop, returning to the beginning of the developmental stage or the process of finalising the model and resulting maps.

Model finalisation and final classification

The knowledge-based digital decision-tree model with the highest producer's and user's accuracy was used to develop the final classifications. This model best reflected the pedogenic concepts in the study area when compared with aerial photography (Fig. 27.3). The final classification resulted in a 10-m-resolution pixel-based predictive soil map that could be used in a GIS (Plate 27, see Colour Plate Section).

Neighbourhood analysis, vectorisation and polygon editing

A traditional-looking polygon map that would meet Soil Survey Geographic Database (SSURGO) quality standards was developed using the raster-based output of the final classification. Using *Imagine*, a 3×3 majority filter reduced random pixel noise. Groups of like pixels were identified using a clump function and pixel groups <4.0 ha in rangeland or <1.6 ha in potential cropland were eliminated. The resulting pixel-based map had map units that appeared to fit soil-landscape units and appeared similar to a traditional polygon soil survey map. This processed pixel map was used as a base for a vector map product. The raster layer was opened in *ESRI ArcGIS* and converted to polygon coverage. Polygons could then be edited by hand as needed.

Final field verification

Final field verification should occur when all mapping and iteration is complete, and, in the case of NRCS soil surveys, when all map units have been certified. Verification can include qualitative review and quantitative accuracy assessment of the final product. Because this pilot project is part of the larger, ongoing soil

Figure 27.3. *Digital orthophotograph of the southeastern quarter of the Juniper Draw 7.5-min quadrangle of the study area.*

survey project of Northern Johnson County, WY, this will likely not occur for several years.

27.4 Discussion and conclusions

Mapping soils following the PURC methodology allowed for all phases of the pilot project soil survey – pre-mapping, model development and refinement and map compilation – to be completed in a digital environment. Digital, spatially explicit data derived from DEMs and Landsat imagery represented one or more of the soil-forming factors. These various data layers were complied and classified using easy-to-use, commercially available GIS and image processing software to accurately predict the occurrence of map units on the landscape.

Unsupervised, supervised and simple rule-based classifications of data layers were used as pre-mapping tools for recognising soil-landscape patterns and targeting field data collection. However, knowledge-based decision-tree models that incorporated expert knowledge were best for classifying soil-landscape relationships and predicting the distribution of soils throughout the pilot project area. Knowledge-based models allowed for integration of pedogenic understanding of soil and landscape relationships. The quantitative rules used to define the thresholds and ranges of data related to soil patterns on the landscape could be stored and accessed for future evaluation and model refinement.

The PURC methodology using knowledge-based modelling should produce soil maps that are more consistent across the survey area than mapping by traditional methods. All map units were predicted using quantitative rules based on the hypotheses explaining soil occurrence shared by all project members, thus maintaining a close, interactive relationship between field soil scientists and the model developer.

Knowledge-based classification relies heavily on the availability of experienced soil scientists to develop conceptual relationships between soils and environmental factors. If experienced soil scientists were not available in an area that lacks prior soil survey information, it would be difficult to develop a knowledge-based classification. However, if limited soils data were available, conceptual modes captured in hierarchical decision trees could be extrapolated over like areas using data techniques such as classification tree analysis (e.g. Chapter 28). As new classification tools and higher resolution data layers become available, our ability to accurately and efficiently predict soil distribution will improve.

Acknowledgments

This research was partly supported by the Utah Agricultural Experiment Station, Utah State University, Logan, UT. Approved as paper no. 7740.

References

Amen, A., Blaszczynski, J., 2001. Integrated Landscape Analysis. U.S. Department of the Interior, Bureau of Land Management, National Science and Technology Center, Denver, CO, pp. 2–20.

Beven, K.J., Kirby, M.J., 1979. A physically based, variable contributing area model of basin hydrology. Hydrol. Sci. Bull. 24 (1), 43–69.

Chavez, P.S., 1996. Image based atmospheric corrections – revisited and revised. Photogramm. Eng. Remote Sens. 62 (9), 1025–1036.

Congalton, R.G., Green, K., 1999. Assessing the Accuracy of Remotely Sensed Data: Principles and Practices. CRC Press, Boca Raton, FL.

Flores, R.M., Ethridge, F.G., 1985. Evolution of intermontane fluvial systems of Tertiary Powder River Basin, Montana and Wyoming. Rocky Mountain Section–S.E.P.M Cenozoic Paleography of the West-Central United States, pp. 107–125.

Jenny, H., 1941. Factors of soil formation. McGraw-Hill, New York.

Leica Geosystems GIS & Mapping LLC, 2003. ERDAS field guide, 7th Edn. Leica Geosystems GIS and Mapping Atlanta Georgia.

Manis, G., Lowry, J., Ramsey, R.D., 2002. Pre-classification: an ecologically predictive landform model. Remote Sensing/GIS Laboratory, College of Natural Resources, Utah State University. (Available online at http://www.gis.usu.edu/%7Eregap/download/landform) (Last verified April 13, 2004).

Moore, I.D., Grayson, R.B., Landson, A.R., 1991. Digital terrain modeling: a review of hydrological, geomorphological, and biological applications. Hydrol. Process. 5, 3–30.

Rouse, J.W. Jr., Hass, R.H., Schell, J.A., Deering, D.W., 1973. Monitoring vegetation systems in the Great Plains with ERTS. Third ERTS Symposium, NASA SP-351. NASA, Washington, DC, pp. 309–317.

Soil Survey Division Staff, 1993. Soil Survey Manual. United States Department of Agriculture, Washington, DC.

Soil Survey Staff, 2003. Keys to Soil Taxonomy, 9th Edn. USDA Natural Resources Conservation Service, US Government Printing Office, Washington, DC.

Turner, M.G., 1989. Landscape ecology – the effect of pattern on process. Annu. Rev. Eco. Sys. 20, 171–197.

United States Department of Agriculture, Natural Resources Conservation Service, 1998. Soil Survey Geographic (SSURGO) database for Campbell County, Wyoming, Southern Part. United States Department of Agriculture, Natural Resources Conservation Service, Fort Worth, TX (Available online at ftp://ftp.ftw.nrcs.usda.gov/pub/ssurgo/online98/data/wy605/) (Last verified 15 April 2004.)

Western Regional Climate Center, 2000. Dead Horse Creek, Wyoming, Climate Summary. (Available online at http://www.wrcc.dri.edu/cgi-bin/cliMAIN.pl?wydead) (last verified 3 May 2005).

Zeng, X., Dickinson, R.E., Walker, A., Shaikh, M., DeFries, R.S., Qi, J., 2000. Derivation and evaluation of global 1-km fractional vegetation cover data for land modeling. J. Appl. Meteorol. 39 (6), 826–839.

Zhu, A.X., 1997. A similarity model for representing soil spatial information. Geoderma 77, 217–242.

Zhu, A.X., 1999. A personal construct-based knowledge acquisition process for natural resource mapping. Int. J. Geogr. Inf. Sci. 13, 119–141.

Zhu, A.X., 2001. Soil mapping using GIS, expert knowledge, and fuzzy logic. Soil Sci. Soc. Am. 65, 1463–1472.

Developments in Soil Science, volume 31
P. Lagacherie, A.B. McBratney and M. Voltz (Editors)

Chapter 28

INCORPORATING CLASSIFICATION TREES INTO A PEDOGENIC UNDERSTANDING RASTER CLASSIFICATION METHODOLOGY, GREEN RIVER BASIN, WYOMING, USA

A.M. Saunders and J.L. Boettinger

Abstract

We demonstrate that the pedogenic understanding raster classification (PURC) approach can be transferred to a new soil survey project area and that classification tree analysis is a viable alternative to a knowledge-based decision tree for predicting the distribution of soil classes. The new area is in the Green River Basin of Wyoming, USA, which has greater variability in parent material, lower erosion rates and a colder climate than the Powder River Basin of Johnson County, Wyoming, where the PURC approach was developed. There is also a lack of existing soil survey data within or adjacent to the new study area. Topographic data derived from digital elevation models (DEMs) and various band combinations and ratios of Landsat 7 remotely sensed spectral data were selected to represent soil-forming factors. Unsupervised classification techniques were used in the preliminary stage of the methodology to recognise existing soil-landscape patterns and to develop an initial sampling plan. Knowledge-based classification and classification tree analysis were used to develop models predicting soil delineations and quantifying soil map unit concepts. The output images generated from the knowledge-based and classification tree models produced similar predictions of soil patterns across the landscape, with slightly different predictions of soil map units. However, the knowledge-based model was much more time-intensive than the classification trees, and failed to classify all pixels within the study area even after multiple iterations of the model. Classification tree analysis was successfully integrated into the methodology, and is more objective than the knowledge-based classification. Although finals maps were slightly different, both models were transparent and could be further refined as additional data becomes available or as land-use needs change.

28.1 Introduction

Many areas of the world lack completed or updated soil survey data. Traditional soil survey methods are time-consuming, labour-intensive and sometimes result in inconsistent and inaccurate prediction of soil distribution across the landscape. With the goals of making soil survey methods more quantitative and streamlining the cartography process, Cole and Boettinger (Chapter 27) developed a protocol in which pre-mapping, model development, validation and final map generation were completed in a digital environment. This pedogenic understanding raster classification methodology (or, PURC) incorporated spatially explicit topographic and Landsat spectral data to represent Jenny's (1941) soil-forming factors in a geographic information system (GIS) to map soils on rangelands in the Powder River Basin of Wyoming, USA. The final map was created using a knowledge-based tools allow for more efficient and effective transfer of the soil scientist's 'mental model' to the final soil survey product (Bui, 2004). Classification tree analysis is a knowledge-engineering tool that uses both categorical and continuous data in a non-linear approach to classification, which is useful in predicting discrete soil classes that do not necessarily follow a linear trend (McBratney et al., 2003). Classification tree models iteratively partition data into binary homogenous groups based on some specified variable. They are transparent and relatively easy to interpret without statistical expertise. Classification trees that used spatial input data (i.e. topographic data, existing soil maps, etc.) have successfully predicted soil classes and generated soil maps in a variety of physiographic environments under different land uses (e.g. Lagacherie and Holmes, 1997; Bui et al., 1999; Moran and Bui, 2002; Bui and Moran, 2003).

Our objectives in this chapter are to demonstrate that the PURC approach can be transferred to a new soil survey project area and that classification tree analysis is a viable alternative to a knowledge-based decision tree for predicting the distribution of soil classes. The new area is in the Green River Basin of Wyoming, which has greater variability in parent material, lower erosion rates and a colder climate than the Powder River Basin. There is also a lack of existing soil survey data within and adjacent to the study area.

28.2 Study area

The study area is located in a portion of the Green River Basin in Sublette County, just west of the Wind River Range (Fig. 28.1). It covers an area of approximately 600 km^2 and consists of subsets of ten 7.5-min quads. The mean annual temperature of the area is approximately 2.1 °C, with mean annual precipitation of approximately 29 cm, based on data from Pinedale, Wyoming, located just north of the study area (Western Regional Climate Center, 2004). The soil temperature regime for the area is frigid, while the soil moisture regime is ustic aridic (25–36 cm). The study area is sparsely vegetated, dominated by shrubs, grasses and forbs. The terrain consists mainly of gently sloping rolling plains, with ridge, sideslope and drainage landform components.

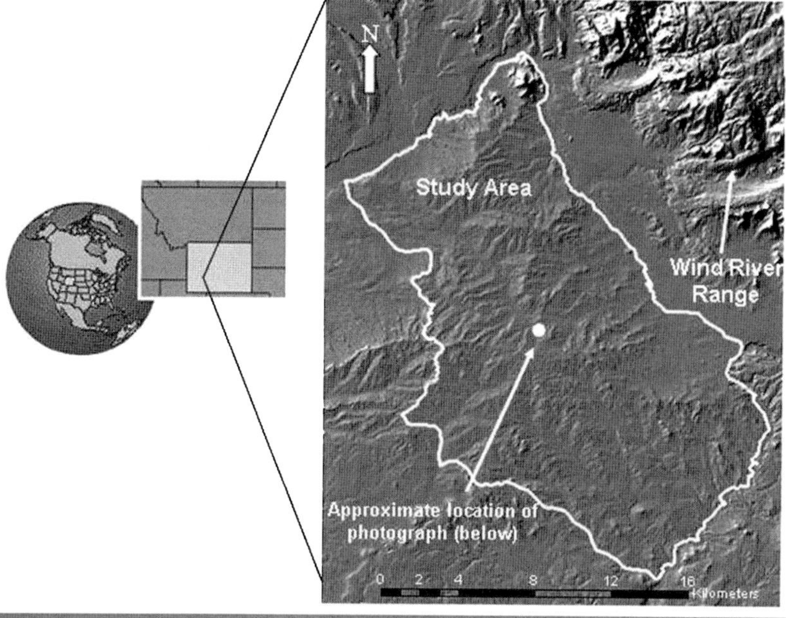

Figure 28.1. *Top: location of study area in relation to Wyoming and the United States and hillshade of study area and surrounding vicinity. Bottom: vegetation and relief of study area, looking east towards the Wind River Range.*

Characteristic vegetative cover and relief of the study area are shown in Figure 28.1. Bedrock geology is predominately inter-layered shale and sandstone from the Eocene Green River (lacustrine) and Wasatch (fluvial) formations (Roehler, 1992).

28.3 Methods and materials

28.3.1 Data acquisition, pre-processing and proxy identification

Raster data layers derived from U.S. Geological Survey (USGS) digital elevation models (DEMs) and various band combinations and ratios of Landsat 7 remotely sensed spectral data were selected to represent soil-forming factors. Data were analysed with ERDAS Imagine and ESRI ArcGIS data and image processing software. All data layers were projected into Universal Transverse Mercator (UTM) Zone 12 North, datum NAD83 and clipped to the extent of the study area. All data layers were also converted into Imagine file format (.img) for use in layer stacking and classifications.

Eleven DEMs with 10-m spatial resolution were mosaiced together in ArcInfo to provide full coverage of the study area. Data layers were derived from the DEMs using ArcGIS, and include slope, aspect, hillshade and compound topographic index (CTI). The CTI quantifies catenary landscape position, where small values of CTI correspond to areas of erosional loss, while larger values relate to areas of sediment accumulation (Gessler et al., 2000).

The Landsat 7 image (path/row 3730 taken July 6, 1999) covering the study area was analysed using Imagine. The image was resampled from 30 to 10-m spatial resolution using cubic convolution. The resampled image was used to derive a soil enhancement ratio and a fractional vegetation cover from the normalised difference vegetation index (NDVI). The soil enhancement ratio uses three band ratios (3/2, 3/7 and 5/7) to identify different parent material types within the study area (Amen and Blaszczynski, 2001). The NDVI is a normalised ratio of the near infrared and red bands of a multi-spectral image, which results in values ranging from −1.0 to 1.0, where higher values indicate higher vegetation density (Rouse et al., 1973). Fractional vegetation cover normalises the values of a vegetation index to provide an estimated percentage value of vegetative cover (Zeng et al., 2000).

Digital data derived from the DEMs and Landsat 7 imagery represented various soil-forming factors (Table 28.1). No raster data layers were used to represent climate or time because climate was roughly uniform throughout the study area and it is difficult to represent time digitally. These digital data were

Table 28.1. *Digital data layers representing soil-forming factors and respective data sources.*

Soil-forming factors	Representative data layers	Source data
Organisms (vegetation)	NDVI, fractional vegetation cover	Landsat 7 imagery
Relief	Elevation, slope, aspect, CTI	DEMs
Parent material	Soil enhancement ratios	Landsat 7 imagery

combined into individual multi-band images using the layer stacking function in Imagine.

28.3.2 Preliminary classifications

Digital data layer stacks representing the soil-forming factors were classified using different methods to identify landforms and predict soil delineations across the landscape. Classification methods included a combination of data-driven and user-defined methods, which were incorporated into digital models of the study area. All of the classifications were performed in Imagine. Combinations of the three soil enhancement ratios, fractional vegetation, slope and CTI data layers seemed to best represent the soil-landscape relationships in the study area, and, therefore, were used the most in the classification methods.

Unsupervised classification of different layer stacks was used in the preliminary stage of the methodology to recognise patterns that could represent existing soil-landscape-vegetation relationships. Classifications were generated with five and ten classes for different layer stacks. The number of classes was chosen by a trial and error approach in an attempt to find the level of detail best suited for initial field reconnaissance, given no prior knowledge of the study area. Five classes proved to be too general, while ten classes provided a desirable level of detail for identifying soil-landscape relationships in the field (Fig. 28.2). Unsupervised classification is an ISODATA clustering technique that takes an iterative approach to classifying pixels by using feature space. Pixels are grouped into classes on the basis of their distance from a central mean, and all pixels must meet a 95% convergence, that is less than 5% of pixels can switch classes between each iteration. The user specifies the number of clusters or classes desired. This is an unbiased classification that is data-driven, allowing objective pattern recognition.

The PURC methodology described by Cole and Boettinger (Chapter 27) included a supervised classification method based on training sites identified from a neighbouring soil survey. Since no soil survey data for the study area or adjacent areas were available, a supervised classification was not performed.

28.3.3 Field data collection

An initial sampling plan for collecting field data was developed using results of the unsupervised classification with 10 classes derived from a layer stack containing soil enhancement, fractional vegetation, slope and CTI (Fig. 28.2). Expert soil scientists determined that this set of proxies most accurately represented the soil-landscape patterns in the study area based on field observations. The unsupervised classifications provided predictions of how soils were distributed across the landscape. The sampling strategy was modified as field data provided

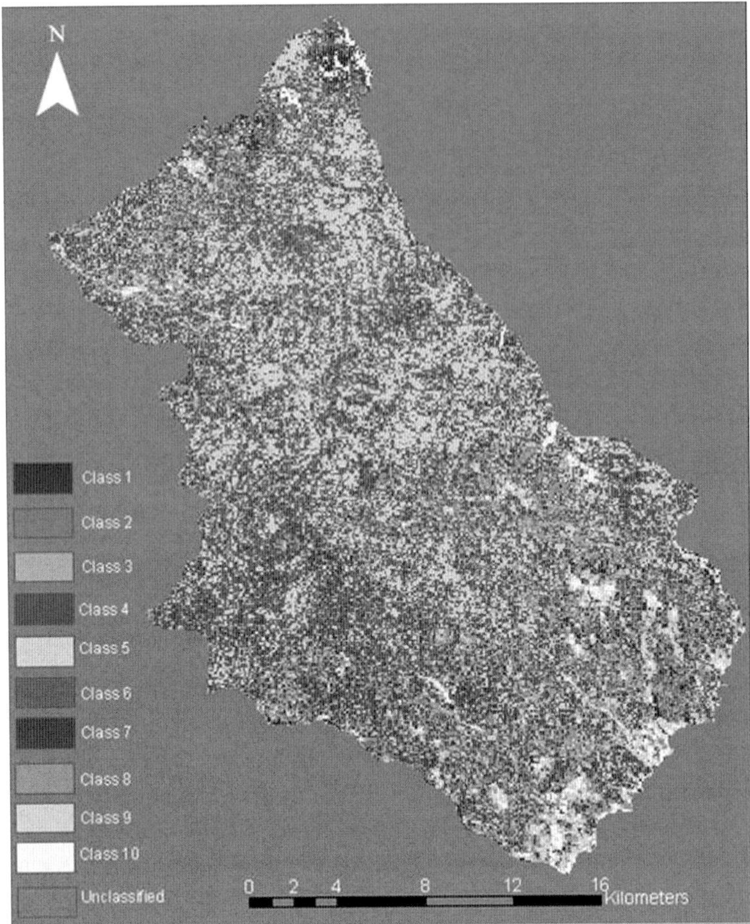

Figure 28.2. *Unsupervised classification of study area with 10 classes derived from a layer stack containing soil enhancement, fractional vegetation, slope and CTI.*

more information about the survey area and soil map units were developed by the soil survey staff. Transects were completed across areas of all of the predominant patterns in the classification, where each transect consisted of 10 points set apart by predetermined intervals (Soil Survey Division Staff, 1993). Hand-excavated or augered holes (approximately 0.5 or 0.1 m^2, respectively) at each transect point were used for observing soil profiles and describing and recording soil properties (i.e. colour, texture, structure, ped and void features, rock fragments and vegetation). The UTM coordinates of all data points were marked with a Garmin GPS unit for easy georeferencing and incorporation with the spatially explicit digital data proxies and classifications. Subsequent sampling was completed using traverses, where spacing is determined by the complexity of the soil pattern, to confirm the soil map unit composition of areas identified with patterns similar to existing transects (Soil Survey Division Staff, 1993). A total of 28 soil map units identified from 380 observation points were collected May through August 2004.

28.3.4 Knowledge-based classification

Knowledge-based classifications were introduced after initial fieldwork was completed and factors driving map unit distinctions were defined. Knowledge-based classification is a decision tree approach in which the user defines classes that are usually rule-based. Knowledge-based classification was created in the Knowledge Engineer in Imagine. The rules specify ranges and/or thresholds for environmental covariates such as slope, vegetation type and percent cover and elevation. Landforms and soil map units were defined on the basis of rules established by a range of values associated with the digital data proxies of environmental covariates (i.e. slope, elevation, CTI, soil enhancement ratios and fractional vegetation cover).

The knowledge of soil scientists familiar with the survey area was incorporated into the knowledge-based model. This included pedogenic concepts that applied to the study area, such as soil-landscape relationships reflecting the influences of local soil-forming factors (climate, organisms, relief, parent material and time). As field data collection progressed, developing conceptual models and an increasing number of field data points were incorporated into the model. The final model incorporated all 380 data points to predict 28 map units.

28.3.5 Classification tree analysis

Classification trees were used to generate additional classifications from the same digital data layers used for the knowledge-based model. These digital data layers served as the independent predictor variables, while the 28 soil map units identified at 380 field data points were the dependent variables. Rulequest Research See 5.0 data mining software was used to generate classification trees. See 5.0 creates classification trees and sets of "if-then" rules that are used to predict data categories (i.e. soil map units).

Boosting was used to enhance the accuracy of the classification trees. In boosting, multiple models are generated that form a "committee" where each member is assigned a different voting weight based on its individual accuracy. The resulting tree will contain predictions of the "most probable" outcomes, based on the input trees. Ten boosting trials were performed for this analysis.

Cross-validation was used to validate trees generated from training data because the number of data points collected in the field was limited. In an *f*-fold cross-validation, evaluation cases (i.e. data points) are divided into f blocks of roughly equal class size and distribution, where f is a number set by the user. Each case is used only once for evaluation, and the remaining cases are used as training data for each fold. The value of f was set at 10 for this analysis.

28.4 Results and significance

Descriptions of each of the soil map units developed and used in the knowledge-based and classification tree models are shown in Table 28.2. The areas

Table 28.2. *Descriptions of soil map units predicted by the knowledge-based and the classification tree models.*

Map unit	Soil family classification[a]
1101	Fine, mixed, superactive, frigid Ustic Haplargids
1201	Fine to coarse-loamy, mixed, superactive, frigid Ustic Haplargids
1202	Fine-loamy, mixed, superactive, calcareous, frigid Oxyaquic Torrifluvents
2201	Fine-loamy, mixed, superactive, frigid Ustic Haplargids
5201	Coarse-loamy, mixed, superactive, frigid Ustic Haplargids
5202	Fine-loamy, mixed, superactive, frigid Ustic Haplargids
5301	Fine-loamy, mixed, superactive, frigid Ustic Haplargids
5302	Coarse-loamy, mixed, superactive, frigid Ustic Haplargids
5303	Fine-loamy, mixed, superactive, frigid Ustic Haplargids and Calciargids
5304	Coarse-loamy, mixed, superactive, frigid Ustic Haplocambids
5305	Coarse-loamy, mixed, superactive, frigid, Ustic Haplargids
5306	Fine-loamy, mixed, superactive, frigid Ustic Haplargids
5307	Fine-loamy, mixed, superactive, frigid Ustic Haplargids
5308	Fine-loamy, mixed, superactive, frigid Ustic Haplargids
5311	Fine, smectitic, frigid Ustic Haplargids and Paleargids
5312	Fine-loamy, mixed, superactive, frigid Ustic Haplargids and Paleargids
5313	Fine, smectitic, frigid Ustic Haplargids
5402	Fine-loamy, mixed, superactive, frigid Ustic Haplargids
5403	Rock outcrop
5405	Fine-loamy, mixed, superactive, frigid Ustic Haplargids and Calciargids
5406	Fine-loamy, mixed, superactive, frigid Ustic Haplargids
5407	Fine-loamy, mixed, superactive, frigid Ustic Haplargids
5501	Clayey, smectitic, calcareous, frigid, shallow Ustic Torriorthents
5502	Fine-loamy, mixed, superactive, frigid Ustic Haplargids
5503	Fine-loamy, mixed, superactive, frigid Ustic Haplargids
5504	Fine-loamy, mixed, superactive, frigid Ustic Haplargids
5601	Loamy, mixed, superactive, frigid, shallow Ustic Haplargids and Badlands
5602	Fine-loamy, mixed, superactive, frigid Ustic Haplargids

[a]Soil family classification (Soil Survey Staff, 2003) listed is for the dominant soil component; map units with the same classification differ in proportion and type of components.

Table 28.3. *Comparison of area predicted for each soil map unit by knowledge-based and the classification tree models.*

Soil map unit	Knowledge-based	Classification tree
	Area (ha)	
1101	48	1192
1201	559	324
1202	2844	483
2201	661	3374
5201	12,306	14,542
5202	495	528
5301	2704	2807
5302	3980	4447
5303	1657	3047
5304	1002	351
5305	102	1033
5306	1686	3182
5307	114	6920
5308	60	0
5311	2376	0
5312	28	0
5313	5480	0
5402	2256	2931
5403	1059	1587
5405	5	1984
5406	122	1605
5407	109	0
5501	541	1840
5502	822	1316
5503	809	362
5504	0	2068
5601	719	617
5602	54	2045
Unclassified	15,986	0
Total area	58,584	

predicted to occur in each map unit by the knowledge-based and classification tree models are shown in Table 28.3.

The knowledge-based classification model was continuously modified as more data were collected and as the soil map unit rule sets needed to be refined for more accurate prediction. The knowledge-based model constantly evolved as more knowledge of the study area became available. However, even after eight modifications of the knowledge-based model, approximately 10% of the survey area failed to classify into one of the predicted classes, which resulted in "black holes" in the output image (Plate 28 (see Colour Plate Section)). Despite the knowledge-based classification allowing for the incorporation of expert knowledge of soil scientists into the decision tree, the overall process is extremely time-intensive and may not result in each pixel being classified.

Rule-based classification trees were generated with See 5.0 using a trial and error approach, which included a combination of boosting, pruning and cross-validation methods. The output image from the classification tree analysis is shown in Plate 28. Multiple classification trees were generated using different values for boosting, pruning confidence-level and cross-validation folds. The lowest mean error (31.8% with 2.3% standard error) of the classification trees generated was obtained with 10 boosting trials, 10 cross-validation folds and 50% pruning confidence-level, which produced an average of 45 rules for predicting 28 soil map unit classes. The classification trees classified all pixels within the given prediction area based on extrapolation, which could be useful for predicting soil map units in inaccessible areas. However, since classification trees are data-driven, they require large amounts of input for all desired output classes in order to accurately predict soil map units within a given area.

Digital maps generated from the final knowledge-based classification and classification tree showed overall similar patterns with slight differences. Both models were transparent and could be further refined as additional data becomes available or as land use needs change. Both digital maps could also be modified to meet the needs of the end user. Soil scientists familiar with the study area felt comfortable that both maps reasonably represented their conceptual models of soil-landscape relationships (J. Karinen, Project Leader, Sublette County Soil Survey, personal communication, 2004). However, creation, maintenance and update of the classification tree model were less time-consuming than the knowledge-based model.

The study validated the application of the PURC methodology to a 600-km^2 study area different than that in which it was developed. The PURC methodology worked successfully without the use of existing soil survey data for training the model. Classification trees were also successfully integrated into the methodology as a less time-intensive and more objective alternative to knowledge-based classification. Classification trees could potentially replace knowledge-based classification in inaccessible areas and in areas where expert teams of soil scientists are not available. However, classification trees do require user input, so field data would have to be collected if it did not already exist for a given area. Further development of a hybrid model incorporating both knowledge-based and classification tree methods could be useful. Small-scale predictions developed by classification trees could be refined for large-scale mapping of detailed soil delineations using knowledge-based models. Alternatively, intensive field data collection and development of knowledge-based models in multiple, small areas could provide data necessary for mapping soil classes in the larger encompassing area.

Acknowledgements

This research was partly supported by the Utah Agricultural Experiment Station, Utah State University, Logan, UT. Approved as paper no. 7741.

References

Amen, A., Blaszczynski, J. 2001. Integrated Landscape Analysis. U.S. Department of the Interior, Bureau of Land Management, National Science and Technology Center, Denver, CO, pp. 2–20.

Bui, E.N., 2004. Soil survey as a knowledge system. Geoderma 120, 17–26.

Bui, E.N., Moran, C.J., 2003. A strategy to fill gaps in soil survey over large spatial extents: an example from the Murray-Darling basin of Australia. Geoderma 111, 21–44.

Bui, E.N., Loughhead, A., Corner, R., 1999. Extracting soil-landscape rules from previous soil surveys. Aust. J. Soil Res. 37, 495–508.

Gessler, P.E., Chadwick, O.A., Chamran, F., Althouse, L., Holmes, K., 2000. Modeling soil-landscape and ecosystem properties using terrain attributes. Soil Sci. Soc. Am. J. 64, 2046–2056.

Jenny, H., 1941. Factors of Soil Formation. Mc-Graw Hill, New York.

Lagacherie, P., Holmes, S., 1997. Addressing geographical data errors in a classification tree for soil unit prediction. Int. J. Geogr. Inf. Sci. 11 (2), 183–198.

McBratney, A.B., Mendonça Santos, M.L., Minasny, B., 2003. On digital soil mapping. Geoderma 117, 3–52.

Moran, C.J., Bui, E.N., 2002. Spatial data mining for enhanced soil map modeling. Int. J. Geogr. Inf. Sci. 16, 533–549.

Roehler, H.W., 1992. Correlation, Composition, Aerial Distribution, and Thickness of Eocene Stratigraphic Units, Greater Green River Basin, Wyoming, Utah, Colorado. U.S. Geological Survey Professional Paper 1506-E, 49pp.

Rouse Jr. J.W., Hass, R.H., Schell, J.A., Deering, D.W., 1973. Monitoring vegetation systems in the Great Plains with ERTS. Third ERTS Symposium, NASA SP-351, NASA, Washington, D.C., pp. 309–317.

Soil Survey Division Staff, 1993. Soil Survey Manual. United States Department of Agriculture, Washington, D.C.

Soil Survey Staff, 2003. Keys to Soil Taxonomy, 9th edition. U.S. Department of Agriculture, Natural Resources Conservation Service, U.S. Government Printing Office, Washington, D.C.

Western Regional Climate Center, March 31 2004. Daily Temperature and Precipitation. June 8 2004. http://www.wrcc.dri.edu/cgi-bin/cliMAIN.pl?wypine.

Zeng, X., Dickinson, R.E., Walker, A., Shaikh, M., DeFries, R.S., Qi, J., 2000. Derivation and evaluation of global 1-km fractional vegetation cover data for land modeling. J. Appl. Meteorol. 39 (6), 826–839.

Developments in Soil Science, volume 31
P. Lagacherie, A.B. McBratney and M. Voltz (Editors)

Chapter 29

RULE-BASED LAND UNIT MAPPING OF THE TIWI ISLANDS, NORTHERN TERRITORY, AUSTRALIA

Ian D. Hollingsworth, Elisabeth N. Bui, Inakwu O.A. Odeh and Phillip McLeod

Abstract

We have applied a decision tree analysis (DTA) to map soil, plant community and land unit classes across the Tiwi Islands (7320 km^2), located in northern Australia. Our survey substituted environmental analysis and DTA for traditional air photo interpretation to provide a continuous land unit coverage over the islands. DTA was used to derive mapping rules for land units, their component vegetation classes and soil families from secondary survey site observations and distributed environmental data. The mapping was tested on a legacy data set. The environmental variables used are: elevation, slope, latitude, longitude, landform pattern class, wetness class, static wetness index, erosion or deposition index, Landsat TM band 5:7 and vegetation cover class extracted from digital topographic 1:50,000-scale mapping.

We needed to reduce the number of the resulting land unit classes derived from historical surveys into nine classes so as to produce a more meaningful mapping. This was achieved by generalising the component vegetation and soil units of the land unit classes thus producing the broader land unit classification. This procedure dovetails well with the current survey approach in which surveyors often develop more detailed classification systems than can be accurately mapped. We recommend a predictive mapping approach based on explicit mapping rules that integrate available land resources information and facilitate production of upgradeable maps.

29.1 Introduction

29.1.1 Background

Thematic map products of land resources survey in Australia have been criticized for not making use of the range of distributed environmental information that is now available and for not producing the land information of a quality required by modern users (Cook et al., 1996; McBratney et al., 2003; McKenzie and Austin, 1993). Land resource assessment in the Northern Territory (NT) is based on air photo interpretation of map units and descriptive surveys (McDonald et al., 1996) to represent soil, landform and habitat diversity in a landscape where the native vegetation is largely intact. The land unit map product of this process is widely used for development and conservation planning. However, there are gaps and inconsistencies in the resulting coverage because much of the survey work has been project based.

Opportunities now exist to upgrade and integrate the land resource maps using extensive and ubiquitous digital elevation models across northern Australia. Digital soil mapping (DSM) methods could potentially be used to improve the quality and extent of land resource mapping in an economic environment where public funding of field survey work is declining. Several DSM techniques have been reviewed extensively by McBratney et al. (2003). A most relevant technique to our work here is, decision tree analysis (DTA), which could be used to generate rule-based mapping analogous to expert systems used by traditional land resource surveyors. Guisan (2000) describes a broad range of applications using this technique for quantitative modelling of ecosystem. DTA has been used to map vegetation properties (Franklin et al., 2000) and to model habitat distribution (Guisan and Zimmermann, 2000) but it is prone to error when extrapolated to unfamiliar landscapes (Gahegan, 2000). DTA was used in Australia to extract soil mapping rules from extensive geology and DEM-derived attributes (Bui et al., 1999) and to produce continuous soil property maps from disparate soil survey data (Bui and Henderson, 2003; Henderson et al., 2005). However, although the results of the traditional land resource approach are generally thematic maps, DTA has not been used at local-to-regional landscape scales to map land resources including soil.

The underlying main aim of the work reported here was to make a continuous and consistent assessment of land resources of the Tiwi Islands, comprising Bathurst and Melville Islands (7320 km^2) for the purpose of strategic development planning. The islands are located in tropics at 12° S, 130° E (Fig. 29.1). Disparate land unit maps had been created over parts of the Tiwi Islands in the 1970s to assess land capability for forestry and agriculture (Olsen, 1980; Van Cuylenburg and Dunlop, 1973; Wells and Cuylenburg, 1978; Wells et al., 1978). Since this time, digital 1:50,000 topographic mapping (capable of deriving a 50-m DEM) over the Tiwi Islands has become available and a range of DSM techniques based on the analysis of DEM data has been developed.

In this chapter, we utilised a DTA mapping approach in an analogous manner to conventional land resource assessment as practiced in the NT making use of legacy soil survey data, digital topographic mapping and Landsat TM data. The aim was to demonstrate the technique in a way that could be readily used in the routine survey program to rationalise map units and extrapolate existing mapping. Because climate and lithology are relatively uniform across Tiwi Islands (Nott, 1994) we focused on sampling physiographic factors controlling soilscape diversity. We utilized digital mapping methods in a first-pass

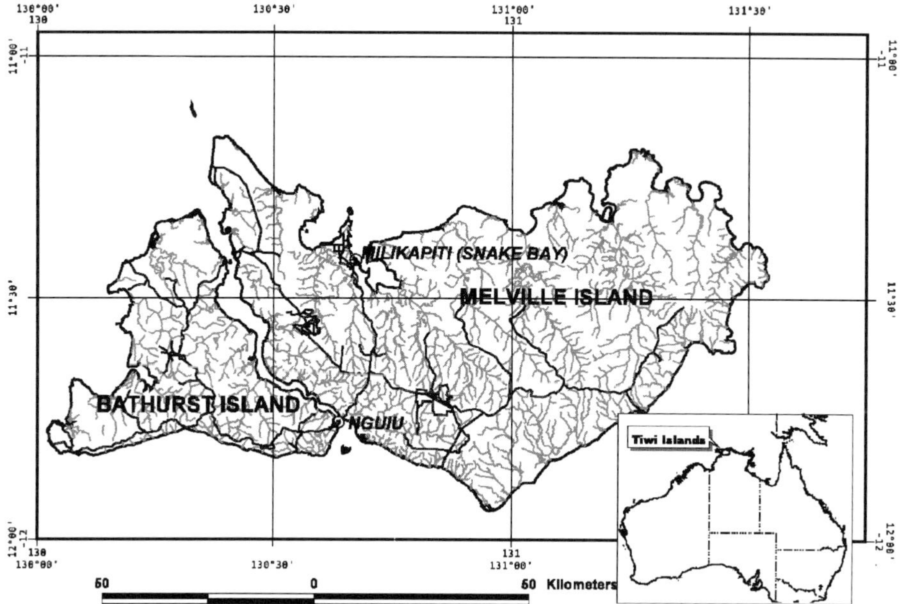

Figure 29.1. *Location of the Tiwi Islands in Australia.*

assessment of the land resources over the full extent of Tiwi Islands, a remote region where traditional land resource mapping has been difficult to apply.

29.1.2 Environment

The Tiwi Islands have an equatorial savannah climate that is dominated by northwest monsoon and the southeast trade weather patterns. The islands experience seasonal drought during the southeast trade season between May and September. The surface hydrology exerts considerable control over habitat diversity in this climate because water supply is a limiting factor for plant growth (Eamus, 2003).

The geology is relatively simple and consists of quaternary alluvium, flat bedded, tertiary Van Diemen sandstone and the underlying Cretaceous Mookinu mudstone and Wangarlu mudstone members of the Bathurst Island formation. Tertiary Van Diemen sandstone is a fine-to-medium grained quartzose sandstone of fluvial and partly littoral origin. The sandstone covers most of Melville and Bathurst Islands and varies in thickness to a maximum of 80 m and dips gently to the northwest. The origin of the clastic sediments in the sandstone were highlands, located to the south of the Tiwi Islands during the Tertiary Age (Nott, 1994). Dissected plateau remnants 150-m high form the current highlands in the centre of the islands, which are fringed by coastal mangroves and cheniers. Soils on sandstone are typically Oxisols (Ustoxs) with deep sandy to sandy loam surface horizons. Soils originating from mudstone are typically Ultisols (Ustults) with strongly acid subsoils overlain with a layer of pisolithic gravel.

29.2 Materials and methods

29.2.1 Definition of land units

Land units, as described in the NT land resources survey, represent unique subclass combinations of geomorphic unit, lithology, vegetation and soil classification that occur as a repetitive pattern within broader land systems. The land units are delineated from stereographic air photo interpretation and field survey observations onto 1:100,000-scale topographic base maps.

The complexity of the unit concept and the subjectivity of subclass definitions in different surveys often lead to disparity in the quality and types of unit being used in different surveys. We correlated historical soil and vegetation survey reports of the Tiwi Islands to produce a legend of 27 land units describing 10 soil family classes and 26 plant community types (Brocklehurst, 1998; Olsen, 1980; Wells and Cuylenburg, 1978; Wells et al., 1978) an excerpt of which is produced in Table 29.1. This approach was consistent with the current land unit classification system used in the NT, which concatenates landform class, soil class and vegetation class to construct a unique land unit class.

29.2.2 Landscape analysis

The ANUDEM, program, was used to generate 50-m grid by fitting a splined elevation surface to 1:50,000 digital contours interspersed with spot heights, drainage network and coastline data. TAPES-G (Gallant and Wilson, 1996; Wilson and Gallant, 2000) was used to derive slope, steady-state wetness index and a sediment–erosion index to characterise the spatial distribution of soil water content and erosion–deposition processes. The steady-state wetness index is a derivative of a specific catchment area that could be used as a surrogate for subsurface flow. This is most appropriate to a humid environment (Troch et al., 1993). The steady-state wetness index (ω) is defined as:

$$= \ln\left(\frac{A_s}{\tan\beta}\right)$$

where A_s is the specific catchment area (catchment area draining across a unit width of contour; m^2/m) and β is the slope angle (in degrees). This index is similar to the specific catchment area, or upslope area per width of contour that has been used widely in soil property mapping from digital terrain data (Wilson and Gallant, 2000). The wetness index was divided into six quantiles to

Table 29.1. *Land unit, vegetation and soil classes used.*

Mapped land unit	Land unit legend code	Landform element	Mapped vegetation unit	Described vegetation community	Soil family
(a) 85	Gec084	Foot slopes	(a) Eucalypt forest	1a: *E. miniata*, *E. tetrodonta* and *E. nesophila* open-forest with Chrysopogon fallax grassland understorey	K11 Koolpinyah: deep, gravelly, imperfectly drained, yellow sandy loam over sandy clay loam
	Gaq085	Hill crests	(a) Eucalypt forest	1a: *E. miniata*, *E. tetrodonta* and *E. nesophila* open-forest	K9 Hotham: deep, gravelly, well drained, red, sandy loam over sandy clay
	Gfc085	Slopes	(a) Eucalypt forest	1a: *E. miniata*, *E. tetrodonta* and *E. nesophila* open-forest	K9 Hotham: deep, gravelly, well drained, red, sandy loam over sandy clay
	Udf085	Slopes	(a) Eucalypt forest	1a: *E. miniata*, *E. tetrodonta* and *E. nesophila* open-forest	K9 Hotham: deep, gravelly, well drained, red, sandy loam over sandy clay
(b) 86	Gaq086	Summit surfaces	(a): Eucalypt forest	1b: *E. miniata* and *E. tetrodonta* open forest/ woodland	K8 Berrimah: very deep, well drained, red, sandy loam over acidic, sandy clay
	Laq086	Summit surfaces	(a): Eucalypt forest	1b: *E. miniata* and *E. tetrodonta* open forest/ woodland	K7 Berrimah: deep, slightly or nongravelly, well drained, red, sandy loam over sandy clay loam
(c) 88	Lfc087	Plains	(a) Eucalypt forest	1b: *E. miniata* and *E. tetrodonta* open forest/ woodland	K8 Berrimah: very deep, well drained, red, sandy loam over acidic, sandy clay
	Gec088	Fan	(a) Eucalypt forest	1b: *E. miniata* and *E. tetrodonta* open forest/ woodland	T6 Cockatoo: deep, nongravelly, well drained, red sandy
	Uec088	Fan	(a) Eucalypt forest	1b: *E. miniata* and *E. tetrodonta* open forest/ woodland	T6 Cockatoo: deep, nongravelly, well drained, red sandy soils

Figure 29.2. *Static soil wetness index derived from 50-m DEM and classified into six quantiles.*

represent the distribution of runon and runoff elements in the landscape (Fig. 29.2). While no direct association with traditional landform elements was made, the visible distribution of wetness index quantiles in the landscape indicated a close association with the surface drainage system and hillslope topography.

The erosion–deposition index (ΔT_c), a dimensionless sediment transport capacity, was also computed using TAPES-G as a nonlinear function of specific discharge and slope, expressed as

$$\Delta T_{cj} = \left[A_{sj-}^{m}\left(\sin \beta_{j-}\right)^{n} - A_{sj}^{m}\left(\sin \beta_{j}\right)^{n}\right]$$

where β is the slope (in degrees) and A_s is the specific catchment area or drainage area per unit width orthogonal to a flow line (m^2/m). ΔT_c represents the change in sediment transport capacity across a grid cell, and can be used as a measure of the erosion or deposition potential in each grid cell (Wilson and Gallant, 2000). Uniform excess rainfall conditions were assumed.

The published 1:250,000-scale geological map did not accurately discriminate outcrops of mudstone from extensive flat-bedded sandstone. As an alternative, we used a clay index calculated as a ratio of Landsat TM bands 5 and 7 to assess the variation in lithology. Vegetation cover was assessed from vegetation layer in the 1:50,000 digital topographic mapping. The mapped classes are: medium-density woodland, scattered woodland, saline coastal flats and marine swamps, pine plantation, dense vegetation, mangrove, inland water, intertidal flat foreshore, bare areas, lakes, dunes and cheniers.

29.2.3 Survey design

The stratified survey design covering 120 sites included 4 replicate observations in each of 6 elements of 5 landscape patterns. Landform element (50-m grid) and landform pattern (100 ha) variation was premapped using classification of the static soil wetness index at fine and coarse levels of resolution. First, six quantiles of the frequency distribution of static wetness index was used to represent landform element variation. Secondly, a network of regular 100-ha hexagons was overlaid on the wetness class grid and the areas of each wetness index class were cross-tabulated for each hexagon. Using an agglomerative classification, ALOC in the PATN software program (Belbin, 1987) five landform classes representing coarse (100-ha resolution), landform pattern variation were created. The landscape pattern variation and the survey site coverage are depicted in Figure 29.3. Prior to field work, survey site locations were selected in elements of each landscape pattern. Access considerations, landform element size and overall coverage of the islands also influenced wherever survey sites were placed.

29.2.4 Survey analysis

DTA models were fitted using the See5 program (http://www.rulequest.com/) to predict the land unit, and their component soil family and vegetation classes from latitude, longitude, elevation, slope, wetness index, wetness class, landform class, erosion–deposition index, vegetation cover and clay index (Landsat TM band 5:7). To restrict the size of the tree that was generated, at least four observations were required for each the DTA leaf (final node). DTA models were developed on a training dataset (120 sites in the recent survey) and tested on

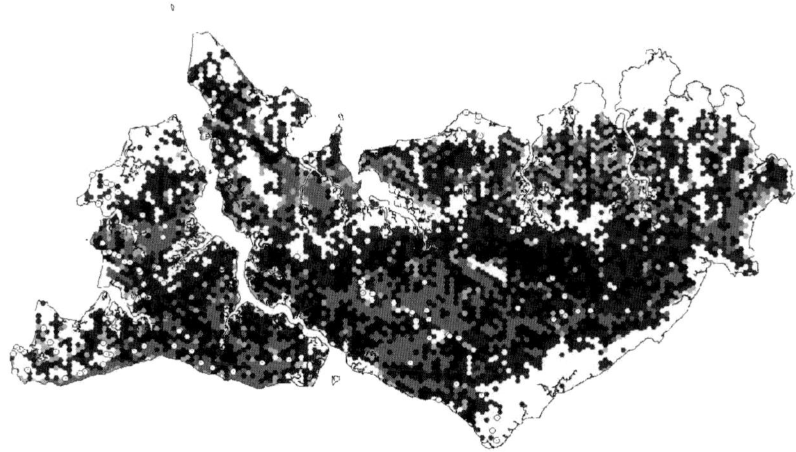

Figure 29.3. *Landform pattern variation and field survey site coverage.*

legacy survey data (108 sites) that had been collected to support forestry development projects in the southeast of Bathurst Island and the western end of Melville Island in the 1970s. The quality of the resulting maps was assessed based on the frequency of correct and incorrect allocations of these sites to classes.

29.3 Results

The 9 aggregated classes, extracted from 27 land unit classes obtained from historical survey reports, exhibit a reasonable mapping result. We used a simplified vegetation classification (7 classes instead of 26) to achieve generalization in this case. A selection of the resulting land unit classes, vegetation classes and soil classes are presented in Table 29.1 along with the more detailed classifications that were extracted from the historical survey legends.

The tabulated results for training and test datasets from models for land unit classes, vegetation classes and soil classes are shown in Tables 29.2–29.4. There is a considerable increase in the error of prediction between the training data and the test data. This is due to the fact that the test data selected from previous surveys were less accurately located (± 100 m compared with ± 10 m) and were concentrated in a part of the islands (Plate 29a, see Colour Plate Section) not well represented in the training dataset, thus adding to errors in prediction. For example, in the test survey area, land unit 109 comprises undulating landscapes with underlying cretaceous clay sediments that were misclassified as the more extensive land unit 88, a similar landscape but underlain with unconsolidated sandstone and sand colluvium. These misclassifications were manually corrected in the map so as to agree with site observations.

The land unit map is shown in Plate 29b (see Colour Plate Section). The red and orange colours indicate extensive areas of Ustoxs. The soil prediction model could not distinguish between gravelly and nongravelly phases of the soils (described as Hotham and Berrimah soil families in the previous land unit mapping). The grey colours represent Ustults and the blue colours Aquic soils. In the instance of the test survey area, the mapping model did not distinguish Ustoxs formed in residual sandstone from Ustults formed in cretaceous clay sediments in a reliable way. These inaccuracies in the mapping were ascribed to shortcomings in the ability to discern lithological boundaries using the clay index imagery. Woodland vegetation with a thick understory and grass cover would have obscured soil spectral signature except where land had been recently burnt.

Deep Ustoxs on flat-to-undulating terrain occur in land units 85, 86, 88 and 90 showed surface horizon texture variation from sandy (Kiluppa) to sandy loam (Berrimah) and gravel content variation from nongravelly (Berrimah) to

Table 29.2. *Land unit classification tree evaluation.*

Training data											
Total	Errors	(a)	(b)	(c)	(d)	(e)	(f)	(g)	(h)	(i)	Land unit classified as
33	3	23	1	2			5		2		(a) 85
11	2	2	7		1		1				(b) 86
25	1	1		22			1		1		(c) 88
5	1	1			4						(d) 90
0	0										(e) 95
18	1				1		17				(f) 105
3	0							3			(g) 112
23	8		2	1	1		2	2	15		(h) 109
2	0									2	(i) 114
120	16										Error rate = 13%

Test data											
Total	Errors	(a)	(b)	(c)	(d)	(e)	(f)	(g)	(h)	(i)	Land unit classified as
15	9	6		3	3		1	1	1		(a) 85
8	7	2	1		2		1	2			(b) 86
35	25	6		10	8		4	5	2		(c) 88
0	0										(d) 90
12	12	1		1	1		2	2	5		(e) 95
23	21	6	1	8	1		2	3	2		(f) 105
15	10	1					1	5	8		(g) 112
0	0										(h) 109
0	0										(i) 114
108	84										Error rate = 77%

Table 29.3. *Soil classification tree evaluation.*

Training data												
Total	Errors	(a)	(b)	(c)	(d)	(e)	(f)	(g)	(h)	(i)	(j)	Soil family classified as
34	11	22	9	1	1					1		(a) Hotham
44	9	4	35	1	4							(b) Berrimah
9	1		1	8								(c) Mirrikau
9	3		1		6		2					(d) Ramil
2	2	2										(e) Irgil
6	3		2				3			1		(f) Killuppa
3	0							3				(g) Wangitti
2	2							1		1		(h) Rinnamatta
10	5	1	2	1				1		5		(i) Marrakai
1	1									1		(j) Koolpinyah
120	37	32	50	13	13	0	5	7	0	9	0	Error rate = 31%

Test data												
Total	Errors	(a)	(b)	(c)	(d)	(e)	(f)	(g)	(h)	(i)	(j)	Soil family classified as
11	7	3	3	4						1		(a) Hotham
28	10	7	14	1	2			2		2		(b) Berrimah
13	12	4	1	1				0		7		(c) Mirrikau
14	13	5	3	1	1			1		3		(d) Ramil
2	2	1	1					0				(e) Irgil
14	13	6	4	1	1		1	0		1		(f) Killuppa
1	0							1				(g) Wangitti
1	2	1						0				(h) Rinnamatta
13	9	6	2					1		4		(i) Marrakai
11	11	6	4					1				(j) Koolpinyah
108	79	13	32	8	4	0	1	6	0	4	0	Error rate = 73%

Table 29.4. *Vegetation classification tree evaluation.*

Training data

Total	Errors	(a)	(b)	(c)	(d)	(e)	(f)	(g)	(h)	(i)	(j)	Vegetation classified as
92	2	90			2							(a) Eucalypt forest
7	6	2	1	1	3							(b) Melaleuca forest
6	2	2		4								(c) Mangrove forest
10	0				10							(d) Vine Forest
0	0											(e) Sparse woodland
4	0						4					(f) Grassland/sedge
1	1	1										(g) Coastal woodland
120	11											Error rate = 9%

Test data

Total	Errors	(a)	(b)	(c)	(d)	(e)	(f)	(g)	Vegetation classified as
81	14	67		3	11				(a) Eucalypt forest
14	14	11		1	2				(b) Melaleuca forest
0	0								(c) Mangrove forest
9	7	7			2				(d) Vine Forest
3	3	2			1				(e) Sparse woodland
0	0								(f) Grassland/sedge
1	1	1							(g) Beaches/cheniers
108	39								Error rate = 36%

gravelly (Hotham). The Hotham, Berrimah and Kiluppa soils mapped relatively consistently with most of the misclassifications occurring as a result of confusion between each of these units.

29.4 Discussion and conclusions

The traditional soil survey method produces a knowledge base that is expressed in soil map legends and can be formalized in terms of sets of mapping rules based on distributed environmental attributes (Bui, 2004). DTA produces sets of rules that are analogous to traditional survey methods. However, with this DSM technique, it is possible to upgrade the mapping, or knowledge base, as more field survey data become available. This approach would appear to have many advantages in the digital age: the ability to integrate with the knowledge accumulated over the years being not the least. However, statistical mapping methods such as those applied in the current Tiwi Islands study may not readily reproduce the level of map legend complexity used in routine survey work undertaken in the NT.

This is because traditional surveys have a tendency to over specify land unit classes in relation to the field survey datasets, even in low relief landscapes with fairly monotonous lithology such as the Tiwi Islands. This has also been identified as a map quality issue in Australian land resources mapping (McKenzie and Austin, 1993). Consequently, the expectations of end users of land resources mapping may need to be modified or attuned to the need for further field survey work to meet specific information requirements. Also, alternative approaches to the one used in our study to defining test and training datasets could be considered.

The location accuracy of legacy soil survey data is a recognized issue when this data have been used to develop test predictive models (Bui and Moran, 2001). Our test dataset comprised site information extracted from legacy soil surveys conducted 20–30 years ago. A cross-validation testing method would keep more information in the training set. However, because of the different methods used to select and locate sites, we decided to separate legacy and current survey data in our analysis. A general finding was that there is less error in the test and training datasets for vegetation class predictions than for soil and land unit class predictions (Tables 29.2 and 29.3). Plant communities tend to compensate for edaphic variation by changing their topographic positions (Guisan and Zimmermann, 2000).

The simplification that we needed to make to apply data-based mapping techniques on the Tiwi Islands demonstrates the level of inference that can be drawn reliably from the amount of available survey information. The mapping that we have produced is general relative to the land unit classification system in use in the NT and a review of the field survey database in the NT in relation to map unit descriptions and some test area surveys may be warranted to rationalize the land unit framework and to check and develop formalized mapping rules. The advantages of developing explicit rule-based mapping systems are that they would formalize current knowledge of soil and ecosystem variation and would integrate the current, disparate land resources information base using continuous environmental coverages, to create continuous land resources coverages and measurements of their reliability. We also found a much higher error of prediction for the test area than for the more extensive training area, which is probably explained by the poor delineation of different lithologies at the local scales. High-resolution lithological information and additional sampling of different lithologies will be needed to map land units according to the system in use elsewhere in the NT.

The key outcome of our work is to demonstrate the use of digital topographic data (1:50,000) for land resources assessment and environmental analysis in the NT and to point out the limits to mapping inferences that can be made from field survey work. No assessment of the reliability of the traditional land unit survey information has been made. However, it is likely that land unit map legends overstate the inferences that can be made. A review of the NT database and map legend structure to produce an explicit set of knowledge-based mapping rules that can be tested and upgraded by the ongoing survey program may be worthwhile. DSM methods such as DTA could be integrated into this approach.

Acknowledgements

We need to acknowledge the support and guidance of the Tiwi Land Council, Kate Haddon, Sylvatech Pty Ltd. and the Northern Territory Department of Business Infrastructure and Resource Development for this work.

References

Belbin, L., 1987. The Use of Non-Hierarchical Allocation Methods for Clustering Large Sets of Data. Australian Comp. J. 19, 32–41.

Brocklehurst, P. 1998. The History and natural Resources of the Tiwi Islands, Northern Territory. Chapter 4 – Vegetation, Parks & Wildlife Commission, N.T.

Bui, E.N., 2004. Soil survey as a knowledge system. Geoderma. 120 (1–2), 17–26.

Bui, E.N., Henderson, B.L., 2003. Vegetation indicators of salinity in northern Queensland. Austral Ecol. 28, 539–552.

Bui, E.N., Loughhead, A., Corner, R., 1999. Extracting soil-landscape rules from previous soil surveys. Australian J. Soil Res. 37, 495–508.

Bui, E.N., Moran, C.J., 2001. Disaggregation of polygons of surficial geology and soil maps using spatial modelling and legacy data. Geoderma 103, 79–94.

Cook, S.E., Corner, R.J., Grealish, G., Gessler, P.E., Chartres, C.J., 1996. A rule-based system to map soil properties. Soil Sci. Soc. Amer. J. 60, 1893–1900.

Eamus, D., 2003. How does ecosystem water balance affect net primary productivity of woody ecosystems? Functional Plant Biol 30, 187–205.

Franklin, J., McCullough, P., Gray, C., 2000. Terrain variables for predictive mapping of vegetation communities in Southern California. In: J. Williams and J. Gallant (Eds.), Terrian Analysis: Principals and Applictions. John Wiley and Sons, New York, p. 381.

Gahegan, M., 2000. On the application of inductive machine learning tools to geographical analysis. Geogr. Anal. 32, 113–139.

Gallant, J.C., Wilson, J.P., 1996. TAPES-G: a grid-based terrain analysis program for the environmental sciences. Comp. Geosci. 22, 713–722.

Guisan, A., Zimmermann, N.E., 2000. Predictive habitat distribution models in ecology. Ecol. Model. 135, 147–186.

Henderson, B.L., Bui, E.N., Moran, C.J., Simon, D.A.P., 2005. Australia-wide predictions of soil properties using decision trees. Geoderma 124, 383–398.

McBratney, A.B., Mendonca Santos, M.L., Minasny, B., 2003. On digital soil mapping. Geoderma 117, 3–52.

McDonald, R.C., Isbell, R.F., Speight, J.G., Walker, J., Hopkins, M.S., 1996. Australian Soil and Land Survey Field Handbook. Inkata Press, Melbourne, 250 pp.

McKenzie, N.J., Austin, M.P., 1993. A quantitative Australian approach to medium and small scale surveys based on soil stratigraphy and environmental correlation. Geoderma 57, 329–355.

Nott, J., 1994. Long-Term Landscape Evolution in the Darwin Region and Its Implications for the Origin of Landsurfaces in the North of the Northern-Territory. Australian J. Earth Sci. 41, 407–415.

Olsen, C.J., 1980. A Report on the Land Resources of South East Bathurst Island. LC 80/2, Land Conservation Unit. Conservation Commission of the Northern Territory, Darwin, N.T.

Troch, P.A., Detroch, F.P., Brutsaert, W., 1993. Effective Water-Table Depth to Describe Initial Conditions Prior to Storm Rainfall in Humid Regions. Water Resour. Res. 29, 427–434.

Van Cuylenburg, H.R.M. and Dunlop, C.R., 1973. Land Units of the Seventeen Mile Plain, Melville Island. 14, Animal Industry and Agricultural Branch Department of the Northern Territory, Darwin NT.

Wells, M.R. and Cuylenburg, H.R.M.v., 1978. Land Units of Areas Adjacent to the Tuyu and Yapilika Forestry Plantations, Melville Island, N.T. LC78/9, Land Conservation Unit, Territory Parks and Wildlife Commission.

Wells, M.R., Cuylenburg, H.R.M.v. and Dunlop, C.R., 1978. Land Systems of the Western Half of Melville Island, NT. LC78/10, Land Conservation Unit Report, Territory Parks and Wildlife Commission, Darwin.

Wilson, J.P., Gallant, J.C., 2000. Terrain Analysis Principles and Applications. John Wiley & Sons, New York, 469 p.

Developments in Soil Science, volume 31
P. Lagacherie, A.B. McBratney and M. Voltz (Editors)

Chapter 30

A TEST OF AN ARTIFICIAL NEURAL NETWORK ALLOCATION PROCEDURE USING THE CZECH SOIL SURVEY OF AGRICULTURAL LAND DATA

L. Boruvka and V. Penizek

Abstract

Artificial neural networks (ANN) can be used for the development of models for automated soil allocation to predefined soil units. This chapter tests a minimum input data number for reliable ANN model development, and allocation improvement by including terrain data in the model. Results of the Soil Survey of Agricultural Land (SSAL) carried out in the Czech Republic in the period 1960–1972 were used as soil data. Primary terrain attributes (altitude, aspect, and slope) were used as covariates. Increasing the number of training data leads to better allocation results. Nevertheless, a number of 20–30 input profiles showed to be sufficient for most soil units under study; increasing this number did not bring an important improvement in allocation performance of the models. For a good allocation, the classes should be clearly defined and distinguished from each other. Similarities between soil units (e.g. between Luvisols (LV) and Albeluvisols (AB) in some characteristics) increase the proportion of incorrectly allocated soils. Using auxiliary data should improve the allocation results. Nevertheless, the predictors (both soil attributes and covariates) and their structure should be selected according to what is the most important for soil classes to be predicted. Development of a useful ANN allocation model requires good training-data selection, suitable model structure selection and thorough training and exhaustive validation

30.1 Introduction

A wide range of pedometric techniques for analysing and processing soil data has been developed, originating in the sphere of conventional statistics, geostatistics and other mathematical methods (Goovaerts, 1999; McBratney et al., 2000, 2003). In soil classification, fuzzy methods are becoming more and more important. The task is to find a suitable and reliable method for soil allocation to soil groups. In addition to soil data, environmental covariates can be exploited as the input. It thus represents a specific case of *scorpan* models (McBratney et al., 2003). An early work comparing traditional soil survey with a computer allocation method was published by Norris and Loveday (1971). A lot of work dealing with soil fuzzy allocation into soil units was done using the Australian soil classification (e.g. Mazaheri et al., 1995, 1997). McBratney (1994) reviewed different objective methods of unknown soil allocation to continuous soil classes. Unsupervised fuzzy clustering can yield more or less homogeneous soil classes reflecting the natural structure of the landscape (Triantafilis et al., 2001). However, they do not necessarily need to correspond to traditional soil classes that are still in use. Nevertheless, Bragato and Lulli (1995) found fuzzy classes connected to traditional soil units; fuzzy clustering then only modified the allocation of several soil individuals. Albrecht et al. (Chapter 31) attempt to create a fuzzy allocation procedure based on the German Soil Systematics. Zhu (2000) used neural networks for allocation of soils into prescribed soil classes; the derived soil information provided greater spatial detail than that derived from the conventional soil maps.

Data from traditional soil survey present an important source of information about soils. In the Czech Republic, a systematic Soil Survey of Agricultural Land (SSAL) aimed at detailed mapping of all agricultural land in the country was done in the period 1960–1972 (Nemecek et al., 1967). Each selected soil profile that will be used in this study corresponds to 70–180 ha according to the heterogeneity of the area (more heterogeneous areas were surveyed with higher density). Resulting materials still provide an essential source for the evaluation of soil conditions. There are soil maps resulting from this survey, and databases of soil properties determined on samples from described soil profiles. However, only agricultural land was included in the survey and the maps are formed by crisp polygons.

Several attempts to apply new pedometric methods on the data from the SSAL were made. Penizek and Boruvka (2004) tested several ways to reduce the nugget effect of the SSAL data variograms. Exclusion of a few outlying values after detailed data supervision was more efficient than simple exclusion of all data representing local extremes (Fluvisols (FL), Gleysols (GL)). Boruvka et al. (2002) applied unsupervised fuzzy classification of soils. Different inputs were tested and the results were compared with the traditional soil classification. The best result was obtained when a semiquantitative approach was used, with the degree of principal pedogenetic processes or morphological features of the whole profiles as the input data. Nevertheless, the correspondence to conventional soil classes was in some cases low.

The aim of the artificial neural network (ANN) technique is generally to create a model to describe a system, while there is no assumed structure of the model (Gershenfeld, 1999). This contribution presents a preliminary study of the ANN application as a supervised fuzzy classification method. Point soil data from the SSAL are used as the input. It should provide a fuzzy alternative to crisp polygon maps, and it should be also applicable for allocation of unknown soil individuals. Principal objectives were:

1) To propose a minimum input data number for a reliable ANN model.
2) To assess allocation improvement by including terrain attributes in the model.
3) To compare ANN as a method of soil allocation to classes with traditional soil classification and to compare different soil units with respect to their predictability by ANN models.

30.2 Materials and methods

30.2.1 Input dataset

The chosen district of Tabor is located in Southern Bohemia. Its total area is 1327.3 km^2. 59.4% of the area is agricultural land. Forests cover is 29.2% and 3.4% are water bodies. The altitude ranges from 360 to 720 m.

Data from 586 soil profiles from the SSAL were available (Nemecek et al., 1967). The set consisted of 14 Fluvisols (FL), 23 Luvisols (LV), 132 Albeluvisols (AB; 27 haplic – ABh and 105 stagnic – ABs), 301 Cambisols (CM; 212 haplic – CMh, 28 luvic – CMl, 22 stagnic – CMs and 39 stagnoluvic – CMsl), 35 Stagnosols (SG) and 81 Gleysols (GL). All of them represented only agricultural land.

Soil profiles of each soil group were selected randomly to training and validation subsets. FL were excluded from calculation due to low number of observations. Soil pH and clay content (particles <0.01 mm) of the surface and subsurface soil horizons, and the ratio between clay content in the subsurface and surface horizons were used as the input data. The reason why just these attributes were selected is that there is a much larger set of profiles in the SSAL database where only these characteristics are available. Finding a suitable allocation procedure based on these soil attributes would therefore enable to include these profiles into the prediction of soil classes spatial distribution and into a new map creation.

30.2.2 Part 1: minimum input data number testing

For the assessment of the effect of the input data number on the accuracy of the ANN model prediction, two series of training datasets were used. The first one included soil profiles of all soil groups (or subgroups in case of AB and CM); separate training datasets contained 5, 10, 15, 20 and 30 profiles of each group (except LV, ABh, CMl and CMs – maximum 15 profiles, and CMsl and SG – maximum 20 profiles). The second series included profiles of only the most frequent units: ABs, CMh and GL; separate training datasets contained 5, 10, 20 and 40 profiles of each group, and also 60 profiles of ABs and 60 and 100 profiles of CMh. For testing the models of the first series, a validation file containing data of 20 profiles of ABs, CMh, GL; 19 profiles of CMsl; 15 profiles of SG; 13 profiles of CMl; 12 profiles of ABh; 8 profiles of LV and 7 profiles of CMs was used. For testing the models of the second series, a file containing data of 40 profiles of each of the three soil groups (ABs, CMh and GL) was used.

30.2.3 Part 2: including terrain attributes as auxiliary data

Terrain data were obtained from the Fundamental Base of Geographic Data of the Czech Republic at the scale 1:10,000 (ZABAGED; LSO, 2001). Three types of primary terrain data were selected: altitude, aspect and slope. All combinations of the terrain attributes with soil data were tested. Training datasets consisted of 15 profiles of LV, ABh, CMs and CMl, and 20 profiles of all other groups (or subgroups in case of AB and CM); in total, it contained 160 soil profiles. Testing dataset included data of 7 profiles of CMs, 8 profiles of LV, 12 profiles of ABh and CMl, 15 profiles of SG, 19 profiles of CMsl and 20 profiles of ABs, CMh and GL; in total, it contained 133 soil profiles.

30.2.4 Data processing and software

The program Neuropath version 1.2 (Minasny and McBratney, 2002a) was used. It is a multilayer perceptron type, but it enables only one layer of maximum 10 hidden units. It applies bootstrap aggregating (or bagging) method to enhance the accuracy of prediction; details see in Minasny and McBratney (2002b). In our study, 30 bootstrap datasets with 100 iterations were used in all cases. The number of hidden units was 6 to keep the structure reasonably simple and to avoid overfitting. The output provided similarity values. The success of allocation and correspondence to traditional soil classes was based on the maximum similarity value for each profile.

30.3 Results and discussion

30.3.1 Statistical description of the dataset

Table 30.1 presents basic statistical parameters of soil pH and clay content in the surface and subsurface horizons and altitude ranges for each soil group in the dataset. It is apparent that CM, GL and AB range to higher altitudes than LV and SG. In case of LV, AB (particularly stagnic) and CM it corresponds to soil zonality. The differences in slope and aspect between soil groups were less evident, except higher maximum slope for CM (up to 10 degrees). According to variance values of clay content, GL and SG were more heterogeneous compared with other soil groups; LV and AB were relatively most homogenous.

On the level of subgroups, no big difference in the basic parameters was found between ABh and ABs. In case of CM, CMh and CMs contained less clay and more sand and provided slightly lower topsoil pH values than CMl and CMsl.

Table 30.1. *Basic parameters of soil groups in the whole dataset.*

Attribute(unit)		ABh	ABs	CMh	CMl	CMs	CMsl	GL	LV	SG
Number		27	105	212	28	22	39	81	23	35
Altitude (m)	Mean	483	475	508	517	496	483	487	450	444
	Min.	424	410	364	424	410	420	406	404	412
	Max.	560	622	686	650	648	614	644	516	564
Max. slope (degrees)		4.0	5.0	10.0	6.0	6.0	8.0	7.0	5.0	4.0
Topsoil										
pH	Mean	5.8	5.8	5.4	5.8	5.3	5.7	5.2	6.0	5.9
	Variance	0.4	0.4	0.4	0.6	0.7	0.3	0.4	0.4	0.5
	Min.	4.1	4.2	3.8	4.1	3.9	4.2	4.1	5.0	4.5
	Max.	7.1	7.4	7.3	7.4	7.4	7.0	7.3	7.3	7.8
Clay content (particles <0.01 mm) (%)	Mean	26.3	26.6	19.9	28.5	22.2	25.3	31.1	29.8	26.6
	Variance	24.7	21.9	30.9	28.6	41.1	24.6	102.0	18.5	59.4
	Min.	11.9	15.3	3.1	14.3	6.3	13.2	10.8	22.4	6.4
	Max.	35.1	43.4	36.2	38.2	34.8	32.7	58.7	41.6	43.8
Subsoil										
pH	Mean	5.5	5.4	5.3	5.3	5.2	5.2	5.1	5.4	5.6
	Variance	0.6	0.4	0.5	0.8	0.8	0.5	0.7	0.5	0.6
	Min.	4.0	4.0	3.8	3.8	3.8	4.1	2.3	4.0	3.9
	Max.	6.7	7.1	7.3	7.5	7.0	7.2	7.3	6.7	6.9
Clay content (particles <0.01 mm) (%)	Mean	29.1	31.6	15.7	35.2	21.3	29.0	32.4	36.3	36.6
	Variance	26.2	39.2	58.6	109.7	75.7	64.9	234.6	15.9	185.8
	Min.	14.3	16.9	0.2	15.3	2.1	8.3	1.6	27.9	12.2
	Max.	38.5	47.2	39.8	56.8	38.8	43.6	67.0	44.1	68.5

30.3.2 Part 1: minimum input data number testing

Generally, as was expected, the success of prediction increases with the number of profile data in training datasets (Fig. 30.1). The percentage of success was determined as the percentage of agreement between real soil classification (from SSAL) and soil classification by the ANN model that is the soil group with maximum similarity (or "membership") value at the point, with no respect to the absolute value of the similarity value. The graph combines results of the two series of training datasets. The CMh and GL curves reached fairly stable values at about 15–20 training profiles; then a further increase with increasing the training data number was very slow. ABs reached the stable level for approximately 30 training profiles. The maximum allocation success differed between soil groups. CMh reached values higher than 90%, ABs 70–80%, and GL approximately 50%. Allocation success for SG was low, around 20%. This is probably related to the fact that ABs and CMh are subgroups where a higher homogeneity can be expected, while the other soils are groups. Moreover, the distinction of SG based on the selected attributes is worse because it can be similar to stagnic subunits of AB, CM, LV, etc. A potential contribution to the error in the distinction between SG and stagnic subunits of other soil groups

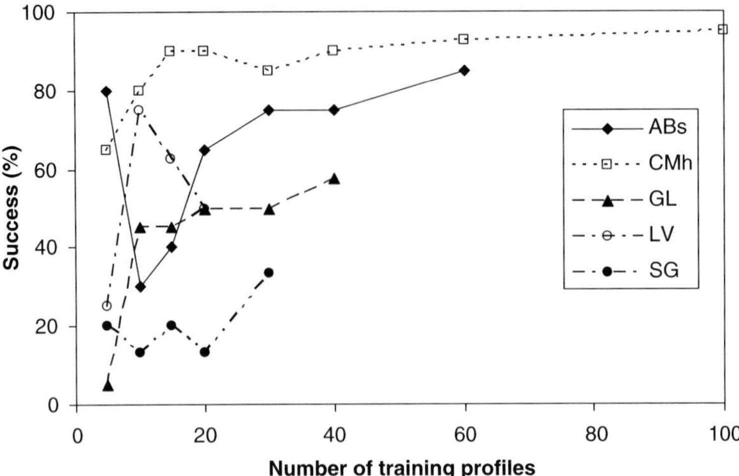

Figure 30.1. *Graph of the success of allocation in relation to the number of training profiles for each soil group (% of agreement between real and allocated soil groups; combined results of the two input file series).*

due to the subjective allocation in SSAL cannot be excluded, either. LV gave very inconsistent results and the stable level was not achieved due to a lack of data. In summary, 20–30 input profiles can be considered as a minimum for training the ANN soil allocation model. Five and ten input profiles provided rather poor allocation and cannot be considered sufficient, but for ABs the ANN with five training profiles performed pretty well; it is probably a random case-specific situation that cannot be generalised.

30.3.3 Part 2: including terrain attributes as auxiliary data

Figure 30.2 summarises the allocation success of all combinations of soil data with terrain attributes. For AB and CM the results for individual subunits were combined so that allocation of a subunit into another subunit of the same soil group was also considered as successful. The distinction between the subunits was rather poor in some cases, as it will be discussed further on. Nevertheless, the ANN model using AB and CM subunits performed better than a model based on the whole groups only (results not shown). The best allocation success was obtained for CM, followed by GL and AB. Allocation was worse for SG, as it is mentioned above, and LV. LV were most often incorrectly allocated among AB (Fig. 30.3), which can be explained by the fact that both soil groups developed through illimerisation process; tonguing of bleached soil material into the illuviation horizon, typical for AB, was not taken into account by the ANN models

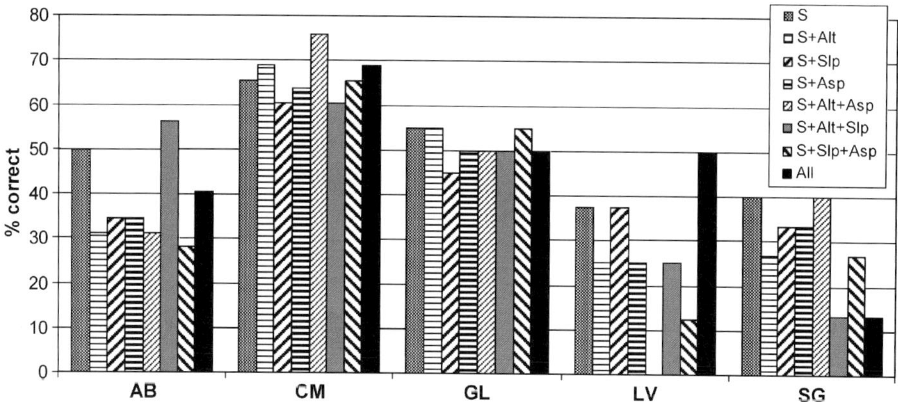

Figure 30.2. *Graph of the success of allocation with different model input data combinations; results of subgroups are combined in the case of AB and CM (S – soil attributes, Alt – altitude, Slp – Slope, Asp – aspect).*

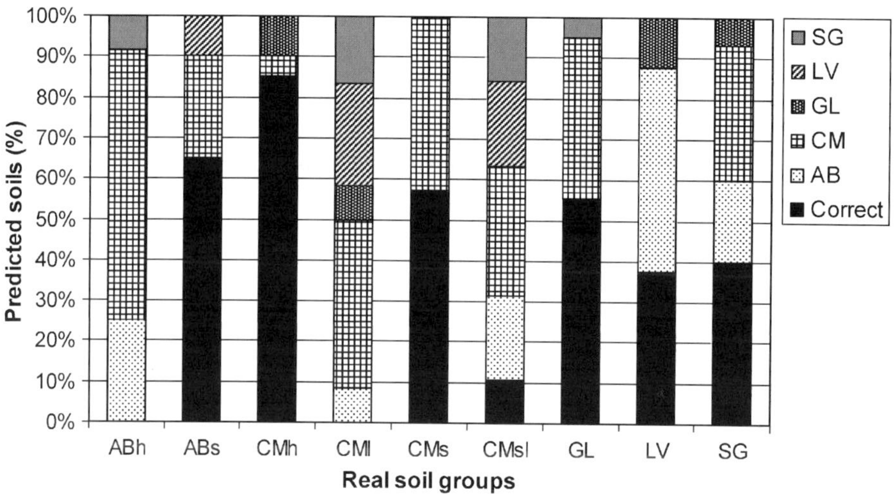

Figure 30.3. *Graph of the share of predicted soil groups for each real soil group (model using only soil attributes as the input data).*

as a criterion. Another reason can be a smaller number of training data for LV (15). Though 15 training profiles were used also for ABh, CMl and CMs, these soils were allocated often to other subgroups of the same soil groups (i.e. AB or CM respectively), which are represented in Figure 30.2 as correct allocation.

The terrain attributes did not present a strong improvement; in many cases the allocation results were even worse for combination of soil and terrain data compared with soil data alone. AB allocation was better when both altitude and slope were added. CM were better allocated by ANN models including altitude and its combinations with aspect or aspect and slope. LV allocation was improved when all three terrain attributes were added. The reason why the terrain attributes did not play a more important role in the allocation may consist in the fact that their range was relatively low and comparable between soil groups (see Table 30.1); a different result may be obtained in more hilly or mountainous regions with Podzols, Leptosols, etc. To summarise, models including terrain attributes provided better results for "zonal" soils allocation (LV, AB and CM), while for "non-zonal" soils (GL and SG) the terrain attributes did not yield any improvement. A better result could be obtained using some secondary, more sophisticated terrain attributes as the covariates.

Figure 30.3 shows the share of correct soil class allocation in the ANN model using only soil attributes. Allocation results using combination of soil data and terrain attributes were generally similar. For AB and CM, the allocation to correct soil subgroups and to other subgroups of the same soil group is distinguished. The best allocation success was achieved for CMh (85%). Other 5% of CMh were allocated to other CM subgroups. CMs profiles were allocated correctly in 57% of cases; all the other CMs profiles were allocated to other CM subgroups. The intermediate nature of luvic and stagnoluvic CM, where both cambic processes and clay migration take place, was shown by the allocation results. These soils were classified at similar numbers of cases as CM and as soils where clay migration occurs, that is AB and LV. Approximately 15% of both CMl and CMsl were classified as SG, which in case of CMsl makes sense because hydromorphic features and processes are another characteristic there. In contrast, it is rather surprising that no CMs (and no ABs) were classified as SG. Allocation of ABh was in most cases incorrect; 67% were allocated to CM and 8% to SG. In contrast, ABs profiles were allocated correctly in 65% of cases. 25% of ABs was classified as CM, which shows similar clay content and pH range in these two soil units. 10% of ABs was allocated to LV. The ANN cannot indicate the transition between eluvial and illuvial horizons specific for AB, as mentioned above. The properties of less eluviated AB can be therefore very similar to the properties of LV. This gradual transition between LV and AB is reflected also in high proportion of LV allocated to AB (50%). GL profiles were allocated correctly in 55% of cases. As the GL can be rather newly developed from other soils due to high groundwater level, they can keep some properties of the original soils. This may explain the 40% of GL allocated to CM. Allocation of SG also reflects probable origin of these soils. They developed most often from CM and AB, and these two groups manifested a high share of the allocated SG profiles. It should be noted here that for many profiles the difference between the highest similarity value and the second highest value was rather small, which even further confirm the intermediate nature of the soils. Continuous soil classes reflect therefore better the soil heterogeneity than crisp classes.

30.4 Concluding remarks

Artificial neural networks can be a good tool for supervised continuous soil classification. Data from legacy soil surveys can be successfully exploited by this technique. However, there are still many questions and problems to be solved in further research, as was shown in this preliminary study. They may be grouped into the following key points:

1) *Soil classes.* There are differences between soil classes in their allocation success, depending on their heterogeneity. For a good allocation, the classes should be clearly defined and distinguished from each other. It must be decided which classification level can be predicted, with respect to the heterogeneity of the classes and to available number of profile data.
2) *Input data structure.* A wide range of input data can be used, including soil attributes, terrain data, remote sensing, etc. (McBratney et al., 2003). Though it is not very convincing in our results, adding auxiliary data should improve the allocation results. However, a large number of predictors do not necessarily mean better prediction. The predictors (both soil attributes and covariates) and their structure should be selected according to what is the most important for soil classes to be predicted. For example, altitude can be decisive for some soils (especially the "zonal" ones) while for other soils it has low importance.
3) *Training data number.* The number of training samples (profiles) necessary for a reliable model generation depends on the structure of the data (their predicative value) and on the soil classes themselves. Certainly, it depends also on the number of available data. Generally, increasing number of training data should lead to better prediction. However, after a certain point further improvement is not significant. This threshold should be determined specifically in each case. An improvement can be achieved by a selection of really representative training profiles for each class (Zhu, 2000).
4) *Neural network structure.* In this study, a simple one-layer network structure was used. More sophisticated neural network models have potential for performance improvement (McBratney et al., 2000). Number of layers, number of hidden units in each layer, weights etc. are parameters that should be thoughtfully tested and validated.
5) *Generalisation of the models.* A good model should not be specific for a region or for a small group of soil classes, but it should be applicable to a wide range of conditions. Therefore it requires good training data selection, thorough training and exhaustive validation. Meeting all these tasks can provide fruitful results.

Acknowledgement

This study was supported by grant no. 526/02/1516 of the Grant Agency of the Czech Republic.

References

Boruvka, L., Kozak, J., Nemecek, J., Penizek, V., 2002. New approaches to the exploitation of former soil survey data. 17th World Congress of Soil Science, Bangkok, Thailand Paper no. 1692.

Bragato, G., Lulli, L., 1995. Application of fuzzy clustering to improve intensive soil survey. Quaderni di Scienza del Suolo 6, 19–31.

Gershenfeld, N., 1999. The Nature of Mathematical Modelling. Cambridge University Press, Cambridge, UK.

Goovaerts, P., 1999. Geostatistics in soil science: state-of-the-art and perspectives. Geoderma 89, 1–45.

LSO. 2001. Fundamental Base of Geographic Data of the Czech Republic (ZABAGED). Land Survey Office, Czech Republic.

Mazaheri, S.A., Koppi, A.J., McBratney, A.B., 1995. A fuzzy allocation scheme for the Australian Great Soil Groups classification system. Eur. J. Soil Sci. 46, 601–612.

Mazaheri, S.A., McBratney, A.B., Koppi, A.J., 1997. Sensitivity of memberships to attribute variation around selected centroids and intergrades in the continuous Australian Great Soil Groups classification system. Geoderma 77, 155–168.

McBratney, A.B., 1994. Allocation of new individuals to continuous soil classes. Aust. J. Soil Res. 32, 623–633.

McBratney, A.B., Odeh, I.O.A., Bishop, T.F.A., Dunbar, M.S., Shatar, T.M., 2000. An overview of pedometric techniques for use in soil survey. Geoderma 97, 293–327.

McBratney, A.B., Mendonca-Santos, M.L., Minasny, B., 2003. On digital soil mapping. Geoderma 117, 3–52.

Minasny, B., McBratney, A.B., 2002a. Neuropack. Neural Networks Package for Fitting Pedotransfer Functions. A User's Guide. Australian Centre for Precision Agriculture, Sydney, Australia.

Minasny, B., McBratney, A.B., 2002b. Neuropack. Neural Networks Package for Fitting Pedotransfer Functions. Technical Note. Australian Centre for Precision Agriculture, Sydney, Australia.

Nemecek, J., Damaska, J., Hrasko, J., Bedrna, Z., Zuska, V., Tomasek, M., Kalenda, M., 1967. Pruzkum zemedelskych pud CSSR (Soil Survey of Agricultural Lands of Czechoslovakia). 1st part. MZVZ, Prague (in Czech).

Norris, J.M., Loveday, J., 1971. The application of multivariate analysis to soil studiesII. The allocation of soil profiles to established groups: a comparison of soil survey and computer method. J. Soil Sci. 22, 395–400.

Penizek, V., Boruvka, L., 2004. Processing of conventional soil survey data using geostatistical methods. Plant Soil Environ 50, 352–357.

Triantafilis, J., Ward, W.T., Odeh, I.O.A., McBratney, A.B., 2001. Creation and interpolation of continuous soil layer classes in the lower Namoi Valley. Soil Sci. Soc. Am. J. 65, 403–413.

Zhu, A.X., 2000. Mapping soil landscape as spatial continua: the neural network approach. Water Resour. Res. 36, 663–677.

Developments in Soil Science, volume 31
P. Lagacherie, A.B. McBratney and M. Voltz (Editors)

Chapter 31

COMPARISON OF APPROACHES FOR AUTOMATED SOIL IDENTIFICATION

C. Albrecht, B. Huwe and R. Jahn

Abstract

Pedometric methods are often used for the classification of soils. One application is the allocation of individual soil profiles to the taxonomic units of a national soil taxonomy which we call identification. We compare three identification approaches, one from Australia, one from the USA and one from Germany, on their ability to reproduce the conventional soil identification process. We also investigate their data requirements as well as the significance of the results.

31.1 Introduction

During the last few decades much effort has been invested in developing software program systems which would automate the classification and identification of soils. Most of them are unfortunately only valid for specific purposes or for narrowly defined areas. Common examples of such systems would be the classification of soils according to their heavy metal concentration by Markus and McBratney (1996) or the classification of soils according to textural and internal drainage classes on an alluvial plain in Greece by Kollias et al. (1999). Only two approaches exist that model national soil taxonomies. The first is the fuzzy allocation scheme for the Australian Great Soil Groups (AGSG) Classification system developed by Mazaheri et al. (1995a). The second is the Expert System for Soil Taxonomy developed by Galbraith et al. (1998). At present, a third system for automated identification is under development. It is applicable to the German Soil Systematics and is based on fuzzy 'If–Then' rules.

Each of these three approaches has a different structure and requires different amounts of data to operate. These characteristics are not a scientific result from testing the advantages of different methods neither by aiming to achieve the highest accuracy, but rather are a consequence of the varying conditions and prerequisites of each national soil taxonomy.

The approaches considered here do not enable the creation of whole soil maps as described in McBratney et al. (2003). The goal of the work reported in this chapter is to compare approaches for automatic soil profile identification (cf. Chapter 30). The advantages of such tools are manifold. For example, it is possible:

- to combine modern techniques with conventional nomenclatures,
- to easily translate old data into modern nomenclatures,
- to provide high-quality standardised input data for other approaches and
- to accelerate the soil identification process.

Another field of application would be the quality assessment of soil data.

31.2 Approaches for automated soil identification

31.2.1 The fuzzy allocation scheme for the Australian Great Soil Groups Classification system

The AGSG Classification system is based on the concepts developed by Dokuchaev. It possesses different hierarchical levels, whereas all levels except the Great Groups find little practical use. Despite certain disadvantages, it remains very popular in Australia (Moore et al., 1983). It has now officially been replaced by the new Australian Soil Classification system (Isbell, 1996).

The allocation scheme was published by Mazaheri et al. (1995a). The Great Soil Groups are characterised by a set of 20 properties such as soil colour and carbonate content. The Groups are defined as a whole profile without a regard to diagnostic horizons or information pertaining to the depth. Fifty prototypes of the Great Groups were determined while taking into account the modes and means of the parameter values as well as information from the general description of the Great Groups. New individuals may be allocated by computing the distance between the individual and the prototypes in a multi-dimensional space. Figure 31.1 shows the identification procedure in principle.

The mathematical background rests on the clustering algorithm fuzzy *k*-means with extragrades, which is described in detail in a number of publications, for example De Gruijter and McBratney (1988) or McBratney and De Gruijter (1992). The allocation scheme was programmed in a spreadsheet with a user interface (Mazaheri et al., 1995b). It allows the identification of one soil profile at a time.

31.2.2 The Expert System for Soil Taxonomy

The US Soil Taxonomy (Soil Survey Staff, 1998) is one of the most well-known soil ordering systems in the world. The taxonomic units are defined by the occurrence of diagnostic horizons or properties. A sophisticated system for the combination of syllables is used for the labelling of these units at different

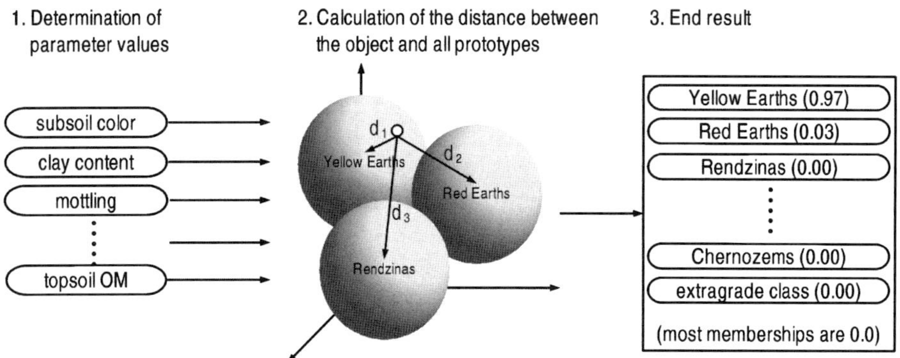

Figure 31.1. *Identification of individual soils with the fuzzy allocation scheme for the Australian Great Soil Groups (AGSG) Classification system, schematic illustration.*

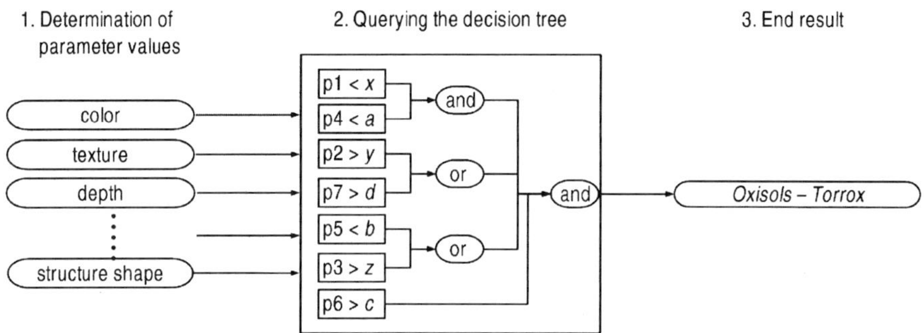

Figure 31.2. *Identification of individual soils with the Expert System for Soil Taxonomy, schematic illustration.*

hierarchical levels. Soil profiles are identified by querying a sequence of single-access, eliminatory keys which is a classification key.

The computational design of the Expert System adapts this structure (Galbraith et al., 1998). The definitions of the Keys to Soil Taxonomy are translated into decision trees. Added heuristic knowledge in the form of expert rules prevents indecision. Seventy independent properties must be known to identify 27 subsections from the Histosol, Spodosol, Andisol and Oxisol soil orders. A minimum data set is required when using only the expert rules, which consists of 13 field description properties, a set of predetermined default values for 20 other properties and estimated values for three particle size properties.

The Expert System is intended to query soil databases while considering a large amount of data. A simplified illustration of the system is provided in Figure 31.2.

31.2.3 The identification system for the German Soil Systematics

The German Soil Systematics is a morphogenetic soil ordering system encompassing all six hierarchal levels (AG Boden, 1998; Scheffer and Schachtschabel, 1998). The central category consists of the soil types (*Bodentypen*) which are roughly equivalent to the Great Groups in the US Soil Taxonomy. The identification process is accomplished in two distinct steps. First the single horizons are identified. Afterwards the unit of the Soil Systematics is derived from the horizon sequence considering the depth of the diagnostic horizons.

The focus of each horizon definition lies on the processes which determine the soil properties due to the morphogenetic character of the systematics. Thus, the majority of horizons are descriptively and verbally defined (as opposed to a deterministic definition) since pedogenetic processes can hardly be precisely with only a few, easily measured parameters. In addition, a certain number of vague threshold values are given to aid in the soil identification. They are treated as assisting or guiding values.

Currently, we are developing an identification software program, the first component of which is fully described in Albrecht et al. (2005). The two-step procedure identifies soil horizons and profiles separately. A set of 89 'If–Then' rules with 34 'If' components is defined for the first step (horizon identification), one for each possible horizon. The horizon identification forms the core of the software program and is crucial for the correct soil type delineation. Figure 31.3a describes the procedure in principle. The first step is the fuzzification of parameter values, that is information like 'the humus content is 2%' is translated into 'the humus content is small to medium with a certain membership'. At this stage verbal information can be easily incorporated when no measured values exist. Afterwards, the predefined rules are queried using the fuzzy parameter values. The memberships of single 'If' components are connected using a simple 'max–min' operator. The rule responses are normalised using the weighted sum combination (Bardossy and Duckstein, 1995) returning the best compromise for all rules.

We developed two special techniques for ensuring proper identifications:

- Single parameter values may be set as 'unknown' to prevent outliers in the profile descriptions.
- Single 'If' components may be skipped in order to handle inconsistent definitions and to prevent the influence of absent values.

The skipping method is also used for intergrade searching. Skipping three parameters in tandem ensures the correct identification with partially varying properties. The large number of parameters in many cases prevents misidentifications from occurring. The organic surface layers are peculiarly susceptible to wrong allocations. They are thus double-checked with respect to their position in the profile.

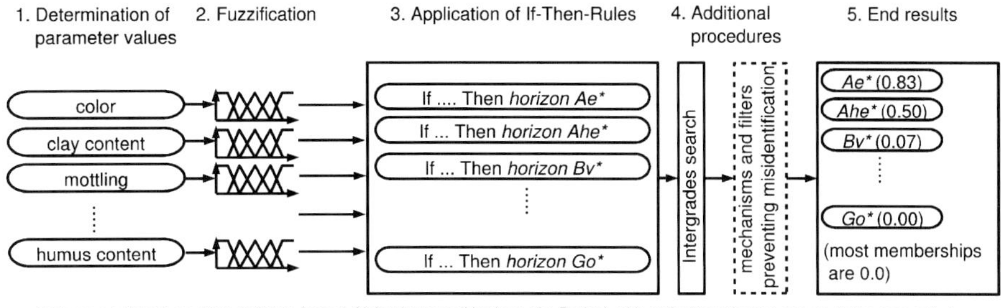

* German soil horizon abbreviations: *Ae* and *Ah* for the topsoil horizon of a Podzol, with and without humus accumulation respectively; *Bv* for the mineral horizon of a cambic soil; *Go* for the mineral horizon of a gleyic soil with oxidative properties

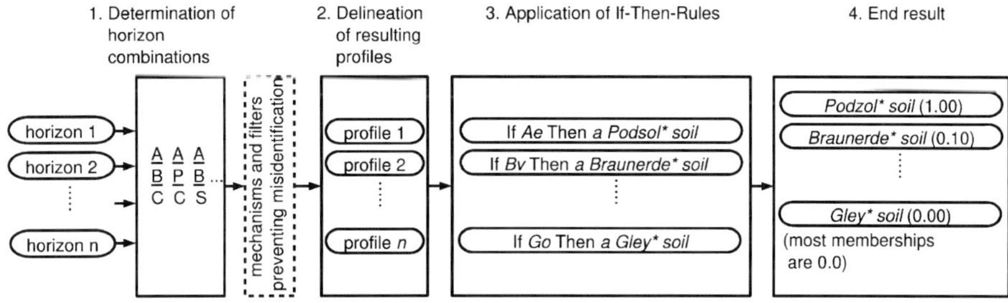

* The German soil types are roughly equivalent to the following WRB soil units: *Podsol* – Podzol, *Braunerde* – Cambisol, *Gley soil* - Gleysol

Figure 31.3. *Identification of soils with the Identification System for the German Soil Systematics, (a) soil horizon identification (above), (b) delineation of soil types (below), schematic illustration.*

The delineation of soil types is done in a similar way (Figure 31.3b). The various combinations of soil horizons are compiled and a set of 'If–Then' rules queries these combinations for diagnostic horizons. In many cases, the horizon identification has multiple results for each single horizon. In that case, the number of resulting profiles increases dramatically. Therefore, it is advantageous to insert a preprocessing step which eliminates illogical results and optimises computational efforts (e.g. by incorporating information on the pedogenetic processes). These tools are still under development and testing and preliminary results have shown that the soil type delineation results are often difficult to obtain and require the most effort on our part.

The identification system is designed to operate both on single soil profiles and on large soil data bases. The first tests with the horizon identification show a satisfying performance whereas the delineation of soil types is suboptimal.

31.3 Differences between the identification approaches

31.3.1 The structure of the underlying soil ordering system

The identification schemes for the soil taxonomies possess a different structure. The fuzzy allocation scheme requires a unique parameterised description of the taxonomic units. This requirement is achieved using the descriptions of the Great Groups in Stace et al. (1968). Taking into account the structure of the underlying soil ordering system, the fuzzy allocation scheme is not a viable choice in reproducing either the German Soil Systematics or the US Soil Taxonomy. For the first system, the lack of quantifiable information is the fundamental drawback. The US Soil Taxonomy is structured as a key, therefore, it is easier to determine which properties a soil does *not* have than to determine its typical appearance.

In Germany and in the USA, a sufficient set of soil profile descriptions could form the basis of a cluster analysis which would become the basis for a fuzzy allocation scheme. This goal would be difficult to achieve since the single soil profile descriptions differ in their set of required parameters. This difference leads to irregular datasets and violates the demand for unique profile descriptions. Furthermore, single parameters have a different weight in different definitions which may result in wide ranging parameter values. This result would have a strong influence on the identification. Lastly, a likely problem exists when one wishes to obtain a data set both unique and large enough to achieve statistical validity.

As stated above the US Soil Taxonomy is designed as a classification key. Adapted to a programmable version it becomes a decision tree. The other possible programming approaches are not expected to be better suited since an adequate reproduction of the US Soil Taxonomy requires a sequence of queries on whether particular conditions are matched or not. The use of a rule-based system is disadvantageous since some queries are dependant on others.

For the German Soil Taxonomy, a set of 'If–Then' rules is the only appropriate tool in reproducing the identification process. The given definitions focus heavily on the soil-forming processes and their differentiating properties. Therefore, any computational identification process would only be able to test the appearance of selected parameter values since neither a complete and detailed soil type parameterisation nor a classification key with quantitative parameters exist.

The single taxonomic units do possess a stand-alone definition. That means the relationships between different soil types are easy to describe, yet are neither part of the definition nor of the soil profile identification. Therefore, the If–Then rules must be queried in tandem or parallel. The use of decision trees would lead to poor results due to the non-hierarchical structure of the identification instruction. In contrast to the allocation scheme of the AGSG, the fuzzy approach is not applicable to model the continuity of the pedosphere but rather to only handle the verbal descriptions.

31.3.2 Required parameters

The consideration of required parameters may be divided into three components; the nature and the number of parameters and the quality of the data sets.

The Australian approach places the highest demand on data quality. The discriminant analysis equations become insolvable when a single gap occurs in the soil profile description. This fact is most likely a problem since soil databases are often incomplete. During field work, the soil surveyor could falsely use a default parameter value preliminarily provided by the software instead of choosing the correct one. Such misuse of the software would lead to misidentifications.

The other identification systems have built-in prevention mechanisms for the case of missing values. The continuation of the computation is guaranteed by assuming default values or skipping unknown parameters, respectively. The American approach requires a minimum of 13 known parameters, whereas the German identification system has no prerequisites regarding the completeness of the parameter set. The successful completion of the identification must be paid with a higher uncertainty or a lower accuracy.

As stated above, the identification procedures require 20, 70 or 34 parameters to be known, respectively. In the first two cases (AGSG and US Soil Taxonomy), the number is valid for a whole profile, whereas the parameters must be determined for each soil horizon separately in the latter case. The small number in the Australian approach represents the minimum data set required to adequately separate the Great Soil Groups. The number of parameters required for the other identification systems results from the aggregation of all parameters used in the definitions. The American identification system requires much more information, which is a direct consequence of the extensive and detailed definition of the taxonomic units.

Soils are usually described with both quantitative and qualitative parameters. The incorporation of qualitative information into a computer program is often restricted to the underlying mathematical approaches handling quantitative data only. The main advantage of the German approach is the direct access to both kinds of parameters via fuzzy sets, whereas the other identification systems are fixed to quantitative information.

31.3.3 Significance of the identification result

The American Expert System provides the best results when one takes into account its ability to reproduce the underlying soil ordering system. The decision trees used are practically an exact duplication of the structure of the US Soil Taxonomy. The main disadvantage of both the Australian and the German approaches is the higher likelihood of subjective interpretations made by different soil surveyors. This subjectivity can hardly be implemented in a computer program. Aside from this drawback, the identification process is well translated in all approaches.

Another aspect of interest is the comparison between the computed results and the results determined by a soil surveyor. The fuzzy allocation scheme for the AGSG clearly distinguishes itself from the other approaches. The calculation of the distance between an individual soil and the prototypes allows the quantitative characterisation of gradual transitions. Therefore, the information content of the computed results is higher and more precisely than the information content of the results determined conventionally. The Expert System for Soil Taxonomy cannot produce better results than that done by a soil surveyor since the identification rules are identical in the Expert System and in the Keys to Soil Taxonomy. The advantages lie in the fast identification and the consequent consideration of all threshold values by the computer program. The results of the German identification program may achieve the quality of the Australian program. The parallel querying of several fuzzy If–Then rules results in several outcomes with different membership grades. It is thus possible to characterise intergrades between soil types, although the quantification of similarity is restricted.

31.4 Conclusions

Several approaches exist which generally automate soil identification. They differ in their structure and required parameters. It is not possible to compare the identification systems directly since they are optimised to operate on different soil ordering systems under varying conditions. The results of one approach may not be readily adapted to those of another approach.

Scientists who aim to develop a soil identification system should carefully analyse the structure of the underlying soil ordering system as well as the availability of soil data. The former is meaningful for the decision concerning the appropriate approach and the latter is important in assessing the quality of the results. The use of fuzzy sets is strongly recommended in order to incorporate vague verbal descriptions which are almost always a component of soil identification schemes.

Acknowledgements

We thank the *German Research Foundation* (*Deutsche Forschungsgemeinschaft*, DFG) for financial support. Chris Tarn helped to increase the quality of the paper.

References

AG Boden, 1998. Systematik der Böden der BRD. Mitteilungen der Deutschen Bodenkundlichen Gesellschaft 86 (in German).

Albrecht, C., Huwe, B., Jahn, R., 2005. Automatische Horizontidentifikation mit einem regelbasierten Fuzzy-Ansatz. Mitt. DBG, 107, 291–292 (in German).

Bardossy, A., Duckstein, L., 1995. Fuzzy Rule-based Modelling with Applications to Geophysical, Biological and Engineering Systems. CRC Press, Boca Raton, New York, London, Tokyo.

De Gruijter, J.J., McBratney, A.B., 1988. A modified fuzzy k-means method for predictive classification. In: H.H. Bock (Ed.), Classification and Related Methods of Data Analysis. Elsevier, Amsterdam, pp. 97–104.

Galbraith, J.M., Bryant, R.B., Ahrens, R.J., 1998. An expert system for soil taxonomy. Soil Sci. 163, 749–758.

Isbell, R.F., 1996. The Australian Soil Classification. CSIRO, Melbourne.

Kollias, V.J., Kalivas, D.P., Yassoglou, N.J., 1999. Mapping the soil resources of a recent alluvial plain in Greece using fuzzy sets in a GIS environment. Eur. J. Soil Sci. 50, 261–273.

Markus, J.A., McBratney, A.B., 1996. An urban soil study: heavy metals in Glebe, Australia. Aust. J. Soil Res. 34, 453–465.

Mazaheri, S.A., Koppi, A.J., McBratney, A.B., 1995a. A fuzzy allocation scheme for the Australian Great Soil Groups Classification system. Eur. J. Soil Sci. 46, 601–612.

Mazaheri, S.A., Koppi, A.J., McBratney, A.B., Constable, B., 1995b. Australian Soil Identification Spreadsheet (ASIS): a program for allocating soil profiles to Australian Great Soil Groups (GSG), version 1.1. http://www.usyd.edu.au/su/agric/ACSS/sphysic/asis/asis1.html

McBratney, A.B., de Gruijter, J.J., 1992. A continuum approach to soil classification by modified fuzzy k-means with extragrades. J. Soil Sci. 43, 159–175.

McBratney, A.B., Mendonça-Santos, M.L., Minasny, B., 2003. On digital soil mapping. Geoderma 117, 3–52.

Moore, A.W., Isbell, R.F., Northcote, K.H., 1983. Classification of Australian soils. In: Soils: An Australian Viewpoint. Division of Soils, CSIRO, pp. 253–266. CSIRO, Melbourne/Academic Press, London.

Scheffer, F., Schachtschabel, B., 1998. Lehrbuch der Bodenkunde. Ferdinand Enke Verlag, Stuttgart (in German).

Soil Survey Staff, 1998. Keys to Soil Taxonomy, 8th edition. United States Department of Agriculture/ Natural Resources Conservation Service, Washington, DC.

Stace, H.T.C., Hubble, G.D., Brewer, R., Northcote, K.H., Sleeman, J.R., Mulcahy, M.J., Hallsworth, E.G., 1968. A Handbook of Australian Soils. Rellim Technical Publications, Glenside, South Australia.

F.ii. Examples of predicting soil attributes

Chapters 32–38 gives examples of predicting soil properties or attributes from Australia, Brazil, England & Wales, France, Hungary and the UNITED STATES. Soil properties predicted include available phosphorus, hydromorphic features, organic matter, salinity, available phosphorus, drainage status and texture.

The techniques used comprise geostatistical methods such as ordinary kriging and multiple indicator kriging, and statistical methods for dealing with covariates such as multiple linear regression, discriminant analysis and pricinipal component analysis. The hybrid method of regression-kriging or kriging with external drift is also used. Among the examples, there are neither uses of co-kriging or conditional simulation, nor of data-mining tools such as regression trees and neural networks. There is also little formal use of existing pedological knowledge.

The sample sizes vary from hundreds to thousands of observations. There are no examples with very small (<100) or very large (>10000) sample sizes. The resolution of the final maps vary from 30 m to 5 km for areas between 770 km^2 and 334,000 km^2.

The most common predictors are terrain attributes derived from DEM and remotely-sensed images. Consideration of soil-forming factors and processes at multiple scales seems to help predictability.

Three of the examples show the use of digital soil mapping for producing quantitative data in monitoring programs. This kind of application has significant policy implications.

Future work will need to consider the problems of very small or very large sample sizes, more sophisticated prediction procedures and an improved inclusion of pedological knowledge and multi-scale information. Demonstrations of how soil maps of this kind can be used for enhanced soil assessment and policy making need to be made.

Developments in Soil Science, volume 31
P. Lagacherie, A.B. McBratney and M. Voltz (Editors)

Chapter 32

DIGITAL MAPPING OF SOIL ATTRIBUTES FOR REGIONAL AND CATCHMENT MODELLING, USING ANCILLARY COVARIATES, STATISTICAL AND GEOSTATISTICAL TECHNIQUES

Inakwu O.A. Odeh, Mark Crawford and Alex. B. McBratney

Abstract

The application of statistical techniques to spatially predicting soil attributes from ancillary variables evolved from Jenny's factors of soil formation, termed "Climate, Organisms, Relief, Parent material and Time" or "*corpt*". The *corpt* approach was recently extended to include other soil attributes (*s* as surrogates) and space (*n*), and thus it is termed *scorpan* with time factor *t* in *corpt* replaced as age (*a*). The main objectives of this chapter are to collate and integrate various land feature digital layers to the same resolution and coordinate system, and to develop spatial prediction models based on *scorpan* and *scorpan*-kriging methods, for predicting selected soil attributes.

Existing analogue maps were first digitised and transformed into digital maps. These, along with other existing digital information, were reprojected in the same coordinate system, the Geocentric Datum of Australia (1994) and Map Grid of Australia (1994), namely GDA-94 and MGA-94. Further these digital map layers, along with Landsat bands and digital terrain attributes, were used to predict soil attributes and thus producing different soil attribute maps for a number of soil depths. The spatial prediction methods used were *scorpan* methods, such as multiple linear regressions (MLR) and *scorpan*-kriging (SK), which combines simple kriging with MLR. While MLR was good enough model to predict a number of soil attributes, SK was more appropriate for electrical conductivity for two layers: 0–10 and 70–80 cm layers and was equally good, if not slightly better than the *scorpan* technique of MLR. However, the results of *scorpan*-kriging exhibit more detailed variation across the extent of the study area compared with kriging, are displayed as geographical information systems (GIS) layers, which could potentially be useful for various environmental and catchment modelling.

32.1 Introduction

It is now widely recognised that the natural resources (including land and water) of this planet are not inexhaustible. However, the demand imposed upon these resources is huge and increasing on a scale more than natural cycles can replenish. Thus the sustainability and the management of these resources are vital for the survival of all living things, including the human race. The management of these resources requires knowledge and information about them. In Australia and worldwide, the importance of updating regional land resource maps has come to the forefront of many government and nongovernment organizations (Bui and Moran, 2001). This is due to increased awareness of environmental impact of land use and the fact that land is an important component of the agroecosystems that needs to be conserved for the future generations. The knowledge of variability of land features over large areas has become increasingly more important for numerous scientific and policy purposes (Cihlar et al., 2000). Soil is an important component of the land. Therefore, digital mapping of soil can greatly meet the needs of land feature information necessary for land use planning and resource management.

Digital soil maps can be created using interrelationships of soil with ancillary variables such as digital elevation models (DEMs) (McBratney et al., 2000), remotely sensed data (Odeh and McBratney, 2000), chemical and physical characteristics obtained through laboratory analysis of soil samples. Additionally, existing natural resource maps are useful surrogates for predicting soil attributes (Bui and Moran, 2001). For a useful comprehensive soil database obtained from different sources or derived from different ancillary variables, it is of utmost importance that the different data layers are appropriately georeferenced in a common coordinate system and projection; they should also be of the same spatial resolution and format. Integration of land cover information derived from old topographic maps with digital land cover maps derived from satellite imageries is difficult because the geometric and thematic characteristics of historical maps are not always well known (Petit and Lambin, 2001). One of the processes involved in digital map integration is the transformation of analogue maps into a digital form, but because of their nature, digital maps are inevitably linked to geographical information systems (GISs). GISs are essential for collating and analysing data on environmental processes in space and time (Oldak et al., 2002). GIS, as a tool for collating all kinds of spatial information (Burrough and McDonnell, 1998), is in itself incapable of soil mapping, as the latter requires an intellectual framework (McBratney et al., 2003). However, GISs are especially useful in working with most geographic phenomena over large areas when the volume of information is large and environmental variables are interrelated (Burrough, 1986). Indeed, GISs are becoming more sophisticated as they continue to incorporate more complex modelling capabilities.

The main objectives of this chapter are to collate and integrate various land feature digital layers to the same resolution and coordinate system, and to develop spatial prediction models, based on *scorpan* techniques (McBratney et al., 2003), for predicting selected soil attributes important for further hydrological and environmental modelling. The prediction models are based on correlation with existing land resource maps obtained from Australia's New South Wales (NSW) Department of Natural Resources (NSW DNR). Additional land feature databases such as Gamma radiometrics, DEMs, aspect, slope, plan curvature and topography wetness index (TWI) were also utilised to improve the accuracy and, therefore, reduce the level of uncertainty of the digital soil attribute maps.

32.2 Materials and methods

32.2.1 Study area

The location of the study area is in the northwestern part of NSW of Australia, within the Namoi river catchment (Fig. 32.1). The Namoi catchment comprises more than 43,000 km^2, which is about 6% of NSW (NSW EPA, 1997). The study area is part of a larger survey project under the auspices of the Australian Cotton Cooperative Research Centre. The primary industry for the region is agriculture, which is predominantly irrigated cropping. The main crops grown are cotton, oil seeds and cereals (Banks, 1995). The proportion of land use for

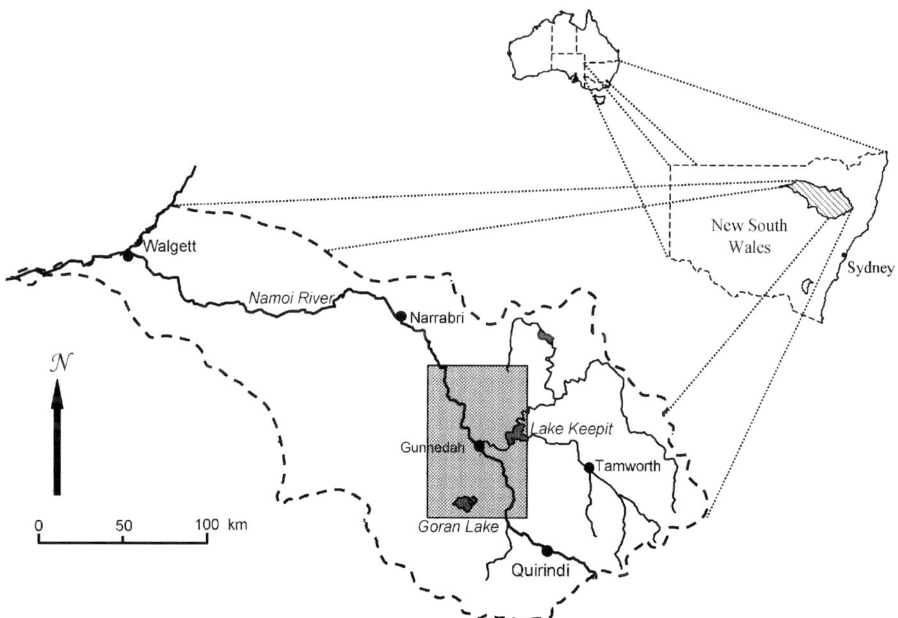

Figure 32.1. *The Namoi river catchment showing the study area (in grey) within Australia and New South Wales.*

crop production, particularly irrigated cotton, has increased since the 1950s, when it became possible to cultivate the heavy textured soil on the plains.

A relatively detailed previous soil survey was carried out within the study area by the NSW DNR. The extent of this survey is the Curlewis 1:100,000 map sheet. Soils in the area are defined in terms of soil parent materials (Banks, 1995) and also based on the standard classification systems of the Great Soil Groups (Stace et. al., 1968) and the Northcote Key (Northcote, 1979). According to this soil map the study area consists mainly of clay plains interspersed with sharply elevated parts that in turn consist predominantly of older sediments capped with younger basalts (Duggin and Allison, 1984). The dominant geologic formation (approximately 80% of the study area) is the Liverpool Plains. Physiographically, the area is characterised by relatively flat landscape, with shallow slopes dipping towards the north (Duggin and Allison, 1984) with slopes generally less than 3° (Crawford, 1976) (Table 32.1). Other areas with slopes of less than 3° include the watercourses of the Namoi river, Mooki river and Cox's Creek. Flooding occurs mainly in the flat plains from January to March and occasionally in the month of July (Wiles, 1996). There are some local areas where the slope can be up to 80°; this is usually associated with the Hunter–Mooki thrust fault system (Wiles, 1996).

Climatically, the study area is defined as having a dry subhumid climate with annual average rainfall of 642 mm (Bureau of Meteorology, 2003). It has a dominant summer rainfall due to summer thunderstorms (Banks, 1995), although there is somewhat even spread of precipitation throughout the year. Rainfall tends to decrease from east to west and from the hills to the plains (Hird, 1976). In terms of vegetation cover, the hill slopes of the upper Namoi have been cleared except for the particularly steep ridges and the NSW State Forests (Banks, 1995). The uncleared areas are dominated by white cypress pine (*Callitris glaucaphylla*) and a number of eucalypts as subdominant species.

Table 32.1. *Average root mean square error (RMSE) and R^2 of prediction of percent clay content using scorpan and scorpan-kriging techniques.*

Depth	*Scorpan* (multiple linear regression)		*Scorpan*-kriging	
	R^2	RMSE	R^2	RMSE
0–10	0.69	13.59	0.66	12.15
10–20	0.51	13.81	0.57	12.85
30–40	0.61	12.32	0.58	12.11
70–80	0.47	12.59	0.65	9.73

32.2.2 Data capture and acquisition

In order to make digital soil maps for the study area, field soil sampling was required to determine the target soil attributes to be mapped. Because the study area is not a perfect rectangle (Fig. 32.2), we first divided it into northern and southern sections. The soil samples locations (Fig. 32.2) were obtained using a stratified random sampling system. Each section was subdivided into four equal-size blocks (as strata) to ensure that samples were taken relatively evenly across the entire region. The sites were then randomly selected within each block. Soil sampling involved taking of a soil core at each site and the core was subdivided into six depth categories: 0–10, 10–20, 30–40, 70–80, 120–130 and 190–120 cm.Samples were taken back to laboratory for chemical and physical analysis.

All soil samples were first air-dried and ground to less than 2 mm, with the gravel fraction removed and its proportion determined. All subsequent analyses were based on the so-called earth potion (<2-mm diameter). Particle-size fractions were determined using the micropipette method (Odeh et al., 2003). Effective cation exchange capacity (CEC) was determined using the method described in Rayment and Higginson (1992). Exchangeable bases (Ca^{2+}, Mg^{2+}, Na^{+} and K^{+}) were the only cations extracted, as the soils in the surveyed area contain negligible amounts of other cations (Fe and Al). Both pH and electrical conductivity (EC) were determined using a 1:5 soil–water suspension, using deionised water, following the method of Rayment and Higginson (1992). The total carbon content was determined for the top two layers (0–10 cm and 10–20 cm) using the Dumas total combustion method in a CHN-100 elemental analyser (Leco Inc., St. Joseph, MI, USA). For lack of space in this chapter, only a few of the soil attribute maps are shown here.

32.2.3 GIS analysis of existing soil and ancillary variables: satellite data, digital elevation and terrain analysis

GISs have been used extensively for the integration of spatial and thematic data. Development of digital land feature maps involved the conversion of the existing soil information, in addition to acquiring the more needed land feature data. It was necessary to integrate lithology, land management units (LMUs) and soil descriptions, obtained from NSW DNR and the existing soil maps from CSIRO Division of Land and Water. This is required for them to be of a uniform geographic projection and of the same coordinate system. We used ArcInfo/ ArcGIS® and ERDAS IMAGINE® for the integration. For example, the soil data acquired from NSW DNR was in geographic coordinate system not suitable for the spatial prediction models we were to carry out. The data had to be converted into the coordinate systems and projection used in this study, namely map grid

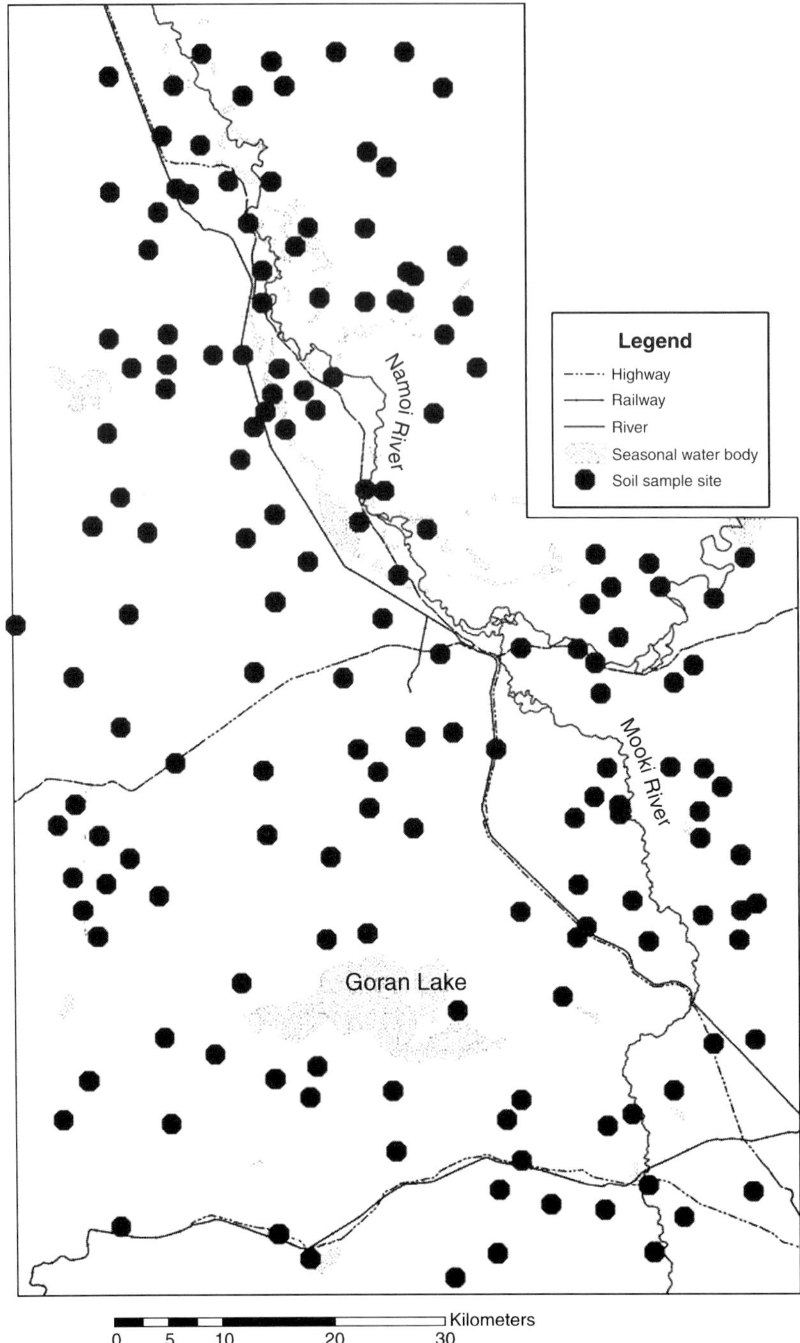

Figure 32.2. *The study area showing the sampling pattern.*

of Australia (MGA) based on geocentric datum of Australia (GDA), 1994. This coordinate system is based on Universal Transverse Mercator (UTM) projection. A DEM, with a spatial resolution of 250 m, acquired from CSIRO Land and Water, provided the base map for this study. The DEM came in the form of GDA MGA 94 that was our working coordinate system. All of the vector layers of roads, river network and other related infrastructure were also reprojected into the same coordinate system. Once all coordinate conversion and reprojection were done, the maps were visualised using ArcInfo/ArcGIS (ESRI, 2004).

It was hypothesised that the utilisation of satellite data and DEM would improve the accuracy of spatial prediction of soil features (Odeh and McBratney, 2001). However, this process requires the transformation of the satellite images into a data format usable for the prediction process. Landsat TM imagery acquired on 22 May, 1999, obtained from Geoscience Australia in Canberra, consists of several bands: 3 (blue, green and red) in the visible section of the electromagnetic spectrum, 2 in the near infrared and 2 in the midinfrared. These are respectively referred to as b1, b2, b3, b4, b5, b6 and b7. The raster image was resampled into our working grid of a resolution of 250 m.

Specific geomorphometry (Evans, 1980) was utilised in this study to determine the point terrain attributes from DEM. Odeh et al. (1991) defined specific geomorphometry as the measurement and analysis of specific land surface features that are defined according to clearly defined criteria. For this study, a number of primary terrain attributes were determined, including slope, aspect, plan curvature and profile curvature. These attributes, defined as primary terrain attributes, have considerable influence on the hydrological patterns of the surrounding area. For example, slope influences the rate of water and sediment flows (Odeh et al., 1994). Aspect defines the direction of flow and thus determines the upslope area of the catchment in which it originated. Plan curvature and profile curvature is the rate of change aspect and gradient, respectively. The effect of the changing aspect on the surrounding area influences the flow convergences or divergences, whereas the changes in the gradient affect the rate of flow acceleration and deceleration that influences soil aggradation or degradation (Odeh et al., 1991).

We also used the secondary terrain attribute, topographic wetness index (TWI), sometimes referred to as compound topographic index (CTI). TWI quantifies the position of a point in the landscape in terms of water and sediment movement, and is a hydrological index that is related to zones of surface saturation (Moore et al., 1993). TWI is derived using the expression:

$$\mathrm{TWI} = \ln\left(\frac{A_s}{\tan\beta}\right) \tag{32.1}$$

where A_s is the upslope area and β is the slope radian. Combined with the *flow accumulation* and *flow direction* as derived using the modules of ARC/INFO (ESRI, 2004) and primary terrain attributes such as slope and aspect, the attributes provide not only good GIS layers for environmental and hydrological modelling but also useful input data for spatially predicting the difficult-to-measure soil properties.

32.2.4 Spatial prediction of soil attributes

In order to spatially interpolate some of the soil variable methods such as kriging, regression models – also known as *scorpan* (McBratney et al., 2003) and *scorpan*-kriging were used. Kriging can generally be described as a moving weighted averaging that estimates property values at unsampled points based on the relative distance of the neighbouring sampled point (Webster and Oliver, 1990).

MLR models have been previously used for the purpose of deriving relationships between soil attributes and ancillary variables. MLR models are based on the linear equation (Eq. 32.1), which has been widely used because of its ease and availability (McBratney et al., 2003).

$$y = \beta_0 + \beta_1 x_1 + \ldots + \beta_n x_n \tag{32.2}$$

where y is the predicted attribute; β_0 the intercept; $x_1, \ldots, x_n$; and $\beta_1, \ldots, \beta_n$ are the regression coefficients. With the use of the Splus program, MLR prediction formulas were derived using a stepwise regression technique that determined the best combination of soil predictors of the sampled sites and thus used for the prediction of the unknown areas of the survey site. For the soil attributes that produced a small R^2 value using MLR models, a combination of MLR followed by ordinary kriging of the regression residuals and summing them (known as *scorpan*-kriging) was applied.

Scorpan-kriging, as used here, can be described as the combination of a multiple linear regression model with simple kriging of the regression residuals (Odeh et al., 1995; McBratney et al., 2003). The *scorpan*-kriging method was carried out using the combination of simple kriging pf the residuals resulting from MLR, and the MLR models. The *scorpan*-kriging is based on the general expression:

$$S(\mathbf{x}) = m(\mathbf{x}) + e'(\mathbf{x}) + e \tag{32.3}$$

where $S(\mathbf{x})$ is the predicted soil value; $m(\mathbf{x})$, the regression component; $e'(\mathbf{x})$, the locally varying component or regression residuals, interpolated by simple kriging (McBratney et al., 2003), and e, spatially incoherent error term.

32.2.5 Validation of the prediction methods

The primary validation criterion used to test the quality of spatial prediction of soil attributes was the root mean square error (RMSE) (Voltz and Webster, 1990). RMSE is expressed as:

$$\text{RMSE} = \sqrt{\frac{\sum_{i=1}^{n} (z_i - \hat{z})^2}{n}} \tag{32.4}$$

where z_i is the measured value and $\hat{z}$ is the predicted value. RMSE is a measure of precision and bias. As RMSE is sensitive to both systematic and random errors, it could be used to estimate the accuracy of prediction (Atkinson and Foody, 2002) and could be based on a validation sample set selected independent of the training set. However, in validating a spatial prediction model a large sample size is required when simple dichotomous sets, one for validation and the other for training are used (Bishop and McBratney 2001, Odeh et al., 2003) are used. But when the sample size is small as is in the case here – 163, it is difficult to have sufficient (and separate) sample datasets for modelling and validation. Moreover, it has been suggested in the literature that a sample size of 100 is the barest minimum required for variogram estimation (Webster and Oliver, 1992). For this reason, a modified jack-knifing technique (Good, 1999) was used to resample the base sample data repeatedly 10 times for the purpose of validation. The resampling size for validation was maintained at approximately 1/5 of the available data, that is, 30 out of 163. RMSE (Eq. 32.4) as a measure of precision and bias was estimated, for each of the resampled validation sets. The prediction quality of each prediction method was determined by the averaging of RMSEs of the 10 jack-knifed samples. Additionally, we used the coefficient of determination (R^2) of the predicted versus the measured value of the soil attributes as a measure of performance of each technique.

32.3 Results and discussion

32.3.1 Land feature maps

Figure 32.2 shows the perspective view of the study area based on DEM. It shows the complexity of the landscape varying from the relatively flat alluvial plains along the Namoi–Mooki river, to the steep slopes in the eastern fringe and central and southwest portion of the study area. As stated previously, both primary (slope, aspect, plan curvature, profile curvature) and secondary terrain attributes (TWI), derived from the DEM, can provide vital information on the hydrological patterns of the survey area. Figure 32.3 shows some of these layers. Flow direction, flow convergence and sediment accumulation, just to mention a few, are all derived from DEM and play a vital role in the development of digital

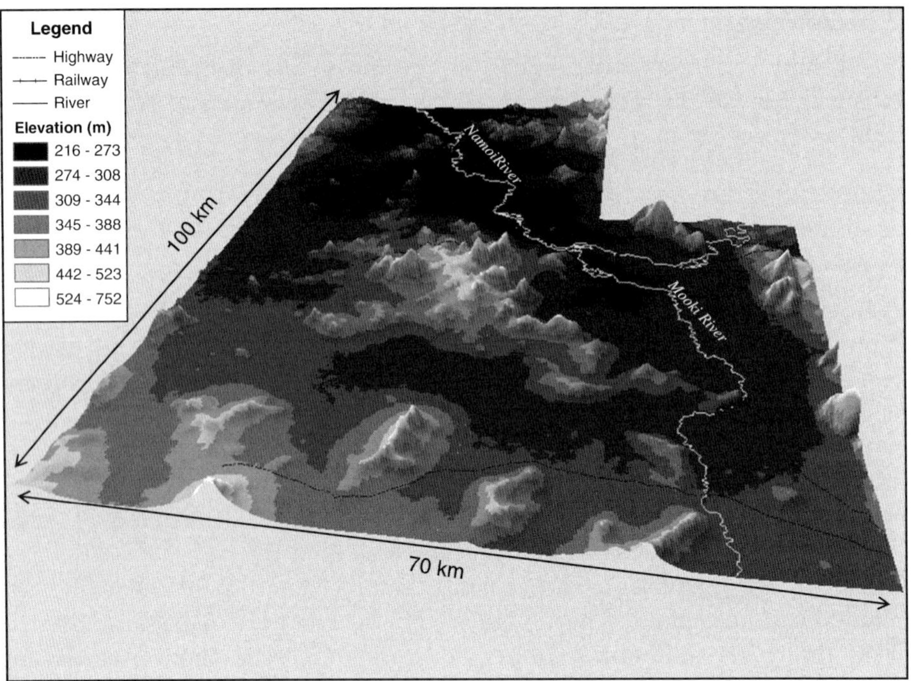

Figure 32.3. *Perspective view of digital elevation model (DEM) of the study area.*

land feature maps. TWI provides valuable information on the flow patterns and accumulation areas of water. This is demonstrated in Figure 32.3(b), which illustrates the high TWI values as the likelihood of water accumulation. Figure 32.3(a) shows the spatial distribution of plan curvature that also illustrates the hydrological patterns of the survey area; plan curvature is highly influential on the rate of flow acceleration or flow deacceleration. Therefore, understanding the hydrological properties and physical parameters as depicted by these digital layers is useful for catchment modelling and planning, and thus, can greatly benefit the management of natural resources.

Based on the fundamental understanding of the relations between the soil attributes on the one hand, and terrain and landscape features on the other, soil attributes, especially the more difficult-to-measure ones, could be predicted more accurately. In pursuing this endeavour, both datasets require a common coordinate systems and projection, which could allow for the spatial interpolation of sampled site parameters onto the unknown locations. Secondly, the number of predictor variables used in the *scorpan* model more often determines the degree of accuracy; thus the more ancillary variables are used in the prediction model the greater is the reduction of the model uncertainty. Applying attributes derived from DEM, Landsat images and previously soil and landscape layers for the study area would evidently improve the accuracy of the resulting digital soil attribute maps.

32.3.2 Digitisation of existing soil data

'The existing digital land feature maps should not only be of the same coordinate system but also coincident spatially with the prediction base map. But for them to be used in consonance with the soil attribute layers, they need to be in the same coordinate system and projection. The digitisation of the existing data such as lithology, LMUs and Murray Darling Basin soil groups (MDB soil) was carried out in order to develop the digital land features for the designated soil attributes. As described in subsection 32.3, the majority of work is centred on conversion of the variety of geographical coordinate systems to a uniform system, thus reducing the spatial errors in the prediction models. However, the importance of these features is based not only on their valuable source of information for management purposes but also their added benefit in enhancing the prediction of the soil attributes at unknown locations. All the terrain attribute layers and lithology, MDB soil and LMU layers were transformed and projected to the same coordinate system of the sample base map. Two of these layers, TWI and lithology, are illustrated in Figure 32.4.

As mentioned above, one of the existing data sources was LMU, which was obtained from NSW DNR. LMU were derived by dividing the landscape into areas based on slopes and soil types, as a reflection of how they can be managed. The LMU, therefore, constitute useful GIS layers that can be used for predicting soil attributes, as they are very versatile. However, the LMU coverage of the study area is incomplete, covering the Liverpool Plains in the southwest of the study area. Another useful and important data was lithology (or geology) layer (Figure 32.4b), which was extracted from Tamworth 1:250,000 k geological map sheet obtained from Geoscience Australia. The lithology layer provides one of the most valuable secondary information for predicting soil attributes.

Another existing secondary sourced GIS layer used here is the MDB soil (not shown here because of lack of space) – which contains a complex combination of many soil types as extracted from the whole soil map of Australia; it is accompanied by an elongated legend as obtained from its source. It is based on detailed Northcote soil classifications (Northcote, 1979) with tens of soil classes for NSW region. As there are over 50 soil types in the study area alone, it became too complex for the modelling process. To overcome this problem an abridgement of minor soil types with similar major soil classes was carried out. This reduced the original 50 to 28 abridged classes, consisting mainly of Duplex, Uniform and Gradational soil type classes. As soil class is a very important factor in the

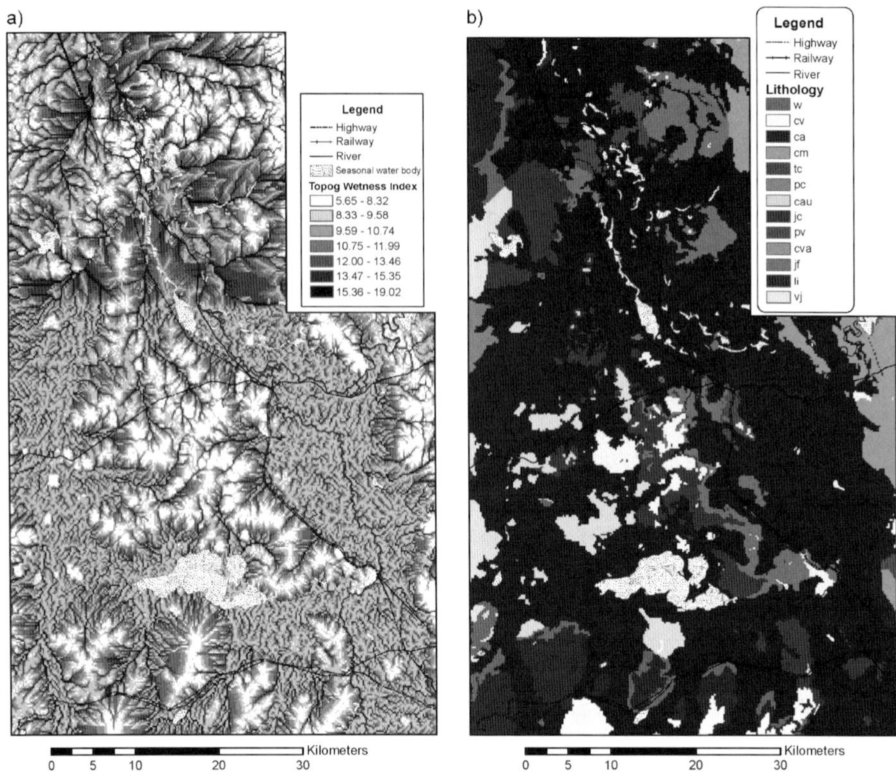

Figure 32.4. *(a) Topographic wetness index and (b) Lithology of the study area.*

prediction of soil attributes, the abridgement process were validated using a range of soil types at the sampled locations.

32.3.3 Validation of prediction techniques

The average values of RMSE (derived using Eq. (32.4)) and the coefficient of determination (R^2) were used to compare the quality of the prediction. Tables 32.1 and 32.2 show the results for predicting the clay content and EC at various soil depths. In the case of clay content, *scorpan*-kriging only shows a marginal improvement over *scorpan* technique of multiple linear regression in terms of both RMSE and R^2. However, for the 70–80 cm depth, both validation criteria indicate *scorpan*-kriging as far more superior to *scorpan* multiple linear regression. The main reason for this is the high influence of the predictor variable, the landscape configuration encapsulated by digital terrain models, as soil texture tends to be genetically related to land surface hydrological processes. In the case of spatially predicting EC for all the soil layers, *scorpan*-kriging appears to be

Table 32.2. *Average root mean square error (RMSE) and R^2 of prediction of electrical conductivity (EC) using scorpan and scorpan-kriging techniques.*

Depth	*Scorpan* (multiple linear regression)		*Scorpan*-kriging	
	R^2	RMSE	R^2	RMSE
0–10	0.40	0.10	0.56	0.13
10–20	0.21	0.17	0.52	0.14
30–40	0.35	0.18	0.67	0.16
70–80	0.37	0.83	0.72	0.32

better than multiple linear regression as indicated by RMSE and R^2, with the exception of topsoil layer (0–10 cm) (Table 32.2). Again, the main cause of spatial variability of EC are mainly the natural movement of salt leading to dryland salinity, as influenced by land use (land clearing), landscape variability (hydrological factors as influenced by slope, surface curvatures and drainage areas). All these factors were used as the predictor variables, but appear to be less influential at depths than at or close to the surface layers.

32.3.4 Digital maps of soil attributes

The results of our spatially predicting soil attributes are digital maps of several soil properties. These maps are displayed in an integrated geographic information system that can be viewed, combined or queried for useful information for catchment management and modelling. For lack of space, here we present the maps for clay content and EC created for the two sampled depths – 0–10 and 70–80 cm.

Maps of particle-size fractions – the clay content: Particle-size proportion is regarded as one of the important factors in the development of hydrological models. The particle-size fractions, especially clay content, provide some indicators of physical characteristics of the soil landscape, especially water infiltration, overland flows, flux of material through the soil and into the groundwater. Figure 32.5 show the resulting maps of soil layer clay content (%) as produced using *scorpan* for topsoil (0–10 cm depth) and *scorpan*-kriging for subsoil (70–80 cm depth). On close look at both soil layers, it seems that there is slight increase in the clay content with depth, especially in the river plains. It also indicates the predominance of high-to-medium clay zones in the flood plains of the Namoi and Mooki rivers. These zones are traditionally used for irrigated agricultural production; thus for management and further agricultural practices this particular land feature map would be highly beneficial. Figure 32.5(b) depicts the topsoil % clay exemplifying the natural topography of the study area. It is also evident from the Murray Darling soil landscape map (not shown

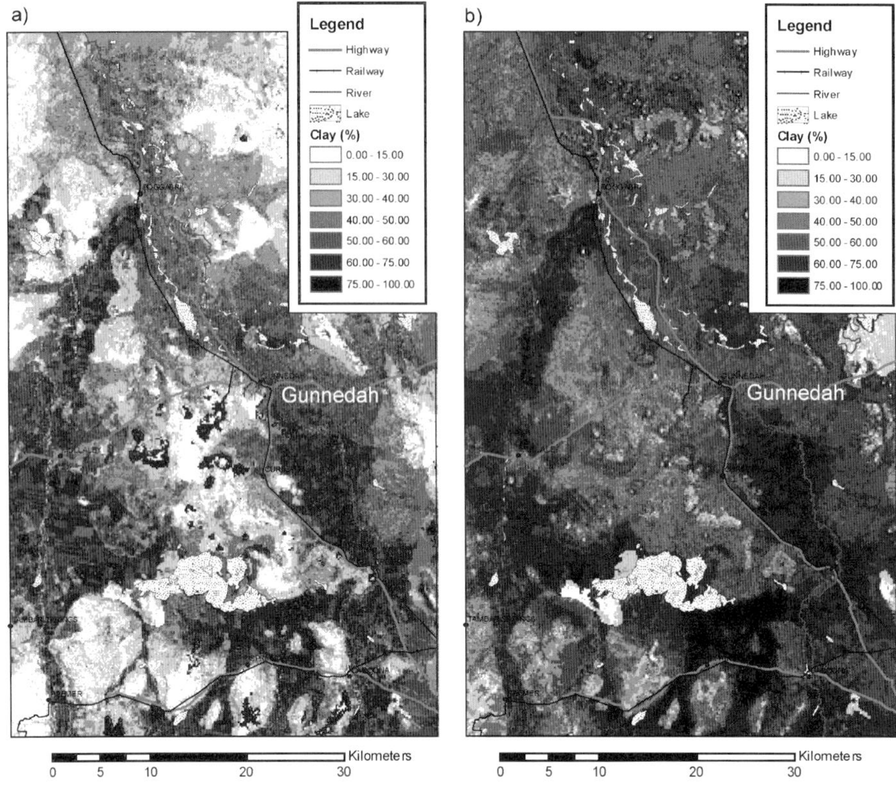

Figure 32.5. *(a) Topsoil Clay % 0–10 cm, (b) Subsoil Clay % 70–80 cm (scorpan-kriging).*

here) that there is a direct correlation between the locations of uniform soil types (Uf in the old Australian soil classification) and areas with medium-to-heavy clay; this is true for all of the percentage clay for the two other subsoil layers (not shown here due to lack of space), which is consistent with uniform clay content throughout the profile.

The natural drainage lines and accumulation points for both sediments and water are illustrated in Figure 32.5(a) and 32.5(b) with clearly defined stream lines apparent in both maps. Slope, elevation and profile curvature were all used in the multiple linear regression models and were well the dominant predictors of the particle-size fractions and it shows very well in the resulting digital attribute maps.

Salinity map layers: As is the case with physical properties, chemical properties play significant roles in the development of hydrological and catchment models. In Australia, efforts have been focused on understanding the relationship of salinity in relation to landscape using information about the extent and severity of the problem. Many studies have investigated salinity and hydrology simultaneously in order to understand their interactions, especially the incidence of dryland salinity in relation to landscape. In the examples provided here, the several chemical properties investigated, analysed and spatially predicted for the survey area were EC, Exchangeable Sodium Percentage (ESP), CEC, organic carbon, P and pH. We present the results of salinity as measured by EC. Since previously stated MLR predictions EC resulted in low R^2 values, the method was adopted for the chemical attributes as well. This method, however, was not good enough for EC with low R^2 ($R^2 = 0.11$ for 0–10 cm layer and 0.17 for 70–80 cm layer). However, in spite of relatively low R^2 for layers 10–20 and 30–40 cm, the MLR residuals exhibited sufficient spatial correlation for *scorpan*-kriging to be used. The MLR results were accepted for these layers, however. For layers 0–10 and 70–80 cm, *scorpan*-kriging was used to produce good EC maps. The resulting EC layers for two depths, 0–10 cm and 70–80 cm, are shown in Plate 32 (see Colour Plate Section). Obviously, the top layers (Plate 32a, see Colour Plate section) exhibit less variation than the subsoil layer as shown in Plate 32b (see Colour Plate Section). Additionally, it is obvious from that that EC increases with soil depth, which is not surprising. The increased salinity with depth could be ascribed to groundwater accretion due to land clearing, especially on the hill slopes. These data layers thus provide valuable information on the hydrological processes taking place and therefore could be used for the management of salinity in the area. It is also noteworthy to point out the relatively high salinity in the subsoil (70–80 cm) in the plains north of Gunnedah and around Boggabri and further south in the Liverpool plains around Curlewis and Breeza (Plate 32b, see Colour Plate Section). These areas are major agriculturally productive farmsteads of the upper Namoi region. It points to the importance of targeting resources to managing salinity in these high spots.

32.4 Concluding remarks

The development of more sophisticated hydrological models in recent years has been a cause of a mixture of optimism and debate amongst hydrologists, physical geographers and environmental scientists. Models have been developed from a range of hypotheses based on conceptual type and distributed models to the latest generation of physically-based distributed and parsimonious physically-based models. Many of these models rely heavily on empirical or statistical relationships among component characteristics of the catchment. It is, therefore, required that the physical characteristics of the catchment that influence the physical processes are accurately observed and spatially modelled to reflect the physical variability within the catchment. The work reported here is a contribution towards this requirement. This is particularly important in Australia where management of natural resources are increasingly being considered within each catchment, with the catchment communities given greater responsibility for the management of their resources. This has largely being organised through Catchment Management Authorities. This study that covers part of the Namoi river catchment in northern NSW, has undoubtedly contributed towards achieving the goal of having accurate soil and related land resource information for catchment planning and modelling for forecasting the impact of resource use. As potential input to catchment and hydrological models, the digital soil information resulting from this study, readily accessible as GIS layers, has the potential to enhance the modelling process and would lead to informed decision regarding resource management within the catchment.

References

Atkinson, P.M., Foody, G.M., 2002. Uncertainty in remote sensing and GIS: fundamentals. In: G.M. Foody and P.M. Atkinson (Eds.), Uncertainty in Remote Sensing and GIS. Chichester, Wiley.

Banks, R.G., 1995. Soil Landscapes of the Curlewis 1:100 000 Sheet Report. New South Wales Department of Conservation and Land Management, Sydney.

Bishop, T.F.A., McBratney, A.B., 2001. A comparison of prediction methods for the creation of field-extent soil property maps. Geoderma 103, 149–160.

Bui, E.N., Moran, C.J., 2001. Disaggregation of polygons of surficial geology and soil maps using spatial modelling and legacy data. Geoderma 103, 79–94.

Bureau of Meteorology., 2003. http://www.bom.gov.au/climate/averages.shtml.

Burrough, P.A., 1986. Principles of Geographical Information Systems for Land Resources Assessment. Clarendon Press, Oxford.

Burrough, P.A., McDonnell, R.A., 1998. Principles of Geographical Information Systems. Oxford University Press, Oxford.

Cihlar, J., Latifovic, R., Chen, J., Beaubien, J., Li, Z., Magnussen, S., 2000. Selecting representative high resolution sample images for land cover studies. Part 2: Application to estimating land cover composition. Remote Sensing Environ. 72, 127–138.

Crawford, K.L., 1976. Topography, in Gunnedah District Technical Manual. P.E.V. Charman (Ed.). Soil Conservation Service of NSW, Sydney.

Duggin, J. and Allison, P.N., 1984. The Natural Grasslands of the Liverpool Plains, NSW. Department of Ecosystem Management, University of New England, Armidale.

NSW Environmental Protection Authority, 1997. Proposed Interim Environmental Objectives for NSW Waters: Namoi Catchment. Environmental Protection Authority, Sydney.

ESRI, 2004. ARC/INFO User's manual. Environmental Systems Research Institute, Redlands, Ca. Supplemented by ARC/INFO Online help.

Evans, I.S., 1980. An integrated system of terrain analysis and slope mapping. Geomorphol. Supply 36, 274–295.

Good, P.I., 1999. Resampling Methods: A Practical Guide to Data Analysis. Birkhauser, Boston.

Hird, C., 1976. Climate, in Gunnedah District Technical Manual. P.E.V. Charman (Ed.). Soil Conservation Service of NSW.

McBratney, A.B., Mendonca Santos, M.L., Minasny, B., 2003. On digital soil mapping. Geoderma 117, 3–52.

McBratney, A.B., Odeh, I.O.A., Bishop, T.F.A., Dunbar, M.S., Shatar, T.M., 2000. An overview of pedometric techniques for use in soil survey. Geoderma 97, 293–327.

Moore, I.D., Gessler, P.E., Nielsen, G.A., Peterson, G.A., 1993. Soil attribute prediction using terrain analysis. Soil Science Society of America. J. 57, 443–452.

Northcote, K.H., 1979. A Factual Key for the Recognition of Australian Soils, 4th ed. Rellim Technical Publications, Adelaide.

Odeh, I.O.A., Chittleborough, D.J., McBratney, A.B., 1991. Elucidation of soil landform interrelationships by canonical ordination analysis. Geoderma 49, 1–32.

Odeh, I.O.A., Chittleborough, D.J., McBratney, A.B., 1994. Spatial prediction of soil properties from landform attributes derived from digital elevation model. Geoderma 63, 197–214.

Odeh, I.O.A., McBratney, A.B., 2000. Using AVHRR images for spatial prediction of clay content in the lower Namoi valley of eastern Australia. Geoderma 97, 237–254.

Odeh, I.O.A., McBratney, A.B., Chittleborough, D.J., 1995. Further results on prediction of soil properties from terrain attributes: heterotopic cokriging and regression-kriging. Geoderma 67, 215–225.

Odeh, I.O.A., Todd, A.J., Triantafilis, J., 2003. Spatial prediction of soil particle-size fractions as compositional data. Soil Sci. 168, 501–515.

Oldak, A., Jackson, T.J., Pachepsky, Y., 2002. Using GIS in passive microwave soil moisture mapping and geostatistical analysis. Int. J. Geogr. Inf. Sci. 16 (7), 681–698.

Petit, C.C., Lambin, E.F., 2001. Integration of multi-source remote sensing data for land cover change detection. Int. J. Geogr. Inf. Sci. 15, 785–803.

Rayment, G.E. and Higginson, F.R., 1992. Australian Laboratory Handbook of Soil and Water Chemical Methods, Vol. 3. Inkata Press, Melbourne.

Stace, H.C.T., Hubble, G.D., Brewer, R., Northcote, K.H., Sleeman, J.R., Mulcahy, M.J., Hallsworth, E.G., 1968. A Handbook of Australian Soils. Rellim Technical Publications, Glenside, Australia.

Voltz, M., Webster, R., 1990. A comparison of kriging, cubic splines and classification for predicting soil properties from sample information. J. Soil Sci. 41, 473–490.

Webster, R., Oliver, M.A., 1990. Statistical Methods in Soil and Resource Survey. Oxford University Press, Oxford.

Webster, R., Oliver, M.A., 1992. Sample adequately to estimate variograms for soil properties. J. Soil Sci. 43, 177–192.

Wiles, L., 1996. Coal Resource Audit of the Gunnedah Basin. New South Wales Department of Mineral Resources, Sydney, 319 pp.

Developments in Soil Science, volume 31
P. Lagacherie, A.B. McBratney and M. Voltz (Editors)

Chapter 33

COMPARING DISCRIMINANT ANALYSIS WITH BINOMIAL LOGISTIC REGRESSION, REGRESSION KRIGING AND MULTI-INDICATOR KRIGING FOR MAPPING SALINITY RISK IN NORTHWEST NEW SOUTH WALES, AUSTRALIA

J.A. Taylor and I.O.A. Odeh

Abstract

In Australia, soil salinity is one of the most devastating forms of land degradation facing agricultural production. In northern New South Wales (NSW), where irrigated-cotton production is a dominant agricultural commodity, salinity even though is not currently such a serious problem, could pose a potential threat to cotton production. This chapter explores the use of discriminant analysis (DA) and other alternative models for using ancillary data to create threshold-based risk maps (rather than continuous maps) of soil salinity and compares these maps with more traditional indicator predictions. The opportunity for incorporating an error analysis into DA is also explored. The hypothesis for including the error analysis in DA and related models is that a certain level of information remains in the residuals of the initial model and that this information could be extracted and analysed to improve the final prediction. The ancillary-based models used in this chapter did not produce better predictions than a Multi-Indicator Kriging (MIK) approach that did not include the ancillary data. The DA models produced better prediction than the Multi-Variate Stepwise Linear Regression with the Kriging (MSLR-K) and Binomial Linear Regression models. A benefit of the DA approaches is the generation of probability maps, similar to MIK, that may be used in future decision making.

33.1 Introduction

In Australia, irrigated cotton is predominantly grown in the northwest regions of New South Wales (NSW), Australia. This is primarily due to the prevalence of heavy clay soils and the availability of irrigation water from dams located in the Northern Tablelands. Traditionally flood irrigation has been used as it is most cost-efficient when water is cheap and plentiful. Flood irrigation, however, is an inefficient means of water usage as it may lead to a large loss of water into groundwater. Throughout Australia considerable secondary salinisation has occurred in irrigated regions due to rising watertables bringing subsoil salts into the vadose zone. While salinity is not currently widespread in the northwest of NSW, the presence of some salt outcrops in the river valleys has raised concerns within the cotton industry. Over the past decade, the industry has funded numerous soil surveys within the cotton production areas of NSW (including the Namoi and Gwydir valleys) to test for soil and production sustainability. One of the primary objectives was to map soil properties, particularly salinity (ECe), across these regions to facilitate management and minimise the environmental footprint of the industry. The aim of this chapter is to investigate alternative options for constructing soil ECe (salinity) hazard maps across two neighbouring river valleys, the Gwydir and Namoi (~1.2 million ha), at a resolution of 200 m^2.

33.2 Methods

33.2.1 Available data

A 200-m resolution digital elevation model (DEM) constituting part of the Murray–Darling Basin Soil Information System, was obtained for the survey area from CSIRO Land and Water. This DEM formed the basis of the interpolation grid of attribute maps for the study area. The elevation data was used to derive primary (slope, aspect, flow direction, upslope area, downslope area) and secondary (Compound Topographic Index, Topographic Wetness Index) terrain attributes.

Landsat imagery was purchased from the Australian Centre for Remote Sensing (ACRES) for the two catchments as a single scene captured in January 1999. An aerial γ-radiometric survey was also obtained from NSW Department of Mineral Resources that comprises several private surveys. This contained information on the Th, Ur, K and total counts covering the catchments. A large number of ancillary variables were available – some of which were strongly correlated and some had missing values. Several of the available ancillary variables were culled and the predictors used in this study are listed in Table 33.1.

Soil surveys have been conducted in the two catchments at various times over the past 20 years. The Edgeroi dataset was derived initially from an equilateral triangular grid survey conducted in 1985 (McGarry et al., 1989). The data

Table 33.1. *Ancillary variables selected for the prediction of ECe risk.*

Coordinates	DEM attributes	Landsat attributes	γ-Radiometric
Eastings	Elevation	Blue	K radiometrics
Northings	Slope	Green	
Eastings × Eastings	Aspect	Red	
Northings × Northings	Topographic Wetness Index	Near-infra red	
Eastings × Northings		Short-wave IR	
		Thermal IR	

was subsequently supplemented by a further 100 site-directed soil samples. In 1997, a site-directed survey was conducted in the lower Namoi valley to the west of the Edgeroi area comprising 125 sites (Odeh and McBratney, 2000). Similarly 153 sites were identified in the Gwydir valley and sampled in 1996 (Singh et al., 2003). While the depth of sampling varied between surveys (and sometimes between sites), all samples were surveyed to a minimum depth of 110 cm. For this study, data from the topsoil (0–10 cm) and subsoil (100–110 cm) was utilised. Various soil properties were measured in the laboratory including particle size analysis, pH and EC in 1:5 soil:water extract. For this investigation, a predicted hazard map of ECe (not EC), with areas above and below a threshold value of 6 dS/m that is the tolerant level for growing cotton, was the desirable outcome. Thus the conversion of the EC 1:5 soil:water value into ECe was needed. This was done using the texture-based lookup table in SOILpak that was developed for the Australian Cotton CRC (Daniells and Larsen, 1991). The data was analysed for unusual or erroneous values at each sites and trimmed if such data was found. This resulted in 573 sites of which 50 were randomly removed to form a validation set leaving 523 in the prediction data set. (Fig. 33.1)

33.2.2 Interpolation approaches

To map threshold values of a spatial process, Indicator Kriging (IK) (Goovaerts, 1997) or Multi–Indicator Kriging (MIK) (Cattle et al., 2002) is usually the preferred method of interpolation. A MIK approach using a global variogram was performed on the data with ECe thresholds set at increments of 2 from 2 to 20, increments of 5 from 30 to 70, increments of 10 from 70 to 100 then at 150 and 200. A monotonic regression was fitted to derive the conditional cumulative distribution function (CCDF). The CCDF was solved for the probability of an ECe value of 6 dS/m (see Cattle et al. (2002) for a full description of the method). The probability map was converted into a threshold map by assigning locations with a probability 0.5 the nominal value '>6' and for probabilities <0.5 the value '<6'.

However, as previously stated, a number of ancillary data layers are available for the study area and it may be preferable to incorporate these ancillary data into the interpolation process. This may be done by co-kriging, indicator co-kriging or regression-kriging. Co-kriging techniques (including indicator co-kriging) are computationally large and usually restricted to only a few variables (Odeh et al., 1995; Goovaerts, 1997)). Previous studies (Odeh et al., 1995; Knotters et al., 1995; Goovaerts, 1999; Bishop and McBratney, 2001) have highlighted the benefits of regression kriging techniques when large amounts of ancillary data are available. With up to 13 ancillary variables available for this study (Table 33.1) regression kriging should be preferable to co-kriging or indicator kriging techniques.

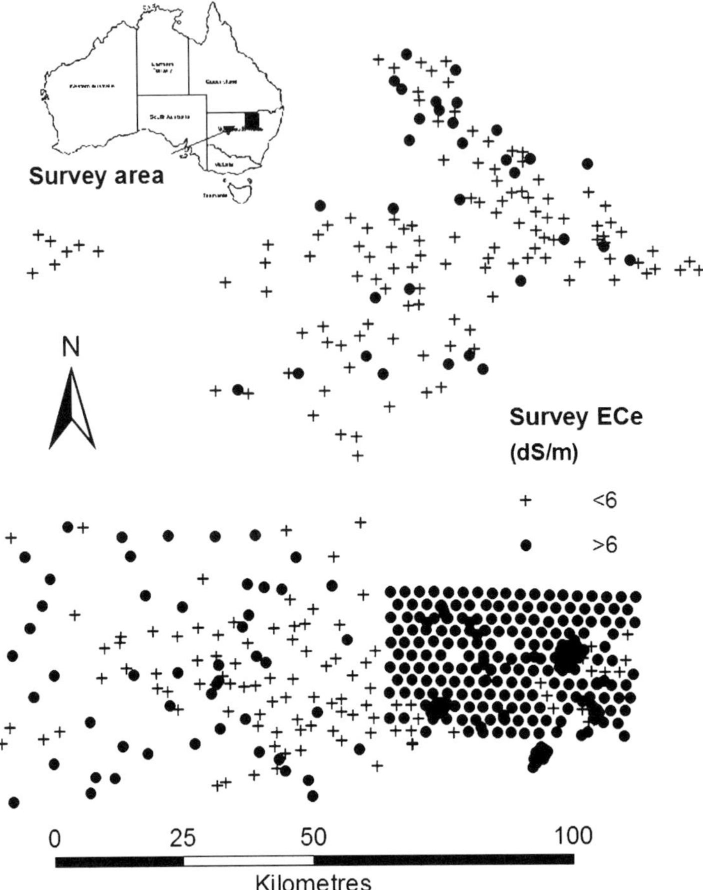

Figure 33.1. *Location of soil survey area and location of soil survey points within study area with mark symbols differentiation sites <6 dS/m and >6 dS/m.*

Initially the regression kriging approach of Odeh et al. (1995) involving Multi-Variate Stepwise Linear Regression with Kriging of the regression residuals (MSLR-K), was applied to the data. This was done with the raw ECe data, which produced a continuous map of EC across the study area that was subsequently converted into a threshold map. The resultant continuous and threshold maps are incoherent with expert expectations of the study area. The raw ECe data is positively skewed and the data was log-transformed and the process repeated on the transformed data (logMSLR-K) before being back-transformed and mapped. This produced a more coherent result, however there are limitations associated with this approach as

(1) the sampling scheme was not designed to adequately characterise the range of environmental variation (leading to over- and under-prediction in the regression equations). This may be overcome by improved sampling designs (e.g. Chapter 32) or
(2) the data is too sparse and the kriged residuals do not exhibit enough detail to correct the regression errors.

An alternative approach to utilise the ancillary data, to increase the effective data density, is proposed using discriminant analysis (DA) combined with an analysis of the errors from the DA. It is hypothesised that the error (residuals) of the initial model still contains information that can help improve our prediction. The ECe data was reorganised from a continuous dataset into a new dataset (Y_1) with two nominal data ranges, <6 dS/m and >6 dS/m. DA was performed and the inverse discriminant function (IDF) applied to the entire ancillary data set to produce a prediction (y_1) of the respective areas of <6 dS/m and >6 dS/m and an associated probability of prediction ($p(y_1)$).

A second nominal variable (Y_2) was derived indicating which samples sites were correctly and incorrectly classified, that is an error dataset from the DA. Y_2 contains three classes,

Type I errors (Y_{1i} (<6 dS/m), y_{1i} (>6 dS/m)) denoted as $Y_{2\ \mathrm{I}}$,
Type II errors (Y_{1i} (>6 dS/m), y_{1i} (<6 dS/m)) denoted as $Y_{2\ \mathrm{II}}$ and
Type III a correct class (Y_{1i} (<6 dS/m), y_{1i} (<6 dS/m)) or (Y_{1i}(>6 dS/m), y_{1i} (>6 dS/m)) denoted as ($Y_{2\ \mathrm{III}}$).

DA of Y_2 and the application of the resultant IDF to the ancillary data produced a misclassification prediction (y_2) and associated probabilities ($p(y_{2\ \mathrm{I}})$, $p(y_{2\ \mathrm{II}})$ and $p(y_{2\ \mathrm{III}})$).

Since a probability of classification ($p(y_1)$) and probability of misclassification ($p(y_2)$) at each site exist, these were analysed in combination to improve the initial salinity hazard prediction. To achieve this, an arbitrary threshold probability value of 0.65 was used; however, this value could be optimised. Thus if the probability of misclassification is greater than the probability of classification ($p(y_2) > p(y_1)$) and the probability of misclassification exceeds a defined threshold (in this case $p(y_2) > 0.65$) then the initial classification is deemed incorrect and the prediction reversed. In all other cases, the initial classification is considered correct. The incorporation of the misclassification (residual) analysis with the original analysis, we term discriminant and error analysis (DAEA).

A Binomial Logistic Regression (BLR) was also applied to the data. This was performed in *JMP*™ using the same predictors as the DA approach. Similarly to DA the BLR approach provides a probability of correct classification. This permitted the error analysis (described above) to be applied to the BLR output again using a threshold probability of 0.65. This approach was termed Binomial Logistic Regression with Error Analysis (BLREA).

To validate the results of the MSLR-K, logMSLR-K, MIK, DA, DAEA, BLR and BLREA models an independent validation subset of 50 soil sample points, with associated ancillary data, was randomly omitted from the initial model

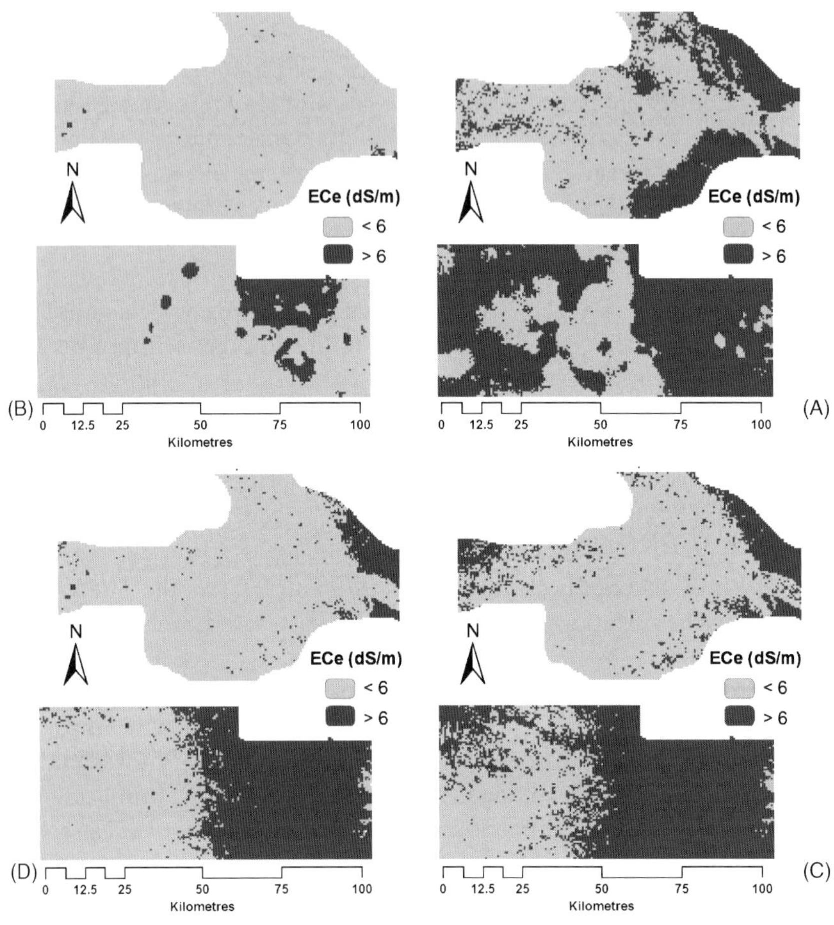

Figure 33.2. *Predictions of ECe threshold (6 dS/m) using (A) Multi-Variate Stepwise Linear Regression with the Kriging (MSLR-K), (B) logMSLK-R, (C) discriminant analysis (DA), (D) Binomial Logistic Regression (BLR), (E) discriminant and error analysis (DAEA) (error threshold 0.65), (F) Binomial Logistic Regression with Error Analysis (BLREA) (error threshold 65), (G) DAEA (error threshold 0.60) and (H) Multi-Indicator Kriging (MIK).*

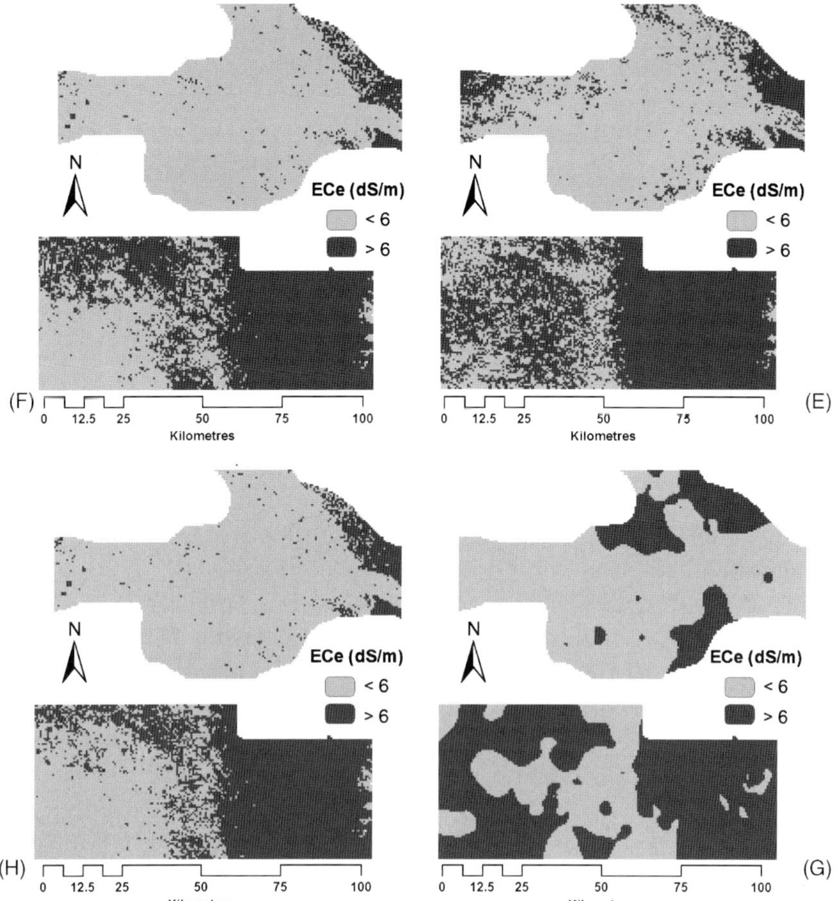

Figure 33.2 (*Continued*)

development. The models were subsequently applied to the validation points to determine the accuracy of prediction.

33.3 Results and discussion

The topsoil samples did not exhibit any sign of salinity and all the calculated ECe values were below the threshold value of 6 dS/m. No maps of topsoil ECe were produced. The subsoil exhibits some salinity risk with 62% of the soil cores showing ECe > 6 dS/m at 1 m.

Threshold maps of subsoil ECe produced using MSLR-K, logMSLR-K, DA, BLR and MIK are shown in Figure 33.2. The DAEA and BLREA maps adjusted for the error analysis are also shown in Figure 33.2. Table 33.2 shows the

Table 33.2. *Percentage misclassification of ECe using a different prediction models.*

Model	Misclassification (%)	
	Prediction set	Validation set
MIK		18.0
MSLR-K		42.0
logMSLR-K		24.0
DA	21.3	22.0
DAEA*	19.5	22.0
BLR	23.5	26.0
BLREA*	21.3	28.0

Note: MIK, Multi-Indicator Kriging; MSLR-K, Multi-Variate Stepwise Linear Regression with the Kriging; DA, discriminant analysis; DAEA, discriminant and error analysis; BLR, Binomial Logistic Regression; BLREA, Binomial Logistic Regression with Error Analysis.
*Threshold of 0.65 used for DAEA and BLREA error analysis.

misclassification % exhibited by the different approaches in both the prediction and validation data sets.

While the MIK analysis produced the lowest misclassification in the validation set, the MSLR-K on the untransformed data has the highest misclassification % indicating the effect of the skewed raw data on the regression analysis. When log-transformed the MSLR-K analysis misclassification % was nearly halved (logMSLR-K = 24%). The misclassification results for DA and DAEA are both better than BLR, BLREA, MSLR-K and logMSLR-K. The addition of the error analysis did not improve the DA misclassification and in the case of BLR actually led to an increase in the misclassification in the validation set. However, in the original model the error analysis improved misclassification in the prediction set.

The MIK analysis, which does not involve the use of ancillary variables, produced the lowest misclassification of 18% (Table 33.2). It is possible that indicator co-kriging may further improve the predictions, however, while this is valid in theory but it does not necessarily happen (Goovaerts, 1997, p. 298). The selection of suitable co-variates is also important as co-kriging approaches are computational expensive and limited to only a few variables (Odeh et al., 1995). The MIK analysis produces probability maps which are shown in Plate 33A (see Colour Plate Section), in addition to the threshold maps (Fig. 33.3) that can be used for risk assessment.

The DA approaches were marginally better than the logMSLR-K; however, the BLR approaches were the worst models (excepting the untransformed MSLR-K). The improvement in DA over MSLR-K indicates that there is some opportunity for the use of these functions in delineating threshold maps for soil properties, such as ECe. The inclusion of the proposed error analysis needs further investigation to identify if it is of benefit to the model. Jack-knifing or bootstrapping the data is the next step to test the validity of the DA approaches, especially the benefit of incorporating the error analysis. An arbitrary error threshold probability of 0.65 has been selected for this preliminary investigation and further work is required to identify the optimum threshold. In Figure 33.2, images E and G show the effect of varying this threshold value. Another contentious point is the selection of the predictor variables. Without the benefit of stepwise processes in the DA and BLR approaches "expert" opinion has been used in the selection of the predictor variables. This was a subject of the DSM workshop and while not discussed here it is acknowledged that this is an important factor in optimising predictions.

A benefit of the DA approaches is the generation of probability maps (Plate 33B (see Colour Plate Section)), similar to MIK, that may be used in future decision making. The error analysis also provides maps of where the initial model is predicting poorly or well. This may aid in additional sampling or the selection of predictor variables in future models. From the misclassification map (not shown), it appears that the coordinates and γ-radiometric data are strongly influencing the error analysis. The significant influence of coordinates (Easting and Northing) is indicative of some trend in the data. This trend is obvious in Fig. 33.2 and Plate 33, with salinity generally increasing from west to east. The approaches used here are all on the basis of the same set of predictor variables in the initial and error analysis. This may not be ideal and again the proposed methodology may benefit from the proper selection of predictor (ancillary) variables at all stages of the process. Alternative DA algorithms, for example Optimal Discriminant Analysis for Ordinal responses, have been shown to outperform normal discrimination (Coste et al., 1997) and may have an application in DSM.

33.4 Conclusions

This chapter explores alternative models for using ancillary data in creating threshold-based risk maps (rather than continuous maps) of soil properties using the example of soil ECe. The hypothesis for including the error analysis is that a certain level of information remains in the residuals of the initial model and that this could be extracted and analysed to improve the final prediction. The ancillary-based models used in this chapter did not produce better predictions than MIK approach that did not include the ancillary data. The DA models produced better prediction than the MSLR-K and Binomial Linear Regression models.

References

Bishop, T.F.A., McBratney, A.B., 2001. A comparison of prediction methods for the creation of field-extent soil property maps. Geoderma 103, 149–160.

Cattle, J.A., McBratney, A.B., Minasny, B., 2002. Kriging method evaluation for assessing the spatial distribution of urban soil lead contamination. J. Environ. Qual. 31, 1576–1588.

Coste, J., Walter, E., Wasserman, D., Venot, A., 1997. Optimal discriminant analysis for ordinal responses. Stat. Med. 16, 561–569.

Daniells, I., Larsen, D., 1991. SOILpak: A Soil Management Package for Cotton Production on Cracking Clays. NSW Agriculture, Orange NSW Australia.

Goovaerts, P., 1997. Geostatistics for Natural Resources Evaluation. Oxford University Press, New York.

Goovaerts, P., 1999. Using elevation to aid the geostatistical mapping of rainfall erosivity. Catena 34, 227–242.

Knotters, M., Brus, D.J., Oude Voshaar, J.H., 1995. A comparison of kriging, co-kriging and kriging combined with regression for spatial interpolation of horizon depth with censored observations. Geoderma 67, 227–246.

McGarry, D., Ward, W.T., McBratney, A.B., 1989. Soil Studies in the Lower Namoi Valley: Methods and Data: The Edgeroi Data Set Vols. 1 & 2. CSIRO, Australia.

Odeh, I.O.A., McBratney, A.B., 2000. Using AVHRR images for spatial prediction of clay content in the lower Namoi valley of eastern Australia. Geoderma 97, 237–254.

Odeh, I.O.A., McBratney, A.B., Chittleborough, D.J., 1995. Further results on the prediction of soil properties from terrain attributes: heterotopic cokriging and regression kriging. Geoderma 67, 215–226.

Singh, B., Odeh, I.O.A., McBratney, A.B., 2003. Acid buffering capacity and potential acidification of cotton soils in northern New South Wales. Aust. J. Soil Res. 41, 875–888.

Developments in Soil Science, volume 31
P. Lagacherie, A.B. McBratney and M. Voltz (Editors)

Chapter 34

FITTING SOIL PROPERTY SPATIAL DISTRIBUTION MODELS IN THE MOJAVE DESERT FOR DIGITAL SOIL MAPPING

D. Howell, Y. Kim, C. Haydu-Houdeshell, P. Clemmer, R. Almaraz and M. Ballmer

Abstract

We developed models from soil profile descriptions and GIS landscape analysis to estimate the spatial distribution of soil properties to assist soil scientists with soil-landscape information. Soil profile descriptions were obtained within soil survey projects in the Mojave Desert of southeastern California, USA. Sites were located on broad alluvial fans. Soil development varied from young soils with little or no soil development to well-developed soils on older alluvial fan remnants.

We obtained a set of profile descriptions ($n = 264$) from the traditional ongoing fieldwork. The location of these sample points was determined by soil scientist judgment of combinations of soil-forming factors. The project area is sparsely vegetated and access is relatively unimpaired in most areas. We feel that these purposive samples represent the range of the soil-forming factors and that sample location bias will be low. Although this bias is not measurable. We wanted to see if we could make use of these data. We developed models from these data and evaluated the performance of the models using the measured values at randomly located sites not used to fit the models. The models estimated selected soil characteristics continuously in a 30-m raster over the project area. The response variables that we modelled were soil genetic features that are used as diagnostic properties in USDA Soil Taxonomy, for example particle-size class, presence or absence of argillic horizon.

Soil profiles and landscape features were described at 97 randomly located field sites within a portion of the active soil survey project. Explanatory variable information was developed for each of these sites through GIS extraction from digital elevation model data, landform derivatives, band-ratio satellite images and geomorphologic data.

Model estimates for particle-size class were correct or within one class of the correct class for 73% of sample points. Models for depth to soil features had a range of performance. The best fitting model estimated the depth to secondary carbonates within 20 cm of actual depths for 71% of sample points, which contained carbonates. The model for depth to calcic horizon performed less well, while the model for depth to argillic was slightly less reliable. The model for presence or absence of calcic horizon was the most reliable logistic model.

Soils on millions of hectares will be mapped in this general area in the future and we are trying to increase mapping efficiency and depth of understanding of soil-landscape relationships. Model development techniques will be adapted and applied to adjacent areas in the future. Further work will require more field data (to document the response variables) and more complete soil-forming factor spatial data. New soil survey products may result from these continuous raster estimates of soil properties. These model outputs are intended to augment and guide field soil survey data collection, not replace it.

34.1 Introduction

In the western United States the spatial and attribute resolution of digital soil-forming factor data is still quite coarse. The explicit relationships between these explanatory variables and the resulting individual soil properties are not well understood in most areas. Despite several decades of worldwide research and development of GIS soil modeling methods, the outputs from these models are rudimentary information.

The spatial resolution of input data is on the order of 10–>30 m, while soil variation occurs at a scale as fine as 0.5 m. In addition, some of these data have been converted to raster format from large polygons, for example geology, which may to lead to an incorrect assessment of resolution. Attribute resolution is also out of sync, for example geologic attributes describe entire formations rather than individual rock types. It is beyond the resolution of the input soil-forming factor data and our explicit understanding of soil-forming relationships to attempt to create detailed, taxonomic (or even multi-property) soil maps directly from explanatory variable data. At this stage, we feel the appropriate goal for GIS soil-landscape modeling should be to produce maps showing estimates of important individual soil genetic features in order to increase understanding of soil-landscape relationships and to guide field data collection by soil scientists working on soil survey projects.

In this project, we have attempted to develop models to estimate the spatial distribution of individual soil genetic features. These features are defined by objective criteria in Soil Taxonomy (Soil Survey Staff, 1999). These genetic features are commonly used to classify soil pedons using a soil taxonomic system. These genetic features also serve as markers in the stages of development, or genesis, of soils. We have made no attempt to model these genetic features, as they would be combined in a taxonomic system. We feel that these individual genetic features are important measurable soil properties that influence soil use and management. Also, the relationships between each individual genetic feature and the soil-forming factors will be more direct than developing one relationship to model the variation of all of the soil properties together. Our models allow each feature to vary independently and continuously (at separate scales of variation) across the landscape as described by McKenzie, et al. (2000).

Soil survey depends on developing relationships between the soil-forming environment and the resulting soil properties. Dokuchaiev (Glinka, 1927), Hilgard (1914) and Jenny (1941) spoke of relationships between the soil-forming factors and soil properties. It is our interpretation that they were speaking specifically of soil properties rather than taxonomic classes (based on combinations of soil properties). Jenny in particular spoke of developing quantitative relationships. Modelling soil properties directly seems more appropriate than modelling taxonomic classes when using quantitative statistical models based on physical soil-forming processes. There have been many papers published on the use of GIS and statistical inference to develop these relationships. We refer readers to the recent complete review and framework proposed by McBratney et al. (2003) and in Chapter 1.

Our focus was on the development of GIS tools for production soil survey, not research. Our statistical modelling methods draw heavily on the work of others, Webster and Burrough (1972), McKenzie and Austin (1993), McSweeney et al. (1994), Gessler et al. (1995) and McKenzie and Ryan (1999).

34.2 Materials and methods

34.2.1 Study area

The study site is located in the western Mojave Desert approximately 160 km northeast of Los Angeles, California, USA (Fig. 34.1). The study site receives 76–127 mm of rain per year with the majority falling between November and March. Summer precipitation is common after convection storms. Elevation ranges from 180 to 1425 m. Sampling sites were located on broad alluvial fans and associated landforms. Soil development varied from young soils with little or no soil development (Typic Torriorthents) to soils consisting of older well-developed fan remnants underlying younger, more recently deposited alluvial material (Argidic Argidurids). Vegetation communities are dominated by arid climate shrubs such as *Larrea tridentata* (Sessé & Moc. ex DC.), Coville (creosote bush) and *Ambrosia dumosa* (Gray) Payne (white bursage) with *Yucca schidigera* Roezl ex Ortgies (*Mojave yucca*) and *Yucca brevifolia* Engelm. (Joshua tree) occurring in some areas (USDA, 2004a).

34.2.2 Attribute selection

The soil attributes we modelled were soil genetic features such as: presence or absence of argillic horizon, secondary carbonates, calcic horizon, durinodes, duripan and separate (continuous) models estimating the depth to the occurrence of these features. We also estimated particle-size class.

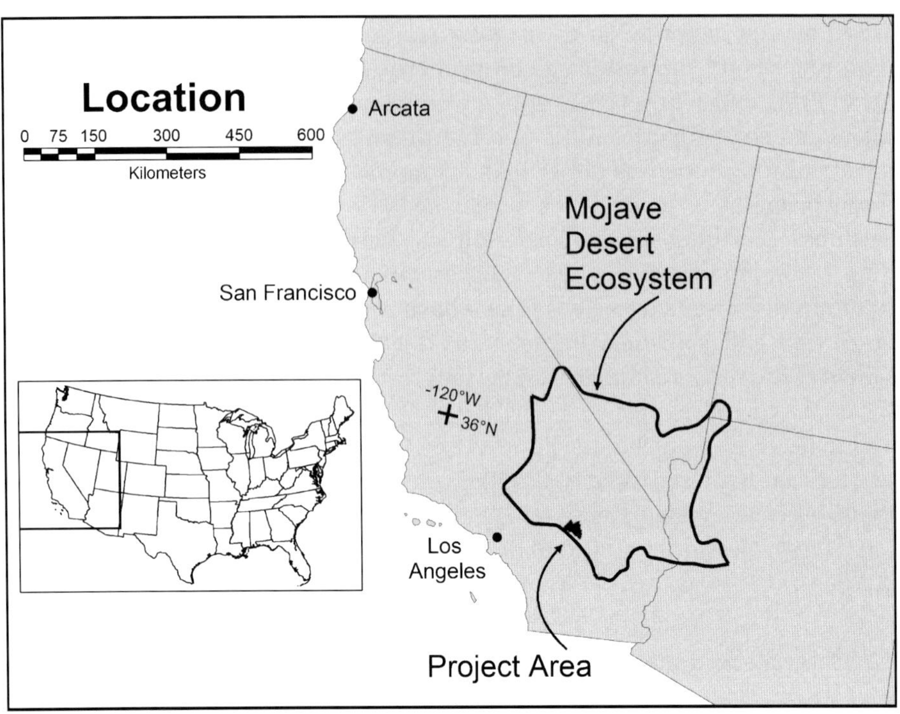

Figure 34.1. *Location of the project area in the western United States.*

The resolution of the model spatial input and output data is 30 m. The entire study area is approximately 77,280 ha.

34.2.3 Digital spatial data

The data layers used to represent the soil-forming environment were DEM and derivatives, band-ratioed Landsat Thematic Mapper (TM) imagery (Clemmer, 2003) and geomorphology (U.S. Army Topographic Engineering Center and Louisiana State University, 2000). See Table 34.1.

For taxpartsize the classes are indicated by a number code used in the National Soil Information System (USDA, 2004b). The classes are: 30 sandy-skeletal, 33 loamy-skeletal, 40 sandy, 44 loamy, 46 coarse-loamy, 50 coarse-silty, 54 fine-loamy, 59 fine-silty, 63 clayey and 69 fine. These class numbers are ordinal. Low numbers are coarse and high numbers are fine textures.

The compound slope shape data were derived from DEM data. The plan and profile curvatures were calculated as floating point values using the CURVATURE routine in ArcInfo (Environmental Systems Research Institute, 2002). These curvature numbers were evaluated against a digital raster graphic

Table 34.1. *Response and explanatory variables.*

Variable names	Description[1]
Response	
Argillic	Argillic (clay accumulation) horizon in the soil: Yes = present, No = absent.
Argillic depth	Depth to the top of argillic horizon
Calcic	Calcic (carbonate accumulation) horizon: Yes = present, No = absent.
Calcic depth	Depth to the top of calcic horizon
Carbonates	Secondary carbonates: Yes = present, No = absent.
Carbonate depth	Depth to the top of accumulation of secondary carbonates
Durinodes	Durinodes (silica masses): Yes = present, No = absent.
Durinode depth	Depth to the top of accumulation of durinodes
Duripan	Duripan (silica cemented layer): Yes = present, No = absent.
Duripan depth	Depth to the top of duripan
Taxpartsize	Particle-size class: 30, 33, 40, 44, 46, 50, 54, 59, 63, 69
Explanatory	
Gisaspect	Slope direction: −1 to 360 DEM derivative
Giselev	Elevation above sea level (in metres) DEM derivative
Gisplan	Plan slope curvature (across the slope) DEM derivative
Gisprof	Profile slope curvature (up and down the slope) DEM derivative
Gisshape	Compound slope shape class (*categorical*) DEM derivative reclassification 9 classes for combinations of concave, linear and convex
Gisslope	Slope steepness in percent DEM derivative
Ratio_band1	Reflectance value for band 1 TM band ratio band 3/band 2
Ratio_band2	Reflectance value for band 2 TM band ratio band 3/band 7
Ratio_band3	Reflectance value for band 3 TM band ratio band 5/band 7
Landform1	Geomorphic landform general (*categorical*)

[1]See Soil Taxonomy (Soil Survey Staff, 1999) for soil definitions.

topographic map to assign these values to three classes. Plan curvature was reclassified to concave, linear or convex based on a subjective comparison with the contour lines. The same process was carried out for the profile curvature. The three shape classes for each direction were added together to form nine possible classes of compound curvature.

The soil enhancement band ratio product was processed using Landsat TM imagery acquired in August of 1993. There was some cloud contamination of the image, however this only appeared to effect the results in localised areas. The ratio composite was developed from research conducted previously in arid areas in Utah by the USDI Bureau of Land Management, National Science and Technology Center.

Although extensive accuracy assessment has not as yet been accomplished, soil scientists in the Utah studies found this product to be useful in delineating and pre-mapping soil polygons. The product, along with other ancillary data, helped to plan field sampling, find discrete changes in soil make up, and was very useful in helping to map remote and more inaccessible areas in difficult terrain. Although vegetation is directly linked to soil type and setting, this methodology appears to be most useful in arid areas where there is little interference from vegetation canopies and where more bare soil is exposed.

The indexing ratio uses bands 2, 3, 5 and 7 of the image and is usually displayed in the following colour gun assignments: (Red) 3/2, (Green) 3/7 and (Blue) 5/7.

In the Utah studies, the 3/2 component was indicative of carbonate radicals (e.g. caliche and limestone); the 3/7 component seemed to indicate ferrous iron; while the 5/7 component was indicative of hydroxyl radical (e.g. clay).

The geomorphology and earth material data have been developed for the entire Mojave Desert region. This will form an important consistent data layer for modeling in this area. In some areas, it does not register well with the apparent landforms. It was developed at a smaller scale (1:100,000) than we are using it at for soil survey work (1:24,000). The models derived from these data have the same apparent misplacements in some areas.

34.2.3 Point data

Two sets of soil point data were obtained from the project area. Field soil profile descriptions were used to characterise soil properties at each location. In order to simplify the models all sample points on mountain landforms were excluded and the models were not estimated for those areas.

Soil profiles were described at 97 randomised locations (randomly generated UTM coordinates) within a portion of an ongoing soil survey project. We refer to these as random data points.

Profile descriptions were obtained from the ongoing soil survey ($n = 264$) (Haydu-Houdeshell, 2004). The locations for these data were selected in a traditional manner by the judgment of the soil scientists to represent particular sets of soil-forming factors. These purposively located sample points are sometimes called judgment samples. We will refer to these as purposive data points (Fig. 34.2).

The project area is sparsely vegetated and access is relatively unimpaired in most areas. We feel that these purposive samples represent the range of the soil-forming factors and that sample location bias will be low. Although this bias is not measurable. We wanted to see if it was possible to use these purposive data points to fit models for the entire project area. We wanted to make use of these extensive data. We refer to these models as models fit to the purposive data.

We used the UTM location coordinates of the random data points to extract the estimated soil property values from the models fit to the purposive data. A

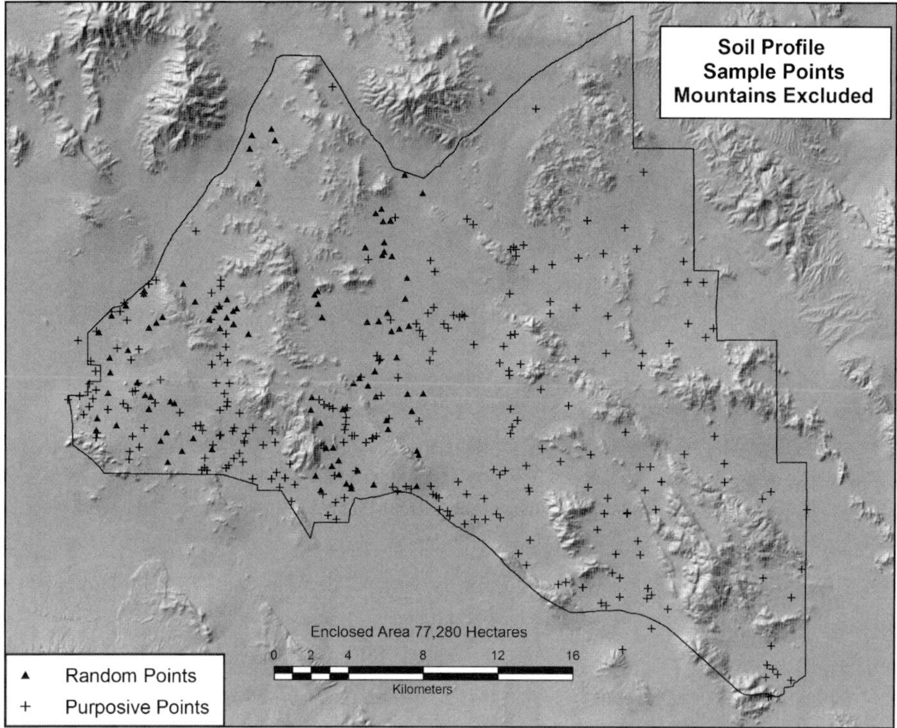

Figure 34.2. *Location of soil profile sample points.*

comparison was made of these extracted purposive model estimates to the actual measured values at the random data point locations (Environmental Systems Research Institute, 1998).

A range of methods was applied to develop optimal models. For continuous response variables, such as depth from the surface to a feature, generalised linear models were used after thorough investigation of optimal Box–Cox transformations on variables and multi-collinearity structures among variables (SAS, 2001). We then compared performance of models on the purposively collected dataset via maps, graphs and statistical summary tables. Various model selection criteria and diagnostic measures were used for these comparisons. We also compared the performance of logistic models on the binomial (presence/absence) variables after transformation and multi-collinearity checks.

34.3 Results and discussion

The resulting models and significant terms are listed below. Box–Cox transformation routines determined that the square root transformation for Gisslope,

Table 34.2. *Models and significant terms.*

GLM models

	Overall F	R^2	Significant terms
Summary of GLM models for the *purposively* collected dataset			
Model 1: carbonates-depth$^{-1/2}$	3.60***	0.128	Giselev***, Ratio_band1***, Landform1**
Model 2: durinodes-depth	3.25***	0.163	Ratio_band2, Gisshape***
Model 3: argillic-depth	3.68***	0.114	Ratio_band2, Landform1***
Model 4: calcic-depth$^{-1/2}$	3.23***	0.268	Gisshape***, Landform1***
Model 5: taxpartsize	21.48***	0.432	Giselev, Ratio_band1***, Landform1***

Logistic models

	Overall χ^2	% Concordant	Significant terms
Summary of logistic models for the *purposively* collected dataset			
Model 1: calcic	10.31***	59.8%	Gisplan**, Gisslope$^{-1/2}$**
Model 2: argillic	20.92***	66.0%	Giselev***, Ratio_band1**, Ratio_band2***
Model 3: duripan	15.24***	69.0%	Ratio_band1***, Ratio_band2***
Model 4: durinodes	40.29***	72.7%	Giselev*, Gisplan***, Ratio_band1***, Ratio_band2***
Model 5: carbonates	33.43	78.4%	Giselev***, Gisplan***, Gisprof*, Gisslope$^{-1/2}$**

***p-value < 0.01.
**p-value < 0.05.
*p-value < 0.1.

carbonates depth and calcic depth is optimal. We found that model fitting, model assumptions and various diagnostics were better after the transformation (Table 34.2).

The models with the most reliable estimates are given in the Table 34.3.

The particle-size class comparison is for class assignment. The model estimate was assigned to the nearest class. For comparison purposes a particle-size class model was also fit to the randomly located points. The assigned class was correct or within one class of the correct class for 73% of the sample points using the model fit to the purposive points. The estimates were within two classes of the correct class for 92–97% of the sample points over an area of 77,280 ha. See Plate 34 (see Colour Plate Section) for a map displaying the output from the purposive data model and a graphical comparison of the model fit to the random points and the model fit to the purposive points.

Table 34.3. *Comparison of model estimates to actually measured soil properties.*

				Number of Classes Estimate Missed By					
Model Fit on Data Points	*n*	Compared to Actual Values at Points	Correct Class %	1 %	2 %	3 %	4 %	5 %	6 %
Particle-size Class									
Purposive	264	Random	24	49	21	3	2	1	0

Model Fit on Data Points	*n*	Compared to Actual Values at Points	% Estimates within 0 to 10 cm	% Estimates within 10 to 20 cm	% Estimates within 20 to 30 cm	% Estimates missed by >30 cm
Carbonate Depth						
Purposive	230	Random	48	23	17	12

Note: *n* controlled by occurrences of each feature in the dataset.

The evaluations of models for continuous estimates of depth to a certain genetic feature show a range of model performance. The model for depth to secondary carbonates performed best. The estimates were within 20 cm of actual measured values at the random sample points for 71% of these points.

The models based on logistic regression for the presence or absence of features are harder to evaluate. The best performance was for calcic horizon. For model probability estimates equal to or greater than 0.5 (4 points) 75% contained a calcic horizon. For model probability estimates less than 0.5 (93 points) 62% of these points did not have calcic horizons. The model for probability of argillic horizon performed next best, where 56% of points with estimated probability of an argillic horizon equal to or greater than 0.5 actually had one. Space limitations do not allow us to show graphs and maps displaying these results. More work is needed to improve these models.

34.4 Conclusions

We feel that the results of these models based on these initial data show promise as a tool for pre-mapping products for production soil survey. They will serve as a framework for visualising soil-landscape relationships. They will also serve as a guide for fieldwork that will increase efficiency.

In the future as we proceed to map the Mojave Desert area, we will gather more point data. We hope to improve these models as the increased data better represents soil-forming factor variation. The models will be applied to subsequent mapping areas as pre-mapping estimates of the distribution of soil properties.

We also think that additional soil-forming factor proxies may improve the models. We hope to test hyperspectral imagery, LiDAR elevation data and gamma-ray imagery as described by Wilford in Chapter 16.

The partnership of a statistician with GIS and soil science workers is a practical team for developing explicit digital soil-landscape relationships. We feel that it is critical for soil scientists to guide the development of these statistical soil-landscape models as a tool for soil survey. But we do not feel that the modelling effort has to be designed so that only soil scientists are working on it. The statistical modelling expertise provided by a statistician is a valuable contribution and can be found at universities throughout the USA near most soil survey offices. Implementing the models in GIS should be performed by soil scientists with appropriate academic preparation and experience in GIS. The application of digital mapping tools will become universal in soil survey.

The application of raster data for soil survey pre-mapping analysis is useful. As explicit digital soil-landscape models improve some of these raster products may soon stand on their own as new soil survey products.

Future work will focus on increasing the number of soil properties evaluated for the larger purposive point dataset. We will also look into combining the binomial models (presence/absence) of features with the estimates of depth for those features, for example to produce a map estimating areas with >50% probability of the presence of an argillic horizon with estimates of depth in those areas. We also hope to improve the models for depth estimates so that several soil feature surfaces can be visualised in 3-dimensional perspective view draped over a land surface. These combinations of continuous estimations of soil features could form new soil survey products, when the models perform better. In this study each response variable was modelled separately, but future study needs to include modelling taking multiple response variables into account, that is from a multivariate point of view.

Acknowledgments

This work was accomplished through the hard work of field soil scientists in difficult, remote conditions (who collected the 'real data') and by a statistician, a GIS specialist/soil scientist and a remote sensing specialist. We appreciate the support of Dave Smith, State Soil Scientist, USDA Natural Resources Conservation Service, California.

References

Clemmer, P., 2003. Band-ratio Landsat 5 Thematic Mapper Imagery. Personal communication. USDI Bureau of Land Management.

Environmental Systems Research Institute, 1998. GridSpot70. [Online] Available: http://arcscripts.esri.com/details.asp?dbid=11037.

Environmental Systems Research Institute, 2002. ArcInfo v8.3 and v9. Redlands, CA.

Gessler, P.E., Moore, I.D., McKenzie, N.J., Ryan, P.J., 1995. Soil-landscape modelling and spatial prediction of soil attributes. Int. J. Geogr. Inf. Systems 9, 421–432.

Glinka, K.D., 1927. Dokuchaiev's ideas in the development of pedology and cognate sciences. U.S.S.R. Academy of Science. Russian Pedological Investigations, I.

Haydu-Houdeshell, C., 2004. Soil profile descriptions Johnson Valley Off Highway Vehicle Area Soil Survey Project. Personal Communication. USDA Natural Resources Conservation Service.

Hilgard, E.W., 1914. Soils. The McMillan Company, New York.

Jenny, H., 1941. Factors of Soil Formation. A System of Quantitative Pedology. McGraw-Hill Book Company, Inc., New York.

McBratney, A.B., Mendonca Santos, M.L., Minasny, B., 2003. On digital soil mapping. Geoderma 117, 3–52.

McKenzie, N.J., Austin, M.P., 1993. A quantitative Australian approach to medium and small scale surveys based on soil stratigraphy and environmental correlation. Geoderma 57, 329–355.

McKenzie, N.J., Cresswell, H.P., Grundy, M., 2000. Contemporary land resource survey requires improvements in direct soil measurement. Comm. Soil Sci. Plant Anal. 31, 1553–1569.

McKenzie, N.J., Ryan, P.J., 1999. Spatial prediction of soil properties using environmental correlation. Geoderma 89, 67–94.

McSweeney, K., Slater, B.K., Hammer, R.D., Bell, J.C., Gessler, P.E., Petersen, G.W., 1994. Towards a new framework for modeling the soil-landscape continuum. In: R. Amundson, J. Harden and M. Singer (Eds.), Proceedings of Factors of Soil Formation: A Fiftieth Anniversary Retrospective Symposium. Denver, Colorado. 28 October 1991. Soil Science Society of America Special Publication Number 33, Madison, WI, pp. 127–145.

SAS, 2001. SAS statistical software Release 8.02. SAS Institute Inc., Cary, NC, 1999–2001.

Soil Survey Staff, 1999. Soil Taxonomy, 2nd edition. Agriculture Handbook No. 436. United States Department of Agriculture, Natural Resources Conservation Service, Washington, D.C.

U.S. Army Topographic Engineering Center and Louisiana State University, 2000. Earth Materials and Landform Mapping Project. [Online] Available: http://www.mojavedata.gov/datasets.php?qclass=geo.

USDA Natural Resources Conservation Service, 2004a. The PLANTS Database, Version 3.5. [Online] Available: http://plants.usda.gov [June 4, 2004]. National Plant Data Center, Baton Rouge, LA 70874-4490 USA.

USDA Natural Resources Conservation Service, 2004b. National Soil Information System (NASIS) Choice list report. [Online] Available: http://nasis.nrcs.usda.gov/documents/.

Webster, R., Burrough, P.A., 1972. Computer-based soil mapping of small areas from sample data. I. Multivariate classification and ordination. II. Classification and smoothing. J. Soil Sci. 23, 210–234.

Developments in Soil Science, volume 31
P. Lagacherie, A.B. McBratney and M. Voltz (Editors)

Chapter 35

THE SPATIAL DISTRIBUTION AND VARIATION OF AVAILABLE PHOSPHORUS IN AGRICULTURAL TOPSOIL IN ENGLAND AND WALES IN 1971, 1981, 1991 AND 2001

S.J. Baxter, M.A. Oliver and J.R. Archer

Abstract

The Representative Soil Sampling Scheme (RSSS) has monitored the soil of agricultural land in England and Wales since 1969. Here, we describe the first spatial analysis of the data from these surveys using geostatistics. Four years of data (1971, 1981, 1991 and 2001) were chosen to examine the available phosphorus (P). At each farm, four fields were sampled, but for the earlier years, coordinates were available for the farm only and not for each field. The average data for each farm were suitable for analysis and the variograms showed spatial structure even with a small sample size. Temporal change was evident in the kriged maps and also in the results of a Student's *t*-test. The decline in values over time, particularly in the east of England, is associated with a reduction in the use of fertilizer P.

35.1 Introduction

The importance of soil monitoring has now been recognised with the European Union Soil Monitoring Directive proposed for 2004 and a soil strategy for the United Kingdom that incorporates a soil monitoring scheme for England. The latter should be implemented in the near future. Several schemes exist to monitor soil properties in England and Wales. The National Soil Inventory (NSI) recorded a wide range of soil physical and chemical characteristics at 5692 sites, in 1981 (Oliver et al., 2002). The Environmental Change Network (ECN) was established to develop permanent sites where information including soil characteristics could be monitored over medium-to-long term (Burt, 1994). As part of the Countryside Survey 2000, 531 samples of soil were obtained from 115 sites on different environmental strata (Firbank et al., 2003). The British Geological Survey is also surveying the soil in some parts of Great Britain. The Representative Soil Sampling Scheme (RSSS) has monitored the soil of agricultural land annually in England and Wales since 1969. The properties measured include, pH and available nutrients, phosphorus (P), potassium (K) and magnesium (Mg). These data have been analysed annually using a range of classical statistics. The results have shown changes in pH, P and K over time, in different regions (West and Wales, Southern, Midlands, Eastern and Northern regions) with different land use (Skinner and Todd, 1998; Skinner et al., 1992; Church and Skinner, 1986).

This chapter describes the first spatial analysis of the RSSS data using geostatistics. The variogram gives an unbiased description of the scale and pattern of the variation of the properties and can be used with the data to predict by kriging (Webster and Oliver, 2001). The results should enable the spatial and temporal variation of the properties to be investigated further. They will also provide an assessment of these data as a basis for monitoring.

Four years of data from 1971, 1981, 1991 and 2001, which embrace the period monitored, have been chosen to examine the spatial variation of pH, Mg, K and P. For this chapter we have selected P to illustrate the analysis.

35.2 Methods

Church and Skinner (1986) describe the methods of sampling P for the RSSS. The soil sampling sites were on farms that were a selected subsample of those used for the Survey of Fertilizer Practice (Chalmers et al., 1990), and they represent the major types of farm in the United Kingdom. At each farm, four fields were selected randomly and were sampled. Phosphorus was determined as sodium bicarbonate soluble P (Ministry of Agriculture, Fisheries and Food, 1985).

Most of the sites of 1971 were sampled again in 1981, but there were also some additional sites in 1981. In 1991, the sites sampled were new, and in 2001 some of these sites were sampled again together with some new sites. The year of sampling is indicated by a superscript, for example, P measured in 1991 is referred to as P^{91}. The number of farms and fields for each year is given in Table 35.1, and the sampling scheme for each year is shown on the maps in Plate 35a,b (see Colour Plate Section). The data for 1971 and 1981 had coordinates for the farms, but not for each field. This meant that only one value for each soil property could be used for each farm; this could be either one of the four values or their average. We chose the latter and first had to determine the effect of averaging the data on the spatial analyses. The full data for 1991 and that obtained by averaging the values within a radius of 5 km were analysed. This, in effect, averages the values for each farm: the number of sites for the average data is close to the number of farms sampled, Table 35.1.

The summary statistics show that P had a skewed distribution and it was transformed to its common logarithm, $\log_{10}P$. Experimental variograms for $\log_{10}P^{91}$ were computed before and after averaging the data, and compared. The results indicated that it was feasible to base the analysis on the averaged data. This meant that the data for 1971 and 1981 could be analysed in the same way as those for 1991 and 2001. Several authorized models were fitted by least squares approximation to the experimental semivariances of $\log_{10}$P using GenStat (Payne, 2000). The variogram model parameters were used with the data for ordinary punctual kriging. The predictions were made at the nodes of a 2.5-km grid and then used for mapping.

Since the kriged predictions are affected by differences in the proportion of nugget:sill variance, we computed a pooled within-class variogram from the 4 years of data for $\log_{10}$P. This entailed taking the mean value for each year (class) from all the values in that year and the residual values were used to compute a single variogram. In this way, a single variogram could be used with the data for each year to krige $\log_{10}$P. This removes the effect of the annual variogram on the kriged estimates. Cross-validation was used to assess the reliability of the predictions using the pooled within-class variogram compared with those using the variogram for each year. This method removes each sample point in turn, and then estimates the value there from the remaining data by punctual kriging. The mean squared error (MSE) was calculated and the method of interpolation with the smaller MSE was considered to be more accurate. Temporal change in P was also assessed using an unpaired Student's *t*-test with the raw data.

35.3 Results

35.3.1 Assessment of averaging values within a radius of 5 km

Table 35.1 gives the summary statistics for P based on the raw data and the averaged data for 1991. There is little difference in their mean values. The standard deviations are smaller for the averaged data as one would expect because the sample support has increased and the local sampling effects removed. Table 35.1 also gives the summary statistics for P for all selected years using the averaged data. The mean decreases over time from 31 mg l^{-1} in 1971 to 26 mg l^{-1} in 2001. The standard deviation and skewness also decrease suggesting

Table 35.1. *Summary statistics of P (mg l^{-1}).*

Year and type of data	No. of fields	No. of farms	No. of averaged values	Mean	Min	Max	Standard deviation	Skewness
1991 raw	716	181	0	26.50	4.00	114.5	16.15	1.85
1971 averaged	417	107	106	31.09	6.25	152.8	24.00	2.30
1981 averaged	881	229	229	28.06	2.33	115.3	16.75	1.91
1991 averaged	716	181	177	26.79	6.50	76.3	12.09	1.41
2001 averaged	660	167	183	25.96	5.00	81.5	11.68	1.36

a decline in the number of sites with large concentrations of P over the period of time examined.

The experimental variograms computed from the raw and averaged data for $\log_{10}P^{91}$ and their best fitting models, that is, those for which the residual mean squares were minimized, are shown in Figure 35.1. Table 35.2 gives the parameters of these models. The nugget:sill variance is slightly smaller for the averaged data for $\log_{10}P^{91}$; for the full data it is 75% and for the averaged data 68%. The range of spatial dependence for the raw data is 71 and 125 km for the averaged data. The variograms of both the raw and averaged data show a clear structure for $\log_{10}P$, which suggests that there has been little loss of detail in spite of the considerable reduction in the number of values used for the latter. To compare the variograms for different years they were computed from the averaged data: this means that all years have the same sample support, which aids the interpretation and comparison between years.

35.3.2 Spatial analysis

Figure 35.1 shows the experimental variograms of $\log_{10}P$ with their best fitting models for each of the selected years, and Table 35.2 gives their model parameters. There is a considerable variation in both the nugget:sill variance and the correlation range for the years examined. A notable difference between these

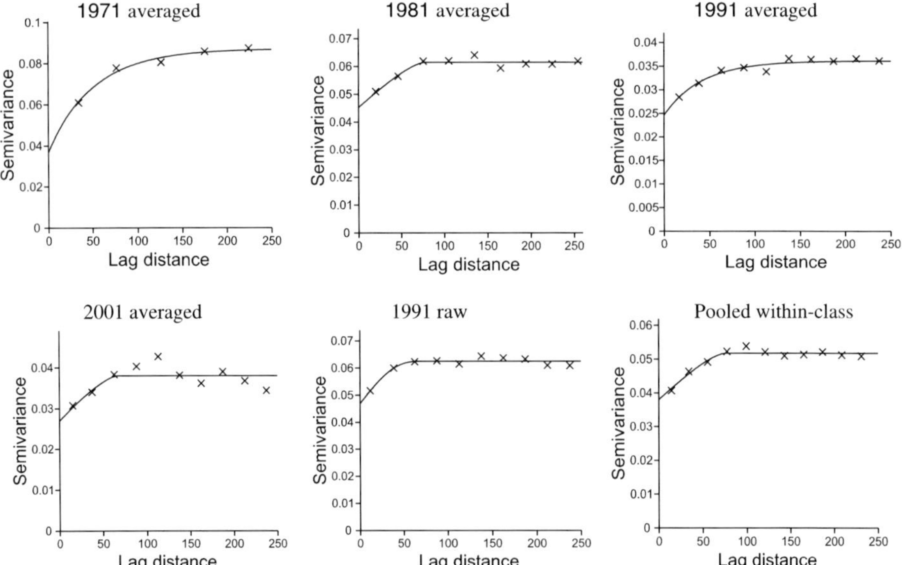

Figure 35.1. *Variograms of $\log_{10}$ P; the experimental values are indicated by the symbols, the model by the line, and the lag distance is in kilometres.*

Table 35.2. *Variogram model parameters for log_{10} P.*

Variable	Model	C_0	C	a (km)	r (km)	*Nugget:sill variance* (%)
1971 averaged	Exponential	0.0366	0.0504	51.0	153.0	42.1
1981 averaged	Circular	0.0453	0.0162	78.1		73.6
1991 averaged	Exponential	0.0246	0.0115	41.7	125.1	68.2
2001 averaged	Circular	0.0269	0.0111	66.4		70.9
1991 raw	Pentaspherical	0.0468	0.0156	71.0		75.0
Pooled within-class	Circular	0.0381	0.0137	75.5		73.5

Where c_0 is the nugget variance, c is the sill of the autocorrelated variance, a is the range of spatial correlation, r the effective range of the exponential model ($r = 3 \times$ distance parameter(a)).

Table 35.3. *Cross-validation for log_{10} P.*

Year and variogram used	MSE
1971 pooled within-class	0.07630
1971 averaged	0.07640
1981 pooled within-class	0.05723
1981 averaged	0.05720
1991 pooled within-class	0.03559
1991 averaged	0.03568
1991 raw	0.03559
2001 pooled within-class	0.03488
2001 averaged	0.03486

To calculate the MSE values for the pooled within-class variogram the mean was added back to the values so that they were comparable with those for the raw and averaged variograms.

variograms is the decrease in sill variance over time, which reflects the change in the general variability of P in England and Wales.

The results of the cross-validation showed that when the pooled within-class variogram was used, the MSEs were similar to those for the variograms of the individual years (Table 35.3). The maps of $\log_{10}$P using the variograms for each year and the single pooled within-class variogram were also similar. This suggests that the pooled within-class variograms provided reliable predictions and should enable temporal change in the properties to be observed. The kriged predictions using the pooled within-class variogram were back-transformed to the original scale of measurement for mapping. The back transformation for ordinary kriging with common logarithms was calculated by:

$$\hat{Z}(\mathbf{x}_0) = \exp[\hat{Y}(\mathbf{x}_0) \times \ln 10 + 0.5\sigma^2_{\hat{Y}}(\mathbf{x}_0) \times (\ln 10)^2 - \psi \times (\ln 10)^2] \qquad (35.1)$$

where $\hat{Y}(\mathbf{x}_0)$ is the kriged estimate of the common logarithm at $\mathbf{x}_0$, its kriging variance is $\sigma^2_{\hat{Y}}(\mathbf{x}_0)$ and ψ, the Lagrange parameter. Plate 35a,b (see Colour Plate

Section) shows the kriged maps. The redder areas have the larger P concentrations and the bluer areas the smaller ones. Plate 35a,b (see Colour Plate Section) shows the maps of kriging variance (not back transformed) for each year. The kriging variances are small near the sampling sites and the larger values indicate where the kriged predictions are less reliable. They are around the edge of England and Wales and where the data were sparser compared to the rest of England and Wales. In 1971, there were the fewest samples; therefore, the estimation variances are larger not only around the boundaries, but also in the middle of southern England (Plate 35a, see Colour Plate Section).

The maps of P show that there has been an overall decline in soil P concentrations over time; values above 40 $mg\,l^{-1}$ are not evident in the map for 2001 (Plate 35b, see Colour Plate Section), but they covered quite a large area in 1971, Plate 35a (see Colour Plate Section). This decrease in soil P values over time was also identified by classical statistical analysis (Skinner and Todd, 1998).

Plate 35b (see Colour Plate Section) includes a map showing the difference in P between 1971 and 2001. It shows that P has decreased mainly in east England and West Yorkshire ($>20\,mg\,l^{-1}$), Devon and Cornwall ($>10\,mg\,l^{-1}$) and in parts of south Wales ($>5\,mg\,l^{-1}$). Increases in P are evident in the rest of the west and north of England by around 5–10 $mg\,l^{-1}$. A map of the differences between the kriging variances between 1971 and 2001, Plate 35a,b, (see Colour Plate Section) reflect the differences between the two sampling strategies with approximately 15% of England and Wales with less reliable predictions located where samples were not taken, however, about 40% of England and Wales have more reliable predictions due to the increased number of samples taken in 2001 compared to 1971.

Soil nutrient concentrations are often expressed as soil indices, as an aid to management. Concentrations of P are divided into nine indices ranging from 0 (deficient) to 9 (very large) (MAFF, 2000). The maps for 1971, 1981 and 2001 show that almost half of England and Wales has a P index greater than 3 (the range for this index is 26–46 $mg\,l^{-1}$), and for 1991, it is about two thirds of the country. At this scale none of England and Wales could be regarded as being deficient in P at any time (<index 2, the range for index 1 is 10–15 $mg\,l^{-1}$) for arable and grass land; however, an index of 3 is recommended for vegetables and bulbs.

35.3.3 Temporal analysis

The results of the unpaired Student's *t*-test showed a moderately significant difference in P concentrations of the soil between 1971 and 1991 (probability = 0.088) and between 1971 and 2001 (probability = 0.041). The differences between pairs of years closer together were less significant. Considering that there are relatively few data points compared to the size of the country and the likelihood of noise there are clear differences in the mean values over time. This again supports other observations and results.

35.4 Discussion and conclusion

Although not designed for spatial analysis, the RSSS dataset has proved suitable for such an analysis. As coordinates were unavailable for the individual fields in 1971 and 1981 an analysis was done to determine the effect of averaging the data for 1991. The variogram from the averaged data was similar to that from the raw data (Table 35.2, Fig. 35.1). This suggests that for this type of analysis only one bulked soil sample needs to be taken from each farm. The irregular sample distribution in these data means that a range of sampling intervals is covered. This is advantageous, in particular where the data are sparse, to describe variation at the shorter lag distances.

The variograms of $\log_{10}$P showed spatial structure although the sample size was small (Table 35.2, Fig. 35.1). However, all the variograms have a large nugget variance because of the large sampling interval between farms in general. The short sampling interval associated with the four fields at each farm contributed little to reducing the nugget variance as could be seen from the analysis of the raw data for 1991. Although this sampling scheme misses the local variation in the soil, it seems to resolve that at the national scale, adequately. This is supported by the similarity of the RSSS variograms for $\log_{10}$P compared with that from the NSI, which had a much larger sample size but on a fixed 5-km square grid. The experimental variogram of $\log_{10}$P from the NSI data, however, was fitted by a nested model that described two spatial scales; the first range was about 27 km and the second 113 km (Oliver et al., 2002). The scales of variation identified from the RSSS for $\log_{10}$P correspond more closely with the long-range structure. The RSSS sampling scheme was not detailed enough to identify the short-range structure. Although there were over 5600 sites for the NSI, the variogram of $\log_{10}$P had a larger nugget:sill variance of 83% than those computed for the RSSS. This suggests that if variation at the national scale is of interest then a smaller sample size than that used for the NSI can be adopted. The sampling configuration has a large effect on the patterns of the maps as seen from the maps of kriging variance Therefore, for a relatively small sample size the sites need to be chosen with care and placed to attain a more even cover across England and Wales.

Temporal change in P was identified from the kriged maps and confirmed by a Student's *t*-test. The mean and standard deviation of P have decreased over time (Table 35.1), and the maps of P (Plate 35a,b, see Colour Plate Section) also show a decline in values.

This is most noticeable in eastern England, the main arable farming region of England and Wales, from over 40 mg l^{-1} in 1971 to between 25–30 mg l^{-1} in 2001. This is due to a reduction in the number of fields where the P status was unnecessarily large. There has been some increase in P concentrations in the western part of the country. This is likely to be due to the greater use of livestock manure and insufficient allowance for P in those manures (Haygarth et al., 1998; Defra, 2004). The greater general depth of ploughing since 1969 might also have had the effect of diluting the overall phosphorus concentration (Skinner and Todd, 1998). Phosphorus fertilizer use declined in England and Wales between 1969 and 1993 from 386,000 mg to 311,000 mg P_2O_5 mineral fertilizer (Skinner and Todd, 1998). It has continued to decline; the Department for Environment Food and Rural Affairs (2001) reported that overall phosphate use on arable crops had gradually declined since 1983, from a 5-year mean of 58 kg ha^{-1} in the period 1983–1987 to 47 kg ha^{-1} during the 4-year period 1998–2001. The rate in 2001 of 42 kg ha^{-1} is the lowest since records began in the United Kingdom in 1983. The overall rate of phosphate application on grassland was largest in 1983, at 28 kg ha^{-1}. Mean annual use of 20 kg ha^{-1} over the period 1997–2001 represents a net decline of 5 kg ha^{-1} in overall phosphate rate, compared to the 1983–1987 mean. The decrease in P could also reflect the change in the type of fertilizer used. Since the early 1970s, there has also been a change from insoluble phosphate fertilizers, such as basic slag or ground rock phosphate used on grassland to compounds that are water soluble (Defra, 2001).

Acknowledgments

This work was funded by Defra (the U.K. Department of Environment, Food & Rural Affairs) and we thank Alan Todd of Rothamsted Research for providing the data. The soil samples were collected by ADAS staff.

References

Burt, T.P., 1994. Long-term study of the natural-environment – perceptive science or mindless monitoring. Prog. Phys. Geogr. 18, 475–496.

Chalmers, A.G., Kershaw, C.D., Leech, P.K., 1990. Fertilizer use on farm crops in Great Britain; results from Survey of Fertilizer Practice, 1969–88. Outlook Agric 19, 269–278.

Church, B.M., Skinner, R.J., 1986. The pH and nutrient status of agricultural soils in England and Wales, 1969-83. J. Agric. Sci., Cambridge 107, 21–28.

Department for Environment Food and Rural Affairs (DEFRA), 2001. The British Survey of Fertiliser Practice. Fertiliser use on farm crops for crop year 2001. HMSO, London.

Department for Environment Food and Rural Affairs (DEFRA), 2004. Managing the problem. Risks of diffuse water pollution from agriculture. Defra. Available: http://www.defra.gov.uk_/environment/water/quality/diffuse/agri/pdf/mapping-problem.pdf

Firbank, L.G., Barr, C.J., Bunce, R.G.H., Furse, M.T., Haines-Young, R., Hornung, M., Howard, D.C., Sheail, J., Sier, A., Smart, S.M., 2003. Assessing stock change and change in land cover and biodiversity in GB: an introduction to Countryside Survey 2000. J. Environ. Manage. 67, 207–218.

Haygarth, P.M., Chapman, P.J., Jarvis, S.C., Smith, R.V., 1998. Phosphorus budgets for two contrasting grassland farming systems in the UK. Soil Use Manage 14, 160–167.

Ministry of Agriculture, Fisheries and Food, 1985. The Analysis of Agricultural Materials. Reference book 427. HMSO, London.

Ministry of Agriculture, Fisheries and Food, 2000. Fertiliser Recommendations for Agricultural and Horticultural Crops (RB209). The Stationary Office, U.K., London.

Oliver, M.A., Loveland, P.J., Frogbrook, Z.L., Webster, R. and McGrath, S.P., 2002. Statistical and geostatistical analysis of the national soil inventory of England and Wales. The Technical Report. Defra.

Payne, R.W. (Ed.) 2000. The Guide to GenStat: Part 2 Statistics. VSN International, Oxford.

Skinner, R.J., Church, B.M., Kershaw, C.D., 1992. Recent trends in soil pH and nutrient status in England and Wales. Soil Use Manage 8, 16–20.

Skinner, R.J., Todd, A.D., 1998. Twenty-five years of monitoring pH and nutrient status of soils in England and Wales. Soil Use Manage 14, 162–169.

Webster, R., Oliver, M.A., 2001. Geostatistics for Environmental Scientists. J. Wiley & Sons, Chichester, UK.

Developments in Soil Science, volume 31
P. Lagacherie, A.B. McBratney and M. Voltz (Editors)

Chapter 36

THE POPULATION OF A 500-M RESOLUTION SOIL ORGANIC MATTER SPATIAL INFORMATION SYSTEM FOR HUNGARY

E. Dobos, E. Micheli and L. Montanarella

A pilot study on the derivation of soil organic content information was carried out using MODIS data, digital elevation data (SRTM-30, 1-km resolution), derived digital terrain variables and the profile database of the Hungarian Soil Monitoring system. A regression-kriging procedure was used to spatially predict the organic matter content for the whole of Hungary.

The goal of the study was to characterise the performance of regression analysis using satellite and terrain data as a potential method for soil variable mapping and to derive a soil-organic matter content map for Hungary. Eight-day composites of Terra MODIS were used (1–8 May, 2000 and 1–8 Sept, 2000, with 500-m resolution, channels 1–7 and 21). The satellite image layers were spectrally improved. Numerous terrain variables were derived from the SRTM-30 and were used as input variables to the regression. Logarithmic and square root transformations were done for some of the layers and variables to achieve normality. Via these preprocessing and transformation algorithms a new integrated satellite and terrain database was created. This database was used for estimating the soil organic matter content by regression analysis and kriging. The statistical correlations and the percentages of variation explained by the regression models were always significant. However, their values were relatively low.

36.1 Introduction

The soil resource is faced with a variety of degradative impacts, which are partly human induced and partly natural. A soil protection strategy has been set up in many countries to maintain or improve the soil quality. However, the soil protection strategies of the EU member states vary a lot in terms of their aims and efforts. Thus, the European Union has initiated a program to harmonise the soil protection strategies throughout Europe and enforce the soil protection actions to ensure the sustainable use of soil resources (COM/2002/179 Final Communication toward a Thematic Strategy for Soil Protection, 2002).

One of the major stresses on the soil resource is the decline of organic matter content (SOM). In many of the European countries there is no reliable, up-to-date, spatially defined SOM information. Soil information and monitoring systems were set up in these countries to survey the recent situation and estimate the rate and trend of potential changes of SOM. These monitoring systems are profile-based networks, with regular sampling period, which can provide a limited spatial coverage for a particular country. These data need to be extrapolated to create a continuous coverage of the area in question. In the meantime, staying in line with the European mapping standards is also an important requirement (see Chapter 6).

The SOM content of the soil is strongly related to the land use, vegetation, climate and terrain features, which can be modelled with DEM and satellite data.

The type and the amount of soil organic matter are strongly related to the presence of water and the lateral redistribution of the surface material by erosion. Both of these phenomena are partially controlled by the terrain. Among others compound topographic index (CTI), wetness index, potential drainage density (PDD) (Dobos et al., 2000, 2001), curvature, slope gradient and flow accumulation variables proved to have a significant contribution to the estimation of the depth of A-horizon, soil carbon content (McKenzie and Ryan, 1999; Gessler et al., 2000), soil organic matter content (Moore et al., 1993; see also Chapters 13 and 37) and topsoil carbon (Arrouays et al., 1998; Chaplot et al., 2001). The overall predictive values of these models were around 50–70%. Many previous studies have demonstrated the usefulness of digital data sources and quantitative, spatially predictive models to extrapolate point data and estimate soil properties (McBratney et al., 2003; see also Chapter 32). The data and the procedure are in line with the spatial soil inference systems theory (see Chapter 1).

In this chapter, a method for extrapolating point information based on an integrated digital elevation and satellite dataset and statistical-geostatistical tools to create a SOM map of Hungary is demonstrated. The overall aim of the work is to develop a methodology, which can be used to derive spatially defined SOM information for EU policy support.

36.2 Materials

36.2.1 The study area

The study area covers the Carpathian basin, and represents a great variety of soils, landscape types, land use, climate, topography and vegetation. Figure 36.1 shows the location of the study area.

36.2.2 Satellite data

The Moderate Resolution Imaging Spectroradiometer (MODIS) sensor data was used for the project (Salomonson et al., 2002). In order to represent different environmental conditions, two dates, May and September of 2000 were selected. MODIS provides a lot of spectral bands for numerous applications. In this study, the 1–7 reflective bands and the normalized difference vegetation index (NDVI) with 500-m resolution and one thermal infrared band (band 31) with an original

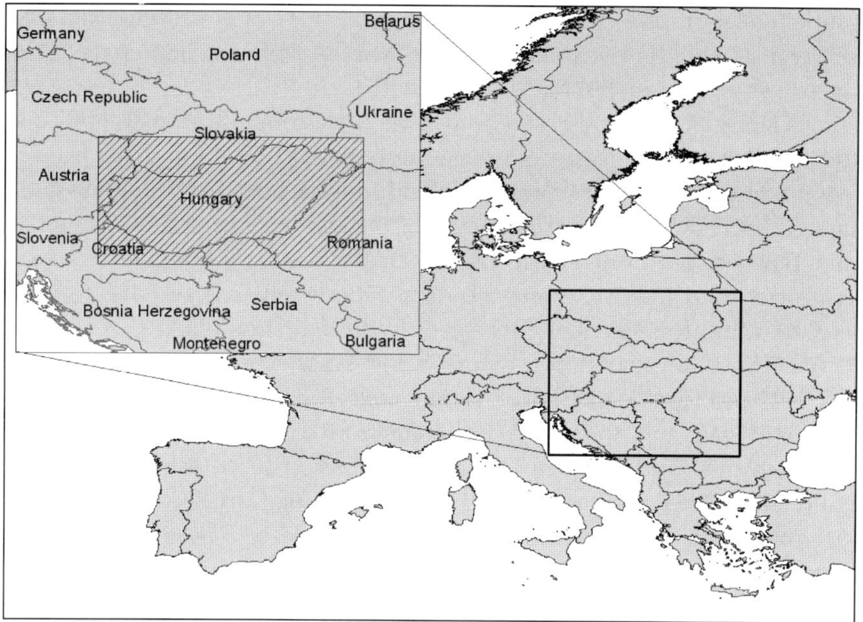

Figure 36.1. *The study area location. The hatched rectangular on the central European map is the pilot area.*

Table 36.1. *The MODIS bands used for the project and their spectral ranges.*

Band	Spectral range (nm)
1	620–670
2	840–870
3	460–480
4	540–560
5	1230–1250
6	1630–1650
7	2100–2150
31	10780–11280

resolution of 1 km were selected. The 1-km layer was later resampled to 500 m. Table 36.1 shows the MODIS bands and the corresponding wavelengths.

36.2.3 SRTM-30 Global elevation data

The SRTM-30 database is an improved version of the GTOPO-30. GTOPO-30 was completed in late 1996, following development over a 3-year period through a collaborative effort led by the staff at the USGS's EROS Data Center (Gesch et al., 1999). Its documentation can be read from the homepage, http://edcdaac.usgs.gov/gtopo30/README.asp#h10.

GTOPO-30 has a spatial resolution of 30 arc-seconds (approximately 1 km). Produced for use in large-area studies, these global DEM data have been generated at a resolution that is compatible with the advanced very high resolution radiometer (AVHRR) sensor. The quality of the dataset varied depending on the original datasets used to compile GOTO-30. The areas with higher relief are in general of a good quality, while the data of low relief areas have significant artefacts that limit its use for any type of hydrologic modelling.

SRTM-30 was a great step forward in low-resolution digital terrain modelling. It is a near-global digital elevation model, covering the Earth's surface between 60° south and north. The data come from a combination of the GTOPO-30 and from the Shuttle Radar Topography Mission (Farr and Kolbrick, 2000), flown in 2000. The basic product has a resolution of 1 arc-second, but it is not publicly available. These data were first generalised to 3 arc-seconds by averaging a 3×3 cell area, and then to 30 arc-seconds by a 10×10 grid averaging of the 3-arc-second product. There are small gaps in the dataset due to the shadowing effect of the radar. These gaps were filled with data from the GTOPO-30. This occurs in ~0.15% of the dataset. The data for the area north and south of the 60° latitudes is obtained solely from the original GTOPO-30.

The data were transformed from geographical coordinates (latitude and longitude) to the standard Lambert Azimuthal Equal-Area projection. Numerous terrain attributes were created and added to the database:

- Altitude
- Specific catchment area (A_s: the ratio of the number of cells contributing flow to a cell and the grid size)
- Profile, planar and complex convexity (see the ArcInfo® online manual)
- Slope percentage (S)(average maximum technique, Burrough, 1986, see also ArcInfo® online manual)
- PDD (Dobos and Daroussin, 2005)
- Aspect
- Flow accumulation (number of cells contributing flow to a cell)
- Relief intensity (difference between the maximum and minimum elevations within a preset sized neighborhood) (Dobos et al. 2005)
- Compound topographic index (CTI: $\ln A_s/S$) (Wilson and Gallant, 2000)

The terrain dataset originally had a 1-km resolution. In order to match the 500-m resolution of the MODIS, the terrain data layers were resampled to 500 m, using the bilinear function of Arc/Info.

36.2.4 Geographic position representing the climatic changes

Two artificial layers were created, one for easting and one for northing to represent the geographic position. The east–west direction represents the transition between the oceanic and the continental climate, and is strongly correlated with the rainfall distribution. The western part of Hungary gets 700–800 mm annual rainfall, while the eastern part gets much less, around 500–600 mm.The north–south direction is also very important in explaining the sub-Mediterranean climate impacts. These trends are very evident in the soil type distribution of Hungary.

36.2.5 Soil monitoring system for Hungary (TIM)

TIM is part of the Hungarian Environmental Monitoring System that was created and maintained since 1995 (Várallyay et al., 1995). This point-vector database consists of 1236 soil profile descriptions. The locations of these points were selected as representative points of the natural landscape units of Hungary, so the database can be considered a realistic characterisation of soil resources of the country. Besides the detailed soil description data, it contains numerous soil physical and chemical measurement data for monitoring the soil changes in time as a result of anthropogenetic and natural processes.

The TIM data served as reference information for regression analysis and kriging (dependent variable). The SOM contents were calculated on a horizon basis in Mg/ha, and the horizon SOM contents were summed up to derive the total SOM content of the area. The variables used to calculate the SOM content were the SOM percentage, bulk density and horizon depths.

36.3 Methods

Regression kriging was used to create the SOM content layer for Hungary. The procedure comprised four major steps, (1) database construction, (2) linear regression, (3) kriging of the regression residuals and (4) summing up the regressed and kriged residual values to derive the final spatial soil inference system.

36.3.1 Database construction

The MODIS bands of the two dates provided 18 layers, representing 18 environmental variables; 10 layers of terrain variables were created as well. Altogether, 30 independent variables were derived including the easting and northing layers. In order to achieve normal or normal-like distribution for all the variables, logarithmic and square root data transformations were carried out. In the end, 45 layers (variables) were created.

The average SOM content from the years of 1992 and 1998 – derived from the TIM – was used as the dependent variable.

All variable layers were sampled for the TIM points and an excel datasheet was created with 46 variables and 1188 records (observations). This database was used as input for the statistical package SPSS 8 for linear regression. Kriging was done with the geostatistical package in ArcGIS.

36.3.2 Regression kriging

Forward regression was used and 12 variables were selected for the regression equation. With the use of the derived regression equation, a continuous layer of estimated SOM content was created. In the second step, the estimation errors were calculated for all the TIM points and were kriged to create a continuous layer. Finally, the regressed and kriged values were summed up and the final version of SOM map was completed.

36.4 Discussion

The forward regression selected 12 variables. The variables and the regression coefficients are given in Table 36.2. The adjusted R^2 was quite low, but significant, 0.238, meaning that there is significant correlation between the SOM content and terrain and spectral variables. The scatterplot of the estimated and original SOM values are shown in Figure 36.2. The root mean square error (RMSE) was 11642.92 g/m^2.

Table 36.2. *The variables selected and their regression coefficients for soil organic matter content estimation.*

Variable	Variable name	Regression coefficients
Intercept		+87921.241
Square root transformed NDVI from Sept. 2000	[Sqrsndvi]	−1829.276
MODIS band 5, May 2000	[May95hu]	1.469
Square root transformed PDD	[Sqrtpdd]	7626.015
PDD	[Pddnd4hu05]	1240.775
Logarithmic transformed altitude	[Lndem]	12097.897
Relief Intensity	[Ridemndhu05]	54.247
MODIS NDVI, May 2000	[Mayndvi]	105.134
MODIS band 3, May 2000	[May93hu]	33.099
Profile convexity	[Prcurv100hu05]	9179.56
Square root transformed MODIS band 1, May 2000	[Sqrm1]	2556.134
MODIS band 4, May 2000	[May94hu]	3.998
Aspect	[Aspecthu05]	10.738
Summary statistics		
$R^2 = 0.238$, significant (0.000), RMSE = 11642.92 $g.m^{-2}$		

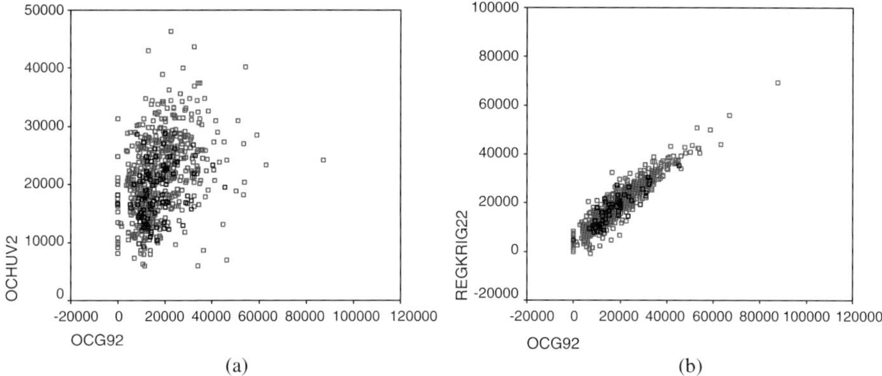

Figure 36.2. *Scatterplots of the original (OCG92) and the predicted SOM values for (a) the regression derived (OCHUV2) and (b) the regression-kriging derived (REGKRIG22) datasets.*

Despite the low statistical correlation, the overall look of the map appears promising. It coincides with our understanding of the spatial distribution of SOM content over Hungary, determined by the climatic, geologic, biotic and human impacts on the soil formation. The low R^2 value and the scatterplot indicate the complex nature of the SOM distribution, determined by important soil-forming factors, which are not significantly represented by the satellite images or the terrain variables. Although the major factors regulating the SOM balance in general were present among the variables, the performance of the regression model was disappointing. The authors identified two potential reasons. The first one arises from the scale issue and the representativity of the training dataset. The independent variables have a 500-m nominal resolution, which is quite low compared with the training dataset. The training points were taken as single borehole samples, which do not necessarily characterise well the entire, 500×500 m grid cell area. The organic carbon was sampled twice previously, first in 1992 and then in 1998. The comparison of the two datasets showed a very high, often unrealistic variation in the SOM content, which is probably due to the sampling design. A block sampling design for the monitoring system would be more appropriate and would result in a much better and consistent SOM database. It would help in the data regionalization as well, which is one of the most important issues at national level. Similar problems for training data characterization were listed and discussed in Chapters 11 and 15.

Besides the representativity question, a well-defined error trend was also identified. The organic carbon content (OCC) of the chernozem areas on loessial parent materials and on the mountainous areas are well estimated or slightly

Table 36.3. *Pearson correlation coefficients for the training data and for the results of the regression prediction and the regression-kriging prediction.*

	Regression kriging	Regression	Training dataset
Regression Kriging	1	0.456[a]	0.934[a]
Regression	0.456[a]	1	0.346[a]
Training dataset	0.934[a]	0.346[a]	1

[a]Correlation is significant at the 0.01 level (2-tailed).

underestimated by the regression model, while the OCC of the sandy and clayey regions of the plain area of Hungary are significantly overestimated. This trend was captured by the ordinary kriging of the regression residuals (Plate 36a, see Colour Plate Section). The combination of the regression and error kriging steps resulted in a refined SOM database, with a much smaller RMSE (4382.7) and higher correlation (Table 36.3, Fig. 36.2b, Plate 36b (see Colour Plate Section)).

36.5 Conclusions

Providing adequate and reliable soil property data for users is one of the most critical tasks of the soil science community. The lack of data, the unknown data accuracy and the spatial inconsistency of the databases are the most limiting factors of this area. These problems are even more crucial on continental and global extents, where harmonised, spatially and thematically consistent database do not exist.

A digital soil mapping procedure was tested here to produce a SOM spatial information system for Hungary. The results of the linear regression procedure is quite promising, however, the statistical quality measures are relatively low. Regression combined with kriging of the residuals, the so-called regression-kriging, produced a much more accurate result with acceptable statistical measures and realistic spatial distribution of the SOM. The method is based on existing digital data sources with global coverage, thus can be repeated anywhere in the world, where soil profile data is available for training. Digital elevation data and remotely sensed (RS) information are among the best environmental descriptors. However, the correlation between these data layers and certain soil properties depends highly on the data quality and the environmental conditions when the data were acquired. In the frame of a well-defined spatial soil inference system, more potential preexisting input data could be used to run the regression model and refine the procedure to better fit our needs and exploit the emerging state-of-the-art IT tools and data.

Acknowledgements

This study was supported by the European Commission, by the Hungarian National Science Foundation (OTKA, 34210) and by the Bolyai Foundation.

References

Arrouays, D., Daroussin, J., Kicin, J.C., Hassika, P., 1998. Improving topsoil carbon storage prediction using a digital elevation model in temperate forest soils of France. Soil Sci. 163, 103–108.

Burrough, P.A., 1986. Principles of Geographical Information Systems for Land Resources Assessment. Oxford University Press, New York, p. 50.

Chaplot, V., Bernoux, M., Walter, C., Curmi, P., Herpin, U., 2001. Soil carbon storage prediction in temperate forest hydromorphic soils using a morphologic index and digital elevation model. Soil Sci. 166, 48–60.

Dobos, E. and Daroussin, J., 2005. Potential drainage density Index (PDD). In: E. Dobos, J. Daroussin, L. Montanarella, (Eds). An SRTM-based procedure to delineate SOTER terrain units on 1:1 and 1:5 million scales. European Commission, EUR 21571 EN.

Dobos, E., Daroussin, J. and Montanarella, L., 2005. The development of a quantitative procedure for building physiographic units for the European SOTER database. In: E. Dobos, J. Daroussin, L. Montanarella, (Eds). An SRTM-based procedure to delineate SOTER terrain units on 1:1 and 1:5 million scales. European Commission, EUR 21571 EN.

Dobos, E., Micheli, E., Baumgardner, M.F., Biehl, L., Helt, T., 2000. Use of combined digital elevation model and satellite radiometric data for regional soil mapping. Geoderma 97, 367–391.

Dobos, E., Montanarella, L., Negre, T., Micheli, E., 2001. A regional scale soil mapping approach using integrated AVHRR and DEM data. Int. J. Appl. Earth Observations Geoinformation, 3, 30–41.

Farr, T.G., Kolbrick, M., 2000. Shuttle Radar Topography Mission produces a wealth of data. Eos, Transactions, Amer. Geophys. Union 81, 583–585.

Gesch, D.B., Verdin, K.L., Greenlee, S.K., 1999. New land surface digital elevation model covers the earth. Eos, Transactions, Amer. Geophys. Union 80, 69–70.

Gessler, P.E., Chadwik, O.A., Chamran, F., Althouse, L., Holmes, K., 2000. Modeling soil-landscape and ecosystem properties using terrain attributes. Soil Sci. Soc. Amer. J. 64, 2046–2056.

McBratney, A.B., Mendonca Santos, M.L., Minasny, B., 2003. On digital soil mapping. Geoderma 117, 3–52.

McKenzie, N.J., Ryan, P.J., 1999. Spatial prediction of soil properties using environmental correlation. Geoderma 89, 67–94.

Moore, I.D., Gessler, P.E., Nielsen, G.A., Peterson, G.A., 1993. Soil attribute prediction using terrain analysis. Soil Sci. Soc. Amer. J. 57, 443–452.

Salomonson, B., Barnes W.L., Xiong, X., Kempler, S. and Masuoka, E., 2002. An overview of the Earth Observing System MODIS Instrument and associated data systems performance. Proc. of the Int. Geoscience and Remote Sensing Symposium. (IGARSS 02) Sydney, Australia.

Várallyay, Gy., Hartyáni, M., Marth, P., Molnár, E., Podmaniczky, G., Szabados, I. and Kele. G., 1995. Talajvédelmi Információs és Monitoring Rendszer. 1 kötet. (In English: Soil Monitoring and Information System) Módszertan. Földművelésügyi Minisztérium, Budapest.

Wilson, J.P., Gallant, J.C., 2000. Secondary terrain attributes. In: J.P. Wilson and J.C. Gallant (Eds.), Terrain Analysis. Principles and Applications. John Wiley & Sons, Inc., New York.

Developments in Soil Science, volume 31
P. Lagacherie, A.B. McBratney and M. Voltz (Editors)

Chapter 37

REGIONAL ORGANIC CARBON STORAGE MAPS OF THE WESTERN BRAZILIAN AMAZON BASED ON PRIOR SOIL MAPS AND GEOSTATISTICAL INTERPOLATION

M. Bernoux, D. Arrouays, C.E.P. Cerri and C.C. Cerri

Abstract

The study area was defined as 6°-square portion of the Brazilian Amazon, from longitudes 60–66° West and latitudes 8–14° South. This region corresponds to ~334,000 km^2 of the western Brazilian Amazon, that is 6.7% of the Brazilian Amazon basin.

A database was constructed from soil profile information reported in published and unpublished soil surveys. Bulk density (BD) values were derived from multiple regressions with other available parameters. Two digitised soil maps were considered: a 1:5,000,000 soil map and 1:1,000,000 soil map of the region that was digitised for the purpose of this study.

Individual C stock for each profile was calculated by two procedures. The first and classical procedure consists in summing C densities by horizon, determined as a product of BD, C concentration, and horizon thickness. The second calculation used a double exponential model of organic C vertical distribution in soils: $(C(X) - C2)/(C1 - C2) = (e^{-b\,X} - e^{-b\,X2})/(e^{-b\,X1} - e^{-b\,X2})$, where X and C(X) are the depth and the C content, X1 and C1 the depth and the C content of a fixed upper position, and X2 and C2 for a fixed deeper position.

Regional C estimates were derived using different map approaches or geostatistical interpolations. Geostatistical interpolations were run not only on the individual C stock, but also on the parameters of the vertical model separately. Two types of validation were conducted, an "internal" validation known as "jackknifing" and an external validation with preserved data (about 10%).

The regional stock was obtained multiplying the carbon content of each taxonomic unit by its area. This calculation leads to similar results using the soil map either at 1:5,000,000 scale or at 1:1,000,000. Regional stocks (0–100 cm) ranged between 2100 Tg using median values, and 2300 Tg, using means, and exhibit an associated error (based on SD) of 900 Tg.

Variographic analysis showed that carbon stocks exhibit a spatial structure at the regional scale. However the high nugget effect reveals that about half of the spatial variability may appear within distances less than a few tens of kilometres. Results of the fitting of the experimental semivariogram for the top metre showed the same pattern with a nugget effect of 4.70 (kg C m^{-2})^2 and a sill value of 4.26 (kg C m^{-2})^2, but the range is reduced to 513 km.

Geostatistical interpolations run either on the C stocks values or independently on each parameters of the vertical model led to similar results.

37.1 Introduction

Soil organic carbon (OC) represents a major pool of carbon within the biosphere. Recent concerns about rising levels of atmospheric CO_2 have directed attention to carbon stocks in soils of the world (Post et al., 1982; Eswaran et al., 1993; Batjes, 1996) and to their role as both a source and sink of carbon. Over short periods of time, changes in vegetation and in land use patterns have a marked effect on topsoil carbon storage. Tropical soils represent at least 32% of the total mass of organic C stored in the soils of the world (Eswaran et al., 1993; Dixon et al., 1994). Among tropical ecosystems, the Amazonian forest is known to play a major role in C sequestration and release (Bernoux et al., 2001a). However, global and regional estimates of C storage in this ecosystem are few (Batjes and Dijkshoorn, 1999; Bernoux et al., 2002; Schroeder and Winjum, 1995). These carbon pools are difficult to estimate because of the limited availability of reliable, complete and uniform data for soils (C concentration and bulk density (BD) down to a sufficient depth) as pointed out by Batjes (1996) and Bernoux et al. (1998a,b). Other principal reasons why C pools are difficult to estimate are the still limited knowledge of the extent of soil types (Sombroek et al., 1993; Batjes, 1996), the high-spatial variability of soil C even within one soil map unit, and the confounding effects of the factors controlling the soil OC cycle (Pastor and Post, 1986; Parton et al., 1987).

The objectives of this study are to compare different techniques in order to obtain reliable estimates of the carbon stocks in the Brazilian Amazon basin. Our goal is to provide more accurate estimates of soil OC under primary undisturbed forest in order to better assess the relative impact of human disturbance caused by clear-cutting and agricultural or grazing use.

37.2 Materials and methods

37.2.1 Study area and soils

The study area was defined as the Brazilian part of a 6°-square, from 60 to 66° West longitude and 8 to 14° South latitude (Fig. 37.1). This region corresponds to 334,100 km^2 of the western Brazilian Amazon that is 6.7% of the legally recognised Amazon basin. Politically this region correspond to nearly all of the State of Rondônia, and small portions of the States of Amazonas and Mato Grosso.

Two main soil divisions of the Brazilian soil classification, Latossolos (Oxisols) and Podzólicos (Ultisols and Alfisols), dominated the studied area covering respectively 40 and 34% of its total (Bernoux et al., 2001b, 2002). These soils are thick

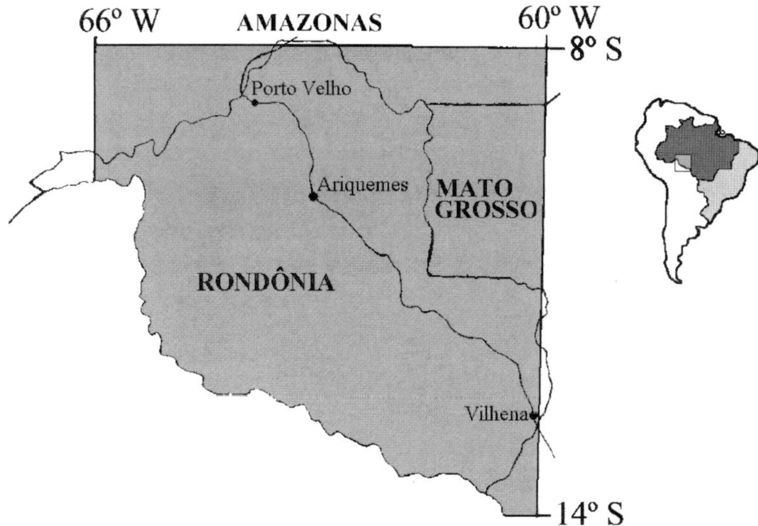

Figure 37.1. *Location of the study area.*

mineral soils with thickness often >2 m. The remainder is distributed among 13 soil divisions, only 4 of which are more than 3% of the superficies: the Plintossolos (Inceptisols, Oxisols and Alfisols) and the Gleissolos (Entisols and Inceptisols) still represent a noticeable extent, accounting for 5 and 3% of the area. The Litólicos (Entisols and Molisols) and the Areais quartzosas (Entisols) show extents representing 7.5 and 5% of the total area. In total, these six soil types cover 95% of the area.

37.2.2 Database

This study used data from the western Brazilian Amazon basin (Rondônia State). A database was constructed from soil profile information for pits surveyed by (i) the Radambrasil project carried out in this region (Ministério das Minas e Energia – Projeto Radambrasil, 1978, 1979) and (ii) by the SNLCS-Embrapa (Empresa Brasileira de Pesquisa Agropecuaria, 1983).

A specific soil database was elaborated from physico-chemical results of soil horizons sampled during surveys realised during the end of the 70 s and the beginning of the 80 s, and published as reports (Ministério das Minas e Energia, 1978, 1979; Rodrigues, 1980; Empresa Brasileira de Pesquisa Agropecuaria, 1983).

In these reports, carbon concentrations (Walkley–Black method) are expressed in g/100 g of fine earth (<2 mm fraction) by horizon. For estimating the carbon storage, a correction was first applied considering that the soil fraction

>2 mm is carbon free. Soil BD was estimated using multiple regressions from other available parameters (e.g. clay content, OC and pH). These regressions were established from data on soil horizons spread over the whole Amazon basin (Bernoux et al., 1998b).

Reported results of soil analyses for 3016 soil horizons, corresponding to 796 soil profiles, were stored in the regional database. But, only 782 profiles were georeferenced, and only 639 of which (2534 soil horizons) having carbon concentration data reported and used in this study.

37.2.3 Soil maps

Two digitised soil maps were considered for this purpose; the first being the 1:5,000,000 soil map (Empresa Brasileira de Pesquisa Agropecuaria, 1981). The second map was obtained by digitisation of the 1:1,000,000 soil maps published in Volumes 16 and 19 of the Radambrasil soil survey report (Ministério das Minas e Energia, 1978, 1979) covering the regional study zone.

37.2.4 Carbon stocks profile by profile

To estimate C stocks and changes, knowledge of the vertical distribution of C in profiles is required. The classical way of calculating C densities (C mass per area) for a given depth consists in summing C densities by horizon, determined as a product of BD, C concentration, and horizon thickness. For each profile, when possible, the corresponding stock C stock (P100 in kg .m^{-2}) was calculated for the first top metre.

A double exponential model was also used in this study. This model was first tested for temperate forest soils (Arrouays and Pélissier, 1994), and also appeared to be appropriate for the tropical soils of this region (Bernoux et al., 1998a). The model equation is:

$$(C(X) - C2)/(C1 - C2) = (e^{-b\,X} - e^{-b\,X2})/(e^{-b\,X1} - e^{-b\,X2}) \qquad (37.1)$$

where X and C(X) are the depth and the C content, X1 and C1 the depth and the C content of a fixed upper position, and X2 and C2 for a fixed deeper position. More detailed regarding this model can be encountered in Arrouays and Pélissier, 1994 and Bernoux et al. (1998a).

X1 was set to 0 cm that is the theoretical contact litter/soil, and X2 was set to 100 cm, therefore C(X) could be written:

$$(CX - C2)/(C1 - C2) = (e^{-b\,X} - e^{-100\,b})/(1 - e^{-100\,b}) \qquad (37.2)$$

For each soil profile of the database, the parameters of the fitted equation were determined with the non-linear regression module Statistica (Statsoft, Inc., 1996). The corresponding C stock (M100 in kg m^{-2}) was obtained integrating CX (eq. 37.2):

$$M100 = 10(C2 - (K1 \,.\, K2) + (K2/b) \,.\, (1 - K1)) \qquad (37.3)$$

Where $K1 = e^{-100\,b}$ and $K2 = (C1 - C2)/(1 - K1)$.

37.2.5 Geostatistical analysis

The geostatistical analyses were run using GS+ software on a PC (Gamma Design Software, 1998). Omnidirectional semivariograms were calculated using a 50 km step (lag distance) to a maximum of 600 km, but using the population after removal of the extremes, that is values beyond three times the interquartile range (IQR) (IQR, difference between the 75th percentile and the 25th percentile of a variable's distribution) from the upper and lower quartiles. Outliers were removed because their inference can be substantial on the estimation of the spatial structure (Cressie, 1993). In fact, very few outliers were removed.

Two types of validation were conducted: an "internal" validation known as "jackknifing" (Cressie, 1993) and an external validation with preserved data (Bourennane et al., 1996). The cross-validation technique ("jackknifing") consists of testing the validity of the semivariogram model by kriging at each sampled location using all other neighbouring samples and then comparing the estimates with the real values. A mean error (ME) close to zero indicates no systematic bias, and a root mean square error (RMSE) close to unity shows a good fit of the semivariogram model and its parameters to the dataset. But this validation is nothing more than a validation of the fitted semivariogram to the original data; it does not validate the reliability of the prediction methods for external data. Before each variographic analysis, 10% of the initial data were kept out (one of ten values of the data after geographical classification by increasing longitude and latitude locations), and the geostatistical analyses were conducted with the remaining data (90% of the information). The 10% of values preserved were used to calculate the mean error (ME2) and the root mean square error (RMSE2). The ME2 should be close to zero for unbiased methods, and the RMSE2 should be small for an unbiased and precise prediction. After cross and external validations, maps of estimated values were obtained by block kriging using 1-km^2 blocks.

37.3 Results and discussion

37.3.1 Classical map approach

It was possible to calculate P100 for the upper metre for only 50% of the original 639 profiles. This is due to the fact that numerous profiles have reported carbon results for topsoil horizons. The M100 calculation was possible for 424 georeferenced soil profiles. Table 37.1 reported the summary statistics on C stocks values calculated using the two methods and on the double exponential parameters model.

Table 37.1. *Statistics on C stock values (P100 and M100 in kg C m^{-2}) and on the parameters of the double exponential depth model.*

Variable	n	Average	Median	Minimum	Maximum	SD
P100	324	7.30	6.30	0.61	41.62	4.51
M100	424	7.32	6.55	1.50	25.19	3.52
C1	424	2.722	2.170	0.336	26.841	2.340
C2	424	0.314	0.288	0.000	1.169	0.168
B	424	0.0577	0.0479	0.0008	0.3202	0.0420

Statistics for M100 and P100 gave similar values. The coefficient of correlation calculated among them was 0.972, for the 304 profiles having both P100 and M100.

A first approximation of the regional stock could be calculated multiplying the values of the mean or of the median by the total soil extent. Using the mean values of P100 would lead to a regional stock of 2400 ± 1500 Tg C stored in the first metre. Using the median, these values are reduced to 2100 Tg in the first metre. In order to refine these values, elementary statistics where calculated by taxonomic unit (Table 37.2).

The regional stock is obtained multiplying the carbon content of each taxonomic unit by its area. This calculation leads to similar results using either the 1:5,000,000 soil map (Embrapa) or the 1:1,000,000 soil map. Regional stocks ranged between 2100 Tg using median values, and 2400 Tg, using means, and exhibit an associated error (based on SD) of 900 Tg. Using the map from Embrapa (1:5,000,000) instead of the Radam (1:1,000,000) leads to an increase of only 25 Tg. It is striking that the results are very similar using (1) the median value of all P100 or M100 values or (2) the medians of M100 and P100 segregated by soil type. But in the second calculation the associated errors based on SD are reduced to less than 50%, whereas it reached 71% in the first calculation.

37.3.2 Classical geostatistical approach

Experimental semivariograms for the variable P100 was best (i.e. most satisfactory validation) modelled using a nugget effect plus a spherical model according to the function:

$$\gamma(h) = c_0 + c_1\left\{\frac{3h}{2a} - \frac{1}{2}\left(\frac{h}{a}\right)^3\right\} \text{ for } 0 < h \leq a$$

$$\gamma(h) = c_0 + c_1 \text{ for } h > a$$

$$\gamma(0) = 0$$

where a is the range of the model, c_0 the nugget and c_1 the structured variance.

Table 37.2. *Statistics of P100 and M100 variables by taxonomic unit. Unit is* $kg\,C\,m^{-2}$.

Soil typel	P100						M100					
	n	Mean	Min	Max	SD	Median	*n*	Mean	SD	Min	Max	Median
PVA	154	6.40	1.44	23.12	3.14	5.84	188	6.51	3.05	1.60	23.44	5.85
LVA	34	7.49	4.40	19.46	3.03	6.97	51	7.95	3.13	4.10	19.60	7.23
LA	26	7.30	4.94	10.71	1.48	7.32	30	7.76	1.61	5.07	11.62	7.80
AQ	17	8.96	4.68	19.13	4.29	6.88	17	8.48	4.02	4.64	19.66	6.71
Ca	13	5.98	2.30	10.69	2.71	5.51	27	6.90	2.60	2.48	10.97	6.74
Pa	12	7.06	1.80	27.69	6.77	5.78	16	7.14	5.06	2.12	24.74	6.16
LVE	10	6.90	3.78	11.52	2.37	6.65	12	7.16	2.29	3.85	12.16	6.59
GPH	9	7.99	3.71	10.86	2.54	7.77	16	7.35	2.18	4.06	11.64	7.06
SA	9	5.57	3.45	11.66	2.45	5.13	10	5.98	2.12	4.10	11.69	5.24
TRE	8	9.52	4.57	15.76	3.80	8.44	15	8.58	3.45	4.54	16.13	7.86
BA	6	11.14	6.61	19.59	4.78	10.37	8	12.15	4.26	7.00	20.15	12.69
Ce	5	10.61	5.19	23.04	7.63	6.23	5	9.48	7.87	5.28	23.54	6.04
PVE	4	5.90	4.16	9.96	2.73	4.74	8	5.96	2.14	3.91	10.13	5.06
PH	4	3.18	1.34	7.62	2.98	1.88	2	8.31	9.63	1.50	15.12	8.31
GH	4	23.32	9.57	33.79	10.25	24.96	5	17.15	5.87	8.76	25.19	16.79
AQH	4	6.02	0.61	13.88	5.78	4.79	3	10.29	6.21	3.22	14.88	12.76
LR	3	19.29	7.12	41.62	19.36	9.14	3	8.34	1.30	7.45	9.83	7.73
Cd	1	13.21				13.21	4	8.55	4.40	4.02	13.42	8.37
SL	1	6.73				6.73	2	7.41	0.36	7.16	7.66	7.41
Pe							1	3.71				3.71
Pd							1	3.60				3.60

Note: PVA = Podzolico Vermelho Amarelo distrofico (Ultisols), LVA = Latossolo Vermelho Amarelo (Oxisols), LA = Latossolo Amarelo (Oxisols), AQ = Areia Quartzosas (Psamments), Ca = Cambissolo alico (Inceptisols), Pa = Planossolo alico (Alfisols, Mollisols), LVE = Latossolo Vermelho Escuro (Oxisols), GPH = Gley Pouco Humico (Aquic suborder), SA = Solo Aluvial (Entisols-Fluvents), TRE = Terra Roxa Estruturada (Ultisols, Alfisols), BA = Brunizem Avermelhado (Chernozems), Ce = Cambissolo eutrofico (Inceptisols), PVE = Podzolico Vermelho Escuro (Alfisols), PH = Podzol Hidromorfico (Spodosols), AQH = Areia Quartzosas Hydromorfica (Psamments), LR = Latossolo Roxo (Oxisols), Cd = Cambissolo distrofico (Inceptisols), SL = Solo Litolico (Lithic subgroup), Pe = Planossolo eutrofico (Ultisols), Pd = Planossolo distrofico (Alfisols, Mollisols).

The experimental semivariogram for P100 was fitted with a nugget effect of 4.70 $(kg\,m^{-2})^2$ plus a spherical model using a sill value of 4.26 $(kg\,m^{-2})^2$ and range of 513 km.This results shows that carbon densities exhibit a spatial structure at the regional scale. However the high nugget effect reveals that about half of the spatial variability may appear within distances inferior to a few tens of kilometres.

Results of the validation procedures showed that the models of the semivariograms are well adapted to the data used to calculate the experimental semivariograms (cross-validation, ME = −0.015 and RMSE = 1.121) and to external data (external validation, ME2 = 0.065 and RMSE2 = 2.668).

Maps of estimated carbon densities and their associated kriged standard deviation (KSD) were calculated by block kriging (Plate 37 (see Colour Plate Section)). Each map is made up of 334,107 1-km^2-square blocks. The P100 kriged values ranged from 4.27 to 10.40 $kg\,m^{-2}$, and KSD from 0.54 to 1.31 $kg\,C\,m^{-2}$. The mean values being 6.665 $kg\,m^{-2}$. The mean of the kriged values is close to the median of the population of the carbon densities without segregation of soil type. This results in a regional soil carbon content of 2220 Tg C, indicating an intermediary level between the previous estimates, but now the associated error is considerably reduced to 295 Tg, that is only 13.5% of the amount.

The geostatistical approach gives regional estimates very similar to those derived from the classical approach based on a soil map, but the advantage of the geostatistical method is to lead to a much lower global associated error and to provide a map of the error.

37.3.3 Combined vertical model and geostatistical approach

The three parameters of the vertical model were estimated for 424 soil profiles. Variographic analysis was done independently for each of the parameters. Results of the variographic analyses showed that the parameters C1 and C2 were spatially structured, but with a high nugget effect. On the contrary the parameter b related to the depth profile curvature show no spatial dependency. Experimental variograms corresponding to parameters C1 and C2 where modelled with a linear model with respective nugget effects of 0.85 and 0.0198 $(kg\,m^{-3})^2$ and slopes of 6.83×10^{-4} and 1.65×10^{-5} $(kg\,m^{-2})^2\,km^{-1}$. Even with such high nugget effects, the validations gave good results (Table 37.3). The ME of the external validation represented respectively for C1 and C2, 2.1 and 4.5% of the mean of each population.

After validation the prediction grid of the parameter values at each location (Plate 37) were obtained by block kriging as described above. Because of the non-linearity in the profile depth model, conditional simulation might be better than kriging, but kriging procedures were chosen because they can be considered as a more generic tool usable by non-specialists. Values varied from 1.61

Table 37.3. *Results of cross-validation and external validation for the parameters C1 and C2.*

Variable	*N*	ME	RMSE	*n*	ME2	RMSE2	Outliers
	Cross-validation			External validation			
C1	351	0.000	1.051	39	−0.047	0.917	34
C2	380	−0.023	1.064	42	−0.014	0.137	2

to 3.19 for parameter C1, with a mean of 2.19. Parameters C2 varied from 0.23 to 0.50, with a mean value of 0.31. Associated kriging error represented in both cases from 9 to 23% with a ME of about 15%.

From the grid of parameter values it is possible to calculate the C stocks corresponding to the first top metre. As parameter b showed no structure, two solutions were possible: the use of a fixed value (mean of median of the population) or a moving window average. The population of parameter b is lognormal, therefore the use of the median was judged more adequate.

This approach was validated in two different ways. First an external validation was done using the values of the 32 reserved data. Those soil profiles had a mean P100 value of 6.914 and mean M100 value of 6.975. The estimates of the stock made punctually for these profiles presented very small ME: −0.12 when compared with P100 values or +0.039 when compared with M100 values.

The second validation was based on comparing the grid obtained by calculating C stock with the grids for parameters C1 and C2 and the median of parameter b population, with the grid corresponding to kriging P100 values. The deviations in percentage relative to the P100 kriged values, ranged from a subestimation of 21.3% to an overestimation of 37.2%, and the mean deviation was an overestimation of 8.2%. Nevertheless the absolute deviations <10% represented 67% of the total area.

37.4 Conclusion

The different approaches used to derive regional C stock estimates furnished similar values. Nevertheless the geostatistical approaches lowered the apparent associated error. The model of vertical variability of C content is well adapted to this region of the Amazon. The use of the model parameters with a geostatistical approach is promising for several reasons: it increases the predictable area of determinable C stock down to 1 m or below, it allows the determination of c stock of a determined region to the bottom of the soil profile, and finally each model parameter has its own significance and can be studied independently.

Acknowledgments

This research was supported partly by the Fundação de Amparo a Pesquisa do Estado de São Paulo (FAPESP) with contracts 94/6046-0 and 95/1451-6, by the Fundação Coordenação de Aperçoamento de Pessoal de Nível Superior (CAPES-MEC) grant number 2129/95, and by the Global Environment Facility (project number GFL/2740-02-4381).

References

Arrouays, D., Pélissier, P., 1994. Modeling carbon storage profiles in temperate forest humic loamy soils of France. Soil Sci 157, 185–192.

Batjes, N.H., 1996. Total carbon and nitrogen in the soils of the world. Eur. J. Soil Sci. 47, 151–163.

Batjes, N.H., Dijkshoorn, J.A., 1999. Carbon and nitrogen stocks in the soils of the Amazon region. Geoderma 89, 273–286.

Bernoux, M., Arrouays, D., Cerri, C.C., Bourennane, H., 1998a. Modeling vertical distribution of carbon in Oxisols of the Western Brazilian Amazon (Rondônia). Soil Sci 163, 941–951.

Bernoux, M., Arrouays, D., Cerri, C.C., Volkoff, B., Jolivet, C., 1998b. Bulk densities of Brazilian Amazon soils related to other soil properties. Soil Sci. Soc. Am. J. 62, 743–749.

Bernoux, M., Carvalho, M.C.S., Volkoff, B., Cerri, C.C., 2001a. CO_2 emission from mineral soils following land-cover change in Brazil. Global Change Biol 7, 779–787.

Bernoux, M., Carvalho, M.C.S., Volkoff, B., Cerri, C.C., 2002. Brazil's soil carbon stocks. Soil Sci. Soc. Am. J. 66, 888–896.

Bernoux, M., De Alencastro Graça, P.M., Fearnside, P.M., Cerri, C.C., Feigl, B., Piccolo, M.C., 2001b. Carbon storage in biomass and soils. In: M.E. McClain, R.L. Victoria, and J.E. Richey (Eds.), The Biogeochemistry of the Amazon Basin. Oxford University Press, New York, pp. 165–184, 365pp.

Bourennane, H., King, D., Chéry, P., Bruand, A., 1996. Improving the kriging of a soil variable using slope gradient as external drift. Eur. J. Soil Sci. 47, 476–483.

Cressie, N.A.C., 1993. Statistics for Spatial Data, Revised edition. Wiley Interscience, New York, 900pp.

Dixon, R.K., Brown, S., Houghton, R.A., Solomon, A.M., Trexler, M.C., Wisniewski, J., 1994. Carbon pools and flux of global forest ecosystems. Science 263, 1900–1985.

Empresa Brasileira de pesquisa Agropecuaria, 1981. Mapa de Solos do Brasil, escala 1:5 000 000.

Empresa Brasileira de Pesquisa Agropecuaria, 1983. Levantamento de reconhecimento de média intensidade dos solos e avaliação da aptidão agrícola das terras do Estado de Rondônia. Contrato Embrapa/SNLCS-Governo do Estado de Rondônia. Rio de Janeiro, Brazil, 895pp.

Eswaran, H., van Den Berg, E., Reich, P., 1993. Organic carbon in soils of the world. Soil Sci. Soc. Am. J. 57, 192–194.

Gamma Design Software, 1998. Geostatistics for the environmental sciences. GS+ version 3.1. Gamma Design Software, Plainwell, MI, http://www.gammadesign.com

Ministério das Minas e Energia, 1978. Projeto RADAMBRASIL, programa de integração nacional. Levantamento de recursos naturais. Vol. 16, Folha SC-20 "Porto Velho", 663pp., MME/DNPM, Rio de Janeiro, Brazil.

Ministério das Minas e Energia, 1979. Projeto RADAMBRASIL, programa de integração nacional. Levantamento de recursos naturais. Vol. 19, Folha SD-20 "Guaporé", 368pp., MME/DNPM, Rio de Janeiro, Brazil.

Parton, W.J., Schimel, D.S., Cole, C.V., Ojima, D.S., 1987. Analysis of factors controlling soil organic matter levels in Great Plains. Soil Sci. Soc. Am. J. 51, 1173–1179.

Pastor, J., Post, W.M., 1986. Influence of climate, soil moisture and succession on forest carbon and nitrogen cycles. Biogeochemistry 2, 3–27.

Post, W.M., Emmanuel, W.R., Zinke, P.J., Stangenberger, A.G., 1982. Soil carbon pools and world life zones. Nature 298, 156–159.

Rodrigues, T.A., 1980. Estudo expedito de solos do Território Federal de Rondônia para fins de classificação, correlação e legenda preliminar. Contrato EMBRAPA/SNLCS-Governo do Território Federal de Rondônia. SNLCS, Boletim Técnico no. 73, 145pp., Rio de Janeiro, Brazil.

Schroeder, P.E., Winjum, J.K., 1995. Assessing Brazil's carbon budget: 1. Biotic carbon pools. Forest Ecol. Manage. 75, 77–86.

Sombroek, W.G., Nachtergaele, F.O., Hebel, A., 1993. Amounts, dynamics and sequestration of carbon in tropical and subtropical soils. Ambio 22, 417–426.

StatSoft, Inc., 1996. Statistica® for Windows, Computer program manual. StatSoft, Inc., 2300 East 14th Street, Tulsa, OK 74104.

Developments in Soil Science, volume 31
P. Lagacherie, A.B. McBratney and M. Voltz (Editors)

Chapter 38

IMPROVING THE SPATIAL PREDICTION OF SOILS AT LOCAL AND REGIONAL LEVELS THROUGH A BETTER UNDERSTANDING OF SOIL-LANDSCAPE RELATIONSHIPS: SOIL HYDROMORPHY IN THE ARMORICAN MASSIF OF WESTERN FRANCE

V. Chaplot and C. Walter

Abstract

One of the most important scientific challenges of digital soil mapping is to develop generic models that may allow the soils to be predicted over large areas. Our objective here is to quantify the relationships between the soils and their "environment" to further our understanding of the rules of soil distribution. The study was conducted in the Armorican Massif (30,000 km^2), a metamorphic area of western France where soils formed under conditions of poor drainage due to water accumulation within slopes and/or low infiltration. Field investigations on a 2-ha site and detailed soil map (1:25,000) analysis, were used to compare quantitative relationships between hydromorphic soils and soil-forming factors of topography, climate and geology. Soil surveys at both scales revealed a continuous succession downslope–upslope of soils with fluvic properties, soils with glossic and/or albic features, soils with other redoximorphic features (RFs) and soils without RF. These soils are distributed along slopes in the same order with only variations in their relative extent. At a single site, the soil hydromorphy significantly correlated with terrain attributes extracted from the surface topography of elevation above the stream bank, ES ($r^2 = 0.79$) and the compound topographic index, CTI ($r^2 = 0.62$). Greater correlations were obtained by using attributes from the topography of the saprolite upper boundary, that possibly better-explained water movement within hillslopes. The total variance of regional data accounted for by the first two principal components (PCs) was 74%. The first PC explained 52% of the variance and exhibited a pattern associated with the presence or absence of RFs in soils. Environmental factors having the highest (absolute) values for this factor are not only mean slope gradient but also geologic substrate, net rainfall (NR) and uplift ratio (UR). These results revealed that other factors than topography have to be considered to account for the range of processes leading to soil hydromorphy over a region. Because these processes act at different spatial scales, multi-scale approaches are required not only to detect the patterns of soil distribution but also to develop generic and high-resolution prediction models for the characterisation of large areas.

38.1 Introduction

Since the soil cover is a continuum of loose material that controls the transfers and transformations of solids, liquids, gases as well as of energy, it is of prime importance to increase our knowledge of the spatial distribution of soils within landscapes. Systematic soil surveys initially begun in many countries of the world in the 20th century for natural resources inventory. But because the standard soil surveys are very costly, requiring a large number of individual observations, measurements and analyses, the needs of a worldwide soil characterisation has met with great difficulties. The recent and increasing environmental concerns have renewed the need of soil inventories since accurate information on soil resources is for instance needed to foster increased crop yields and economical growth while limiting environmental damages. Spatial predictions of soils could be achieved using a variety of interpolation methods including kriging, inverse distance weighting, spline, global polynomial, moving polynomial, radial basis and others (e.g. Voltz and Webster, 1990; Deutsch and Journel, 1992). But in many instances, soil measurements or observations are lacking or are too coarsely spaced, so many interpolation schemes are deficient, especially when high-spatial definition is required. One way to address this problem is to correlate the soil observations or measurements with an auxiliary variable whose spatial distribution could more readily be measured. For the most part, these methods have focused on the relationship between the soil properties and topography because of the known role relief plays (e.g. Moore et al., 1993; Odeh et al., 1994; McKenzie and Ryan, 1999; Bourennane et al., 2000; Gessler et al., 2000; Lagacherie and Voltz, 2000) and also the widespread availability of DEMs used for terrain attribute estimation (e.g. Zevenbergen and Thorne, 1987; Garbrecht et al., 2001). Although DEMs have revolutionised soil modelling (Mc Bratney et al., 2003), their direct use in the estimation of the spatial distribution of soils or soil properties over large areas still faces "prediction quality problems" (e.g. Thompson et al., 2001) leading to the development of mixed methods requiring additional soil observations or the use of a number of supposedly cheaper and easier-to-obtain covariates from new technologies such as electrical resistivity, airborne γ-radiometric or hyperspectral imageries (e.g. Godwin and Miller, 2003).

It is believed that a better understanding of soil-landscape relationships at the natural or preferred spatial scales at which natural processes occur (Sivapalan and Karma, 1995) would allow large areas to be characterised with a high-spatial resolution. This implies the implementation of multi-scale approaches: to detect the patterns and processes of soil formation; and to evaluate the impact of scale on modelling.

The study was conducted in the Armorican Massif, an area of 30,000 km^2 of western France with diverse geological substrates, topographic conditions, climate and tectonic regimes. Previous studies have shown that soils, developed either within geological substrates and aeolian loams, mainly differ in their degree of soil saturation by water, varying between those soils that undergo continuous or periodic saturation leading to the formation of redoximorphic features (RFs) to those with no excess of water (e.g. Curmi et al., 1998). In the region, topographic indices derived from Darcy's law, originally developed for hydrological modelling (Beven and Kirkby, 1979) have been used to evaluate the spatial distribution of RFs on limited and single areas (e.g. Mérot et al., 1995). But when applied at other locations of the same region, these predictive models exhibited high prediction errors (Chaplot et al., 2003a). For instance, when applying a regression model to four other sites of the AM differing in topography (mean elevation from 39 to 202 m and slope gradient from 3.4 to 7.9%), parent material (granite and schist) and precipitation (700–900 mm per year) these last authors showed that low prediction errors occurred at the site of model generation and within the same area, even differing by the geologic substrate, whereas the model was biased elsewhere. Among the possible reasons to explain the lack of generality of their model, the authors suggested a poor knowledge of the soil-landscape relationships. In this study, our objective was to compare the empirical studies of Chaplot et al. (2003a,b) at different scales to deepen the understanding of the soil-landscape relationships, and to further developing a coherent theoretical framework for spatial prediction of soils over the region. Field investigations on limited hillslopes and analysis of soil maps (1:25,000) from the Armorican Massif were used for fitting quantitative relationships in the region between hydromorphic soils and soil-forming factors at different spatial scales.

38.2 Materials and methods

38.2.1 The study sites

The Armorican Massif (between longitudes 47.30°N and 48.90°N and latitudes 1.33°W and 4.78°W) is moderately elevated (from 0 to 380 m) with maximum elevations occurring in the western part (Plate 38a (see Colour Plate Section)). A network of small valleys incised into the bedrock characterises its physiography. The Massif is a complex basement of Proterozoic and Paleozoic aged bedrock with a wide diversity of saprolites. Geological substrates include granites, schists, mica schists, orthogneisses and sedimentary rocks (Plate 38b (see Colour Plate Section)). Granites and schists are the two dominant substrata, comprising 80% of the region's surface. During the Variscan orogeny, lithologic facies were mylonitised and strongly fractured and thus predisposed to deep weathering. Paleocene and Eocene ages, produced deep saprolites (>30 m), rich in kaolin and with ferralithic soils (Soil Survey Staff, 1999) subjected to erosion by the powerful climate cycles from periglacial to temperate, and also by a succession of tectonic movements irregularly expressed in the Armorican Massif during the late Miocene (Messinian) and early Quaternary. Afterwards, a variable depth of aeolian loam, blown from west to east over the last 300,000-year mantles the region. These deposits have been reworked on slopes by solifluxion, water erosion and local alluviation. The drainage patterns reflect the bedrock joints and fractures made by tectonic movements (Van Vliet-Lanoe et al., 1998). The climate is temperate with mean annual temperatures ranging from 10.8°C in the west (city of Brest) to 11°C in the east (city of Rennes). The 30-year average precipitation is 700–1300 mm and the mean annual net rainfall (NR, difference between precipitation and potential evapotranspiration, PET) is 150–500 mm (Plate 38c (see Colour Plate Section)). The agriculture is intensive, essentially oriented towards pig production, dairy farming, winter cereals and maize.

Within the Massif, the soils mainly differ in their degree of soil saturation by water varying between those that undergo continuous or periodic saturation by water leading to the formation of RFs to those with no excess of water.

The soil-landscape relationships were analysed using (i) nine areas of the Armorican Massif and with a total area of 371 km^2 for which 1:25,000 maps were available and (ii) detailed field surveys of single hillslopes.

38.2.2 The analysis of the spatial distribution of soils

At the nine areas of the Armorican Massif selected to exhibit the largest possible range of topographic, climatic and geologic conditions found in the region, soil patterns were extracted along 314 hillslopes selected randomly. Transects were perpendicular to the contour lines, from stream bank to slope summit. For each transect, the soil sequence from downslope to upslope was recorded and the extent of each soil in proportion to total transect length was determined. The proportion of soils with redoximorphic, albic and/or glossic, leached features and the proportion of deep soils (i.e. with a depth of structured soil horizons >1 m) were considered.

Detailed soil surveys were performed in the field within a 2 ha hillslope of the eastern Massif located in the upper part of the "La Roche catchment", 20 km east of Fougère. At the study hillslope, soil pits using an auger hole were drilled according to a 10-m grid. Some additional soil observations were added to precisely define soil limits. At each of the 182 auger pit, several soil features were registered for the estimation of a quantitative index of soil hydromorphy, HI (HI $= P/(V \times C_h)$ with C_h the Munsell chroma of the 0–10 cm layer, P, the cumulative thickness of the soil with RFs divided by the total soil thickness). Other soil features considered for all layers of the soil profiles included the RFs of stagnic, gleyic, oximorphic and/or reductomorphic, and eventually albeluvic tonguing.

38.2.3 The characterisation of the soil-forming factors

A description of the soil-forming factors (Jenny, 1941) was performed both at local and regional levels. At the regional level, the variations of climate, tectonic regime and geological bedrock were considered. Geological substrate and mean slope gradient of the 282 transects were extracted from geologic and topographic maps with a 1:50,000 and 1:25,000 scale, respectively. NR (difference between the precipitation, P and the water loss from the soil through PET) was estimated for each map using 30 years of records. The uplift ratio (UR) was extracted from the map computed by Lague et al. (2000) for the whole Massif.

At the study hillslope, two DEMs with a 10-m resolution were generated (Chaplot et al., 2004) The first DEM is a numerical representation of the soil surface topography, generated from the contour lines with a 5-m interval and 643 additional theodolite points whereas the second DEM is a digital elevation model of the upper boundary of saprolites, generated using the set of 643 data points where the depth to saprolite upper boundary was determined by auger hole. Sub-surface topography is considered here since it was shown to affect the soil water content within hillslopes and thus greatly increased the predictions accuracy of HI (Chaplot et al., 2004). From these DEMs, several primary and secondary terrain attributes were estimated at the auger hole points: the elevation above the sea level (Z), the distance to the stream bank (L), the elevation above the stream bank (ES), the downslope gradient (DG), the slope gradient (S), the specific monodirectional catchment area (A_s, Moore et al., 1991), the specific multi-directional catchment area (A_{sm}), the profile (K_p), contour (K_c), and tangential (K_t) curvatures (Zevenbergen and Thone, 1987) as primary terrain attributes and used to compute the secondary attributes of modified compound topographic index (CTI_m), with DG in place of S, and A_{sm} in place of A_s, and the stream power index (SPI, Moore et al., 1993).

38.2.4 The quantification of the soil-landscape relationships

At the local level, the correlation coefficients r between the soil hydromorphy at each auger hole and the environmental factors of surface topography and saprolite shape were calculated.

Principal components analysis (PCA) was applied to the regional data to quantify the impact of the soil-forming factors of topography, geology and climate on the spatial distribution of hydromorphic soils. PCA converts the actual variables (in our case the soil properties) in the so-called factors, or principal components (PCs), which are linear combinations of the actual variables, not correlated with each other (i.e. they are orthogonal) and explaining altogether the total variance of the data (Jambu, 1991). In this multi-variate statistical tool the first and second factors often explains most of the variance (e.g. more than 70% of the total variance) and therefore the most information contained in the data.

38.3 Results

38.3.1 The environmental conditions at the study sites

General statistics (minimum, min.; maximum, max.; median; average, av.; standard deviation, SD; skewness, skew, kurtosis, kurt. and coefficient of variation, CV) of the elevation above the stream bank (ES), the distance to the stream bank (*L*), the mean slope gradient (*S*), UR and NR for the transects investigated in the field or through map analysis are presented in Table 38.1. The study map transects exhibited a great range of topographic situations as shown

Table 38.1. *Statistics for environmental factors. Two data sets are considered here: (i) 8 transects from the study site plus 24 transects of the same area surveyed by Chaplot et al (2003a); (ii) 282 transects investigated from 1:25,000 soil maps analysis (Chaplot et al., 2003b).*

	ES	L	*S*	UR	NR
Map analysis (282 transects)					
Min.	1	215	0.2	1.1	125
Max.	162	2,800	12.4	4.0	550
Median	23	775	3.0	2.6	225
Av.	27	859	3.5	2.7	241
Skew	22	1.3	1.2	–0.3	1.6
Kurtosis	9.7	2.4	2.1	–1.3	2.8
SD	19.1	425	2.1	1.0	108
CV	70	49	60.4	37.4	45
Sum		24,2297			
Detailed field investigations (32 transects)					
Min.	1.15	125	3.2	1.30	225
Max.	25.00	500	12.9	3.50	225
Median	6.70	254	5.4	3.50	225
Av.	6.32	331	6.3	2.44	225
Skew	2.45	0.11	1.16	–0.09	
Kurtosis	9.14	–1.70	0.55	–2.17	
SD	4.8	138	2.5	1.1	
CV	76.2	41.7	40.5	45.9	
Sum		10,607			

Note: General statistics (minimum, min.; maximum, max.; median; average, av.; standard deviation, SD; skewness, skew, kurtosis, kurt. and coefficient of variation, CV) of the elevation above the stream bank (ES), the distance to the stream bank (*L*), the mean slope gradient (*S*), uplift ratio (UR) and the net rainfall (NR) for the study transects.

by CV of 70% for elevation above stream level (ES), and 60% for the mean slope gradient, *S*, of transects (Table 38.1). ES ranged from 1 to 161 m and transects length from 215 to 2800 m. As a consequence, the computed mean slope gradient of transect was between 0.2 and 12%. Geological substrates include granite, schist, mica schist, orthogneisses, sedimentary rocks and eolian deposits. NR ranges between 125 and 550 mm per year and UR from 1.1 (i.e. relative downlift) in the south of the south Armorican shearing zone to 4.0 (i.e. relative uplift) (Table 38.1). Low URs were observed in the east of AM whereas maximum values occur mainly in the western part, between the south and north shearing zones.

The study hillslope is 150–250 m long and with a mean elevation range of ~10 m. Its topography is relatively smooth since slope angles range from 0% at the alluvial plain and hillslope summit to 8–10% at the middslope position. The general statistics from the eight transects of the study hillslopes and from an additional set of 14 transects from the same area and presented by Chaplot et al. (2003a), revealed high CV values for ES (CV = 76%) and *S* (CV = 40%) (Table 38.1). Finally, the mean slope gradient of transects surveyed in the field ranged between 3 and 13% (Table 38.1).

38.3.2 The spatial distribution of soils

Table 38.2 presents the statistics of the percentage of the total length of each catena for soils with (i) RFs, including soils with fluvic properties, glossic and/or albic features, and other RFs; (ii) soils without RFs.

At both scales of analysis, ~30% of the soil cover comprised soils with RFs (Table 38.2). The average value was 28% for transects surveyed in the field and 37% for the map transects. Soils with RF exhibited a CV of ~85% with values from 2 to 80% on the detailed hillslopes and from 0 to 100% from map analysis. A much greater CV characterised the soils with glossic features with 106% on the detailed hillslopes and 290% on maps which proportion ranged between 0 and 19% at the detailed transects and between 0 and 50 on the maps. Finally, the proportion of soils with fluvic features varied from an average 7% at detailed transects to 27% from map analysis (Table 38.2).

38.3.3 The relation between the soil spatial distribution and the soil-forming factors

At the study hillslope, correlation coefficients between the hydromorphic index, HI and the terrain attributes extracted from surface or sub-surface DEMs ranged between $r = 0.08$ for K_t on the surface or K_p on the sub-surface and $r = 0.80$ for *Z* and DG on the surface and the sub-surface, respectively (Chaplot et al., 2004). As regards the surface topography, greater correlation coefficients were observed for *L* ($r = 0.77$), *Z* ($r = 0.80$), ES ($r = 0.79$), CTI_m ($r = 0.62$) and

Table 38.2. *Statistics for soil features. Two data sets are considered here: (i) 8 transects from the study site plus 24 transects of the same area surveyed by Chaplot et al (2003a); (ii) 282 transects investigated from 1:25,000 soil maps analysis (Chaplot et al., 2003b).*

	Soils with fluvic properties	Soils with glossic features	Soils with redoximorphic features	Soils without redoximorphic features
Map analysis (282 transects)				
Min.	0.0	0.0	0.0	0.0
Max.	100.0	50.0	100.0	100.0
Median	25.0	0.0	28.6	71.4
Av.	27.5	3.3	37.5	62.5
Skew	0.5	3.1	0.6	–0.6
Kurtosis	1.4	9.1	–0.9	–0.9
SD	15.6	9.7	32.5	32.5
CV	56.9	290.3	86.5	52.0
Detailed field investigations (32 transects)				
Min.	1.1	0.0	1.7	20.0
Max.	20.0	19.6	80.0	98.3
Median	5.0	4.0	14.8	85.2
Av.	6.9	6.9	28.4	71.6
Skew	1.1	0.6	1.0	–1.0
Kurtosis	0.8	–1.3	–0.4	–0.4
SD	4.5	7.3	24.4	24.4
CV	65.5	106.3	85.8	34.1

Note: General statistics (minimum, min.; maximum, max.; median; average, av.; standard deviation, SD; skewness, skew, kurtosis, kurt.; and coefficient of variation, CV) of the extension in percentage of the total length of each transect for soils with (i) redoximorphic features (RFs), including soils with fluvic properties, glossic and/or albic features, and other redoximorphic features; (ii) soils without redoximorphic features.

A_{sm} ($r = 0.60$). Looking at the sub-surface landform, HI significantly correlated with L ($r = 0.80$), DG ($r = 0.80$), CTI_m ($r = 0.79$) and A_{sm} ($r = 0.82$). The correlation coefficient greatly increased (e.g. from $r = 0.62$ to $r = 0.79$ in the case of CTI) when using the topography of the saprolite upper boundary instead of using the topography of the soil surface.

The scattergram of "factor scores" of the PCA for each soil feature (RFs, including soils with fluvic properties, glossic and/or albic features, and other RFs; soils without RFs) and environmental factors (the geological substrate; the mean slope gradient, S; UR andNR) is presented in Figure 38.1. The total variance of regional data accounted for by these first two PCs was 74%. The interpretation of the factor coordinates of cases done with the help of their contributions to the variance was performed using soil features that have the greatest values of the contributions. The first PC that explained 52% of the

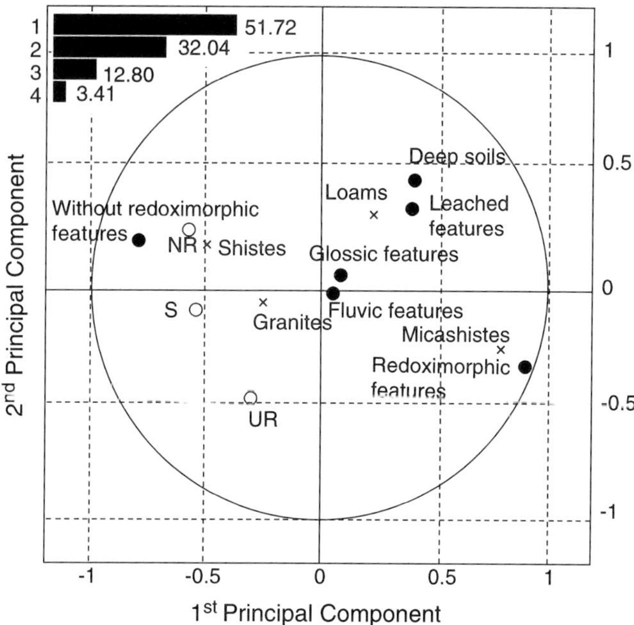

Figure 38.1. *Principal components analysis (PCA) scattergram of "factor scores" for each soil feature (redoximorphic features (RFs), including soils with fluvic properties, glossic and/or albic features, and other RFs; soils without RFs) and environmental factors (the geological substrate; the mean slope gradient, S; the uplift ratio, UR and the net rainfall, NR). Percent of the cumulative variance attributed to axes 1 to 4.*

variance revealed a pattern associated with the soil hydromorphy since soils with or without RFs showed the highest contribution to this factor. Environmental factors that have the highest (absolute) values for this factor loading are the geologic substrate, the mean slope gradient, NR andUR. A greater proportion of soils with RF characterised the micaschists, low NR, UR and mean slope gradient. Lower RF proportions occurred for NRs higher than 300 mm per year, URs up to 3 and, mean slope of transects higher than 9% (Chaplot et al., 2003b). Greater correlation to PC1 occurred for the mean slope gradient of transects ($r = 0.93$), followed by the geological substrate ($r = 0.72$), NR ($r = 0.66$) and UR ($r = 0.41$).

Soil depth and leached features are the variables the most strongly correlated to the second factor. This second component thus revealed a pattern associated with deep and leached soils. The environmental variable scattering suggests an increase of soil leaching with the presence of loamy material ($r = 0.69$) and at lower uplift ($r = 0.76$). Soils with fluvic and glossic properties were not correlated to the two first factors.

38.4 Discussion

This study of the soil spatial distribution using both detailed field study and map analysis revealed that soils in the Armorican Massif mainly differ in their degree of soil saturation by water, from a continuous or periodic saturation to no excess of water thus confirming previous investigations in the region (Walter, 1990; Curmi et al., 1998). Over the region, the soils are distributed along slopes in a consistent catenary sequence from soils with fluvic properties in the bottomlands, soils with glossic and/or albic features, soils with other RFs and soils without RF upslope whose relative extension may vary from site to site. In the Armorican Massif, soil hydromorphy may be because of the water accumulation in hillslopes due to surface and sub-surface runoff (e.g. Beven and Kirkby, 1979) controlled by the gravity and/or by the low infiltration of the soils, the saprolite and the substratum (Curmi et al., 1998) and may thus have a topographic, geologic and/or pedologic origin.

Within hillslopes, the variations of soil hydromorphy due to differences in duration of saturation by water was explained by terrain attributes such as the CTI and the elevation above the stream bank, ES. These factors have indeed a physical basis, explaining respectively the surface and sub-surface runoff (e.g. Beven and Kirkby, 1979), and the variations of the bottomland water level (e.g. Crave and Gascuel-Odoux, 1996). Using these, the predictions of HI, bound on regression models appeared efficient (MAE = 5.3, $P_{\text{level}} < 0.001$, Chaplot et al., 2003a), but more accurate predictions for HI (MAE = 3.9, $P_{\text{level}} < 0.001$, Chaplot et al., 2004) have been however obtained by using attributes from the topography of the saprolite upper boundary. Considering the shape of a subsurface feature, possibly shaped by geologic (e.g. bedrock fracturation, differential alteration) and/or surficial (e.g. differential erosion) processes, improved by 30% the soil prediction accuracy probably due to a better representation of the water redistributions within hillslopes than the surface topography itself (Chaplot et al., 2004). These results revealed that considering the topographic origin of soils hydromorphy was sufficient to predict the spatial distribution of hydromorphic soils at a single hillslope. But the modelling of soil hydromorphy using topography alone was not accurate at the regional level. To account for the other processes of soil hydromorphy, additional information on the geological substrate, NR, and UR should be considered. As previously discussed by Chaplot et al. (2003b), greater extensions of soils with RFs were associated with mica schists, low UR and NR due to the presence of deep and intensively weathered impermeable saprolites.

Since soil processes leading to soil hydromorphy act at different spatial scales, these results revealed the need of multi-scale approaches for the analysis and modelling of the spatial distribution of soils. The knowledge of the processes of soil formation acting at both local and regional levels allowed indeed making hypothesis on the environmental factors to be selected for the quantification of the soil-landscape relationships. In addition, this knowledge determined the modelling process, especially the selection of both scale for soil analysis and areas for data collection. For instance, these results clearly showed that only a preliminary knowledge of the role of tectonics as a soil-forming factor is able to adequately select study areas exhibiting a great variability for this factor.

Further investigations are still needed to develop generic models that may allow the soils to be predicted over large areas. Although at a single hillslope, the spatial distribution of hydromorphic soils was predicted with a high precision by using terrain attributes, these models exhibited high prediction errors when applied to other sites of the region (Chaplot et al., 2003a). In contrast, the models developed to predict the regional variations of the occurrence of hydromorphic soils were inaccurate predictors of detailed soil variations within hillslopes (Chaplot et al., 2003b). High-resolution and generic prediction models taking into consideration the multi-scale aspects of the soil-landscape relationships are not still available. The generation of these could take advantage of the analysis and modelling of the spatial distribution of soils from this multi-scale approach. To inform the detailed soil distribution within hilslopes and its regional variations, predictive algorithms should integrate the rules of soil distribution established at local and regional levels. Among the possibilities, principal components regression (PCR) based on the first PC of PCA from maps data, and the terrain attributes of CTI and ES, may be cited. The regression coefficients of the regression could be defined either through a calibration procedure at limited but detailed study sites exhibiting extreme environmental conditions or during the model generation and considering a large range of environmental conditions at detailed sites. Another possibility is to compute new components (factors or latent variables), that is new explanatory variables, from the set of available map and detailed study sites. In any case, multi-variate statistical methods could be integrated with geostatistical techniques by using, for instance, kriging with as external drift, regression on the PCs, to avoid any bias in the soil prediction. Another possibility is exploiting the quantitative rules of soil distributions established at different scales to determine the conditional probability (Lagacherie and Voltz, 2000) of hydromorphic soils in landscapes using a limited sampling and the soil-forming factor relationships.

38.5 Conclusion

In this chapter on the spatial variation of soils from the hillslope to the regional level our main objective was to quantify the relationships between the soils and their "environment" to further our understanding of soil distribution rules. This issue was addressed in the Armorican Massif (30,000 km^2) where the relationships between the spatial distribution of hydromorphic soils and some soil-forming factors was investigated using field investigations and detailed soil maps (1:25,000) analysis. Two main conclusions could be drawn from this study. First conclusion is that soil modelling could not be achieved by using only processes of soil hydromorphy linked to the duration of soil saturation by water and controlled by topography. To improve the spatial prediction of soils over large areas, a better knowledge of the other processes of soil hydromorphy and their controlling factors (e.g. geology and tectonism) is required.

The second conclusion is that the modelling of the spatial distribution of soils should consider the role of scale in the detection of patterns and processes of soil formation. This gives the opportunity to make models not only range from those *"that emphasis a physically complex description of processes while averaging or lumping landscape heterogeneity, to those that emphasis a representation of the landscape heterogeneity while simplifying process heterogeneity"* (Band and Moore, 1995) but to develop methods that represent both the complexity of soil processes and landscape heterogeneity. Because several processes of soil formation have been identified at different scales, multi-scale approaches should be considered as well for soil modelling.

Finally, this chapter illustrates a method for making the best possible use of (i) soil data from detailed hillslope survey and existing soil maps and (ii) auxiliary information (soil-forming factors) that can help to better describe, understand and quantify the rules of soil distribution within landscapes. From an economic and pragmatic point of view, the impressive developments of technologies such as geographic information systems (GIS) and the wide diffusion of digital databases of all sorts bring favourable conditions for the development and testing of methods to fully address the challenges related to digital soil mapping over large areas.

Acknowledgments

This research was supported by the PNRZH, from the French Ministry of Environment. The authors thank G. Dutin and F. Garnier (INRA Rennes) for their technical assistance in soil science and P. Curmi for his scientific contribution. Authors also gratefully acknowledge Prof. Mc Bratney, Dr. Lagacherie and Dr. Voltz for their useful comments and corrections.

References

Band, L.E., Moore, I.D., 1995. Scale: landscape attributes and geographical information systems. Hydrol. Process. 9, 401–422.

Beven, K.J., Kirkby, M.J., 1979. A physically based variable contributing area model of basin hydrologie. Hydrol. Sci. Bull. 24, 43–69.

Bonnet, S., 1997. Tectonique et dynamique du relief: le socle armoricain au Pleistocene. PhD, Geosciences Rennes, 352 pp.

Bourennane, H., King, D., Couturier, D., 2000. Comparison of kriging with external drift and simple linear regression for predicting soil horizon thickness with different sample densities. Geoderma 97, 255–271.

Chaplot, V., Walter, C., Curmi, P., 2003a. Testing quantitative soil-landscape models for predicting the soil hydromorphic index at a regional scale. Soil Sci 168, 445–454.

Chaplot, V., Van Vliet-Lanoë, B., Walter, C., Curmi., P., 2003b. Soil spatial distribution in the Armorican Massif, western France: effect of soil-forming factors. Soil Sci 169, 856–868.

Chaplot, V., Walter, C., Curmi, P., Lagacherie, P., King, D., 2004. Using the topography of the saprolite upper boundary to improve the spatial prediction of the soil hydromorphic index. Geoderma 123, 343–354.

Crave, A., Gascuel-Odoux, C., 1996. The influence of topography on time and space distribution of soil surface water content. Hydrol. Process. 11, 203–210.

Curmi, P., Durand, P., Gascuel-Odoux, C., Mérot, P., Walter, C., Taha, A., 1998. Hydromorphic soils, hydrology and water quality: spatial distribution and functional modeling at different scales. Nutr. Cycl. Agrosecosys. 50, 127–142.

Deutsch, C.V., Journel, A.G. (Eds.) 1992. Geostatistical Software Library and User's Guide. Oxford University Press, New York, 335 pp.

Garbrecht, J., Ogden, F.L., DeBarry, P.A., Maidment, P.A., 2001. GIS and distributed watershed models. 1: Data coverages and sources. J. Hydrol. Eng. 6, 506–514.

Gessler, P.E., Chadwick, O.A., Chamran, F., Althouse, L., Holmes, K., 2000. Modeling soil-landscape and ecosystem properties using terrain attributes. Soil Sci. Soc. Am. J. 64, 2046–2056.

Godwin, R.J., Miller, P.C.H., 2003. A review of technologies for mapping within-field variability. Biosyst. Eng. 84, 393–407.

Jambu, M., 1991. Exploratory and Multivariate Data Analysis. Academic Press, Boston, 474pp.

Jenny, H., 1941. Factors of Soil Formation: A System of Quantitative Pedology. McGraw Hill, New York.

Lagacherie, P., Voltz, M., 2000. Predicting soil properties over a region using sample information from a mapped reference area and digital elevation data: a conditional probability approach. Geoderma 97, 187–208.

Lague, D., Davy, P., Crave, A., 2000. Estimating uplift rate and erodibility from the area-slope relationship: example from Brittany (France) and numerical modeling. Phys. Chem. 25, 543–548.

Mc Bratney, A.B., Mendonça Santos, M.L., Minasny, B., 2003. On digital soil mapping. Geoderma 117, 3–52.

Mérot, P., Ezzahar, B., Walter, C., Aurousseau, P., 1995. Mapping waterlogging of soils using digital terrain models. Hydrolog. Process. 9, 27–34.

McKenzie, N.J., Ryan, P.J., 1999. Spatial prediction of soil properties using environmental correlation. Geoderma 89, 67–94.

Moore, I.D., Gessler, P.E., Nielsen, G.A., Peterson, G.A., 1993. Soil attribute prediction using terrain analysis. Soil Sci. Soc. Am. J. 57, 443–452.

Moore, I.D., Grayson, R.B., Ladson, A.R., 1991. Digital terrain modelling: A review of hydrological, geomorphological, and biological applications. Hydrolog. Processes 5, 3–30.

Odeh, I.O.A., McBratney, A.B., Chittleborough, D.J., 1994. Spatial prediction of soil properties from landform attributes derived from a digital elevation model. Geoderma 63, 197–214.

Sivapalan, M., Karma, G.A., 1995. Scale problems in hydrology: contributions of the Robertson workshop. Hydrolog. Process. 9, 243–250.

Soil Survey Staff, 1999. Soil Taxonomy. A Basic System of Soil Classification for Making and Interpreting Soil Surveys, 2nd edition. USDA, Washington DC.

Thompson, J., Bell, J., Butler, C., 2001. Digital elevation model resolution: effects on terrain attribute calculation and quantitative soil-landscape modelling. Geoderma 100, 67–89.

Van Vliet-Lanoe, B., Laurent, M., Hallegouet, B., Margerel, J.P., Chauvel, J.J., Michel, Y., Moguedet, G., Trautman, F., Vauthier, S., 1998. The Mio-Pliocene of the Armorican Massive. New data. C. R. Acad. Sci. 326, 333–340.

Voltz, M., Webster, R., 1990. A comparison of kriging, cubic splines and classification for predicting soil properties from sample information. J. Soil Sci. 41, 473–490.

Walter, C. 1990. Estimation de propriétés du sol et quantification de leur variabilité à moyenne échelle: cartographie pédologique et géostatistique dans le sud de l'Ille et Vilaine (France), thèse ENSA Rennes: 172.

Zevenbergen, L.W., Thorne, C.R., 1987. Quantitative analysis of land surface topography. Earth Surf. Processes Landforms 12, 47–56.

G. Quality assessment and representation of digital soil maps

Developmentally, digital soil mapping is closer to infancy than adolescence and a large majority of the research effort, and therefore most of this book has been devoted to clarifying concepts, developing appropriate data layers and predictive methodologies, and showing the first applications of these.

Digital soil maps are an essential part of a soil assessment framework which supports soil-related decision- and policy-making and therefore it is of crucial importance that DSM products are of known quality and are easily comprehended by potential users. Chapters 39–42 are an initial contribution to this area.

In Chapter 39, there is a detailed discussion of the various approaches to measuring the quality of map products both from the producer's and the user's point of view. Digital soil maps are potentially of higher quality than conventional ones, but this remains to be shown. The most important point is that digital soil maps should be of *known* quality. Further work is required in developing more sophisticated quality measures and designing appropriate sampling to estimate them.

Chapter 40 straddles the quality representation divide. Conventionally, soil maps are made up of polygons divided by sharp lines. This is a convenient cartographic and conceptual representation. It is well known that the boundaries between soil mapping units are not uniformly sharp in reality, although they conventionally appear so on a map. Chapter 40 describes how more informative boundaries can be discovered and delineated. This new approach leads to a higher quality of soil information representation.

Visualisation techniques are expected to provide the indispensable insights into the complexity of the soil cover that are required by both digital soil surveyors and end-users. Before the computational era, the choropleth map was probably the best way to summarise on a sheet of paper the complex information resulting from a soil survey. The appearance of computerised techniques dealing with spatial data have dramatically modified this situation by allowing visualisation of more sophisticated conceptualisations of the soil cover such as the ones handled in digital soil mapping. Chapter 41 describes this change through an historical review of developments over the past four decades. A key point is that methods for displaying spatial soil information cannot be dissociated from how the soil cover is conceived by the data producer. Sharing the same idea, Chapter 42 focuses on *scientific visualisation* and *virtual reality techniques* as tools for visualising real soil landscapes. Through the example of spatio-temporal water-table dynamics of a soil-landscape, the interactivity and the availability of multiple visualisations available from such techniques can stimulate our understanding of complex environment systems, and can disseminate digital soil mapping outputs to a large array of potential end-users.

Of all the topics in this book, quality assessment and representation are the least developed and probably require the most research effort.

Developments in Soil Science, volume 31
P. Lagacherie, A.B. McBratney and M. Voltz (Editors)

Chapter 39

QUALITY ASSESSMENT OF DIGITAL SOIL MAPS: PRODUCERS AND USERS PERSPECTIVES

P.A. Finke

Abstract
The assessment of quality of soil maps can be seen from the producer's and the user's perspective. Producers' perspectives have led to several measures of accuracy and precision that describe the intrinsic quality of the produced soil map and information system. These are described in some detail. In conclusion, it seems that adequate measures lack the ability to detect and quantify logical inconsistencies resulting from joining and harmonisation of existing maps. Additionally, indicators need to be developed that assess the semantic quality of maps while accounting for the taxonomic distance between the map units. Users' perspectives lead to a different view on map quality. Some minimal data sets of error are proposed that will enable users to incorporate soil map uncertainty into their applications.

39.1 Introduction

More than half a century of soil inventories has resulted in a great number of soil data sets, collected and presented in various ways via maps and soil information systems. During this period, mapping scales varied, mapping methods changed and mapped areas increased. The resulting conglomerate of soil information systems is being used for a wide range of applications. This opens the question of quality assessment, since ignoring uncertainties associated with the soil data sets in interpretations, may lead to wrong decisions and will also reduce the confidence in soil scientists (Fisher, 1999).

This recognition has lead to studies on the issue of uncertainty in the context of geographical information systems. Zhang and Goodchild (2002) used the term uncertainty as an umbrella for the distinct terms: error, randomness and vagueness. The spatio-temporal soil system that we try to understand and describe during the soil mapping process is associated with randomness, if only because we do not fully understand it and cannot completely measure it. The descriptive model for this system (i.e. the combination of the data acquisition, the derived conceptual model and the applied map inference methods) is associated with error and vagueness, depending on the type of chosen model and its methods.

Quality in relation to cartography has been defined by Moellering (1987) as *the suitability of the data for the intended use*. Others (e.g. Forbes et al., 1982) use terms like *adequacy* and *fitness for use* to describe the quality of soil resource inventories. These approaches have in common that quality is made dependent on the intended use of the inventoried data, which is seldom single-purpose. Any usage of soil resource data is associated with a characteristic sensitivity of the application to variation, due to errors in these data. The same error in basic soil data will cause different uncertainties in different applications (e.g. Finke et al., 1996), and thus the usage aspect of quality is in practice a variable. One objective of this chapter is therefore to indicate some methods to assess the impact of the quality of soil mapping to applications.

Other aspects of quality are constant for the inventory considered because they directly apply to the data. Since 2002, an ISO standard for describing the quality of geographic data exists (ISO 19133), based on a conceptualisation of clear and identifiable objects. Fisher (2003) (Table 39.1) introduced similarities between uncertainty terms and data quality components, while stressing the limitations of the ISO-standard for indeterminate objects. Apparently, the list of data quality components does not reflect all perceived modes of uncertainty and vice versa. Furthermore, indicators of quality are often associated with the methods applied to map soils. Another objective of this chapter is, therefore, to give some quality measures associated with different methods of soil mapping and to different aspects of quality.

In comparison to traditional soil mapping methods, modern predictive methods are better defined, documented and thus less depend on the individual surveyor's style. Also, modern methods come often with an indication of quality. This leads to two questions that deserve elaboration:

Table 39.1. *Similarity relations between components of uncertainty and data quality (Fisher, 2003).*

Uncertainty	Data quality
Error	Positional accuracy
	Attribute accuracy
	Completeness
Vagueness, discord, ambiguity	Semantic accuracy
Error, discord, Vagueness, ambiguity	Currency
Discord	Logical consistency
?	Lineage

Table 39.2. *Confusion matrix with proportions of observations within c mapped classes i and ground truth classes j.*

		Ground truth class j			
		$J = 1$	...	$j = c$	Total
Mapped class i	$i = 1$	$p_{1,1}$	...	$p_{1,c}$	$p_{1+} = \sum_{j=1}^{c} p_{1j}$
	...	...	$p_{i,j}$	...	$p_{i+} = \sum_{j=1}^{c} p_{ij}$
	$i = c$	$p_{c,1}$	...	$p_{c,c}$	$p_{c+} = \sum_{j=1}^{c} p_{cj}$
	Total	$p_{+1} = \sum_{i=1}^{c} p_{i1}$	$p_{+j} = \sum_{i=1}^{c} p_{ij}$	$p_{+c} = \sum_{i=1}^{c} p_{ic}$	$\sum_{i=1}^{c} \sum_{j=1}^{c} p_{ij} = 1$

applicable to soil mapping since classification of remote sensing and digital elevation data plays an important role in modern soil mapping methods.

Some of the classification accuracy presented in the confusion matrix may be due to chance, because some classes may occupy much larger areas than others and thus dominate the validation sample. For these circumstances, the κ-statistic (Cohen, 1968) has been developed. Kappa (<0: the map performs worse than randomly distributed classes; $1 =$ excellent) is calculated by:

$$\kappa = \frac{\theta_1 - \theta_2}{1 - \theta_2} \tag{39.6}$$

where

$$\theta_2 = \sum_{i=1}^{c} p_{i+} p_{+i} \tag{39.7}$$

When full-cover mapping is not yet done but the validation sample is already taken, instead of κ, the τ-statistic (Ma and Redmond, 1995; ranges comparable to those of κ) can be calculated using prior probabilities p_i of class memberships (e.g. estimated from small-scale soil maps or terrain maps):

$$\tau = \frac{\theta_1 - \theta_2'}{1 - \theta_2'} \tag{39.8}$$

where

$$\theta_2' = \sum_{i=1}^{c} p_i p_{+i} \tag{39.9}$$

Precision of fuzzy maps

The maps of the memberships of fuzzy classes can be used to calculate the confusion index (CI) at all map pixels or of the map extent, which is a precision measure calculated by the ratio of the first and second membership m:

$$\mathrm{CI} = \frac{m_{\text{2nd max}}}{m_{\text{1st max}}} \tag{39.10}$$

CI varies between 0 (no confusion) and 1 (maximal confusion), and can be used to indicate transition zones. CI is a precision measure for the defuzzified maps and does not provide quality information on the underlying fuzzy maps. A weakness of this index is that is accounts only for the membership confusion and not for the taxonomic distance between the two fuzzy classes. Hengl et al. (2004) recently proposed a method to combine taxonomic distance by colour separation and confusion by whiteness saturation into one map colour coding system.

Quality of single-value maps

The quality of single-value maps can be expressed by the accuracy measure MSE and by comparable precision measures such as prediction error variance associated with the prediction method (Webster and Oliver, 1990). Additionally, the percentage of variance explained can be calculated and the conditional bias (second regression) can be determined. Since soil mapping does usually not lead to single-value maps, associated quality measures are not extensively treated.

39.2.6 Currency

The currency of a soil map is a function of its age, because the soil system that it describes or the concepts and methods that are used for the description of the soil system may have changed. An outdated map needs to be updated but may keep its value as an historical document, for example to assess historical carbon stocks. The degree of ageing can in some instances be monitored using measures of positional or semantic quality. For example, Finke (2000) gives two parameters to assess map quality of groundwater table class maps in The Netherlands. These parameters are both based on point values of a function G that describes the degree to which two groundwater fluctuation parameters deviate from the definition in the map legend in a particular year. The first parameter is the estimated average value of G for a map sheet:

$$\mathrm{MG} = \sum_{i=1}^{n} g_i \cdot G_i \tag{39.11}$$

where g_i is the weight assigned to a point value of G, depending on the sampling design, and all weights g_i sum up to 1.

The second parameter is the estimated fraction of the area with strong deviations from the map legend:

$$\mathrm{FEXG} = \sum_{i=1}^{n} g_i \cdot I_i \tag{39.12}$$

with $I_i = 1$ if $G_i \geqslant 1$ and $I_i = 0$ if $G_i < 1$.

MG and FEXG can be monitored through G. In Figure 39.1, the evolution of MG and FEXG for one map sheet is shown. The graph also shows empirically derived threshold quality values to support decisions on map updating.

39.2.7 Logical consistency

Logical consistency of a soil map means that no interpretative mistakes have been made in the mapping process that are reflected in the final maps. Interpretative mistakes may (e.g.) occur during field mapping, generalisation,

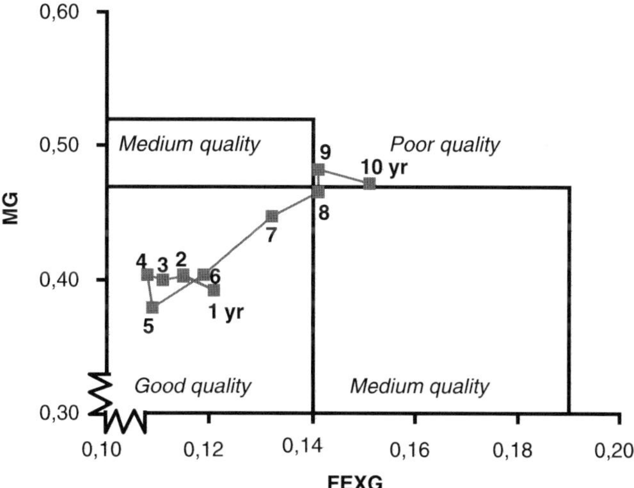

Figure 39.1. *Evolution of map quality parameters (groundwater table class map 1:50,000, sheet 27 East, The Netherlands) for 10 consecutive years after an update, estimated with $n = 52$–73 monitoring wells.*

combination and harmonisation of adjacent maps. It becomes more important to assess this type of error when soil information systems obtain a more composite nature, as they are developed out of regional existing information systems (Van Engelen and Wen, 1995; Deckers et al., 1998; Finke et al., 2003; Lambert et al., 2003; see also Daroussin et al., Chapter 4).

An example of an inconsistency due to generalisation error is the situation where one SMU at the detailed scale is assigned to two SMU that share a boundary at the generalised scale (Fig. 39.2).

Logical inconsistencies may occur as well when results of several mapping projects are combined. An example is given in Plate 39 (see Colour Plate Section), where national boundaries are visible in trans-national soil maps (Lambert et al., 2003; European Soil Bureau Network, 2004). These inconsistencies may occur for conventional soil maps, where different surveyors may have taken different classification or delineation decisions in comparable field situations. It may also occur in pedometric mapping, when clustering methods (e.g. resulting in membership maps derived from fuzzy-*k*-means classification) have been applied, because when training data sets differ, so may the resulting classifications.

This type of logical inconsistency may be recognised visually in thematic maps, but is not easily automatically detected or quantified via precision or accuracy measures. A proposed approach is to use a full-coverage landform classification based on a DEM (the classification possibly being supervised with

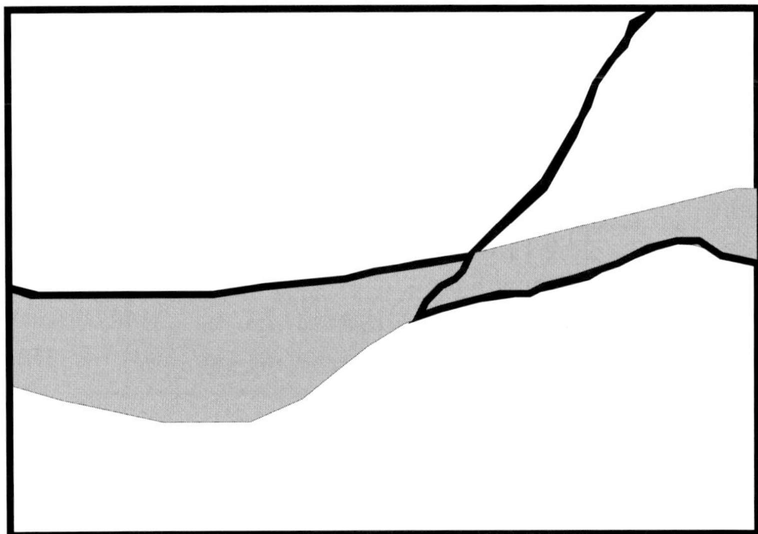

Figure 39.2. *Logical inconsistencies in map generalisation.*

Aerial Photo Interpretation; Hengl and Rossiter, 2003) to detect inconsistencies in the combined soil maps:

(i) The physiographic units (PU) are identified and mapped over the full map extent.

(ii) The PU are split up by the SMU boundaries in each one soil map that contributes to the combined soil map, through the operation $PU2 = PU \cap SMU$. The associated boundary uncertainty can be estimated through identifying AD, a positional accuracy measure. Alternatively, epsilon bands, a positional precision measure, are constructed and their area EBA is counted (Section 39.2.2).

(iii) The AD or EBA polygons are removed from the PU2 trough the operation $PU3 = PU2—(PU2 \cup [AD, EBA])$ and the resulting polygons PU3 are joined at the map boundaries.

(iv) After the combination of the soil maps, those polygons are selected from PU3 ($PU4 \subseteq PU3$), that are again split up due to the SMU boundaries of the combined soil maps. These polygons PU4 are suspect. The precision measure Area of Logical Inconsistency (ALI) is then estimated by:

$$ALI = \sum_{i=1}^{n} \text{Area } (PU4_i) \tag{39.13}$$

39.2.8 Lineage

Lineage is a possible source of uncertainty, because it may involve that errors are introduced when integrating data from different sources, possibly of various ages. A well-known example is the lesser quality of positional data in older topographic maps and information systems, which influences the positional quality of (parts of) the soil map (e.g. Radoševic, 1979). Another example is the usage of recent, detailed soil maps to update parts of smaller-scale soil maps (e.g. Finke et al., 2004). The adaptation of updates from one data set to another leads to the problem of integration of heterogeneous data. If the data model of the different data sets is homogeneous, integration of the data is of a geometric nature. Else, a semantic integration must be done first to avoid faulty comparisons and subsequent logical inconsistencies. Walter and Frisch (1999) evaluate some statistical approaches towards this type of data integration and associated precision measures, but few methods seem to exist that quantify the uncertainty effect of lineage.

39.3 User quality

39.3.1 Attitudes towards uncertainty

Uncertainty often causes dilemmas for the people who are exposed to it, especially when the uncertain data are to be used to support (the development of) policy. Policy makers and stakeholders then become users of uncertain data (and its interpretations). As such, they may experience both advantages (room for improvement, a bandwidth for making decisions and room for argumentation) and disadvantages (loss of public image and the risk to make wrong decisions) from uncertain data. All three types of users benefit by reduction of the disadvantages of uncertainty in the data (Table 39.3). It is, therefore, safe to assume that there will be broad support for activities reducing the disadvantages associated with uncertainty for all three types of users. Quantifying uncertainty (see also Heuvelink and Brown, Chapter 8) is one of these activities, when it is done in such way, that it allows for the identification of the sources of uncertainty to be able to minimise uncertainty. This requests interaction between the data collectors and the data users (Fig. 39.3). Also, uncertainty should be quantified in such manner, that it is useable for methods of decision making (e.g. Raiffa and Schlaifer, 2000). Finally, there is the issue of how to communicate uncertainty and associated risk to stakeholders (e.g. Gutteling and Wiegman, 1996) but this is considered beyond the scope of this chapter.

In the following sections, some examples will be given on the utilisation of uncertain soil and landscape data in policy support studies. Focus will be on the description of uncertainty in an error model relevant to the application. Some of the example studies take the form of an uncertainty analysis. The necessary error model can be considered as the "minimum data set of uncertainty" for these example studies and may differ considerably from the intrinsic data quality measures described earlier as they serve a different purpose.

39.3.2 An error model for evaluation studies using crisp thematic maps

This example is taken from a study by Finke et al. (1999), in which the effect of errors in categorical data (i.e. the generalised soil and vegetation class maps of

Table 39.3. *Impacts of uncertain data on different groups of users.*

User	Positive aspect	Negative aspect	User profits by
Researcher	Window of improvement	Damaged public image	Quantification of U Identification of sources of U Minimising U
Policy maker	Window of decision	Wrong decisions Damaged public image	Quantification of U Minimising U Deciding in the presence of U U or risk communication
Stakeholder	Window of argumentation	Wrong decisions	Quantification of U Deciding in the presence of U U or risk communication

Note: U = uncertainty.

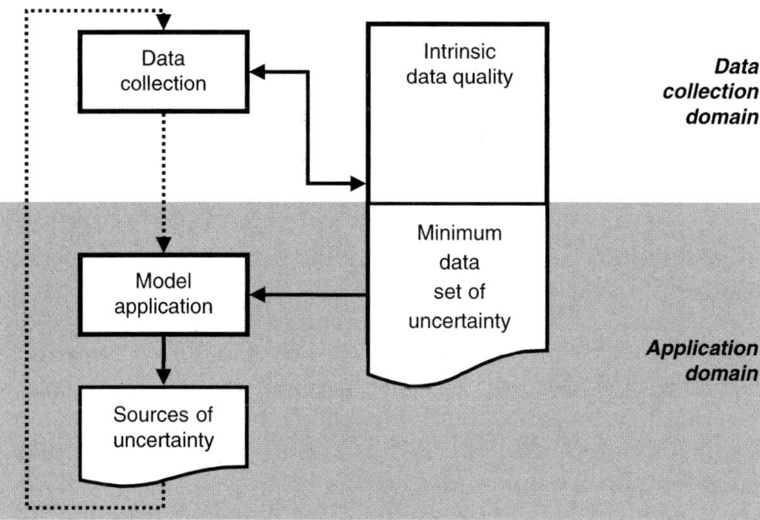

Figure 39.3. *Uncertainty in the data collection and the data application domains. Dotted lines indicate interactions between the domains.*

the EU) on the uncertainty of outcomes of a soil acidification model was analysed. A deposition scenario from the Netherlands environmental outlook (RIVM, 1997) was simulated. To assess the quality of the EU maps, highly detailed maps of soil and vegetation available for the Netherlands (NL) were used as ground truth.

To quantify the degree of error within each EU-category, an indicator variable $I_{t,s}$ for NL soil-vegetation class t within EU soil-vegetation class s was introduced:

$$I_{t,s}(x) = \begin{cases} 1 & \text{if EU}(x) = s \text{ and NL}(x) = t \\ 0 & \text{otherwise} \end{cases} \tag{39.14}$$

where EU(x) is the soil and vegetation class at location x in the EU map and NL(x) the ground truth according to the detailed map. The error model was defined as:

a. The confusion matrix (Lillesand and Kiefer, 1994) with the expectations of I for all combinations of t and s.
b. The binominal variances $s^2(I_{t,s}(x))$ associated with the cells in the confusion matrix.
c. Fitted indicator variograms $\gamma_{I_{t,s}}(h)$, scaled so that the sill equals the binominal variance, for $I_{t,s}(x)$ for all non-zero expectations of I in the confusion matrix. The incorporation of the spatial correlation of the error stemmed from the observation that misclassifications tend to appear in clusters.

The analysis was part of an uncertainty analysis, which also included the effect of uncertainty in continuous data. Because the output of the involved model was known to respond non-linearly to its inputs, the uncertainty analysis was set up as a Monte-Carlo analysis. This approach requires the generation of realisations of model inputs and the error model should allow for this to be done effectively. Thus, this study, reported in Kros et al. (1999), consisted of the following steps:

1. Twenty-five realisations of the EU-map were obtained by sequential multiple indicator simulation (Goovaerts, 1997), using the indicator variograms and a stratification to EU-categories.
2. For each one map realisation, five realisations of soil parameters and five realisations of vegetation parameters were simulated using non-conditional sequential multivariable Gaussian simulation (Pebesma and Wesseling, 1997). For the error model of these continuous data, reference is made to Section 39.3.4.
3. The acidification model was run on all 625 input data sets.
4. The output uncertainty and the relative contributions caused by uncertainty in categorical data, soil parameters, vegetation parameters and an interaction term were quantified in a (nested) ANOVA.

39.3.3 An error model for evaluation studies using fuzzy thematic maps

This example is taken from a study by Gorssevski et al. (2003), in which a continuous landform classification by fuzzy *k*-means is combined with a Bayesian probabilistic modelling approach to obtain probabilistic landslide hazard maps.

The Bayesian approach combines subjective probability with conditional probability. The subjective probability expresses the degree of belief in an event (i.e. the probability of occurrence of a landslide) and is usually called the *prior*. The conditional probability expresses the likelihood of the hypothesis to be true given the evidence (i.e. the probability of occurrence of a landslide given the fuzzy memberships). To calculate the probability of occurrence of a landslide at a location, the following equation applies:

$$P(o|f) = \frac{P(f|o) \cdot P(o)}{P(f)} \quad \text{(Bayes' rule)} \tag{39.15}$$

where o indicates the occurrence of a landslide and f the set of memberships associated with the fuzzy clusters. $P(o)$ is the prior probability for occurrence of a landslide, estimated by the counted occurrences of landslides. $P(f|o)$ is the conditional probability of occurrence of a landslide given the fuzzy memberships, calculated from the relative frequency of association between occurrences of landslide locations and categorised membership values of the fuzzy *k*-means classes at the n locations (grid cells) for which the occurrence (or absence) of landslides is known:

$$P(f|o) = \prod_{i=1}^{n} co_i \tag{39.16}$$

where $\prod_{i=1}^{n} co_i$ is the product of conditional probabilities for occurrence for attributes $i = 1 \ldots n$ of the predictor data sets. $P(f)$ is calculated by

$$P(f) = P(f|o) \cdot P(o) + P(f|a) \cdot P(a) \tag{39.17}$$

where a indicates the absence of a landslide and $P(f|a)$ and $P(a)$ are calculated analogously to $P(f|o)$ and $P(o)$, respectively.

Combination of the above three equations leads to the equation for the Bayesian calculation:

$$P(o|f) = \frac{P(o)\prod_{i=1}^{n} co_i}{P(o)\prod_{i=1}^{n} co_i + P(a)\prod_{i=1}^{n} ca_i} \tag{39.18}$$

The analyses comprised the following steps:

1. Translation of the DEM into landslide-relevant environmental attributes and performing a fuzzy *k*-means classification on training areas. Using the performance indicator FPI (Minasny and McBratney, 2000), the optimal number of classes was found; with this number of classes, the optimal fuzzy exponent ϕ was derived from Odeh et al. (1992) and the memberships at all grid cells of the DEM were calculated.
2. Recording the occurrences of landslide as absence or presence at all grid cells of the DEM.
3. Construction of an error model containing:
   a. Tabulated conditional probabilities of occurrence and absence (expressed as relative frequencies) for each one fuzzy cluster (subdivided in 10 membership subclasses 0–0.1, 0.1–0.2, … 0.9–1.0).
   b. Prior probabilities of absence and occurrence of landslides.
   c. Maps of the memberships of each one fuzzy class.
4. Calculation at each one map pixel of the conditional probabilities based on components a and b of the error model.
5. Calculation of the Bayesian probability of occurrence of landslides at each one map pixel, using Eq. 39.18.

39.3.4 An error model for evaluation studies using (multiple) single-value maps

This example is taken from the same study by Kros et al. (1999) that was briefly described in Section 39.3.3. Below, I focus on step 2 of the over-all procedure described in Section 39.3.3. A commonly applied method to generate realisations of (multiple) single-value maps (Gómez-Hernández and Journel, 1992) are that of joint sequential simulation of Gaussian fields. The necessary error model includes:

a. For each parameter, possibly per stratum, the average value and the variogram.
b. Between parameters, possibly per stratum, the cross-variograms.

To reduce the calculation effort, before starting the simulations, the sensitive parameters were identified. Further simplifications comprised the reduction of the number of parameters that were supposedly correlated, and the assumption that cross-variograms would lack spatial structure so that the covariance would be a constant.

39.4 Concluding remarks

Question 1: Do we expect modern predictive soil maps to be more accurate than conventional soil maps?

- Yes, because the positional quality, currency and lineage of the data used to construct digital soil maps is better than that used to construct the conventional soil maps in the past. Additionally, the fact that precision measures come as by-products of many modern mapping methods allows for the optimisation of these methods. The circumstances are there to make better maps with less field effort, all though this advantage can be lost when too much is economised on the collection of ground truth data. The user of modern predictive soil maps may conclude from the simple presence of precision measures that quality of modern soil maps is less than that of conventional soil maps (because no quality indications were given with these maps). The importance of this aspect should not be neglected and requires communication to the users on quality aspects and on how to manage uncertainty in soil map applications.

Question 2: Do we need new ways of assessing quality?

- Work is needed on the assessment and improvement of uncertainty introduced by combining data from different surveys. There is a need to develop a framework for the harmonisation of soil maps.
- The assessment of semantic quality of soil maps can be improved by including the taxonomic distance as a weight in confusion matrices, such that misclassifications over taxonomically adjacent classes receive less weight.
- The description of intrinsic quality is good for documenting and improving quality but may be insufficient to assess the user quality. Some standardisation in the description of quality in terms of uncertainty or error models is necessary, because it will help bridge the gap between data providers and data users.

This paper has focused on the assessment of several aspects of quality, and on error models to utilise quality for users. It may leave the impression that modern mapping methods may improve quality while conventional soil maps will pass into oblivion. This is certainly not the case. Modern methods may and should be used to improve the quality of existent soil information systems in three ways:

(1) Updating (improving the currency).
(2) Upgrading (improving the completeness).

(3) Corroboration (improving the positional and semantic quality and the logical consistency).

Some experience with (1) and (2) exists (e.g. Finke et al., 2004). The toolbox for (3) is already well filled but applications still seem to be absent.

References

Burrough, P.A., 1990. Principles of Geographical Information Systems for Land Resource Assessment. Clarendon Press, Oxford.

Chrisman, N.R., 1982. A theory of cartographic error and its measurement in digital data bases. Proc. Auto Carto 5, 159–168.

Clarke, S.E., White, D., Schaedel, A.L., 1991. Oregon, USA ecological regions and subregions for water quality management. Environ. Manage. 15, 847–856.

Cohen, J., 1968. Weighted kappa: nominal scale agreement with provision for scaled disagreement or partial credit. Psychol. Bull. 70, 426–443.

Cohen, A.C., 1991. Truncated and Censored Samples: Theory and Applications. Dekker, Inc., New York.

Cohen, M.P., 1996. A new approach to imputation. In: Proceedings of the Survey Research Methods Section of the American Statistical Association, pp. 293–298.

Deckers, J.A., Nachtergaele, F.O., Spaargaren, O.C. (Eds.), 1998. World Reference Base for Soil Resources. Introduction. 84 World Soil Resources Report, FAO, ISRIC and ISSS. Rome, Italy.

Dubois, G., Saisana, M., 2002. Optimizing spatial declustering weights – comparison of methods. In: Proceedings of the Annual Conference of the International Association for Mathematical Geology, Berlin-Germany, September 15–20, 2002, pp. 479–484.

European Soil Bureau Network, 2004. Soil Map Internet Server. http://eusoils.jrc.it/msapps/Soil/SoilDB/SoilDB.phtml (accessed 19 June 2004).

Finke, P.A., Wösten, J.H.M., Jansen, M.J.W., 1996. Effects of uncertainty in major input variables on simulated functional soil behaviour. Hydrol. Process. 10, 661–669.

Finke, P.A., Wladis, D., Kros, J., Pebesma, E.J., Reinds, G.J., 1999. Quantification and simulation of errors in categorical data for uncertainty analysis of soil acidification modelling. Geoderma 93, 177–194.

Finke, P.A., 2000. Updating groundwater table class maps 1:50,000 by statistical methods: an analysis of quality versus cost. Geoderma 97, 329–350.

Finke, P.A., Bierkens, M.F.P., Brus, D.J., Hoogland, T., Knotters, M., de Vries, F., 2004. Mapping ground water dynamics using multiple sources of exhaustive high resolution data. Geoderma 123 (1–2), 23–39.

Finke, P.A., Hartwich, R., Dudal, R., Ibanez, J., Jamagne, M., King, D., Montanarella, L., Yassoglou, N., 2003. Geo-referenced soil database for Europe. Manual of procedures, version 1.1. EUR 18092 EN European Soil Bureau, JRC, Italy.

Fisher, P.F., 1999. Models of uncertainty in spatial data. In: P.A. Longley, M.F. Goodchild, D.J. Maguire, and D.W. Rhind (Eds.), Geographical information systems: Principles, techniques, management and applications. New York, Wiley, pp. 191–205.

Fisher, P.F., 2003. Data quality and uncertainty: ships passing in the night! In: W. Shi, M.F. Goodchild, P.F. Fisher, Proceedings of ISSDQ 2003, Hong Kong. p. 13.

Forbes, T.R., Rossiter, D., van Wambeke, A., 1982. Guidelines for evaluating the adequacy of soil resource inventories. 1987 printing edition of SMSS Technical Monograph 4. Cornell University, Department of Agronomy, Ithaca, NY.

Gómez-Hernández, J.J. Journel, A.G., 1992. Joint sequential simulation of multigaussian fields. In: A. Soares (Ed.), Geostatistics Troia '92, I., pp. 85–94.

Goovaerts, P., 1997. Geostatistics for National Resources Evaluation. Oxford University Press, 483 p.

Gorssevski, P.V., Gessler, P.E., Jankowski, P., 2003. Integrating a fuzzy k-means classification and a Bayesian approach for spatial prediction of landslide hazard. J. Geograph. Systems 5, 223–251.
Gutteling, J.M., Wiegman, O., 1996. Exploring Risk Communication. Kluwer Academic Publishers, Dordrecht, 236 pp.
Hengl, T., 2003. Pedometric mapping. PhD Thesis, ITC, Enschede.
Hengl, T., Rossiter, D.G., 2003. Supervised landform classification to enhance and replace photo-interpretation in semi-detailed soil survey. Soil Sci. Soc.Am. J. 16, 1810–1822.
Hengl, T., Walvoort, D.J.J., Brown, A., Rossiter, D.G., 2004. A double continuous approach to visualisation and analysis of categorical maps. Int. J. Geogr. Inf. Sci. 18, 183–202.
Knotters, M., Brus, D.J., Oude Voshaar, J.H., 1995. A comparison of kriging, co-kriging and kriging combined with regression for spatial interpolation of horizon depth with censored observations. Geoderma 67, 227–246.
Kros, J., Pebesma, E.J., Reinds, G.J., Finke, P.A., 1999. Uncertainty assessment in modelling soil acidification at the European scale, A case study. Journal of Environmental Quality 28, 366–377.
Lambert, J.J., Daroussin, J., Eimberck, M., Le Bas, C., Jamagne, M., King, D., Montanarella, L. (Eds.), 2003, Soil Geographical Database for Eurasia & The Mediterranean: Instructions Guide for Elaboration at Scale 1:1,000,000, Version 4.0. European Soil Bureau Research Report No. 8, EUR 20422 EN, 64 pp. Office for Official Publications of the European Communities, Luxembourg.
Lillesand, T.M., Kiefer, R.W., 1994. Remote Sensing and Image Interpretation, 2nd edition. John Wiley & Sons, New York.
Ma, Z., Redmond, R.L., 1995. Tau coefficients for accuracy assessment of classification of remote sensing data. Photogrammetric Eng. Remote Sensing 61, 435–439.
Minasny, B., McBratney, A.B., 2000. FuzME version 2.1. Australian Centre for precision Agriculture, University of Sydney, NSW 2006. http://www.usyd.edu.au/su/agric/acpa/fkme/FkME.html (accessed 20 June 2004).
Moellering, H., 1987. A draft proposed standard for digital cartographic data. Report no. 8. USA National Committee for Digital Cartographic Standards, American Congress on Surveying and Mapping.
Odeh, I.O.A., McBratney, A.B., Chittleborough, D.J., 1992. Soil pattern recognition with fuzzy c-means: application to classification and soil-landform interrelationship. Soil Sci. Soc. Am. J. 56, 505–516.
Pebesma, E.J., Wesseling, G.C., 1997. Gstat, a program for geostatistical modelling, prediction and simulation. Comput. Geosci. 24, 17–31.
Radoševic, N., 1979. Pre-war military map 1:100 000 (1:50 000) and todays triangulation. VGL, 129–148.
Raiffa, H., Schlaifer, R., 2000. Applied Statistical Decision Theory. Wiley, New York, 356 p.
RIVM, 1997. Nationale milieuverkenningen 1997–2020. RIVM, Bilthoven, The Netherlands.
Rubin, D.B., 1987. Multiple Imputation for Non-Response in Surveys. Wiley, New York.
Van Engelen, V.W.P., Wen, T.T. (Eds.) 1995. Global and national soils and terrain digital databases (SOTER). Procedures Manual (revised edition). UNEP-ISSS-ISRIC-FAO. Wageningen, The Netherlands.
Van Ranst, E., Thomasson, A.J., Daroussin, J., Hollis, J.M., Jones, R.J.A., Jamagne, M., King, D., Vanmechelen, L., 1995. Elaboration of an extended knowledge database to interpret the 1:1,000,000 EU Soil Map for environmental purposes. In: D. King, R.J.A. Jones, and A.J. Thomasson (Eds.), European Land Information Systems for Agro-environmental Monitoring. EUR 16232 EN, 71-84. Office for Official Publications of the European Communities, Luxembourg.
van Reeuwijk, L.P., Houba, V.J.G., 1998. Guidelines for Quality Management in Soil and Plant Laboratories. (FAO Soils Bulletin – 74). Food and Agriculture Organization of the United Nations, Rome.
Walter, V., Fritsch, D., 1999. Matching spatial data sets: a statistical approach. Int. J. Geogr. Inf. Sci. 13, 445–473.

Webster, R., Oliver, M.A., 1990. Statistical Methods in Soil and Land Resource Survey. Oxford University Press, Oxford.

Wösten, J.H.M., Finke, P.A., Jansen, M.J.W., 1995. Comparison of class- and continuous pedotransfer functions to generate soil hydraulic characteristics. Geoderma 66, 227–237.

Zhang, J., Goodchild, M., 2002. Uncertainty in Geographical Information. Taylor & Francis, London.

Developments in Soil Science, volume 31
P. Lagacherie, A.B. McBratney and M. Voltz (Editors)

Chapter 40

USING SOIL COVARIATES TO EVALUATE AND REPRESENT THE FUZZINESS OF SOIL MAP BOUNDARIES

M.H. Greve and M.B. Greve

Abstract

As an alternative to the crisp representation of map units in classical detailed soil mapping, this chapter presents the development of a decision-support system for fuzzification of soil maps. High-resolution auxiliary data such as the soil electrical conductivity (SEC), detailed digital elevation models (DEM), and aerial photographs (APs) and their derivatives carry information on different soil parameters. SEC has very high correlation with soil texture, DEM contain valuable information on soil moisture regime and landscape genesis, and AP carry information on topsoil total organic carbon, TOC. We have developed a system that assists in identifying which auxiliary dataset should be used to determine each map unit boundary width, and thereafter express the boundary as a transition zone between two or more soil types. The method is developed using data from a soil survey on the island of Funen in Denmark and the concept is illustrated in a small strip in this area. A soil database with information on modal soils expresses the degree of drainage, texture and topsoil TOC in each of the mapping units, and we can thereby determine which parameters the bounding map units differ on, and in that way also which dataset should be used for calculating the boundary width.

40.1 Introduction

On choropleth soil maps, all soil variation is positioned exactly on the map unit delineation and in the real world this is not the case. The fuzziness of the soil map delineation is not normally taken into account, except perhaps for general statements in the survey report, and no procedure for classifying or coding map unit delineation is described in the Soil Survey Manual (Soil Survey Staff, 1993).

In spite of pedometrical methods developed for example kriging, fuzzy k-means, clustering (Burrough, 1989) that take the continuous nature of soil variation into account, most detailed soil surveys still use the classical concept of crisp soil boundary representation (McBratney et al., 2003). The process of transforming a traditional choropleth soil map into a map expressing the transition zones between soil types have been described by McBratney and Whelan (1995) and Lagacherie et al. (1996). These methods assign boundary widths to line segments between pairs of mapping units and do not take into account the constantly changing boundary width.

Preliminary investigations on field classification of map unit delineation width in three classes (abrupt, intermediate and gradual) were done by Greve (2004). The experience from the survey showed that it is very difficult and time consuming to classify the map unit delineations into three classes. One of the main problems seems to be the confusion between fuzziness and uncertainty of the delineation (Lagacherie et al., 1996). It was not possible to make a distinction between the two.

The aim of this study is to propose a solution for quantifying fuzziness continuously along soil boundaries and to develop a GIS-based decision-support system for fuzzification of existing soil maps using a range of auxiliary data. The output is a data structure showing soil types/map units and the transition zones between them. We developed the layered MultiGrid data structure presented in Greve and Greve (2004). The method presented uses a range of ancillary data to classify the change in fuzziness along the delineation.

40.2 Materials and methods

40.2.1 Research site

The research area is situated in Denmark in the northern part of the island of Funen and described in detail in Greve (2004). 7.4-ha of approximately 600 ha were selected as a test site for illustrating the principle in the method, see Figure 40.1.

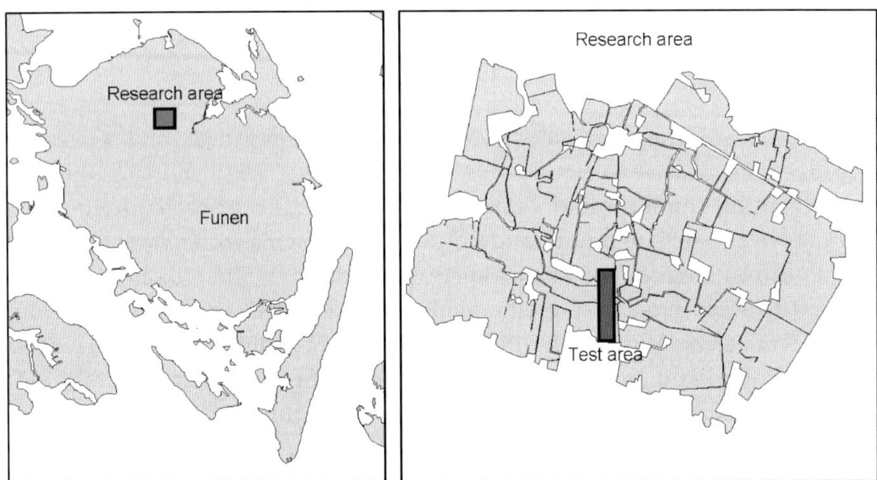

Figure 40.1. *Location of research area and test site.*

40.2.2 The soil survey

Three different data types and their derivatives were used both for the soil survey and in the decision-support system later: aerial photographs (APs), a digital elevation model (DEM) and soil electrical conductivity (SEC). A soil database with information on modal soil horizons contains the soil information.

The data are explained in the following section.

40.2.3 Data used for the survey

Aerial photographs

In this survey, panchromatic B/W AP has been used at a scale of 1:10,000, close to the resolution wanted in the resulting soil database. For post-processing purposes, the AP were scanned in high resolution and rectified.

Digital elevation model

The DEM was generated on the basis of an airborne-laser scanning. The scanner system records the distance to a point every 6 m^2. The points are then interpolated to a DEM in a 2 by 2 m grid with a relative accuracy in elevation of approximately 10 cm.

Soil electrical conductivity

A survey was performed using the non-contact SEC meter EM38. The use of a map of EM38 data during survey operations enables the surveyor to delineate map units manually with very high accuracy, because of the very high correlation between the EM38 map and clay content (e.g. Nehmdahl and Greve, 2001; Durlesser, 1999).

Soil database

A soil database with information on modal soil horizons was used in this study. The database consists of a soil series table and a horizon table. The horizons are defined on the basis of a set of diagnostic properties that are combined into soil series. A soil series has three layers with fixed depths: 0–30 cm, 30–70 cm and 70–120 cm.The diagnostic properties of the system are: geology, drainage, presence of $CaCO_3$, simplified pedology and soil texture. Soil series are distinguished from one another by differences in one or more of the diagnostic properties in one of the three layers. The diagnostic properties are equally important. The properties of the diagnostic horizons were determined on the basis of a statistical analysis of the nationwide soil profile database (Greve, 2004).

40.2.4 The Soil Survey

The mapping area was subdivided into areas with uniform soil conditions and a soil series from the soil database was assigned to each map unit. While traversing the landscape on an all-terrain vehicle, the mapping was performed. As the DEM, SEC and AP were interpreted and strategic augerings were made, the information on soil type was registered in a field computer. The delineation was drawn directly on the computer screen. The area was subdivided into 664 mapping units.

40.3 Developing the method

40.3.1 Selecting the test area

For the development of the method in question, we selected a small area (the test area) in the southern part of the research area (Fig. 40.1). The strip was selected in an area with good relation between the landscape and the soil, and where the soil map is expected to be relatively precise.

Plate 40 (see Colour Plate Section) shows all the available data in the test area in the small strips. The large strip contains the map units. In the central part of the area the gley- and histo-sols are clearly visible on the AP and the profile curvature, but not on the SEC. The SEC and the slope of SEC seem very useful for describing the delineations between the sandy and clayey soils.

40.3.2 Preparing the data

The auxiliary data and the polygons from the soil map were clipped to map the test area. All raster data was resampled to 2 by 2 m pixels, and slope maps in the same resolution were calculated for all three types of auxiliary data. For the DEM, we also calculated the 2nd derivative to express the profile curvature. For each map unit the information on drainage, topsoil total organic carbon (TOC) and texture were simplified into a small number of classes illustrated on Figure 40.2. The map unit boundaries were converted to a dataset containing only the boundary lines.

40.3.3 Register the difference in diagnostic properties on the boundaries and assign each line to a grid

During the survey, a high correlation between the aerial photography and topsoil TOC was found. Topography is known to have high correlation to drainage condition (e.g. Kravchenko et al., 2002) and as stated above, SEC has a high correlation with texture. By calculating the slope in high-resolution raster data, this is an indication of the spatial rate of change in the data, or, in other words, a map of how fast the values in the high-resolution data changes. Therefore, the

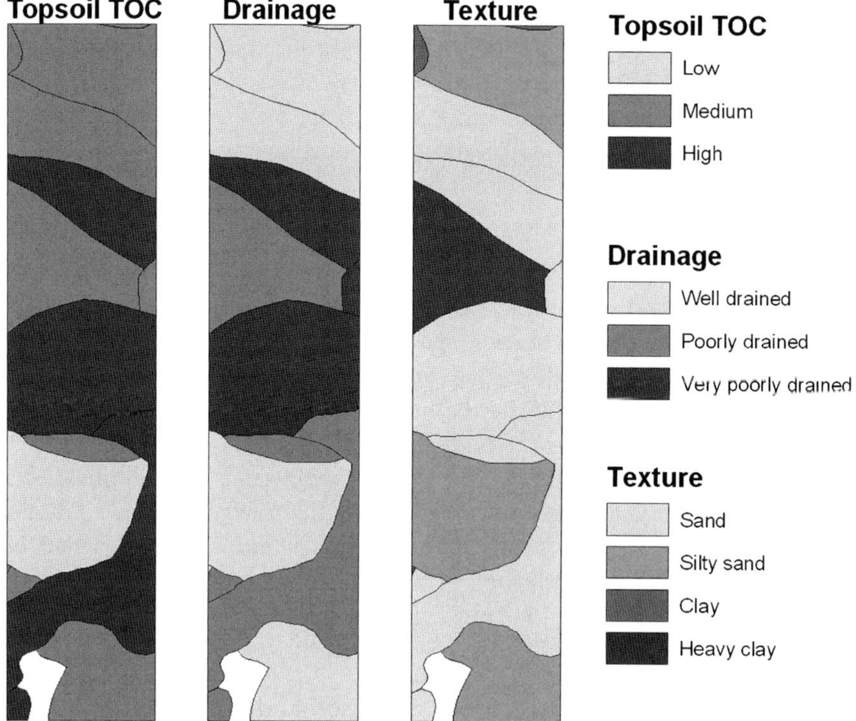

Figure 40.2. *Soil property maps.*

auxiliary data can be used as indicators of the width of the transition zone between two soil types. The data we used in this study were:

- AP: slope.
- DEM: profile curvature.
- SEC: slope.

We have registered the topsoil TOC, the drainage condition and the texture for each of the map units. We then need to find out in which parameters the adjacent polygons differ to determine which dataset/indicator grid should be used for calculating the boundary width. Each line is assigned an indicator grid as sketched below.

In this example we set up priorities between the auxiliary data, so that difference in topsoil TOC take priority over drainage condition, which then take priority over texture. This means, that if two adjacent polygons differ on both topsoil TOC and drainage, only the AP will be used to calculate the boundary width, as topsoil TOC take priority over drainage. The priorities can be changed by the user of the system.

The simple calculation is then, that if the polygons differ on topsoil TOC, the indicator for the line between the polygons will be set to the AP. If the polygons have the same topsoil TOC, but differ in drainage class, the indicator will be set to the profile curvature. If the polygons only differ on the texture, the slope of the SEC grid will be used for calculating the boundary width.

40.3.4 Adjust line position according to grid values

The use of the auxiliary data for compiling the soil map during the soil survey is crucial for the method, as the delineations between mapping units has to be located at the place with the highest spatial rate of change to make use of this criteria in creating the transition zone. Very often the delineation is still a little off, and may be located just beside the high slope. We created an algorithm for moving the lines, so that they would be placed over the steepest cells in the indicator grid, where the rate of change in soil properties is the highest.

We had to select a maximum distance that each line was allowed to move. In our case the distance could be small, as the map unit boundaries had been digitised knowing all the indicator grids. The distance was set to 10 m. We decided not just to move the line, but also alter the shape. This would be done by moving each vertex along the line to the cell in the indicator grid with the highest value in a 10-m buffer zone around the vertex. The indicator grids were resampled to 3 by 3 m pixels for this purpose to smooth out high single-cell values.

We wanted the vertices of each line to be evenly spaced. If the vertices were to close, they would move to the same cell anyhow, and if they were too far apart, we might miss the steep cells on the long stretch. To evenly space the vertices we densified the lines, meaning that extra vertices were added, so that there would be no more than 5 m between two vertices. Then we simplified the lines, only keeping vertices that were at least 20 m apart (twice the maximum distance that the vertex could move). A new line shape file containing the adjusted lines was the output of this step (see Fig. 40.3).

40.3.5 Determine boundary width

After the lines have been moved and/or adjusted, the slope values from the indicator grids are extracted in the cells that the lines run through. The values are then converted to a boundary width. Low-slope values result in broad transition zones and high-slope values in narrow transition zones between the map units. The slope readings in the boundaries are translated to a transition zone width using a linear function assessed by the survey expert. The output here is a map showing the transition zone widths (as numbers) in each line grid cell.

Greve, M.H., 2004. A Danish Soil Reference System for Soil Interpretation on a Detailed Scale. Ph.D. thesis, Institute of Geography, Faculty of Science, University of Copenhagen.

Greve, M.H., Greve, M.B., 2004. Determining and representing width of soil boundaries using electrical conductivity and MultiGrid. Comput. Geosci. 30, 569–578.

Kravchenko, A.N., Bollero, G.A., Omonode, R.A., Bullock, D.G., 2002. Quantitative mapping of soil drainage classes using topographical data and soil electrical conductivity. Soil Sci. Soc. Am. J. 66, 235–243.

Lagacherie, P., Andrieux, P., Bouzigues, R., 1996. Fuzziness and uncertainty of soil boundaries: from reality to coding in GIS. In: P.A. Burrough and A.U. Frank (Eds.), Geographic Objects with Indeterminate Boundaries. Taylor & Francis, London, pp. 275–286.

McBratney, A.B., Mendonça-Santos, M.L., Minasny, B., 2003. On digital soil mapping. Geoderma 117, 3–52.

McBratney, A.B., Whelan, B.M., 1995. Continuous models of soil variation for continuous soil husbandry. In: Robert, P.C., Rust R.H., Larson, W.E. (Eds.), Site-Specific Management for Agricultural Systems, ASA/CSSA/SSSA, Madison, Wisconsin, pp. 325.

Nehmdahl, H., Greve, M.H., 2001. Using soil electrical conductivity measurements for delineating management zones on highly variable soils in Denmark. In: Grenier, G., Blackmore, S. (Eds.), Proceedings of the 3rd European Conference on Precision Agriculture, Montpellier, France, 18–20, June 2001. Vol. 1, pp. 461–466.

Soil Survey Staff, 1993. Soil Survey Manual. United States Department of Agriculture, Handbook no. 18.

Developments in Soil Science, volume 31
P. Lagacherie, A.B. McBratney and M. Voltz (Editors)

Chapter 41

THE DISPLAY OF DIGITAL SOIL DATA, 1976–2004

P.A. Burrough

Abstract

The aim of this chapter is to explore the relations between the perception and observation of soil in the field and the methods used for displaying the information gathered. The main hypothesis is that our insights into the spatial and temporal variation of soil over the landscape are to a large extent controlled by the conceptual models available to us. This range of conceptual models is, in turn, controlled by the methods of display at our disposal, which increasingly make use of digital graphics.

The chapter adopts a historical approach, arguing that developments in the visualisation of soil and other geographical, statistical and cartographic data reflect our understanding of the formation and behaviour of soil in the field. The first digital soil data were merely electronic versions of paper maps and profile descriptions that adopted an exact object approach. Developments in interpolation, especially kriging, use an alternative conceptual model, namely that of the continuous smooth surface, which has been further modified to include notions of noise and short-range variation. Further developments in cokriging have enabled the integration of extra data to improve the interpolation, in particular using data on surface elevation and slope.

The value of elevation data for enhancing understanding soil–landscape relations through improved visualisation was first demonstrated by 3D "draping" of both conventional and interpolated maps of soil data over digital elevation models (DEMs). Further developments in the visualisation of soil types have made direct use of derivatives of DEMs such as slope, elevation, rates of curvature, directly received solar radiation in order to both classify and map soil automatically. The current state-of-the art is the use of 3D in addition to time models of soil formation and distribution that reflect changes in soil patterns over time. The results of these models require computer graphics for adequate display and interaction.

41.1 Introduction

Before the third quarter of the 20th century there were few digital soil maps and no digital soil databases, simply because the technology to manipulate and display large amounts of spatial data digitally was unavailable. Until the 1970s, computers could only handle numerical soil data with difficulty; methods of statistical analysis were limited to analysis of variance, correlation and regression, with respect to hypothesis testing and significance. Field data were collected and classified on intuition, and real maps were manually drawn by skilled craftsmen. Crude computer printouts displaying contoured surfaces were just beginning to become a subject of debate; these "cartograms" were rejected by previous generations of soil cartographers as being completely inferior to the traditional, analogue paper maps that were based on large-scale aerial photographs and high-quality topographic maps.

Then, in the early 1970s, a revolution started. Computers became more powerful, data storage and access leapt from an ability to deal with tens or a few hundreds of soil profiles to thousands and hundreds-of-thousands of soil objects. With the new power to deal with numbers came the desire to act numerically – a new research and application arena was opened up by a new generation of mathematically trained soil scientists: the quantitative revolution had begun.

The ability to act and think digitally created many opportunities that were previously impossible. The development of digital cartography (see MacEachren and Kraak, 1997), and digital soil data in particular, not only meant that one could handle more data, faster, and more efficiently in better and more reproducible ways, but also showed that more information may be gleaned from the data, which in turn leads to the better understanding of the nature and properties of soil. This chapter gives a historical overview of the more important steps that have been implemented from 1976–2004.

41.2 Steps in digital soil data 1976–2004

Like many other pioneering efforts in automated cartography, the first digital maps (including soil) were developed mainly to automate map production and save money (Burrough, 1986). The conceptual model used in these choropleth maps of soil was that of the homogeneous polygon in addition to linked attributes. The location and form of the soil polygon was derived from field survey and aerial photos (later remotely sensed images) and surface geomorphology. Locations of field observations were often only marked approximately on a field map using a thick pencil and hand-written notes.

One of the first, positive effects of digitizing soil maps was not the expected increased efficiency in map production (in fact, in the beginning digital mapping was more expensive than by hand) but the development within soil survey organisations of a common legend for soil surveys made at different scales. Prior to automation there had been little drive to integrate surveys and data. Other unexpected improvements included checks on the quality of map images, such as, all polygons were made closed and labelled and name placement was introduced, and cut-and-paste operations were developed to ensure that the centre of interest could be always kept at the centre of the map.

One important, positive development resulting from the introduction of automation was the creation of national databases of soil polygons and soil profile data. Soil attribute data linked to points, lines and polygons is often in tabular form and permits easy sorting of data according to one or more attributes. Such relational and tabular databases also provide the means for many forms of statistical analysis and empirical or logical modelling. But without computer tools to explore, display and analyse these data, they are not of much use.

41.3 The aim of this chapter

The aim of this chapter is to explore the changes in the ways people have perceived soil and the methods used for communicating that information to improve understanding of how soil is formed, its reactions to use and its distribution over the landscape. The main hypothesis is that our insights into the spatial and temporal variation of soil over the landscape are to a large extent controlled by the conceptual models we have available. This range of conceptual models is, in turn, to a certain extent controlled by the methods of display at our disposal, which increasingly make use of digital graphics.

41.4 Methods for displaying digital map data

The following paragraphs list the main conceptual models that have been (and are still being) used for mapping soil.

41.4.1 The crisp boundary model (1)

Unique soil types are distinguished by hierarchical decision rules leading to crisply defined and sharply bounded units in the soil attribute data space and in the landscape. The conventional theory of soil variation states that for soil series, important changes in soil occur only at crisp boundaries; within boundaries the soil is homogeneous. Soil boundaries can be mapped using features visible in the landscape, on aerial photographs and other remotely sensed information. Unique data on soil properties are linked to a limited number of nonoverlapping soil units. The methods used for displaying soil maps are identical to those used for land use or land cover maps and can be displayed using either vector or raster technology. Pictures of soil profiles and soil descriptions may also be linked to objects and locations to illustrate links between soil formation and place in the landscape (See Fig. 41.1). Today, global positioning systems (GPS) linked to satellites ensure proper geolocation of all geographical objects represented by points, lines and polygons and arrays of grid cells.

Figure 41.1. *Example of crisp polygon map and associated soil sample database. Source: Stichting for Bodemkartering (STIBOKA), Wageningen, Netherlands.*

41.4.2 The crisp boundary model (2)

This model is similar to the crisp boundary model (1), but minor stochastic variations (mean, variance) are tolerated within soil polygons. This increases the size of the attribute database to include means and variances; otherwise these are treated as ordinary, but extra attributes. Soil data attached to polygons, boundaries or points (sample locations) may be displayed in a variety of ways (pie charts, histograms, colour, etc.). The attributes of the polygon, boundary and point objects may be analysed using a wide range of univariate and multivariate statistical methods. See Figure 41.2.

41.4.3 The model of continuous, smooth variation (1)

Much research has demonstrated that the variation of individual soil attributes in space can be modelled by a smooth mathematical function. This smooth function may be a regression or trend surface, or obtained by interpolation from point, line or polygon data. Interpolation may be based on a wide range of deterministic methods (inverse distance, splines, regression) leading to a smoothly varying mathematical field, which is the resulting conceptual object. The combination of data density and the interpolation function used define the form of the field, which is therefore somewhat arbitrary. Continuous variation is

Figure 41.2. *Display of sampling points, elevation surface and concentrations of heavy metals in the topsoil (0–10 cm). Source: Department of Physical Geography, Utrecht University, Nederlands Topografische Dienst.*

displayed using contours, grey scales, colour shading and 2.5D surfaces. See Figures 41.3 and 41.4, and Burrough and McDonnell (1998).

41.4.4 The model of continuous, smooth variation (2) – with added noise

Soil attributes vary in space in a manner that can be modelled by a smooth mathematical function that expresses ideas of optimality and minor uncertainty using the methods of geostatistics – (e.g., Webster and Oliver, 1990; Goovaerts, 1997). The semivariogram is used to provide information on scales and levels of variation and to compute local, optimal values of interpolation weights. Interpolations at unsampled points receive estimates of uncertainty that are

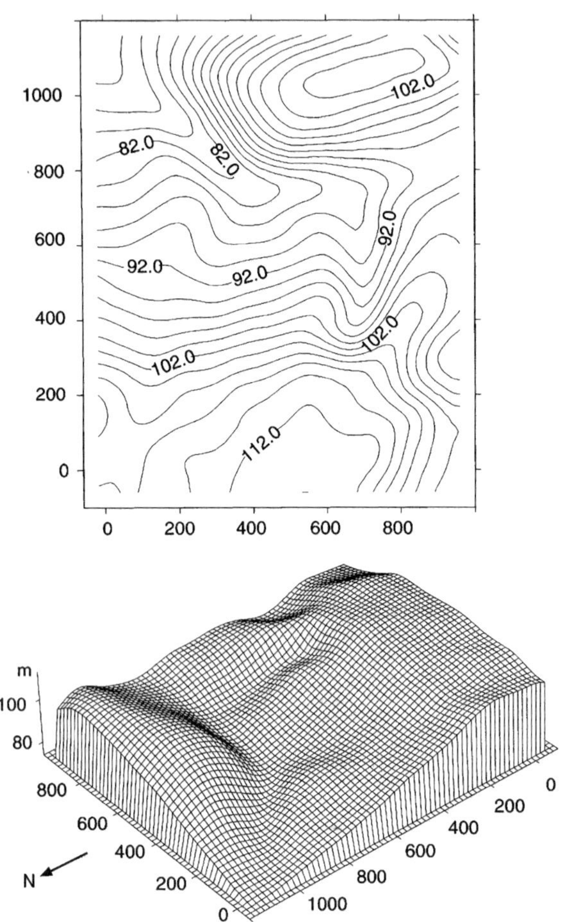

Figure 41.3. *Continuous, smooth variation displayed as contours or gridded surface. Source: Department of Physical Geography, Utrecht University.*

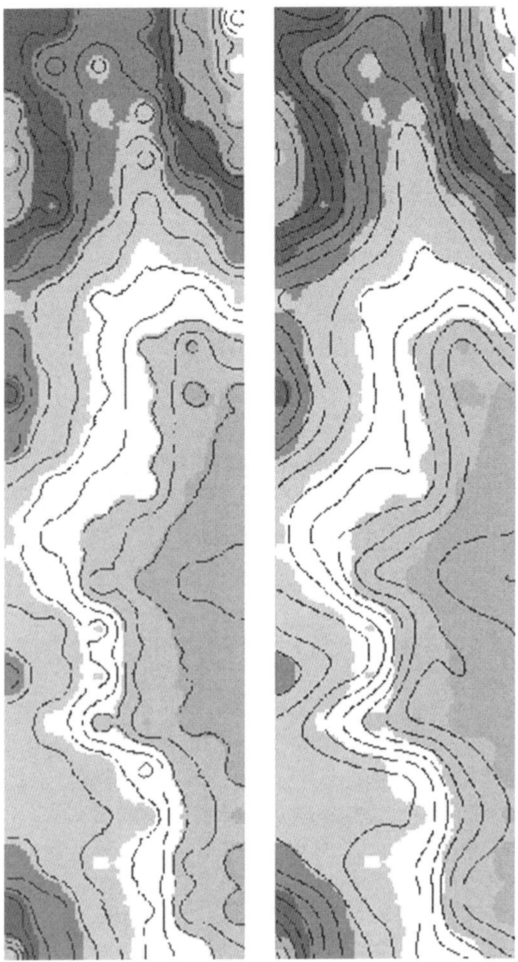

Figure 41.4. *Shows the effect of interpolation method on the resulting contoured surface. Left: inverse square weighting, right: splines. Shading is based on the inverse square weighting to emphasise differences in the methods. Source: Burrough 1969/2004.*

controlled by the form of the semivariogram and the density of data, thereby reducing some of the arbitrary nature of simple interpolation. Because geostatistical interpolation yields two surfaces – an interpolated value and an interpolation error – demands on display increase. (See Figs. 41.5a,b).

41.4.5 The model of continuous, smooth variation (3) – with added information

It is well known that the spatial variation of soil attributes is often spatially correlated, so pedometricians have made use of methods of cokriging (often

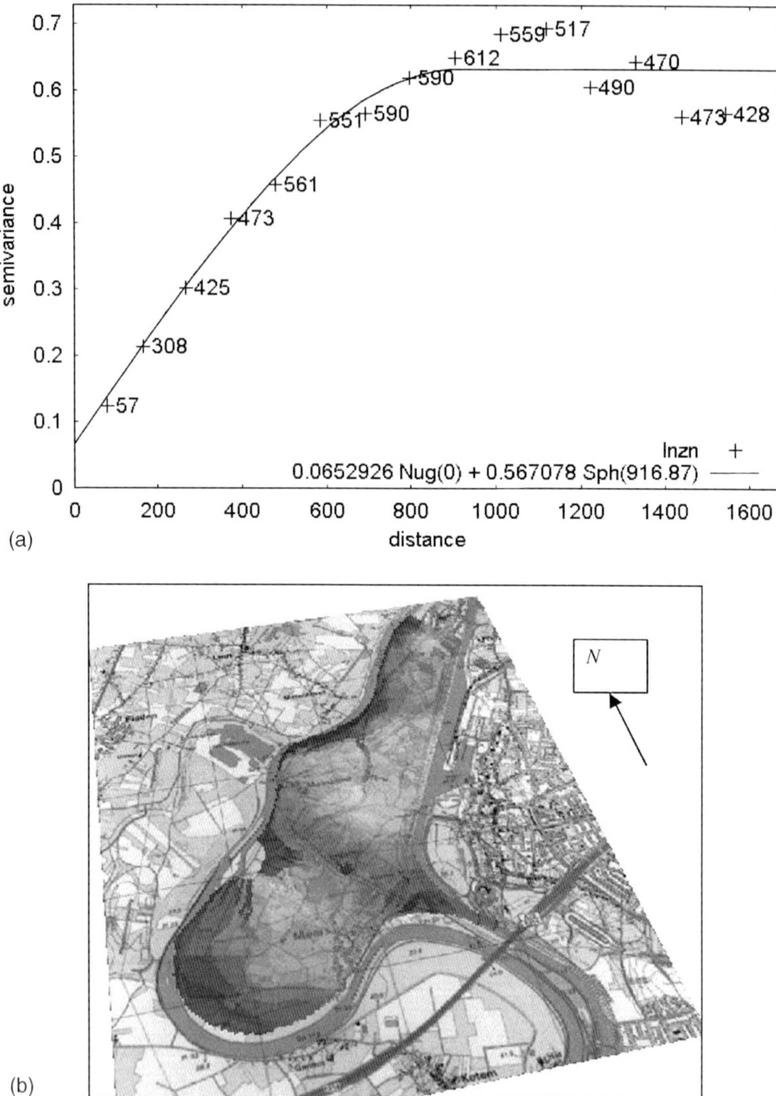

Figure 41.5. *(a) Semivariogram used for computing interpolation weights. (b) Top-soil zinc levels interpolated by co-kriging from point data and flooding frequency zones (shown in blue) to yield a smooth surface which is draped over a semitransparent interpolated elevation surface, displayed over a topographic map. Source: Department of Physical Geography, Utrecht University.*

known as universal kriging) to supplement the information on the variation of costly soil attributes with information gathered from cheap-to-measure but correlated attributes, and by so doing they intend to improve the quality of the interpolation (e.g., Burrough and McDonnell, 1998). Some examples of the application of this methodology are given by Bourennane et al. (1996) (using slope gradient to enhance interpolation of thickness of surface soil layers), Goovaerts (1999) (using elevation to aid the interpolation of rainfall erosivity), Leenaers et al. (1990) (using elevation data to aid interpolation of heavy metal pollution of floodplains), Stein et al. (1989) (using multispectral aerial photography to aid interpolation of soil-available water) and Fisher (1998) (improved modelling of elevation error).

41.4.6 The model of continuous, smooth variation (4): draping thematic data over DEMs

The spatial density of field measured soil observations for both expensive and cheap-to-measure attributes is often very sparse compared to elevation data gathered from aerial photos and GPS-controlled remote sensing. The spatial variations in surface patterns displayed on digital elevation model (DEM) created by photogrammetry or laser altimetry may show much more spatial variation than the one obtained by digitizing topographic maps or interpolation from spot heights. Given the assumption of covariation between landscape form, lithology and composition and soil properties or soil types, a DEM can be used to support 2.5D display of many kinds of soil survey data. This can equally be data that represents crisp polygon data or continuous variation: pictures and text may also be linked to locations on the DEM. Display of soil data "draped" or geolocated accurately over a DEM can be used as an exploratory analysis tool, or as a means of generating or testing the validity of ideas on catenas and other soil–landscape relations. (See Figs. 41.5b and Plate 41a,b, see Colour Plate Section).

41.4.7 The model of soil as a multivariate response to lithology, climate, hydrology and other factors as proposed by pioneering pedologists (such as Hans Jenny)

Unlike the previous concepts given above, this model views soil as a set of multivariate, interacting factors in soil attribute space. No hierarchy is imposed but natural tendencies to cluster in attribute space are investigated using one or more kinds of multivariate analysis. Many kinds of multivariate analysis have been tried out, such as principal components analysis (PCA), numerical taxonomy and clustering, maximum likelihood, neural networks and many more. Methods embodying ideas of class overlap (fuzzy sets) are particularly attractive because they seem to mimic the kinds of complex variation observed in real soil. A multivariate approach to soil development can be carried out using data from derivatives of DEMs. Displaying the results of these overlapping clustering over the original DEMs yields interesting methods for classifying and displaying automatically created maps of soil and vegetation. See Figures 41.6 and 41.7 and Burrough et al. (2001), Pfeffer et al. (2003). The approach adopted in these statistical methods is not that of setting up and testing hypotheses, but of investigating the effects of selecting different subsets of the data to look for clusters and anomalies. The use of exploratory data analysis (EDA) for the interactive analysis of multivariate soil data through several interactive windows, such as a map view, a histogram view, a scatterplot view and a multivariate statsitical view is now commonplace (Gunnink and Burrough, 1997; Bivand and Gebhardt, 2000).

41.4.8 The model of soil as a multivariate response to lithology, climate, hydrology and other factors including temporal change in the landscape

Whereas all conceptual models so far mentioned regard soil as a static phenomenon in space, there are many situations where variations in soil properties change over short intervals of time. Examples include soil moisture percolation over and down the soil profile, surface-water runoff, ground-water level changes, erosion and deposition. The modelling and display of these space–time processes requires suitable computing tools. Among others, this has been developed in the PCRaster toolkit for dynamic modelling (Burrough 1998; Clarke et al. 2002; Mitasova and Mitas, 2002, Wesseling et al., 1996), which provides means for modelling changes of soil properties, interpolation of surfaces and the incorporation of dynamic positive and negative feedback loops (Fig. 41.8). Although Figure 41.9 provides some insights to the modelling of spatial and temporal changes, it is really very difficult to display the change of soil patterns over time without using computer animation.

41.5 Discussion

Display of spatial data on soil and soil–landscape relations has progressed from simple, relatively crude manual techniques based on careful, but necessarily inexact field observations, through precisely controlled use of aerial photographs and exact topographic base maps to techniques for high-quality 2D, 3D and multitemporal display. Increasingly, the need is not just to display map images in digital form, but to create soil databases that are reliable and provide valuable inputs to models of soil formation and soil and ecological change. These models often combine soil information with hydrological, geological and ecological data in order to provide an integrated approach to the better understanding and managing of the earth's surface for people, fauna and flora (see Harmon and Doe, 2001). In this last, soil data cannot be used alone (indeed if we expect that they will only be used independently, then it is quite likely they

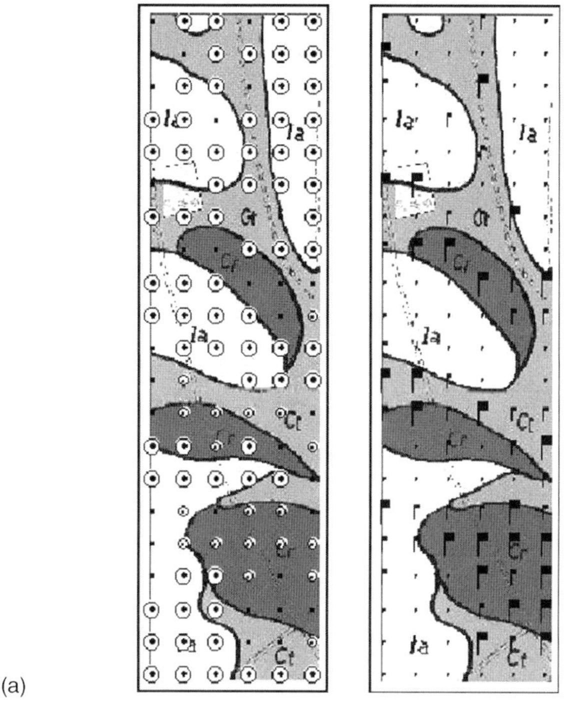

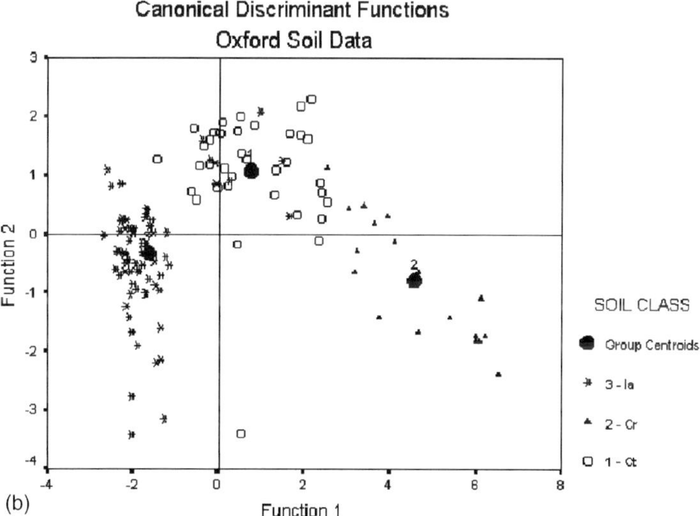

Figure 41.6. *Crisp division of soil landscape and borings into landscape soil units (top). Grey shading and text symbols indicate field mapped soil series; point symbols indicate locations of samples on 100 × 100 m grid. Left: pH, right: percentage clay. Source of data: Burrough 1969.*

Figure 41.7. *Clusters of classified DCA (Detrended Correspondence Analysis) scores for native plants as interpolated by universal kriging within landform classes for Alpine vegetation in Austria, displayed on top of a DEM having 10 m resolution (Pfeffer et al 2003).*

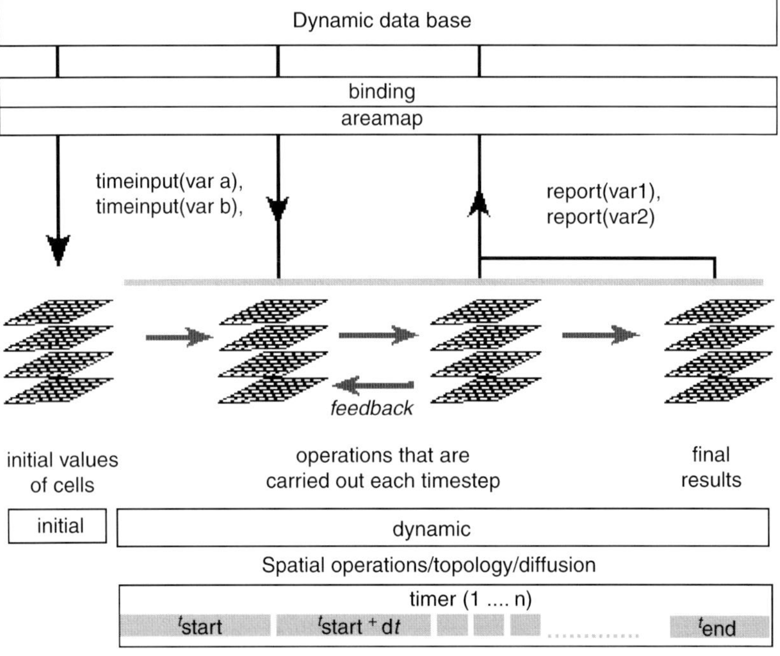

Figure 41.8. *Data structure of PCRaster dynamic modelling tool showing the stacked map layers that are used to created time series of displayed data as in a film (Wesseling et al. 1996).*

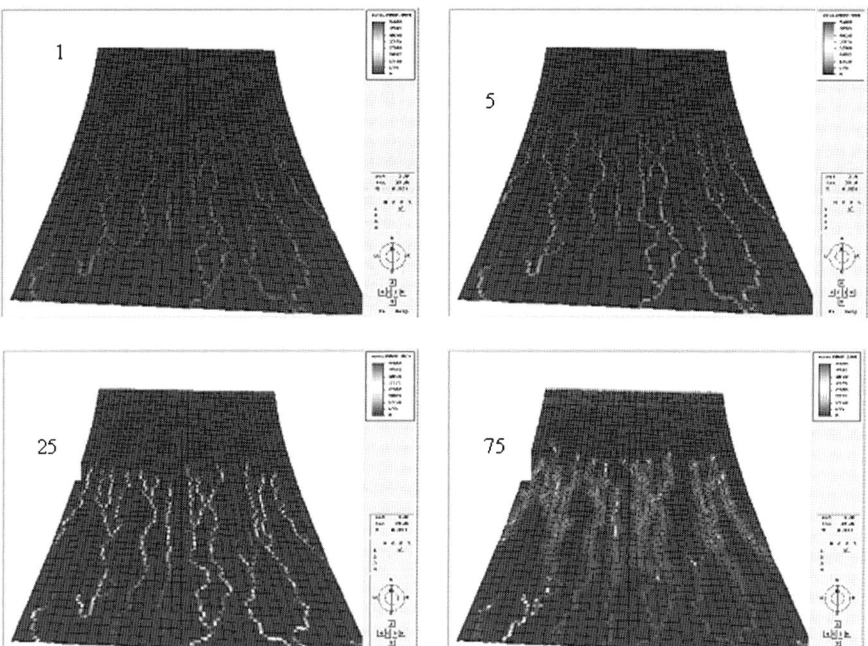

Figure 41.9. *Dynamic display (2): Four frames from a PCRaster simulation of erosion gulleys on a bare paddock. Numbers display model cycle.*

will be ignored by many environmental scientists as being at best difficult to understand, and at worst irrelevant). The storage and display of soil data in digital form is essential for their further use.

One clear lesson that became apparent during the writing of this chapter was the degree to which understanding of the soil as a changing, organic phenomenon has developed. Twenty-five years ago soil was perceived as static, unchanging phenomenon that could be classified by strict hierarchical rules into unchanging groups. Today, there is much more understanding of soil dynamics, and of soil change patterns that have overlapping, and sometimes interfering scales. We are almost reaching a thorough understanding of the complexity and instability of soil with respect to the forces acting upon it, and its ability or inability to resist change.

Much change in soil is seen today as simply degrading in quality. All over the world people have caused soil erosion, soil pollution and the reduction in critical properties of soil that are essential for plant growth. Twenty-five years ago there were few perceived needs for studies on the uncertainty of soil classification, for using local data for local soil classification and assessment, and no major attempts to visualise these changes.

Also, 25 years ago, soil was treated as an object of unique study; today it should be unthinkable to deal with information on soil without at the same time taking account of surface and subsurface moisture, vegetation, land form and land cover. Soil has a central role to play in ecological studies and for studying the processes of land use and land cover change, and also climate change. One would think that soil should take a central place in studying the dynamics of the surface processes operating on our planet, but in fact the reverse is true – interest in soil survey has actually dwindled during recent years, leaving fewer practising soil surveyors and students of soil than 25 years ago. In contrast, interest in modelling urban processes is increasing everywhere (Longley and Batty, 2003). It is high time to reverse this trend, and part of this effort must come through the use of modern information technology for the visualisation and presentation of key data.

References

Acres, B.D., Bower, R.P., Burrough, P.A., Folland, C.J., Kalsi, M.S., Thomas, P. and Wright, P.S., 1975. The Soils of Sabah (report of 1:250,000 soil survey and land evaluation of the whole State in 5 volumes). Land Resources Division, Ministry of Overseas Development, London.

Bivand, R., Gebhardt, D.L., 2000. Using the R statistical data analysis language on GRASS 5.0 GIS data base files. Comput. Geosci. 26, 1043–1052.

Bourennane, H., King, D., Chery, P., Bruand, A., 1996. Improving the kriging of a soil variable using slope gradient as external drift. European J. Soil Sci. 47, 473–484.

Burrough, P.A., 1969. Studies in Soil Survey. D. Phil Thesis, Oxford University.

Burrough, P.A., 1986. Principles of Geographical Information Systems for Land Resources Assessment. Oxford University Press, Oxford, UK.

Burrough, P.A., 1998. Dynamic Modelling and GIS, Chapter 9. In: P. Longley, S. Brooks, R. McDonnell, and W. MacMillan (Eds), Geocomputation: a Primer Proceedings of the Third International Conference on Geocomputation. Wiley, Bristol, UK, 165–192pp.

Burrough, P.A., McDonnell, R.A., 1998. Principles of Geographical Information Systems. Oxford University Press, Oxford, UK.

Burrough, P.A., Wilson, J.P., van Gaans, P.F.M., Hansen, A.J., 2001. Fuzzy k-means classification of topo-climatic data as an aid to forest mapping in the Greater Yellowstone Area, USA. Landscape Ecol. 16, 523–546.

Clarke, K.C., Parks, B.O., Crane, M.P. (Eds) 2002. Geographic Information Systems and Environmental Modeling. Prentice Hall, New Jersey.

Fisher, P., 1998. Improved modelling of elevation error with Geostatistics. Geoinformatica 2 (3), 215–233.

Goovaerts, P., 1997. Geostatistics for Natural Resources Evaluation. Oxford University Press, New York.

Goovaerts, P., 1999. Using elevation to aid the geostatistical mapping of rainfall erosivity. CATENA 34, 227–242.

Gunnink, J., Burrough, P.A., 1997. Interactive spatial analysis of soil attribute patterns using Exploratory Data Analysis (EDA) and GIS. In: M. Fischer, H.J. Scholten and D. Unwin (Eds), Spatial Analysis in GIS. Taylor and Francis, (1996), 87–100.

Harmon, R.S., Doe, W.W., 2001. Landscape Erosion and Evolution Modeling. Kluwer Academic/Plenum Publishers, Dordrecht.

Leenaers, H., Okx, J.P., Burrough, P.A., 1990. Employing elevation data for efficient mapping of soil pollution on floodplains. Soil Use and Management 6, 105–114.

Longley, P.A., Batty, M., 2003. Advanced Spatial Analysis. ESRI Press, Redlands, California.

MacEachern, A.M., Kraak, M.J., 1997. Exploratory cartographic visualisation: Advancing the agenda. Computers and Geosciences 23 (4), 335–343.

Mitasova, H., Mitas, L., 2002. Modeling Physical Systems. In: K.C. Clarke, B.O. Parks, and M.P. Crane (Eds), Geographic Information Systems and Environmental Modeling. Prentice Hall, New Jersey, 189–210pp.

Pfeffer, K., Pebesma, E., Burrough, P.A., 2003. Mapping Alpine Vegetation using vegetation observations and topographic attributes. Landscape Ecol. 18, 759–776.

Stein, A., Bouma, J., Mulders, M.A., Weterings, M.H.W., 1989. Using co-kriging in variability studies to predict physical land qualitites of a level river terrace. Soil Tech. 2, 385–402.

Webster, R., Oliver, M.A., 1990. Statistical Methods in Soil and Land Resource Survey. Oxford University Press, Oxford.

Wesseling, C., Karssenberg, D., Burrough, P.A., van Deursen, W., 1996. Integrating Dynamic Environmental Models in GIS: the development of a Dynamic Modelling Language. Trans. GIS 1, 40–48.

Developments in Soil Science, volume 31
P. Lagacherie, A.B. McBratney and M. Voltz (Editors)

Chapter 42

ARE CURRENT SCIENTIFIC VISUALISATION AND VIRTUAL REALITY TECHNIQUES CAPABLE TO REPRESENT REAL SOIL-LANDSCAPES?

S. Grunwald, V. Ramasundaram, N.B. Comerford and C.M. Bliss

Abstract

Real soil-landscapes are complex consisting of an inextricable mix of patterns and noise varying continuously in the space–time continuum. Soils and parent material show gradual variations in the horizontal and vertical planes forming three-dimensional (3D) bodies that are commonly anisotropic. There is no real beginning and end point in soil-landscapes because environmental conditions are dynamically changed through water flow, biogeochemical processes and human activities. The strengths of soil-landscape modelling lies in hypothesis testing, understanding causal linkages between environmental factors and their interrelationships within a spatial and temporal explicit context. To develop virtual soil-landscape models entails: (i) conceptualisation, that is defining the model framework (e.g. finite space elements); (ii) reconstruction, that is describing and quantifying underlying conditions and behaviour and (iii) scientific visualisation (SciVis), that is abstracting real soil-landscapes into a format that we can comprehend and that helps us to understand the complexity of soil-landscapes. The primary objective in data visualisation is to gain insight into an information space by mapping data onto graphical primitives. Capabilities and limitations of SciVis and virtual reality (VR) techniques are discussed in this chapter. Only recently 3D soil-landscape models have been emerging. We present a case study that translated the spatio-temporal water-table dynamics of a flatwood soil-landscape in Florida into a virtual domain using a geostatistical method to reconstruct the soil-landscape and Virtual Reality Modelling Language (VRML) enhanced with External Authoring Interface (EAI) for visualisation and implementation of interactive functions. Just as maps can visually enhance the spatial understanding of phenomena, interactive spatio-temporal applications can enhance our understanding of complex environmental systems and the underlying transport processes driving soil and water quality.

42.1 Introduction

Scientists have focused on two contrasting concepts to study soil-landscapes and ecosystem processes, which are both equally important. The reductionist approach promotes ever more detailed studies of distributions, soil classes, events and processes, followed by their interpretation. The other approach develops and enunciates an integrative, unifying point of view encompassing and integrating previous observations and results. Both concepts have been employed for quantitative spatially explicit modelling of soil-forming factors evolving through time.

Different kinds of models have been used to translate real soil-landscapes and ecosystem processes into virtual environments. Hoosbeek and Bryant (1994) provide an overview of pedological models using criteria such as the relative degree of computation (qualitative vs. quantitative models), complexity of the model structure (functional vs. mechanistic models) and level of organisation (soil region, pedon to molecular scale). Modelling is about choosing the appropriate metaphor or analogy with which to better understand a phenomenon, for example the spatial distribution of soils, their behaviour and relationship to other environmental factors. In this sense, we create media about phenomena to bridge the gap between what we do not know, and what we are trying to comprehend. Media such as slides, maps, animations, three-dimensional (3D) virtual worlds and digital libraries are also models. Although one might talk of absolutes such as "reality" and "truth", all we have at our disposal are models, which mediate the world for us. We have to acknowledge that different media have different effects on our understanding and interpretation of objects, such as soil-landscapes and ecosystem processes. The following criteria potentially influence to transcend real into virtual soil-landscapes: (i) space, (ii) time, (iii) scale, (iv) ecosystem condition, (v) spatial and temporal variability (vi) interrelationships between environmental factors and (vii) causal linkages or behaviour of the system.

The modelling process can be disaggregated into: (i) conceptualisation, that is defining the model framework (e.g. finite space elements); (ii) reconstruction, that is describing and quantifying underlying conditions and behaviour and (iii) scientific visualisation (SciVis), that is abstracting real soil-landscapes into a format that we can comprehend and that helps us to understand the complexity of soil-landscapes.

Scientific visualisation is defined as the use of the human visual processing system assisted by computer graphics, as a means for the direct analysis and interpretation of information. McCormick et al. (1987) indicated that SciVis transforms the symbolic into the geometric, enabling researchers to observe their simulations and computations. It offers a method for seeing the unseen. It enriches the process of scientific discovery and fosters profound and unexpected insights. In many fields, it is already revolutionising the way scientists do science (McCormick et al., 1987). Senay and Ignatius (1994) point out that the primary objective in data visualisation is to gain insight into an information space by mapping data onto graphical primitives.

Real soil-landscapes are complex consisting of an inextricable mix of patterns and noise varying continuously in the space–time continuum. Soils and parent material show gradual variations in the horizontal and vertical planes forming 3D bodies that are commonly anisotropic. There is no real beginning and end point in real soil-landscapes because environmental conditions are dynamically changed through water flow and biogeochemical processes. In addition, human-induced changes have had remarkable effects on almost all soil-landscapes.

Transforming real into virtual soil-landscapes is based on model predictions and estimations both associated with uncertainty. Estimations use sample data to make an inference about a population whereas prediction refers to a statement made about the future or reasoning about the future. Methods used in science for the derivation of predictions of unknown facts from known facts include (modified from Bunge, 1959):

(1) Logical inference entailing deduction, induction and abduction, the latter one referring to the generation of hypotheses to explain observations.
(2) Structural laws help predict new properties from the known properties of material or formal structures.
(3) Phenomenological laws predict phenomena on the basis of known constant associations.
(4) Functional laws infer properties of a system from knowledge of the functional role of the parts and their interconnections.
(5) Statistical laws help derive collective properties of classes of events from an analysis of such classes.
(6) Mechanical laws extrapolate future (or past) states on the basis of known current states and relations (e.g. Newtonian laws).

42.1.1 Time and space concepts

Frank (1998) provide an overview of time models. Almost all existing soil-landscape models are based on Newtonian time that is focused on a succession of phenomena along a linear time coordinate providing the simplest time concept characterised by causal inertness. Time is viewed as a neutral framework against which independently unfolding events are projected, sorted and measured. Newton argued that time is absolute implying that the universe has a single universal clock capable of determining that two occurrences are simultaneously. The present moment forms the centre point changing constantly. Backcasting and forecasting models exists predicting past and future events (e.g. formation of Spodosols, land use change models) with exponentially increasing prediction errors from the present moment. Other soil-landscape models are "snapshot models" that are limited to describe current environmental conditions. Characteristics of "real" time include that events are non-repeatable and sometimes structured and at other times chaotic.

Generally, it is necessary to divide geographical space into discrete spatial units and the resulting tessellation is taken as a reasonable approximation of reality at the level of resolution under consideration (Burrough and McDonnell, 1998). There are two types of spatial discretisation methods used for soil-landscape modelling: (i) crisp soil map units and (ii) the continuous fields model or pixel-based model (Peuquet, 1988; Goodchild et al., 1992; Burrough and McDonnell, 1998). The crisp model has its roots in empiric observations combined with 19th century biological taxonomy and practice in geological survey. Traditional soil-landscape models use crisp map units which are defined by abrupt changes from one map unit to the other (Voltz and Webster, 1990; Webster and Oliver, 1990). Each soil map unit is associated with a representative soil attribute set (Soil Survey Staff, 1998). Horizons of these soil map units differ from adjacent and genetically related layers in physical, chemical, morphological and biological properties such as texture, structure, colour, soil organic matter or degree of acidity. As such, soil horizons and profiles of these properties correspond to discrete, sharply delineated (crisp) units, which are assumed to be internally uniform. The crisp soil model has been questioned and critically discussed repeatedly (Webster and De La Cuanalo, 1975; Nortcliff, 1978; Nettleton et al., 1991; McBratney and de Grujiter, 1992; Heuvelink and Webster, 2001). An alternative geographic model displays the real world as a set of pixels or voxels (volume element) and is adequate for modelling natural phenomena that do not show obvious boundaries (e.g. soils). This spatial model has the potential to describe the gradual change of soil properties formed by a variety of pedological processes within a domain. The spatial resolution depends on the spatial variability of soil properties. Geostatistical techniques have been introduced to interpolate point observations and construct soil property pixel maps (Goovaerts, 1997; Chilès and Delfiner, 1999; Webster and Oliver, 2001). Challenging is to optimise the density and spatial distribution of observations across a domain to characterise soil-landscape reality without knowing the real spatial distribution of soil properties and operating pedological processes. Crisp and the pixel model have been used extensively to produce two-dimensional (2D) soil maps. This contradicts with the conceptual view of soils as 3D natural bodies. McSweeney et al. (1994) proposed a 3D framework for soil-landscape modelling that has yet to be adopted by soil surveyors and pedologists.

42.1.2 Are current quantitative reconstruction techniques capable to capture spatial and temporal variability of properties and behaviour?

The strengths of soil-landscape modelling lies in hypothesis testing, understanding causal linkages between environmental factors, and their interrelationships within a spatial and temporal explicit context. Goovaerts (1999), McBratney et al. (2000), Heuvelink and Webster (2001), McBratney et al. (2003)

and Grunwald (2005) provided a comprehensive overview of pedometric techniques to model soil spatial and temporal variation. Advanced modelling techniques exist; yet there are major limitations that prohibit their widespread adoption. Geostatistical methods are data intensive requiring a large number of point observations (Webster and Oliver, 2001). Similarly dense frequency datasets are required to describe the change of soil and environmental properties through time. In short, geo-temporal modelling of soil-landscapes is data intensive. Process-based mechanistic models offer alternatives to simulate soil-landscape evolution through time (Minasny and McBratney, 1999). However, most pedological processes operate over long time periods. For those reasons, it is challenging to validate pedodynamic process models. Emerging soil mapping techniques (e.g. soil sensors and remote sensors) provide new opportunities to collect exhaustive datasets that support the reconstruction of soil-landscapes. Yet uncertainties of such datasets are typically high constraining their use.

42.1.3 Are current scientific visualisation and virtual reality techniques capable to represent real soil-landscapes?

The internet, geographic information technology and SciVis provide new education and information delivery capabilities. Numerous studies have shown that SciVis is effective for enhancing rote memorisation and higher-order cognitive skills (Koussoulakou and Kraak, 1992; Barraclough and Guymer, 1998). Stibbard (1997) found that information is absorbed best when using more than one human sense; that is 10% of the information is taken in by reading, 30% by reading and visual, 50% by reading, visuals and sound and 80% by reading, visuals, sound and interaction. Koussoulakou and Kraak (1992) tested the usefulness of different SciVis methods including static maps, series of static maps and animated maps, and found significantly better response times for animated maps. Barraclough and Guymer (1998) reported that advanced visualisation techniques served to better communicate spatial information between people in different fields, such as scientists, administrators, educators and the general public. Just as maps can visually enhance the spatial understanding of phenomena, interactive spatio-temporal applications can enhance our understanding of complex environmental systems and the underlying transport processes driving soil and water quality. According to Fisher and Unwin (2002) visual interfaces maximise our natural perception abilities, improve to comprehend huge amounts of data, allow the perception of emergent properties that were not anticipated, and facilitate understanding of both large-scale and small-scale geographic features of ecosystems.

Only recently 3D soil-landscape models have been emerging. For example, a 3D soil horizon model in a Swiss floodplain was created by Mendonça Santos et al. (2000) using a quadratic finite-element method. Grunwald et al. (2000) presented 3D soil-landscape models at different scales for sites in southern Wisconsin using Virtual Reality Modelling Language (VRML). Sirakov and Muge (2001) developed a prototype 3D Subsurface Objects Reconstruction and Visualisation System (3D SORS) in which 2D planes are used to assemble 3D subsurface objects.

The development of immersive and desktop virtual reality (VR) techniques has been instrumental to develop virtual soil-landscapes and environments. VRML-based models enhanced with Java and External Authoring Interface (EAI) provide capabilities to display real soil-landscapes in 3D and four-dimensional (4D) digital formats (Ramasundaram et al., 2004). Characteristics of VR include: (i) immersion, (ii) navigation (freedom for the user to explore) and (iii) interaction. VR applications are still limited to prototype applications and have not found widespread adoption. Reasons that constrain the extended production of virtual soil-landscape models are due to (i) input data requirements, (ii) labour intensive production (programming is required), (iii) lack in training and education of people to produce such models, (iv) preference of users for traditional 2D maps and (v) lack of realistic abstraction of real soil-landscapes.

42.2 Case study

Our goal was to translate a real into a virtual soil-landscape in Florida. We discuss the limitations of our approach in context of conceptualisation, reconstruction and SciVis.

42.2.1 Objectives

Specific objectives were (i) to reconstruct a flatwood soil-landscape in Florida to describe and display the spatial distribution of soils and topography in 3D format and (ii) to develop space–time simulations that describe and display water flow in 4D format.

42.2.2 Methodology

The study area comprised a 42-ha site in northeastern Florida with hydric and non-hydric soils. About one third of the site was covered by bald cypress (*Taxodium distichum*) and about two-thirds by slash pine (*Pinus elliottii*). In 1994, three silvicultural treatments were administered. While one area was left as a control (uncut), a second area was clearcut. In the third area only the forest on the hydric soils was cut and that on the non-hydric soils left untouched. Morphological and taxonomic soil data were collected at 123 locations. Water table was monitored biweekly at 123 wells from April 1992 to March 1998. Topography was characterised by laser level and ranged from 26.7 to 30.8 m.

We developed a model that characterises soil horizons and terrain across the flatwood site. A digital elevation model (DEM) was developed using the observed point elevation values and ordinary kriging. Soil horizon depths were interpolated in the horizontal plane using ordinary kriging and linear interpolation in the vertical plane to create soil volumes representing the horizons. VRML was used to render face geometry of the soil horizon model (Lemay et al., 1999) and fuse the DEM and soil horizons. The *IndexedFaceSet* VRML class was employed to render polyhedrons. A point-arc geographic data model was used to create *IndexFaceSets*. The appearance of volume objects was coded using the RGB (red–green–blue) colour classification system. An interface is necessary to communicate between a VRML world and an external environment. This interface is called an EAI and it defines the set of functionality of the VRML web browser that the external environment can access. To add interactivity to our web-based model, we developed a Java applet to extend the capability of the Blaxxun3D Java applet, which supports the VRML Java EAI interface.

Dynamic models of hydropatterns were developed from precipitation data and water-table measurements for a period of 6 years (April 1992 and March 1998) using 15-day time increments. Hundreds of semivariograms for water-table depth had to be generated, each representing one specific time period. Water-table levels were interpolated using ordinary kriging. The water-table surface was sliced with the DEM to distinguish inundated from non-inundated areas. The *CoordinateInterpolator* VRML node was used to produce a smooth display of water-table depths between observation periods. Water inundation models for each time period were stored in a digital library. The *IndexedFaceSet* VRML class was employed to render the extent of the study site. The graphical user interface was implemented using a Java Applet that reads the models on-the-fly from the digital library constraint by user-defined start and end times for a inundation simulation.

42.2.3 Results and discussion

The 3D soil horizon model infused with a DEM is shown in Plate 42(a) (see Colour Plate section) and a demo is available at http://3dmodel.ifas.ufl.edu. To access the model, a web browser and VRML plug-in such as Blaxxun3D are required. Users have the ability to view the whole model or each soil horizon independently. VRML sensors enable to rotate, zoom, tilt or pan the model giving users the feeling of emergence in the web-based world. Soil horizons are displayed continuously in 3D geographic space taking into account the displacement from the soil surface.

We implemented the space–time simulations of water-table dynamics in an interactive framework. Users can trigger an event (e.g. inundation simulation) by constraining the border conditions (e.g. time period and geographic domain) of simulations interactively (Plate 42(b) see Colour Plate Section). Such an adaptive simulation framework invites users to study water-table dynamics through observations using a sequence of events: trigger an event – observe ecosystem process – interpretation – assimilation.

Users can study the expansion and contraction of inundated areas over time. This implementation is dynamic and superior when compared with a static 2D GIS "snapshot" map that shows water tables at few specific times. Though elevation changes across the study area are small topography is the major factor driving watertable dynamics on this flatwood site. The upslope drainage area defined the amount of surface water (and most likely some shallow lateral water) draining into the depressional areas. Spodosols and their distinct horizons confounded water flow driven by surface terrain.

We considered the following criteria for the development of our model-simulation environment: (i) global access, that is, web-based implementation, (ii) simulation of a variety of learning mechanisms, (iii) interactivity to engage users, (iv) compartmentalisation and hierarchical organisational structure, (v) abstraction of 2D and 3D geographic objects (e.g. soils, terrain) and dynamic ecosystem processes (e.g. water flow) using geostatistics and SciVis techniques. We used a relatively simple reconstruction method (ordinary kriging) because hundreds of variograms and kriged models needed to be produced. However, in this study we focused on SciVis to gain a better understanding of water-table dynamics at the soil-terrain interface over a long period of time.

Our dataset was unique covering a long time-period with dense observations in time and space. We were able to visualize causal linkages between terrain properties, land use, soils and water movement within a spatial and temporal explicit context. Employing SciVis facilitated multiple views of the content world, which stimulates a greater understanding and insight of the flatwood system. It is this synthesis of geographic datasets that distinguishes the virtual learning environment from conventional instructional media (e.g. 2D GIS maps). The availability of multiple representations – maps, 3D models, space–time models, text – of the same geographical region, each of which offers a different perspective of the soil-landscape, has the potential to improve our understanding of flatwood water dynamics. Desktop VR when combined with other forms of digital media may offer great potential for a cognitive approach to research and education. Scientific visualisation combined with quantitative reconstruction techniques has the potential to translate soil-landscapes and ecosystem processes into a transparent format to enhance our understanding of real-world phenomena and complex environmental systems. Virtual soil-landscape models are beneficial in disseminating geo-referenced soil and landscape data to educators, researchers, government agencies and the general public.

Acknowledgement

All the information concerning the research area was collected by C.M. Bliss and N.B. Comerford. We thank the National Council for Air and Stream Improvement (NCASI) and US Forest Service for funds that allowed the data collection as well as Rayonier, Inc. for allowing the study on their land. We also thank Adrien Mangeot for parts of the coding. This research was supported by the Florida Agricultural Experiment Station and approved for publication as Journal Series No. R-10877.

References

Barraclough, A., Guymer, I., 1998. Virtual reality – a role in environmental engineering education? Water Sci. Tech. 38 (11), 303–310.

Bunge, M., 1959. Causality. Harward University Press, Cambridge, MA.

Burrough, P.A., McDonnell, R.A., 1998. Principles of Geographical Information Systems. Oxford University Press, New York.

Chilès, J.-P., Delfiner, P., 1999. Geostatistics – Modeling Spatial Uncertainty. John Wiley & Sons, New York.

Fisher, P., Unwin, D., 2002. Virtual reality in geography. Taylor & Francis, New York.

Frank, A.U., 1998. Different types of "times" in GIS. In: M.J. Egenhofer and R.G. Golledge (Eds.), Spatial and Temporal Reasoning in Geographic Information Systems. Oxford University Press, New York.

Goodchild, M., Sun, G., Yang, S., 1992. Development and test of an error model for categoryical data. Int. J. Geogr. Inf. Systems 6 (2), 87–104.

Goovaerts, P., 1997. Geostatistics for Natural Resources Evaluation. Oxford University Press, New York.

Goovaerts, P., 1999. Geostatistics in soil science: state-of-the-art and perspectives. Geoderma 89, 1–45.

Grunwald, S. (Ed.) 2005. Environmental Soil-Landscape Modeling. CRC Press, New York.

Grunwald, S., Barak, P., McSweeney, K., Lowery, B., 2000. Soil landscape models at different scales portrayed in Virtual Reality Modeling Language. J. Soil Sci. 165 (8), 598–615.

Heuvelink, G.B.M., Webster, R., 2001. Modeling soil variation: past, present, and future. Geoderma 100, 269–301.

Hoosbeek, M.R., Bryant, R.B., 1994. Developing and adapting soil process submodels for use in the pedodynamic Orthod model. In: Bryant, R.B., Arnold, R.W. (Eds.), Quantitative Modeling of Soil Forming Processes, SSSA Special Publ. No. 39, Madison, WI, USA.

Koussoulakou, A., Kraak, M.J., 1992. Spatio-temporal maps and cartographic communication. Cartographic J 29, 101–108.

Lemay, L., Couch, J., Murdock, K., 1999. 3D graphics and VRML 2. Sams.net Publ., Indianapolis, IN.

McBratney, A.B., de Grujiter, J.J., 1992. A continuum approach to soil classification by modified fuzzy k-means with extragrades. J. Soil Sci. 43, 159–175.

McBratney, A.B., Mendonca Santos, M.L., Minasny, B., 2003. On digital soil mapping. Geoderma 117, 3–52.

McBratney, A.B., Odeh, I.O.A., Bishop, T.F.A., Dunbar, M.S., Shatar, T.M., 2000. An overview of pedometric techniques for use in soil survey. Geoderma 97, 293–327.

McCormick, B.H., DeFanti, T.A., Brown, M.D. (Eds.), 1987. Visualization in scientific computing. Comput. Graphics 21(6), (entire issue).

McSweeney, K., Gessler, P.E., Slater B.K., Hammer, R.D., Peterson, G.W., Bell, J.C., 1994. Towards a new framework for modeling the soil-landscape continuum. SSSA Special Publ. 33, Factors in Soil Formation.

Mendonça Santos, M.L., Guenat, C., Bouzelboudjen, M., Golay, F., 2000. Three-dimensional GIS cartography applied to the study of the spatial variation of soil horizons in a Swiss floodplain. Geoderma 97, 351–366.

Minasny, B., McBratney, A.B., 1999. A rudimentary mechanistic model for soil production and landscape development. Geoderma 90, 3–21.

Nettleton, W.D., Brasher, B.R., Borst, G., 1991. The taxadjunct problem. Soil Sci. Soc. Am. J. 55, 421–427.

Nortcliff, S., 1978. Soil variability and reconnaissance soil mapping: A statistical study in Norfolk. J. Soil Sci. 29, 403–418.

Peuquet, D., 1988. Presentations of geographic space: towards a conceptual synthesis. Ann. Assoc. Am. Geogr. 78 (3), 375–394.

Ramasundaram, V., Grunwald, S., Mangeot, A., Comerford, N.B., Bliss, C.M., 2004. Development of an environmental virtual field laboratory. Comput. Educ. J. 45, 21–34.

Senay, H., Ignatius, E., 1994. A knowledge-based system for visualization design. IEEE Comput. Graphics Appl. 2, 36–47.

Sirakov, N.M., Muge, F.H., 2001. A system for reconstructing and visualizing three-dimensional objects. Comput. Geosci. 27, 59–69.

Stibbard, A., 1997. Warwick University Forum, No. 6.

Soil Survey Staff, 1998. Keys to Soil Taxonomy, 8th edition. Government. Printing Office, Washington, DC.

Voltz, M., Webster, R., 1990. A comparison of kriging, cubic splines and classification for predicting soil properties from sample information. J. Soil Sci. 31, 505–524.

Webster, R., De La Cuanalo, H.E., 1975. Soil transect correlograms of North Oxfordshire and their interpretation. J. Soil Sci. 26, 176–194.

Webster, R., Oliver, M.A., 1990. Statistical methods in soil and land resource survey. Oxford University Press, Oxford.

Webster, R., Oliver, M.A., 2001. Geostatistics for Environmental Scientists. John Wiley & Sons, Chichester, England.

AUTHOR INDEX

A

Abramovich, F. 317–318
Adamchuk, V.I. 168
Adams, M.L. 208
Addiscott, T.M. 309
Ahern, C.R. 26
Ahrens, R.J. 16, 425, 427
Ahuja, L.R. 17
Akaike, H. 273
Albrecht, C. 425, 428
Aldrick, J.M. 193, 195, 202
Allard, D. 288
Allison, P.N. 440
Almaraz, R. 465
Almond, P.C. 335
Althouse, L. 392, 488, 508
Amen, A. 381, 392
Amundson, R.G. 285
Andrade Júnior, O. 48
Andreux, F. 294
Andrieux, P. 10, 288, 543–544
Antoine, J. 3
Appice, A. 7
Archer, J.R. 477
Arnett, R.R. 370
Arrouays, D. 4, 245, 294, 488, 497–498, 500
Ashton, L.J. 31
Ashton, L.U. 208
Asseng, S. 208
Assis, D.S. 47
Astle, W.L. 8
Atkinson, G.L. 147
Atkinson, P.M. 445
Aurousseau, P. 282, 287, 290, 509
Austin, M.P. 27, 29, 327, 373, 401, 412, 467
Avery, B.W. 366
Awater, R.H.C.M. 286
Ayyub, B.M. 98, 103

B

Baca, J.F.M. 48
Baert, G. 288
Baille, M. 16
Baize, D. 227
Ballmer, M. 465
Band, L.E. 293, 518
Banfield, C.F. 282
Banin, A. 225
Banks, R.G. 439–440
Baptista, G.M.d.M. 220
Barak, P. 576
Baran, A. 105
Baran, S. 105
Bardossy, A. 428
Barmore, R.L. 236
Barnes, E.M. 11
Barnes W.L. 488
Barney, M.L. 236–237
Barr, C.J. 477
Barraclough, A. 575
Barrett, J.G. 292
Barrett, L.R. 7
Barry, S.J. 13, 26, 32, 88, 90, 198, 202
Barthès, J.P. 51, 252, 291
Bartsch, H.U. 120
Bas, C.L. 55
Bascomb, C.L. 282
Batjes, N.H. 498
Batty, M. 568
Baumgardner, M.F. 108–109, 111, 289, 488
Baumgartl, T. 286
Bausch, W.C. 11
Baxter, S.J. 477
Beattie, J.A. 334
Beaubien, J. 438
Beckett, P.H.T. 10, 25, 27, 282, 286
Beckman, R.J. 155
Beckstrand, D. 236–237
Bedidi, A. 222

Bedrna, Z. 416–417
Behar, J.V. 138
Belbin, L. 373, 407
Bell, J.C. 467, 508
Ben-Dor, E. 225
Benjamini, Y. 317–318
Bennema, J. 41
Berland, M. 109
Berman, M. 210
Bernoux, M. 488, 497–498, 500
Bertoni, J. 41
Beswick, A.R. 199
Beven, K.J. 290, 381, 509, 516
Bezdicek, D.F. 284
Bhering, S.B. 48
Bie, S.W. 25, 27
Biehl, L. 109, 111, 289, 488
Bierkens, M. 330
Bierkens, M.F.P. 98, 533, 539
Bierwirth, P.N. 208, 338
Biggs, A.J.W. 208
Bishop, T.F.A. 5, 7–8, 27, 29–30, 141, 148, 167, 415, 423, 438, 457, 574
Bivand, R. 564
Blaszczynski, J. 381, 392
Bleys, E. 13, 26, 32, 88, 90, 198, 202
Bliss, C.M. 571, 576
Bockheim, J.G. 281
Boden, AG 428
Bodin, F. 8
Boettinger, J.L. 235, 377, 389
Bogaert, P. 104
Boiffin, J. 285
Bollero, G.A. 546
Bonfils, P. 221, 252
Bonham-Carter, G.F. 202
Bornand, D. 4
Bornand, M. 10, 245, 252, 287–288
Borne, F. 247, 249
Borst, G. 574
Boruvka, L. 415–416
Bouedo, T. 282, 287
Boulaine, J. 281, 286
Boulet, R. 10
Bouma, J. 270, 284, 286–287, 563
Bourennane, H. 498, 500–501, 508, 563
Bourrelly, L. 16
Bouzelboudjen, M. 576
Bouzigues, R. 10, 288, 543–544
Bowler, J.M. 334
Bowyer, J. 32
Box, G.E.P. 104
Boyle, M. 333
Brabant, P. 8–9, 246
Bragato, G. 416
Brakensiek, D.L. 17
Brasher, B.R. 574
Brasil, 41
Breiman, L. 27
Bresson, L.M. 58, 61
Brewer, R. 430, 440
Bristow, K.L. 343–344
Brocklehurst, P 404
Brown, A. 530
Brown, J.D. 97–98
Brown, K.L. 214, 333
Brown, M.D. 572
Brown, S. 498
Bruand, A. 501, 563
Brus, D.J. 16, 137, 153–154, 158, 183, 191, 286, 330, 457, 527, 533, 539
Brutsaert, W. 404
Bryant, R.B. 16, 285, 425, 427, 572
Bui, E.N. 4, 10, 25–27, 29–30, 32–33, 61, 88, 90, 119, 125, 193, 195, 198, 202, 289, 291–292, 329, 331, 390, 401–402, 412, 438
Bullock, D.G. 546
Bullock, P. 78
Bunce, R.G.H. 477
Bunge, M. 573
Bunting, B.T. 370
Burgess, T.M. 168
Burkart, M.R. 8
Burrough, P.A. 7–8, 10, 15, 25, 253, 258, 267, 284, 288, 292, 438, 467, 490, 525–526, 543, 555–556, 560–561, 563–566, 574
Burt, T.P. 290, 477
Butler, B.E. 10, 30, 282, 331–332, 334
Butler, C. 508
Butturini, A. 290

C

Cabello, L. 253
Cacetta, P. 104–105
Cam, C. 11
Camara, G. 120–121

Camargo, M.N. 41
Campbell, L. 236–237
Cannon, G. 195
Carley, J.A. 236–237
Carlile, P. 26, 32–33, 88, 90, 198, 202
Carter, J.O. 199
Carvalho Júnior, W. 48
Carvalho, L.G. de O. 40–41
Carvalho, M.C.S. 498
Catani, R.A. 41
Cattle, J.A. 457
Cattle, S.R. 12–13, 17–18, 285, 287
Cazemier, D.R. 10, 16, 287–288, 293
Cerdan, O. 58, 61
Cerri, C.C. 497–498, 500
Cerri, C.E.P. 497
Cervelle, B. 222
Chabrillat, S. 225
Chadwick, O.A. 7, 25, 27, 30, 327–328, 331–332, 392, 488, 508
Chadwicks, R.S. 236–237
Chagas, C.S. 48
Chalmers, A.G. 478
Chamran, F. 392, 488, 508
Chaplot, V. 290, 488, 507, 509, 511–517
Chapman, G.A. 26, 30, 32, 88, 90, 198, 202
Chapman, P.J. 484
Chappell, J. 334
Chartres, C.J. 31, 292–293, 331, 401
Chauvel, A. 10
Chauvel, J.J. 510
Chavez, P.S. 380
Chen, J. 438
Chen, X.Y. 334
Chery, P. 563
Chilès, J.-P. 574
Chittleborough, D.J. 26–27, 29–30, 141, 443, 457–458, 462, 508, 537
Chopra, P. 30
Chorover, J. 332
Chrisman, N.R. 526
Christakos, G. 138
Christensen, R. 140–141
Christian, C.S. 123
Chéry, P. 501
Church, B.M. 478
Cihlar, J. 438
Clark, R.N. 223, 226
Clarke, S.E. 526
Claudot, B. 282, 287
Clayden, B. 366, 369
Clemmer, P. 465, 468
Cochran, W.G. 286
Coelho, M.R. 42
Cohen, A.C. 307, 310, 313, 527
Cohen, J. 529
Cohen, M.P. 527
Coifman, R.R. 308, 318, 320
Cole, C.V. 498
Cole, N.J. 377
Coleman, D.C. 284
Comerford, N.B. 571, 576
Conacher, A.J. 291, 370
Congalton, R.G. 240, 368, 385
Conover, W.J. 155
Constable, B. 426
Cook, B.G. 26, 28
Cook, S.E. 31, 208, 292–293, 331, 338, 401
Cooke, R.M. 98
Cooper, M. 290
Cormack, R.S. 26
Corner, R.J. 27, 29,31, 208, 291–293, 331, 338, 390, 401, 402
Corwin, D.L. 11, 312–313, 315
Coste, J. 463
Costin, A.B. 334
Couch, J. 577
Coulibaly, L. 8
Courault, D. 288
Couturier, D. 508
Craig, M.A. 207–208
Craig, M.D. 210
Crave, A. 290, 511, 516
Crawford, K.L. 440
Crawford, M. 437
Cressie, N.A.C., 282, 501
Cresswell, H.P. 30–31, 466
Crozier, M.J. 216
Cruz-Lemos, P. de O, 40–41
Curmi, P. 290, 488, 509, 511–517
Cuylenburg, H.R.M.v 402, 404

D

Dabas, M. 7, 283
Damaska, J. 416–417

Daniells, I. 457
Darlympe, J.B. 291
Daroussin, J. 8, 55, 58, 60–61, 64, 77–78, 81, 109, 111, 113, 287, 488, 490, 527, 532
Das, S.N. 236
Daubechies, I. 305, 307, 309–310, 313, 318, 321
Daughtry, C.S.T. 11, 226
Dauth, C. 208
Davis, C. 120–121
Davy, P. 511
Dawes, W. 5, 292–293
De Alencastro Graça, P.M. 498
De Bruin, S. 7, 125
De Deckker, P. 334
De Gruijter, J.J. 10, 16, 137, 153–154, 158, 183, 191, 286, 330, 426, 574
de Jong, E. 208
de Jong, S.M. 220, 270
De La Cuanalo, H.E. 574
de la Rosa, D. 123
de Laat, P.J.M. 286
de Vries, F. 533, 539
DeBarry, P.A. 508
Deckers, J.A. 9, 336, 532
Deering, J.A. 392
Deering, D.W. 381
DeFanti, T.A. 572
DeFries, R.S. 381, 389, 392
Delfiner, P. 574
Dent, D. 327, 331
Dercon, G. 9
Detroch, F.P. 404
Deutsch, C.V. 141
Dickinson, R.E. 381, 392
Dickson, B.L. 208–210, 212
Dietrich, W.E. 285, 290, 295
Diggle, P.J. 149
Dijkshoorn, J.A. 498
Dixon, R.K. 498
Dobermann, A. 154, 167
Dobos, D. 62
Dobos, E. 55, 62, 107, 109, 111, 113, 289, 487–488, 490
Doe, W.W. 564
Doherty, M.D. 373
Dokuchaev, V.V. 281, 285
Donev, A.N. 147
Donnet, A. 252
Donoho, D.L. 308, 317–318
D'Or, D. 104
Doran, J.W. 284
Doraiswamy, P.C. 226
dos Santos, H.G. 39
Dosso, M. 282
Dowling, T.D. 335
Dowling, T.I. 31
Drees, L.R. 8–9, 282
Drew, J.V. 236
Driessen, P. 336
Dubois, G. 527
Duckstein, L. 428
Duda, R.O. 292
Dudal, R. 532
Duggin, J. 440
Dunbar, M.S. 27, 30, 167, 415, 423, 438, 574
Dungan, J.L. 104
Dunkerley, D. 334
Dunlop, C.R. 193, 195, 202, 402, 404
Durand, P. 290, 509, 516
Durlesser, H.P. 545
Dusart, J. 77, 80
Dutton, B. 293
Dwivedi, R.S. 236

E

Eamus, D. 403
Edye, J.A. 30
Egenhofer, M. 120–121
Eimberck, M. 58, 60, 78, 81, 532
Emmanuel, W.R. 498
Engel, B.A. 3
Epema, G.F. 125
Epinat, V. 270
Eswaran, H. 10, 498
Ethridge, F.G. 378
Ettema, C. 175
Evans, I.S. 443
Everitt, B.S. 155
Exposito, F. 7
Ezzahar, B. 290, 509

F

Falipou, P. 17
Farr, T.G. 110, 490
Favrot, J.C. 4–5, 8, 10, 15, 245, 291

Fearnside, P.M. 498
Feigl, B. 498
Fett, D.E.R. 195
Feuerherdt, C. 87
Field, J. 208
Finke, P.A. 104, 523–524, 527, 530, 532–534, 536–537, 539
Finkel, R.C. 285, 290, 295
Firbank, L.G. 477
Fischer, G.W. 3
Fisher, P.F. 523–524, 563, 575
Fitzpatrick, R.W. 208, 220
Flatman, G.T. 138
Fletcher, P. 366
Flitti, M. 253
Flores, R.M. 378
Follain, S. 281
Foltête, J.C. 253
Fonseca, F. 120–121
Foody, G.M. 445
Foote, R.S. 208
Foran, B.D. 28
Forbes, T.R. 124, 524–525
Fort, J.L. 11
Frank, A.U. 7, 573
Franklin, J. 7, 25, 27, 30, 327–328, 331, 402
Fridland, V.M. 10
Friedman, J. 273
Fritsch, D. 533
Fritsch, E. 282
Frogbrook, Z.L. 477, 483
Funtowicz, S.O. 98
Furse, M.T. 477

G

Gaffey, S.J. 222
Gahegan, M. 402
Galbraith, J.M. 16, 425, 427
Gallant, J.C. 25–26, 257, 260, 264–265, 327–328, 335–336, 340, 371, 404, 406, 490
Garbrecht, J. 508
Gascuel, C. 286, 290
Gascuel-Odoux, C. 290, 509, 516
Gashnig, K. 292
Gassman, P.W. 8
Gatehouse, R.D. 334
Gaultier, J.P. 10
Gebhardt, D.L. 564
Gennadiyev, A.N. 281
Gentle, M.R. 28
George, R.J. 208
Gershenfeld, N. 416
Gesch, D.B. 110, 490
Gessler, P.E. 5, 26–29, 31, 154, 208, 290, 292–293, 330–331, 338, 341, 392, 401, 467, 488, 508, 511, 536, 574
Gibbons, F.R. 25
Giblin, A.M. 209
Girard, M.C. 245–247, 253
Gish, T.J. 17
Glinka, K.D. 467
Gómez-Hernández, J.J. Journel, 537
Goddard, T.W. 246, 258
Godwin, R.J. 508
Goetz, A.F.H. 225
Golay, F. 576
Goldrick, G. 30
Gong, P. 249
Good, P.I. 445
Goodchild, M. 523, 574
Goovaerts, P. 103–104, 141, 167, 177, 270, 282, 294, 415, 457, 462, 536, 560, 563, 574
Gorssevski, P.V. 536
Goulding, K.W.T. 309
Gourmelon, J. 252
Govers, G. 9
Goward, S.N. 226
Graetz, R.D. 28
Graps, A. 270
Grasty, R.L. 210, 338
Gray, C. 402
Grayson, R.B. 381
Grealish, G.J. 31, 208, 292–293, 331, 338, 401
Green, A.A. 210
Green, K. 240, 385
Greene, R. 208
Greenlee, S.K. 110, 490
Gregory, L.J. 340
Greve, M.B. 543–545, 550–551
Greve, M.H. 543–545, 550–551
Grohman, F. 41
Grose, C. 26, 32, 88, 90, 198, 202
Groves, P.R. 208, 338
Gruber, T.R. 120

Grundy, M. 26, 30–32, 34, 88, 90, 198, 202, , 328, 466
Grunwald, S. 571, 576
Guenat, C. 576
Guisan, A. 402, 412
Gunnink, J. 564
Gutmans, M. 39, 41
Gutteling, J.M. 534
Guymer, I. 575

H

Haines-Young, R. 477
Hajek, B.F. 286
Hallegouet, B. 510
Hallsworth, E.G. 425, 430, 440
Hammer, R.D. 281, 467
Hammond, E.H. 111
Hansen, A.J. 564
Hardy, R. 4, 10, 245
Harmon, R.S. 564
Harrod, T.R. 310, 317–318
Hart, G.A. 287
Hart, P. 292
Hartigan, J.A. 185
Hartikainen, H. 72
Hartwich, R. 532
Hartyáni, M. 491
Hass, R.H. 381
Hassika, P. 488
Hastie, T. 273
Hatton, T.J. 208
Hauff, P.L. 220
Haydu-Houdeshell, C. 465, 470
Haygarth, P.M. 484
Hazen, R. 8
Hebel, A. 498
Hefting, P. 290
Heimsath, A.M. 285, 290, 295
Heinson, G.S. 208
Helt, T. 109, 111, 289, 488
Henderson, B.L. 26–27, 32–33, 88, 90, 198, 202, 402
Hengl, T. 138–139, 141, 146–148, 153–154, 158–159, 167, 526, 530, 533
Herpin, U. 488
Herriman, R.C. 236
Hesse, A. 7, 283
Heuvelink, G.B.M. 18, 97–98, 103, 105, 137–138, 141, 148, 154–155, 167, 286, 288, 294, 574
Hewitt, A.E. 25, 27, 246, 253, 258, 266–267, 327
Heyligers, P.C. 28
Hiederer, R. 61
Higginson, F.R. 441
Hilgard, E.W. 467
Hird, C. 440
Hénin, S. 247
Hodgson, G. 208
Hofman, G. 288
Hogan, D.V. 310, 317–318
Hole, F.D. 247, 332
Hollingsworth, I.D. 401
Hollis, J.M. 366, 369, 527
Holmes, K. 392, 488, 508
Holmes, S. 291, 390
Hoogerwerf, M. 287
Hoogland, T. 533, 539
Hook, J. 31
Hoosbeek, M.R. 284–285, 572
Hopkins, M.S. 90, 401
Horn, R. 286
Hornung, M. 477
Houba, V.J.G. 526
Houghton, R.A. 498
Houšková, B. 78
Houlder, D. 26
Hovgaard, J. 210
Howard, D.C. 477
Howarth, P.J. 249
Howe, D.F. 26, 32, 88, 90, 193, 195, 198, 202
Howell, D. 465
Hrasko, J. 416–417
Hubble, G.D. 332, 430, 440
Hudson, B.D. 25, 327
Huggett, R.J. 259–260, 332
Humbel, F.X. 10
Hummel, J.W. 11, 168
Humphreys, G.S. 332
Hunt, E.R. 226
Hutchinson, M.F. 26, 31, 264, 328
Hutton, J.T. 334
Huwe, B. 425, 428
Hyvönen, E. 70

I

Ibanez, J. 532
Ignatius, E. 572
Imhoff, M. 26, 32, 88, 90, 198, 202
Isbell, R.C. 90, 214
Isbell, R.F. 271, 273, 333, 401, 426

J

Jabiol, B. 227
Jackson, T.J. 438
Jacquier, D.W. 31, 208, 214, 333
Jahn, R. 425, 428
Jamagne, M. 4, 58, 60–61, 78, 81, 109, 282, 527, 532
Jambu, M. 511
Jankowski, P. 536
Jansen, M.J.W. 324, 524, 527
Jarvis, S.C. 484
Jeansoulin, R. 252
Jeffrey, S.J. 199
Jenkins, G.M. 104
Jenny, H. 43, 281, 285, 366, 377–378, 390, 467, 511
Jensen, E.H. 236–237
Jensen, J.R. 239
Johnson, D.L. 332
Johnston, R.M. 13, 26, 32, 88, 90, 198, 202
Johnstone, I.M. 317–318
Joisten, H. 122
Jolivet, C. 294, 498, 500
Jones, R.J.A. 58, 61, 78, 80, 527
Jordan, G. 323
Journel, A.G. 141

K

Kaffka, S.R. 312–313, 315
Kalenda, M. 416–417
Kalivas, D.P. 425
Karma, G.A. 508
Karssenberg, D. 555, 564, 566
Kavouras, M. 120–121
Kay, J.J. 333
Keig, G. 28
Kele. G. 491
Kershaw, C.D. 478
Kershaw, P. 334
Kicin, J.C. 488
Kiefer, R.W. 528, 535
Kienast-Brown, S. 235
Kiiveri, H.T. 104–105
Kim, Y. 465
King, D. 4, 8, 10, 55, 58, 60–61, 78, 81, 109, 252, 287, 290, 501, 508, 511, 513, 516, 527, 532, 563
Kirkby, M.J. 290, 381, 509, 516
Knotters, M. 330, 457, 527, 533, 539
Knox, E.G. 43
Kok, B. 3
Kokaly, R.F. 226
Kokla, M. 120–121
Kolbrick, M. 110, 490
Kollias, V.J. 425
Koppi, A.J. 17, 416, 425–426
Koussoulakou, A. 575
Kovac, M. 158
Kozak, J. 416
Kraak, M.J. 575
Kravchenko, A.N. 546
Krol, B. 119
Kros, J. 104, 534, 536–537
Krosley, L. 225
Kruk, M. 290
Kruse, F.A. 220
Kues, J. 120
Kupper, H. 39, 41
Kyriakidis, P.C. 104

L

Labib, M.E. 8
Lacaze, B. 252
Lagacherie, P. 3, 5, 10–11, 15–17, 25, 51, 252, 281, 287–293, 390, 508, 511, 513, 516–517, 543–544
Lague, D. 511
Lambert, J.-J. 58, 60–61, 78, 81, 532
Lambin, E.F. 438
Landson, A.R. 381
Lang, M.P. 195
Lanza, A. 7
Lark, R.M. 138, 153, 270–271, 301, 304, 309–310, 312–315, 318, 324
Larsen, D. 457
Latifovic, R. 438
Laurent, M. 510
Laut, P. 28

Lawley, R. 366
Lawrance, C.J. 8
Lawrie, J.W. 158
Le Bas, C. 4, 58, 60–61, 78, 81, 109, 532
Le Bissonnais, Y. 58, 61
Leahy, S. 28, 31
Ledreux, C. 252
Leech, P.K. 478
Leenaers, H. 563
Leenhardt, D. 286
Lees, B.G. 26
Legros, J.P. 10, 15, 17, 25, 282, 284, 292
Lemay, L. 577
Lemos, R.C. 40–42
Leppert, P.M. 208
Lerssi, J. 70
Lesch, S.M. 11, 138–139, 153–154, 313
Leveque, J. 294
Lewis, D.T. 236
Lewis, M. 220
Leys, J. 334
Li, Z. 438
Lilja, H. 67
Lillesand, T.M. 528, 535
Lim, K.J. 3
Lisi, F.A. 7
Loaiza, G. 9
Loffler, E. 28
Léonard, J. 58, 61
Longley, P.A. 568
Loughhead, A. 27, 29, 208, 291, 390, 402
Loveday, J. 415
Loveland, P.J. 61, 477, 483
Lowery, B. 3, 15, 291, 576
Lowry, J. 381
Lozet, J. 228
Lösel, G. 4
Lucas, Y. 10
Lulli, L. 416
Lynch, L.G. 26, 28

M

Ma, Z. 529
Mackey, B.G. 371
Macleod, A.P. 30
MacMillan, R.A. 246, 253, 258, 267
Madeira Netto, J.S. 219–220, 222
Madsen, H.B. 58
Magnussen, S. 438
Maidment, P.A. 508
Maitre, V. 290
Malenovský, Z. 253
Malerba, D. 7
Mallat, S.G. 306, 310
Mangeot, A. 576
Manis, G. 381
Margate, D.E. 220
Margerel, J.P. 510
Margules, C. 28
Mark, D.M. 120
Markewitz, D. 333
Marks, A. 208
Markus, J.A. 425
Marques, J.Q.A. 41
Mars, N.J.I. 120–122
Marsman, B.A. 286
Marth, P. 491
Martin-Clouaire, R. 10, 16, 287–288, 293
Martins, E. 219
Martz, L.W. 208
Maschmedt, D. 26, 32, 88, 90, 198, 202
Masser, I. 3
Masuoka, E. 488
Mathieu, C. 228
Mayr, T.R. 365–366
Mazaheri, S.A. 17, 416, 425–426
McArthur, D. 7, 25, 27, 30, 327–328, 331
McBratney, A.B. 3, 5, 7–10, 12–13, 15–18, 25–32, 43–44, 50–53, 138, 141, 148–149, 153–155, 159, 167–168, 219, 232, 253, 269–271, 283, 285, 287, 290, 292, 294–296, 328, 330–331, 366, 390, 401–402, 415–416, 418, 423, 425–426, 437, 438–439, 443–444, 455–458, 462, 467, 488, 508, 537, 543, 550, 574–575
McCormick, B.H. 572
McCullough, P. 402
McDonald, R.C. 90, 401
McDonnell, R.A. 284, 438, 560, 563, 574
McFarlane, D.J. 208
McGarry, D. 28–30, 271, 287, 456
McGaw, A.J.E. 30
McGrath, S.P. 477, 483
McKay, M.D. 155
McKenney, D.W. 371

McKenzie, N.J. 25–34, 88, 90, 154, 198, 202, 208, 214, 290, 327–328, 330, 333, 338, 340–341, 371, 401, 412, 466–467, 488, 508
McKinlay, C.R. 236–237
McLaren, R.G. 8
McLeod, P.J. 195, 401
McMahon, J. 26
McSweeney, K. 291, 467, 574, 576
McTainsh, G. 334
Mead, R.A. 368
Medina, H.P. 39, 41
Mendes, W. 40–41
Mendonça Santos, M.L. 7, 13, 15–16, 18, 25, 27, 30–31, 39, 43–44, 50–53, 138, 269–271, 283, 290, 292, 328, 330–331, 390, 415, 423, 425, 508, 543, 550, 574, 576
Meneguelli, N.A. 47
Mennis, J.L. 7
Mermut, A.R. 10
Merry, R.H. 31
Metropolis, N. 156
Metternicht, G. 236, 239
Meunier-Caldairou, V. 253
Meyer-Roux, J. 61
Meyers, J.A. 373
Michalski, C.H. 31
Michel, Y. 510
Micheli, E. 109, 111, 289, 487–488
Michelic, E. 62
Milford, H.B. 30
Miller, P.C.H. 508
Milne, A.E. 309
Milne, G. 258
Minasny, B. 7, 12–13, 15–18, 25, 27, 30–32, 43–44, 50–53, 138, 153–155, 159, 167–168, 219, 232, 253, 269–271, 283, 285, 287, 290, 292, 295, 328, 330–331, 366, 390, 401–402, 415, 418, 423, 425, 438–439, 444, 457, 467, 488, 508, 537, 543, 550, 574–575
Minty, B.R.S. 207, 210–211
Mitas, L. 564
Mitasova, H. 564
Mitchell, P.B. 332
Müller, W.G. 138–139, 141, 149, 184
Moellering, H. 524
Moguedet, G. 510
Mokma, D.L. 69, 72
Molenaar, M. 7, 120–122
Molnár, E. 491
Monestiez, P. 288
Montanarella, L. 55, 60–62, 64, 78, 80–81, 107, 109, 487–488, 490, 532
Moodie, K.B. 199
Moore, A.W. 26, 28, 426
Moore, G.E. 5
Moore, I.D. 5, 26–27, 29, 154, 208, 290, 330, 338, 341, 381, 467, 488, 508, 511, 518
Moorman, T.B. 8
Moran, C.J. 4, 10, 26–27, 29–30, 32–33, 61, 88, 90, 119, 195, 198, 202, 289, 291–292, 390, 402, 412, 438
Moreira, M.J. 123
Morgan, C.L.S. 15
Morgan, M.T. 168
Morlat, R. 8, 11
Mougenot, B. 253
Moyeed, R.A. 149
Mérot, P. 290, 509, 516
Muge, F.H. 576
Mulcahy, M.J. 425, 430, 440
Mulders, M.A. 563
Murdock, K. 577
Murphy, C.L. 30
Myers, D.E. 175

N

Nachtergaele, F.O. 5, 336, 498, 532
Nagler, P.L. 226
Nascimento, A.C. 39, 41
Nason, G. 324
Navarro Sanchez, I. 288
Negre, T. 109, 488
Nehmdahl, H. 545
Nemecek, J. 416–417
Nettleton, W.D. 574
Nevalainen, R. 67
Nègre, T. 62
Nguyen-The, N. 51, 252, 291
Nieder, R. 294
Nielsen, G.A. 5, 26, 29, 290, 488, 508, 511
Nimlos, T.J. 281, 293
Nishiizumi, K. 285, 290, 295
Nix, H. 26
Nolan, S.C. 246, 258
Norberg, T. 105
Norman, J.M. 3, 15

Norris, J.M. 415
Nortcliff, S. 574
Northcote, K.H. 30, 90, 425–426, 430, 440, 447
Nott, J. 402–403
Nulsen, R.A. 208
Nunes de Lima, V. 82
Nychka, D. 139, 153, 184

O
O'Callaghan, J.F. 28
O'Connell, D.A. 28–29, 208
Odeh, I.O.A. 26–30, 141, 167, 401, 415–416, 423, 437–438, 441, 443, 445, 455, 457–458, 462, 508, 537, 574
O'Flaherty, S. 309
Ogden, F.L. 508
Ojima, D.S. 498
Okx, J.P. 563
Oldak, A. 438
Oldeman, L.R. 8
Olea, R.A. 138
Oliver, M.A. 172–173, 270, 287, 327, 331, 444, 477–478, 483, 528, 530, 560, 574–575
Olley, J.M. 334
O'Loughlin, E. 5, 292–293
Olsen, C.J. 402, 404
Olsen, H.W. 225
Omonode, R.A. 546
Oude Voshaar, J.H. 457, 527

P
Pachepsky, Y.A. 17, 438
Paiva Neto de, J.E. 41
Palmer, B. 365
Palmer, R. 366
Panissod, C. 283
Park, S.J. 291
Parton, W.J. 498
Pastor, J. 498
Paton, T.R. 332
Payne, A. 30, 33, 193
Pebesma, E.J. 104–105, 155, 534, 536–537, 564, 566
Peedell, S. 82
Peltovuori, T. 67, 69
Penizek, V. 415–416
Percival, D.B. 320
Petersen, G.W. 467
Peterson, G.A. 5, 26, 29, 290, 488, 508, 511
Petit, C.C. 438
Pettapiece, W.W. 246, 258
Pettitt, A.N. 30, 153
Peuquet, D. 574
Pfeffer, K. 564, 566
Philip, S.R. 208
Piccolo, M.C. 498
Pickup, G. 208
Pieters, G. 142
Pillans, B.J. 334
Pinder, G.F. 155
Pélissier, P. 500
Podmaniczky, G. 491
Poesen, J. 9
Pond, B. 333
Post, W.M. 498
Pouget, M. 222
Powell, B. 26, 32, 88, 90, 198, 202
Pracilio, G. 208
Pringle, M.J. 9, 155, 285
Prosser, I. 195

Q
Qi, J. 381, 389, 392
Questiaux, D.G. 334

R
Radoševic, N. 533
Raggatt, T.J. 220
Raiffa, H. 534
Ramasundaram, V. 571, 576
Ramirez, M. 9
Rampant, P.C. 30
Ramsey, R.D. 381
Ranaivoson, A. 253
Rao, B.R.M. 236
Ravetz, J.R. 98
Ravi Sankar, T. 236
Rawls, W.J. 17
Rayment, G.E. 441
Reboh, R. 292
Redmond, R.L. 529
Reich, P. 498
Reinds, G.J. 104, 534, 536–537
Reinsel, G.C. 104
Rhoades, J.D. 138–139, 153–154, 313

Richter, D.D. 333
Richter, J. 294
Richter, R. 221
Ringrose-Voase, A.J. 29, 31, 34, 328
Ritman, K. 26
Robbez-Masson, J.M. 51, 219, 245, 247, 250, 252–253, 289, 291
Roberts, L. 208
Robinson, N. 87
Rodó, X. 301
Rodrigues, T.A. 499
Rodriguéz-Arias, M.A. 301
Roehler, H.W 391
Rosenbluth, A. 156
Rosenbluth, M. 156
Rosenburg, R.J. 40–41
Rosenthal, K.M. 26
Rosén, L. 105
Ross, P.J. 343–344
Rossiter, D.G. 3, 8, 10, 12, 19, 119, 124, 138–139, 146–147, 153–154, 158–159, 524–525, 530, 533
Roudabush, R.D. 236
Rouse, J.W. Jr. 381
Roush, T.L. 223
Rowe, R.K. 334
Royle, J.A. 139, 153, 184
Rubin, D.B. 527
Ruellan, A. 282
Rusco, E. 61
Russ, A.L. 226
Ryan, P.G. 27, 29
Ryan, P.J. 5, 25, 27, 29–30, 154, 208, 290, 292–293, 330, 338, 341, 371, 467, 488, 508
Ryan, S. 220

S

Sacks, J. 139, 184
Saisana, M. 527
Salis, J. de. 193, 195
Salomonson, B. 488
Sanchez, H. 9
Santos, H.G. dos 40–42
Santos, R.D.dos. 42
Saunders, A.M. 389
Sbresny, J. 120
Schachtschabel, B. 428
Schaedel, A.L. 526
Scheffer, F. 428
Scheldeman, K. 288
Schell, J.A. 381
Schellentrager, G.W. 236
Séchet, P. 47
Schiller, S. 139, 184
Schimel, D.S. 498
Schlaifer, R. 534
Schläfer, D. 221
Schmidt, J.S. 246, 253, 258, 266–267
Schneider, J. 120
Schoknecht, N. 26, 30, 32–33, 88, 90, 193, 198, 202
Schott, B. 323
Schroeder, P.E. 498
Schvartz, C. 282, 287
Scott, K.M. 208, 212
Scott, R.M. 28
Scull, P. 7, 25, 27, 30, 327–328, 331
Sebillotte, M. 284
Seevers, P.M. 236
Senay, H. 572
Shaikh, M. 381, 389, 392
Sharma, R.C. 236
Shatar, T.M. 27, 30, 167, 415, 423, 438, 574
Sheail, J. 477
Shenkelaars, V. 7
Shepherd, K.D. 7
Short, D. 5, 292–293
Shrestha, D.P. 220
Si, B. 319
Siderius, W. 119, 123, 130–131, 139, 142, 153, 168, 186
Sier, A. 477
Silva, E.F. 42
Silverman, B.W. 317, 324
Simbahan, G.C. 154, 167–169, 171, 178–179
Simon, D.A.P. 26–27, 30, 32–33, 88, 90, 193, 198, 202, 402
Simons, N.A. 30
Simonson, R.W. 259, 332
Singh, A.H. 236
Singh, B. 457
Sippola, J. 67, 69
Sirakov, N.M. 576

Sivapalan, M. 508
Skidmore, A.K. 5, 292–293
Skinner, R.J. 478, 482, 484
Slater, B.K. 30, 32, 467
Sleeman, J.R. 425, 430, 440
Slocum, J. 292
Slocum, K. 270
Smart, S.M. 477
Smettem, K. 208
Smith, B. 120
Smith, R.V. 484
Solomon, A.M. 498
Sombroek, W.G. 498
Spaargaren, O.C. 336, 532
Speight, J.G. 27, 90, 401
Spiers, R.B. 8
Spooner, N.A. 334
Spoor, G. 61
Spouncer, L.R. 31
Spätjens, L.E.E.M. 191
Squividant, H. 290
Stace, H.C.T. 440
Stace, H.T.C. 430
Staines, S.J. 310, 317–318
Stangenberger, A.G. 498
Steers, C.A. 286
Stein, A. 138–139, 141–142, 146–148, 153–154, 158–159, 167–168, 174–175, 186, 270, 287, 563
Stengel, P. 4, 285
Stewart, B.A. 284
Stewart, G.A. 123, 202
Stibbard, A. 575
Stock, S.S. 236–237
Stokes, H.A. 236–237
Stokes, W.L. 238
Strauss, D.J. 138–139, 153–154, 313
Sudduth, K.A. 11
Sullivan, M.E. 28
Sun, G. 574
Sutinen, R. 70
Switzer, P. 210
Szabados, I. 491

T
Tabbagh, A. 7, 283
Taha, A. 509, 516
Taha, M. 290
Tanaka, A.K. 48
Tandarich, J.P. 281
Tavernier, R. 8, 58, 77–78, 287
Tawn, J.A. 149
Taylor, G. 210
Taylor, J.A. 455
Taylor, J.K. 25
Taylor, M.J. 208
Teller, A. 156
Teller, E. 156
Thammappa, S.S. 236
Thenail, C. 290
Thiry, M. 220
Thom, R. 289
Thomas, M. 208
Thomasson, A.J. 61, 109, 527
Thompson, J. 508
Thorne, C.R. 508, 511
Thwaites, R.N. 257–260, 263, 267
Tibshirani, R. 273
Timlin, D.J. 17
Todd, A.D. 478, 482, 484
Todd, A.J. 441, 445
Tomasek, M. 416–417
Tonkin, P.J. 335
Trautman, F. 510
Trexler, M.C. 498
Triantafilis, J. 28–29, 416, 441, 445
Troch, P.A. 404
Turner, M.G. 381
Turunen, P. 70

U
Uitermark, H.T. 120–122, 125
Unwin, D. 575
Upadhyaya, S.K. 168

V
Valenzuela, C.R. 125
Van Cuylenburg, H.R.M. 402
van de Nes, T.J.M. 286
van Den Berg, E. 498
Van der Meer, F.D. 220
van Deursen, W. 555, 564, 566
van Engelen, V.W.P. 8
van Gaans, P.F.M. 253, 258, 267, 564

van Groenigen, J.W. 139, 142, 153–154, 158, 168, 174–175, 183, 186
van Heesen, H.C. 286
van Holst, A.F. 286
van Loenen, B. 3
Van Meirvenne, M. 288
van Oosterom, P.J.M. 7, 120–122
van Ranst, E. 5, 527
van Reeuwijk, L.P. 526
Van Vliet, B. 290
Van Vliet-Lanoë, B. 509–510, 512, 514–517
Van Wambeke, A. 124, 524–525
Van-Cuylenburg, H.R.M. 195
Vanegas, R. 9
Vanhanen, E. 70
Vanmechelen, L. 527
Vauthier, S. 510
Veitch, S. 32
Veldkamp, A. 125
Venkataratnam, L. 236
Venot, A. 463
Verboom, W.H. 208
Verburg, K. 344
Verdade, F.C. 39, 41
Verdin, K.L. 110, 490
Verwoort, R.W. 12–13, 17–18, 285, 287
Vial, P. 307, 310, 313
Viaud, V. 290
Viscarra Rossel, R.A. 29, 283, 294, 296
Vital, P. 11
Väänänen, T. 70
Volkoff, B. 498, 500
Voltz, M. 286, 288, 292, 508, 517, 574
Von Reibnitz, U. 98
Vossen, P. 61
Vrščaj, B. 55, 64
Várallyay, Gy. 491

W

Wackernagel, H. 141
Walden, A.T. 320
Walker, A. 381, 389, 392
Walker, J. 401
Walker, P.H. 334
Walsh, M.G. 7
Walter, C. 281–283, 286–287, 290, 294, 296, 488, 507, 509, 511–517
Walter, E. 463
Walter, J. 90
Walter, V. 533
Walvoort, D.J.J. 154, 530
Wang, C. 286
Ward, W.T. 28–30, 271, 416, 456
Warrick, A.W. 175
Wassenaar, T. 10, 252
Wasserman, D. 463
Webster, C.P. 309
Webster, R. 8, 10, 30–31, 34, 138, 149, 153, 168, 172–173, 270–271, 282, 286–289, 294, 304, 310, 317–318, 327, 329, 331, 444, 467, 477–478, 483, 508, 528, 530, 560, 574–575
Wells, M.R. 402, 404
Wesseling, G.C. 536, 564–566
Weterings, M.H.W. 563
Wharton, S.W. 247–248
Whelan, B.M. 5, 8, 154, 543, 550
White, D. 526
Wickerhauser, M.V. 320
Wiegman, O. 534
Wielemaker, W.G. 7, 125
Wilding, L.P. 8–9, 282
Wiles, L. 440
Wilford, J.R 207–208, 213, 215, 328, 338
Williams, I.S. 334
Williams, J. 343–344
Williams, M. 334
Wilson, J.P. 26, 260, 264–265, 328, 336, 371, 404, 406, 490, 564
Wilson, S.R. 286
Winjum, J.K. 498
Winter, S. 120
Wisniewski, J. 498
Wladis, D. 104, 534
Wong, M.T.F. 208
Worstell, B. 109, 111, 113
Wright, R.L. 286
Wösten, J.H.M. 17, 524, 527

X

Xing, X. 488

Y

Yaalon, D.H. 282
Yang, C.H. 11
Yang, S. 574

Yassoglou, N.J. 425, 532
Yfantis, E.A. 138
Yli-Halla, M. 69, 72
Young, A. 327, 331
Yue, W. 3, 15

Z

Zavitz, B.L. 371
Zeng, X. 381, 392
Zevenbergen, L.W. 508, 511
Zhang, J. 523
Zhang, Y. 155
Zhu, A.X. 293, 378, 416, 423
Zhu, J. 15
Zimmermann, N.E. 402, 412
Zinck, J.A. 125, 236, 239
Zinke, P.J. 498
Zuska, V. 416–417

SUBJECT INDEX

A

Abstraction rules, 121–123
Accuracy, 98, 104, 360, 523–530, 532–533
Aeolian, 334, 339
Aerial photographs, 556–557, 564
AGSG, 426, 430–432
AIC, 273–274, 276–277
Akaike Information Criterion, 315
Allocation, 415–423
Alluvial landforms, 335
Amazon, 498
Annealing schedule, 156
Antequera, 119, 123–125, 128, 130–131
ANUDEM, 336
Apparent electrical conductivity, 313
Application ontology, 122, 126–128
Argissolos, 42
Armorican Massif, 508–510
artificial neural networks (ANNs), 353, 356, 415–416
Attribute quality, 526
Australia, 25–35, 37
Australian Great Soil Groups (AGSG), 425
Autocorrelation, 97
Autocovariance function, 140
Available Phosphorus, 477
Available water capacity, 343–345
Average kriging variance, 183, 189–190

B

Band depths, 219, 221, 224–226, 231
Bayesian
  belief networks, 372
  inference, 292
  maximum entropy, 104
Billabong catchment, 332–333
Biophysical data, 88

C

$CaCO_3$ contents, 219, 222–224, 226–228
Carbon stocks, 498, 500
Cartograms, 556
Catchment
  analysis, 370
  classification, 370
Catena, 258–259, 261
Cation Exchange Capacity (CEC), 441, 451
Classification, 235, 377, 425–427
  key, 427, 430
  trees, 193–194, 199–201, 353, 356
  tree analysis, 389–390, 396–398
Clay, 440
Clay contents, 219, 221–222, 224–225, 228–230, 269, 273, 275–276, 448–450
Cokriging, 149
Common reference grid, 81
Compact subregions, 183–185
Completeness, 524, 527, 538
Computation speed, 360
Conditional probability, 291–292
Confusion matrix, 103–104
Contingency tables, 368
Continuum removed (CR), 223
Conventional, 328
  surveys, 333
Correlation, 98, 104
Correlogram, 104
Crisp boundary, 557–558
Cross-variogram, 104
Currency, 524, 526, 530, 538

D

Data
  collection, 89, 92
  integration, 119–123, 125, 127–129, 131, 133

Database, 97–98, 105
  structure, 89
Data-mining, 353
Decision support, 543–545, 550–552
DEMs (Digital Elevation Models), 107, 109–116, 137, 264–267, 290, 354, 377, 380–381, 386, 389, 392, 438–439, 443, 445–447, 488, 490, 543, 545–546, 547, 550–551, 555
Denoising, 301, 316–318, 323
Design-based sampling, 153, 191
Detailed, 40–45
Deterministic approach, 290, 294
Digital
  soil mapping, 3, 5–12, 17–19, 55, 57, 59, 61, 63, 65, 252, 269–270, 277, 465
  soil maps, 438
  terrain analysis, 358
Discrete approach, 286
Discriminant analysis, 455, 459–460, 462–463
Domain ontology, 121–123, 125–128, 130–132
D-optimal design, 146
Draping, 555, 563
DTA, 358
Dynamic modeling, 564, 566

E

EM38, 545, 551
Embrapa Solos, 48, 50–51, 54
England and Wales, 477
Environmental, 346
  change, 333
  correlation, 328–329
  variables, 269–270, 277
Erosional uplands, 335, 340
Error, 98, 103–104, 137, 139–141, 145–146, 149
  model, 534–538
Europe, 55–56, 58, 60, 62
Experimental design, 139, 147
Expert system, 292
External validation, 501

F

False discovery rate, 318
Feature space, 154
Fine-resolution maps, 169
Functional approach, 366
Fuzzy
  classification, 257–258, 262, 264, 266
  k-means, 426
  logic, 288, 293
  sets, 98
Fuzziness of soil boundaries, 543

G

Gamma radiometric, 70, 328, 338
Gamma-ray, 371
  spectrometry, 207–210, 217
Generalized linear models, 471
Geographic information systems (GIS), 377, 390
Geological map, 119, 123–124, 128
Geomorphic unit, 336, 340
Geo-pedological system, 125–127, 130
Geostatistical samples, 183–185
Geostatistics, 104, 286–287, 477–478, 560
German Soil Systematics, 425, 428–430
Germany, 354
GIS, 7, 11, 246, 250, 353, 377–378, 383, 385, 387, 544, 551
  analysis, 441
GPS, 557, 563
Grid spacing, 138–139

H

Harmonising and interoperability rules, 81
High-resolution, 507, 517
History, 27, 29
HyMap, 219
Hyperspectral, 219–220

I

If–Then rules, 425, 428–430, 430
Indicator Kriging, 457
Inference system, 12–14, 16–18
Information
  domain, 87
  product, 92
Iniciativa Solos, 39, 50
INSPIRE
  Directive, 78
  legal process, 82

Intermediate scales, 119–120
Interpolation, 555, 558, 560–564

K
K-means clustering, 154, 183, 185, 187, 189, 191
Knowledge-based, 389, 397–398
  classification, 377, 382, 384–385, 387, 389–390, 395–396, 397–398
Knowledge-engineering, 390
Kriging, 137–141, 144–146, 148–149, 282, 287, 437, 444, 478–479, 481–483
  variance, 137–141, 144–146, 149
  with external drift, 141

L
Land
  cover, 235–240, 242, 244
  resource assessment, 25, 27, 32
  unit mapping, 401, 403, 405, 407–409, 411, 413
Landsat, 235, 271, 273–276, 377, 380–381, 384, 386, 389–390, 392
Landscape, 43–44, 52, 245–249, 252–254
  analysis, 370
  classification, 370
  processes, 330
Latin hypercube sampling, 155
Latossolos, 42
Learning vector quantisation, 353, 357
Legend, 556
Lineage, 524, 533, 538
Linear regression, 353, 358
Lithology, 341, 346
Logical consistency, 524, 531, 539
Logistic regression, 473

M
Machine-learning, 27, 29
Map
  purity, 191
  unit, 397
Markov property, 104–105
Mathematical distance, 245
Maximum kriging variance, 183–184, 187, 189–190
Mean, 97, 100, 103–104
  squared shortest distance, 185
Metadata, 83
Minimization criterion, 139
Mixed approach, 286
Model-based sampling, 153–154
Modelling, 508, 516–518
MODIS, 487–492
Mojave Desert, 467
Multicollinearity, 148
Multiple linear regressions, 437
Multiresolution analysis, 301–302, 304, 307–314, 316, 321–322, 324
Multi-scale, 508, 516–517
  approaches, 507
  geographical data, 120
Multi-source
  datasets, 119
  geographical data, 120
Multivariate analysis, 563

N
Namoi river catchment, 439, 452
National database, 557
National Soil Inventory (NSI), 477, 483
New South Wales, 439
Normal distribution, 103
Nugget, 138, 143
  variance, 186–187

O
Ontologies, 119–123, 125, 128, 130–132
Ontology-based data integration, 119, 121
Optimization, 137–139, 141–149

P
Parent material, 67, 72
Partial sill variance, 187
Particle-size fractions, 441, 449–450
Pedogenesis, 332
Pedological knowledge, 281, 285
Pedotransfer, 51
Poisson distribution, 103
Positional quality, 525, 533, 538
Precision, 523, 525–527, 530, 532–533, 538
Predictor, 148
  datasets, 195, 199
  variables, 269–270, 273

Principal component sampling (PCs), 158
Prior points, 183, 187, 190
Probability distribution function (pdf), 97
Production soil survey, 467
Pyramid algorithm, 307–308, 310

R

RADAMBRASIL, 42
Radioactivity erosion, 214–215
Random variable, 100
Range, 186–187
Raster products, 474
Reconnaissance, 41, 43–45
Redness, 219
Reference
  area, 291
  model, 119, 121–123, 128–130, 132
Reflectance spectra, 219
Regional level, 119
Regolith, 257, 262
Regolith-terrain, 257–259, 262, 264–267
Regression, 138–141, 146–149
  kriging, 141, 148
Regular grid sampling, 184
Remote sensing, 104, 137, 207, 225, 235–236, 257, 264
Representative Soil Sampling Scheme (RSSS), 477–478, 483
Risk assessment, 462
RMSE, 440, 445
Root mean square error, 440, 445
Rough sets, 98
Rule-based mapping, 402, 413

S

Saline soils, 235–236
Salinity, 455–457, 459, 461, 463
Sample
  configuration, 137–139, 141, 143–150
  design, 98
Sampling, 193–195, 197, 199, 201, 203–204
  feature space, 154
  geographical space, 154
  optimisation, 167, 174–175
  strategy, 138, 330
Scale dependence, 301, 322, 325
Scientific visualization, 571–572, 575, 578
Scorpan, 50, 52–53, 437–438, 440, 444, 446, 448
Scorpan-kriging, 437, 440, 444, 448–449, 451
Secondary
  data, 119
  data sources, 119
  information, 167, 169–170, 179–180
Segmentation, 246–247, 250, 252–253
Semantics, 119–120, 131
  data integration, 119, 129
  factoring, 119, 121, 126, 128
  integration, 119–121, 123, 125, 128, 130, 132
  matching, 120
  quality, 523, 527–528, 530, 538–539
Semidetailed, 40, 43–45
SiBCS, 42
SigSolos, 39, 48–49, 51, 53
Sill variance, 186
Simple random sampling, 158
Simulated annealing, 139, 142–143, 183, 189–190
SisSolos, 39, 46–48, 53–54
Site-specific management, 258, 260, 262
Small-scale, 39–40, 42
Smooth variation, 558, 560–561, 563
Software, 355
Soil
  attributes, 269, 276–277
  attribute maps, 437, 439, 441, 447
  classes, 269, 271, 273, 353
  classification, 415–417, 419, 423
  database, 499
  database, 50, 88, 97, 102, 104–105, 499
  digital database, 555–556
  genetic features, 466
  geomorphology, 264, 266
  identification, 425–426, 428, 432
  information system, 55, 60–62, 64, 97, 99, 101, 103, 105
  landscapes, 257
    model, 571, 573–576, 578
    rules, 291
    mapping, 119–121
    modelling, 354, 466
    relationships, 508–511, 517

map unit, 382–383, 385, 389, 394–396, 398
mapping, 39–41, 43–44, 55–57, 62–63, 110, 208, 217, 494, 543, 549, 550
maps, 40–42, 45, 50, 52, 282, 286, 387, 390, 500
monitoring, 447
systems, 77
organic
carbon, 167, 169, 171, 178
matter, 487–489, 491–493, 495
patterns rules, 292
portal, 80
prediction, 269
profile classes, 341, 343
properties, 474
surveys, 25–26, 28–31, 33–34, 39–41, 43–46, 49, 51, 53–54, 235–236, 239, 244, 389–390, 393–394, 398, 402, 412
thickness, 340, 343
types, 330
attributes, 42–43
class maps, 191
forming factors, 281, 289
geomorphic map, 119, 123, 126–127, 129
Soilscape, 245, 246–255
Solar radiation, 371
SOTER, 107–109, 111, 113–116
Space time
analysis, 294
variability, 97, 99–100
Space-filling sample, 184
Spatial
and temporal variation, 478
autocorrelation, 104, 142–148
coverage sample, 139, 183–185, 187, 189, 191–192
data infrastructures, 3, 6
decomposition, 270, 273, 277
extension, 331
interpolation, 137–139, 146, 149
prediction, 437–438, 441, 443–445
scale, 301–303, 305, 307, 309–311, 313, 315–319, 321–325
simulated annealing (CSSA), 142, 167, 169
variability, 99
Spherical variogram, 186
SRTM, 107, 109–111, 113, 487, 489–490
Standard deviation, 100, 103
Stationarity, 97, 103–105
Stochastic simulations, 294
Stratification, 330, 373
Stratified spatial sampling, 158
Stratigraphic, 339, 346
Stream incision, 334
Supervised classification, 237–239, 242, 244, 377, 382, 387
Support vector machines, 353, 357
Support, 98, 105

T
3D, 571–578
Temporal variability, 99
Terrain
analysis, 262, 328
attributes, 415, 418, 420, 422, 437, 443–445, 447
characterisation, 109, 115
Topography, 290
Topographic
maps, 556, 563
wetness index (TWI), 336, 341
Training data, 193–196, 199, 201–203
Trend estimation, 137, 141, 143, 145–149

U
Uncertainty, 97–100, 102–105, 347
propagation, 104
Universal kriging, 137–138, 140–141, 144–146, 148–149
Unsupervised classifications, 377, 381–382, 387, 389, 393–394
US Soil Taxonomy, 426, 428, 430–431

V
Validation, 191, 358
Variance quad-tree, 154
Variogram, 104, 138, 140–141, 143, 149, 184, 186–187, 189–191, 270–271, 273, 275, 277, 477–481, 483

Vertical model, 504
Virtual
  landscape, 294
  reality, 571, 575–576
Visible
  SWIR spectra, 222
  near-infrared range, 221
  NIR, 219
Visualisation, 555, 568

W

Water table dynamics, 571, 578
Wavelets, 269–271, 275, 277
Wavelet packet transform, 301, 308, 320
Wavelet transform, 302, 304, 308, 310–311, 316–317, 319, 322–325
Weathering, 207, 212–217
Wetness, 67, 72

Colour Plate Section

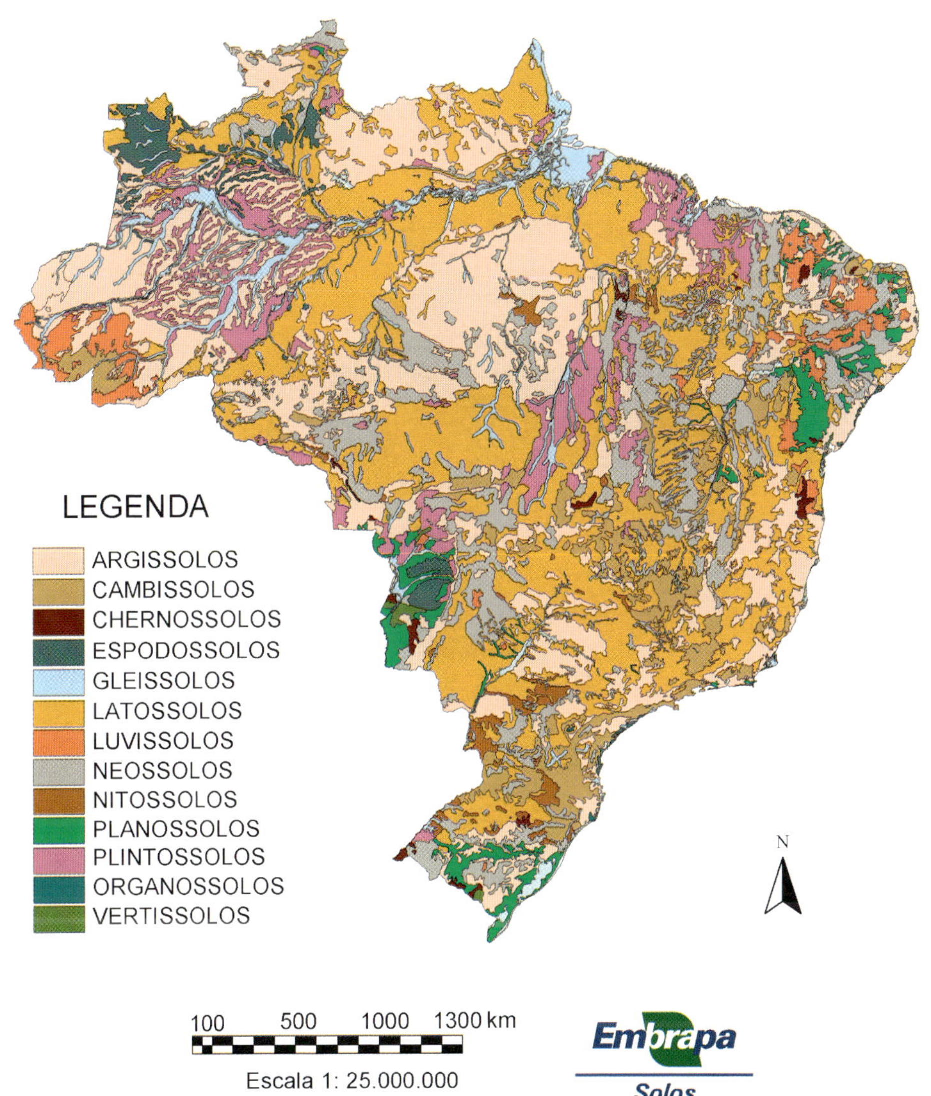

Plate 3. *Map of the soil orders of Brazil (after Embrapa, 2001 based on Embrapa, 1981).*

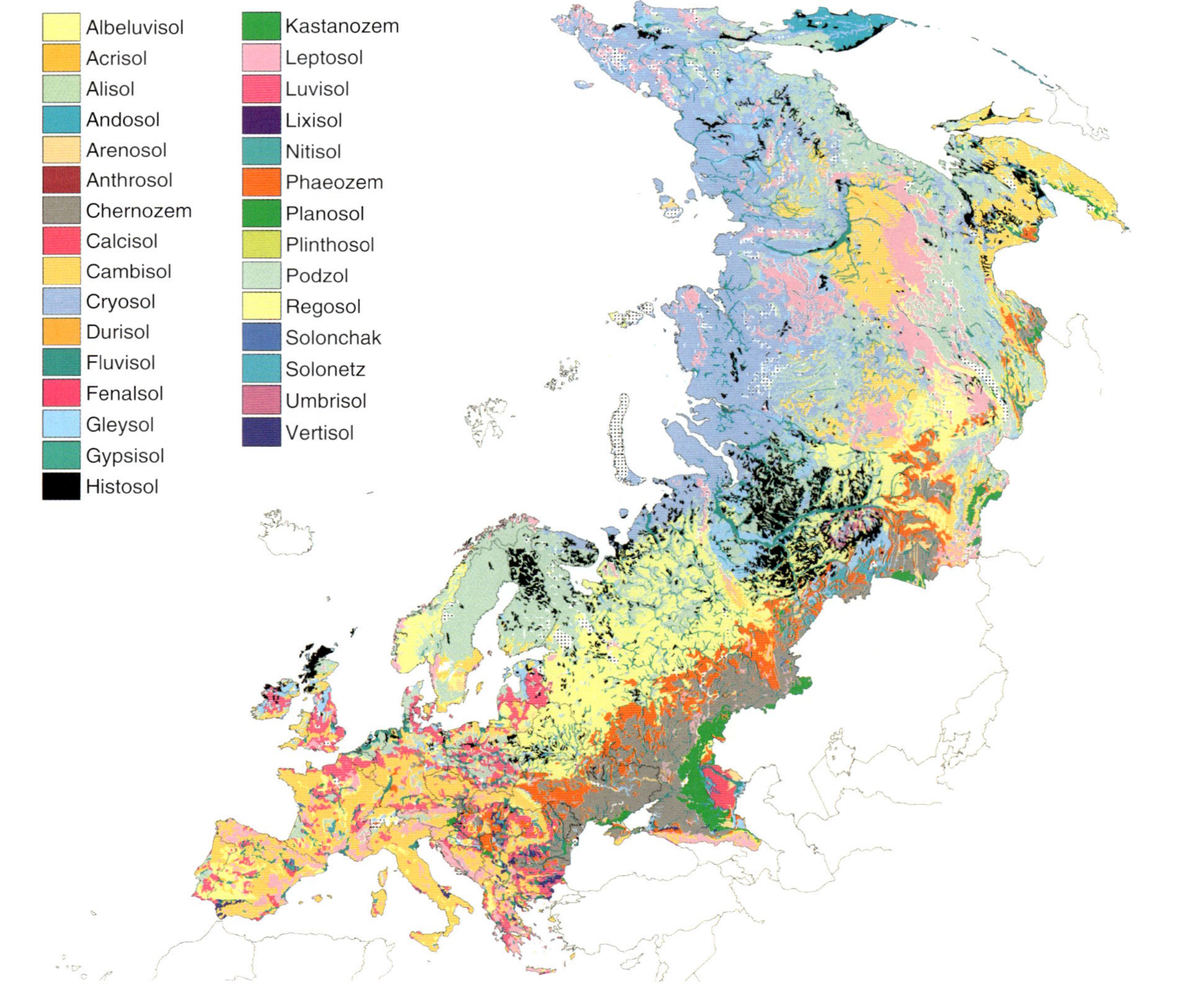

Plate 4. *Eurasia. Soil reference group code from the World Reference Base (WRB) for Soil Resources.*

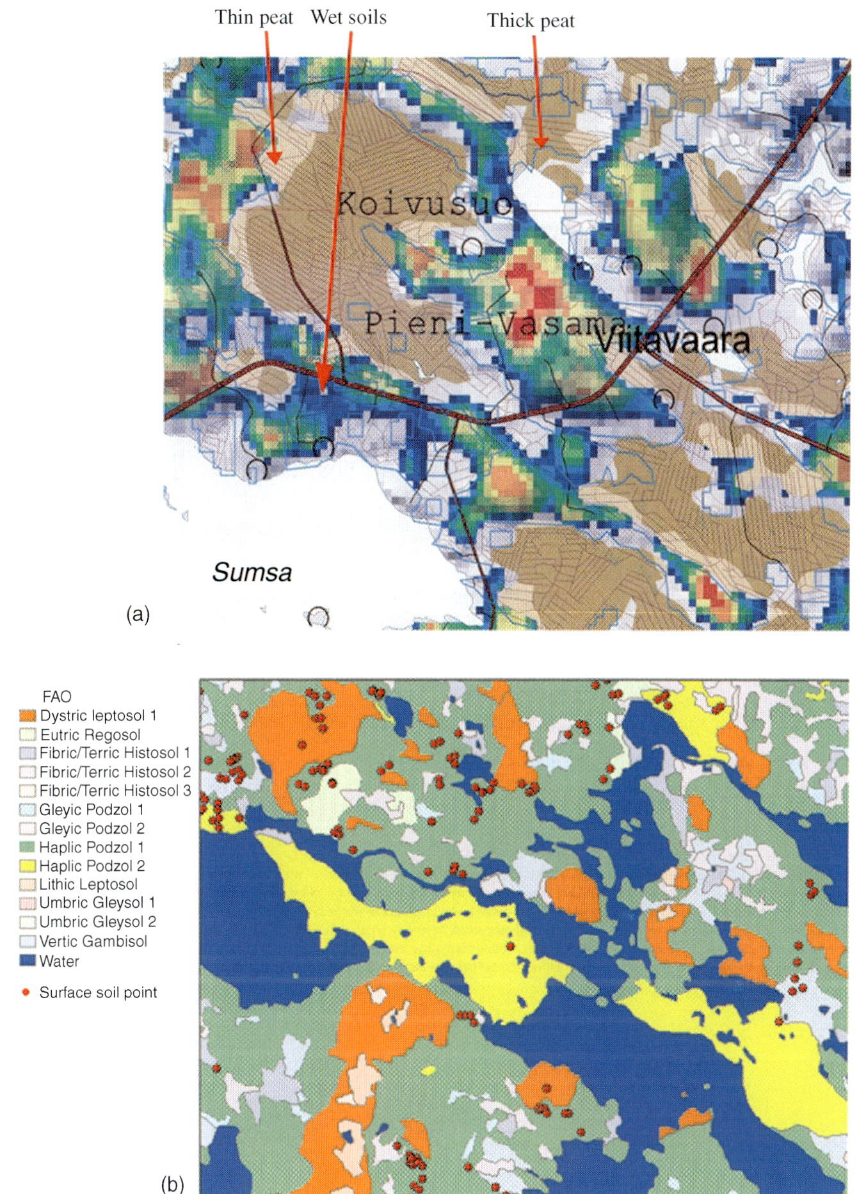

Plate 5. *(a) Thickness of peat mapped with classified airborne radiometric potassium (K). (b) Interpreted soil polygons (soil bodies) in the Sotkamo area of Finland.*

Site code[1] **CLRA48**

Flat plains south-west of Mount Gellibrand

Location	The Sanctuary (Beal Road), Colac district, south-west Victoria
Landform	Lava plain
Geology	Quaternary paludal lagoon and swamp deposits: *silt, clay*
Element	Flat
Slope	0–1%
Aspect	South-west

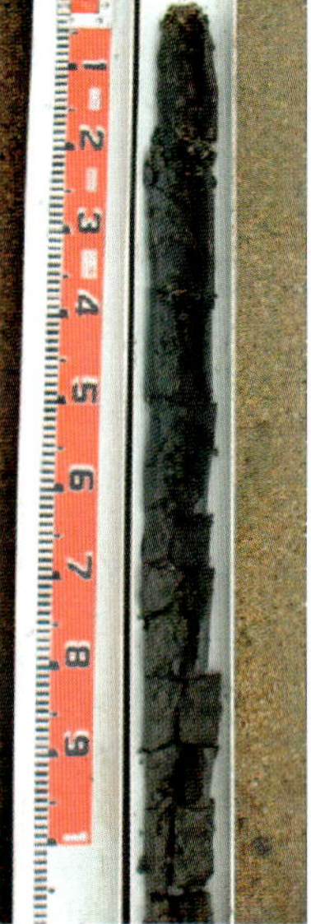

Epihypersodic, Self-mulching, Grey Vertosol

Horizon	Depth (cm)	Description
A	0–35	Black (2.5/N); heavy clay; strong coarse and medium prismatic, parting to medium and fine angular blocky structure; smooth ped fabric; very firm consistence (moist); non-calcareous, pH 7.5; diffuse boundary to:
B21	35–80	Dark grey (2.5Y4/1) with very few fine faint orange mottles; medium heavy clay; strong coarse and medium prismatic, parting to medium and fine angular blocky structure; smooth ped fabric; firm consistence (moist); non-calcareous, pH 9; diffuse boundary to:
B22	80–140+	Dark grey and greyish brown (2.5Y4/1, 2.5Y5/2) with few medium distinct orange mottles; medium clay; very few fine rounded quartz pebbles; strong coarse and medium prismatic, parting to fine angular blocky structure; smooth ped fabric; firm consistence (moist); non-calcareous, pH 9.

[1] Source: Robinson et al (2003) A land resource assessment of the Corangamite region. Department of Primary Industries, Centre for Land Protection Research Report No. 19

Analytical data[2]

Site CLRA48 Horizon	Sample depth cm	pH H_2O	pH $CaCl_2$	EC dS/m	NaCl %	Ex Ca cmol$_c$/kg	Ex Mg cmol$_c$/kg	Ex K cmol$_c$/kg	Ex Na cmol$_c$/kg	Ex Al mg/kg	Ex Acidity cmol$_c$/kg	FC −10kPa %	PWP −1500kPa %	KS %	FS %	Z %	C %
A	0–35	8	7.6	1.3	N/R	6.2	26	2.5	17	N/R	N/R	N/R	N/R	N/R	N/R	N/R	N/R
B21	35–80	8.5	8.1	2.6	N/R	5	26	2.8	38	N/R	N/R	N/R	N/R	N/R	N/R	N/R	N/R
B22	80–140+	8.8	8.3	2.7	N/R	4.5	25	2.8	39	N/R	N/R	N/R	N/R	N/R	N/R	N/R	N/R

Management considerations

This is a structured cracking clay soil. Cracking clay soils vary in their workability depending on their moisture status (highly permeable when dry and impermeable when saturated). These soils are also prone to structure decline particularly when worked wet. They are also generally alkaline with depth and can place stress on roots with their high shrink-swell capabilities. The main priority on these soils is to avoid working when wet (on or below plastic limit).
The alkaline subsoils are associated with a high nutrient capacity but result in an imbalance in nutrient availability (may be restrictive to certain plant species (eg. potatoes). This soil type is often associated with sodic and calcic soil properties, here very sodic, magnesic as well as saline in the subsoil. Growing alkaline tolerant species is a practical option. Soil salinity at depth will affect deeper rooting plants and may indicate water movement restrictions. It is important not to increase the groundwater level bringing the salinity closer to the surface; more efficient use of water by plants and/or deep drainage is suggested.

[2] Source: Government of Victoria State Chemistry Laboratory.

Plate 7. *Australia: example layout of a two-page soil site description.*

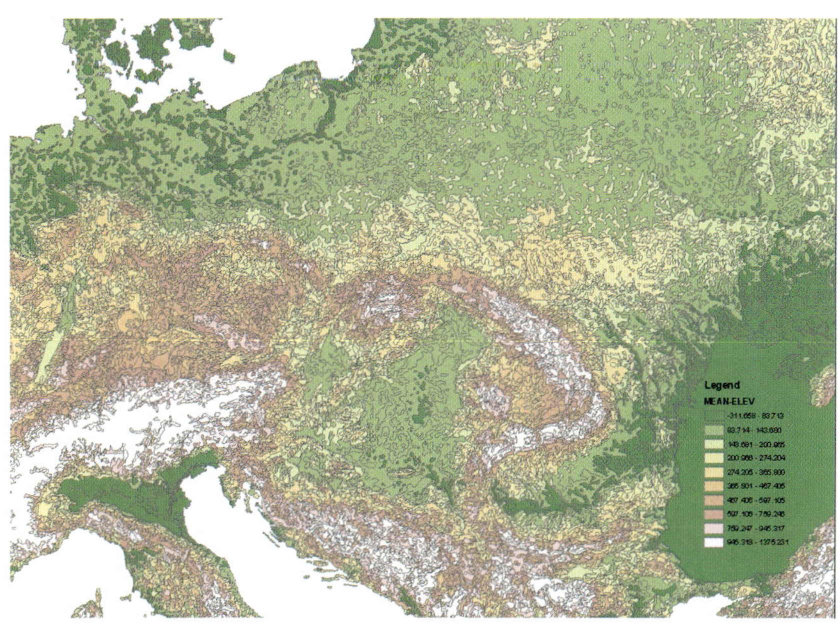

Plate 9. *Central and eastern Europe. The SOTER Unit structure derived from the SRTM30 Digital Elevation Data.*

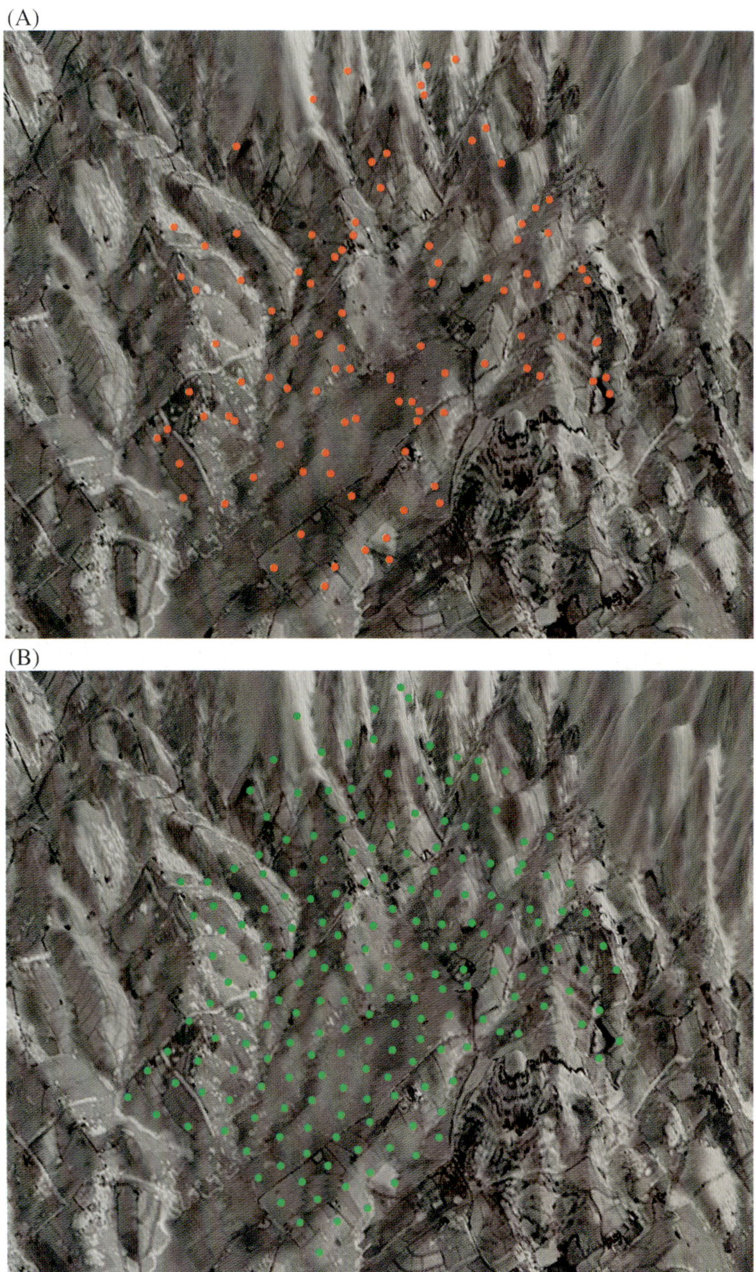

Plate 12. *Hunter valley, NSW, Australia. Location of the sampling units in the landscape, (A) Latin hypercube sampling and (B) equal spatial strata sampling.*

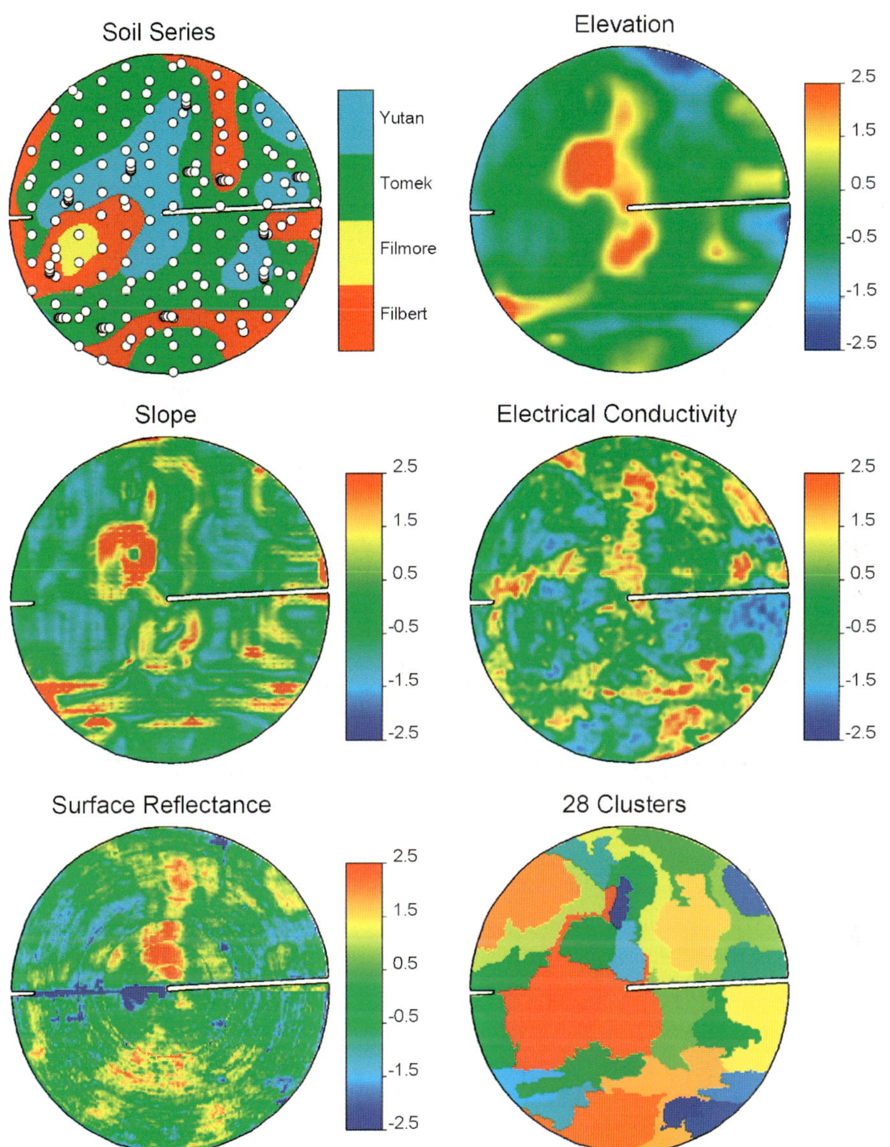

Plate 13. *Data layers in a 52-ha centre-pivot agricultural field near Mead, Nebraska.*

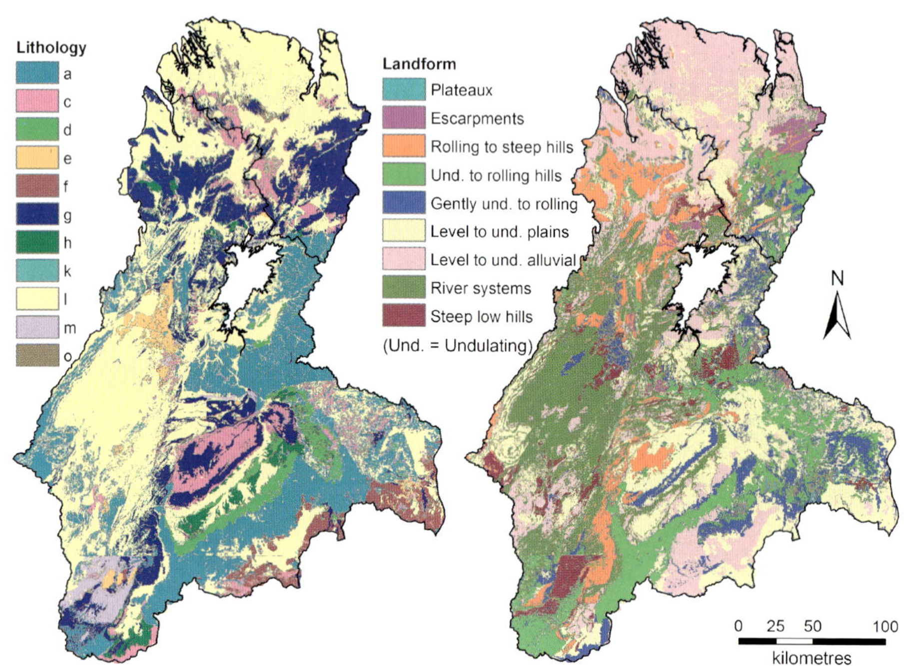

Plate 15. *Ord river basin, Australia. (Left) Mapped modal prediction from 10 classification trees for lithology attribute. (Right) Mapped modal prediction from 10 classification trees for landform attribute.*

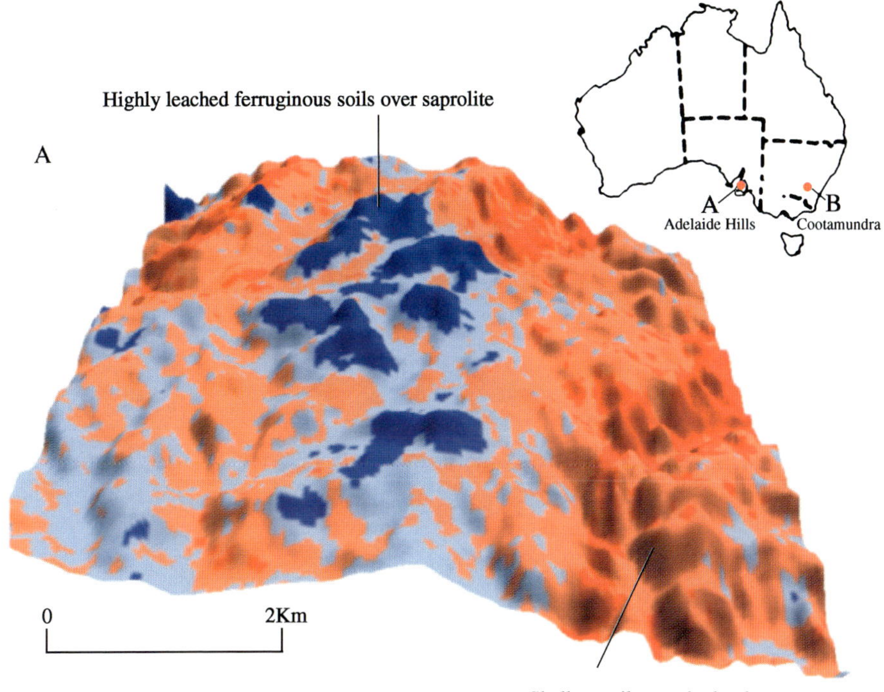

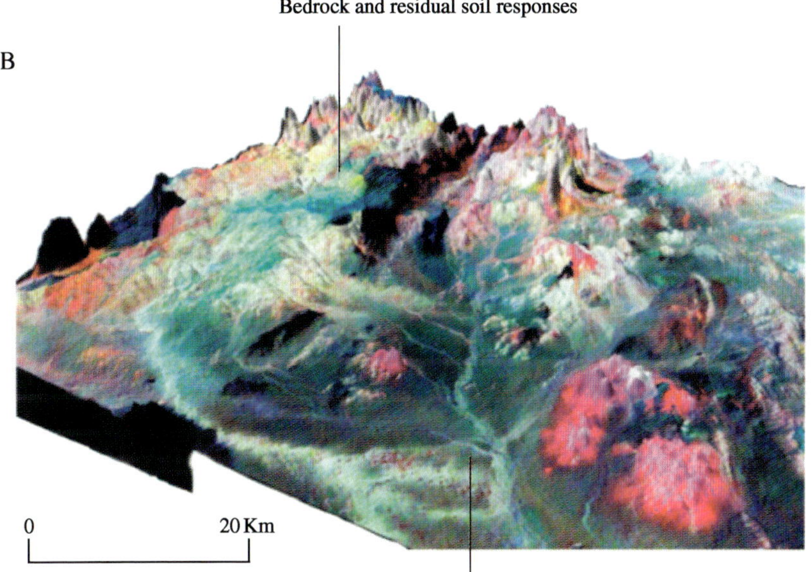

Plate 16a. *Adelaide Hills, South Australia and Cootamundra, New South Wales, Australia. (A) Modelled regolith type: highly leached ferrugionous shallolw soil over slightly weathered bedrock. (B) Ternary gamma-ray spectrometry draped over a DEM (K in red, Th in green and U in blue). This type of display assists in understanding landscape processes and radioelement responses.*

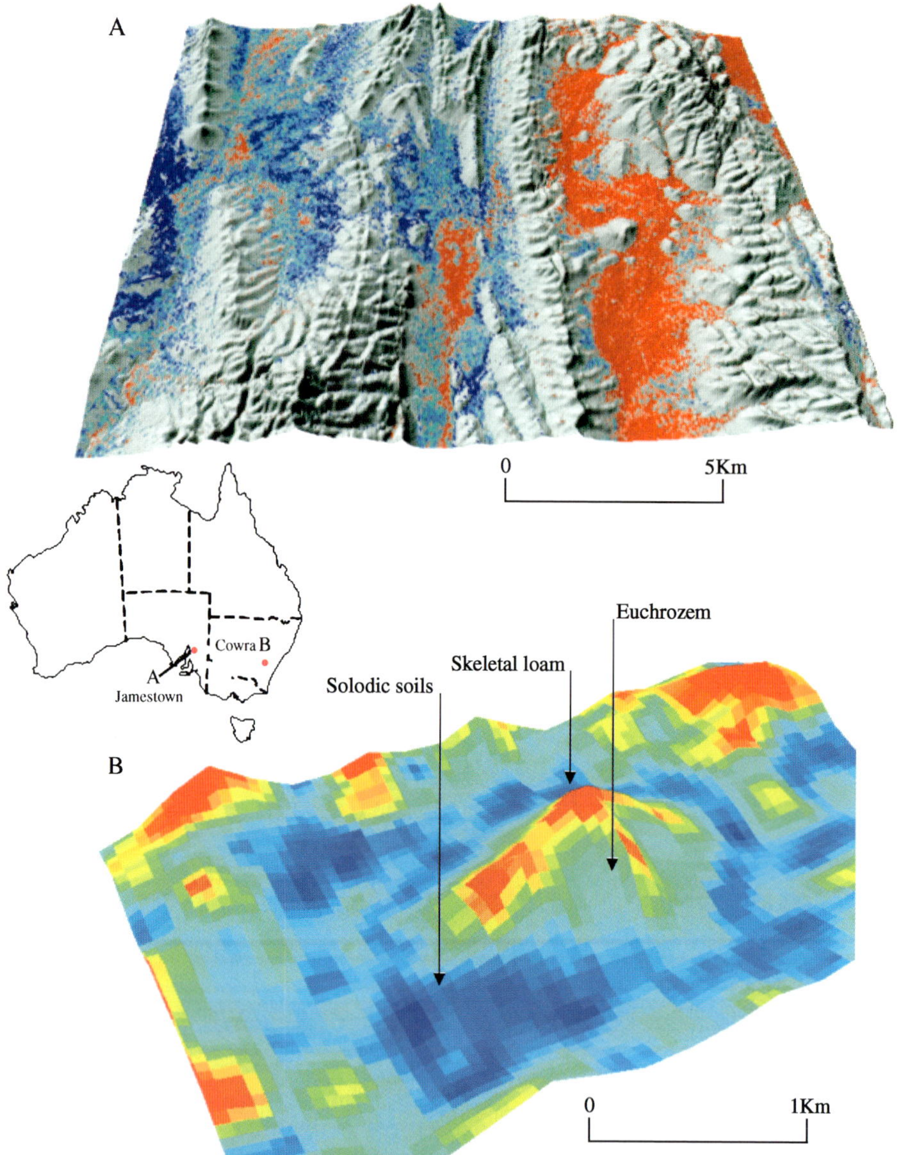

Plate 16b. *Jamestown, South Australia and Cowra, New South Wales, Australia. (A) Prediction of surface soil texture over de-postional landforms using airborne gamma-ray imagery. Silty soil in red, fine sandy soil in blue. (B) Soil catena using K/Th ratio bands and topographic wetness index.*

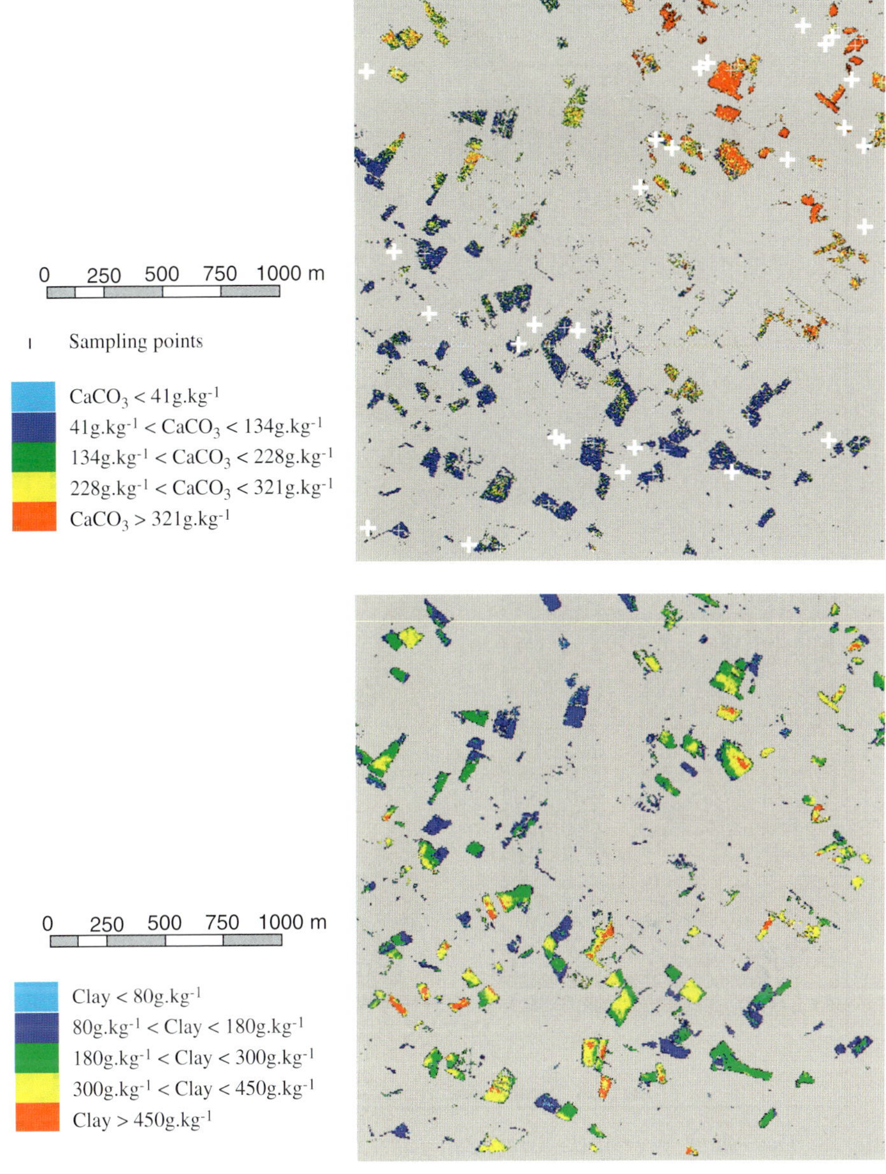

Plate 17. *La Peyne watershed, southern France. Upper: map of estimated CaCO3 content classes at the soil surface using HyMap* $\overline{\mathrm{CR}}_{2341}$. *Classes were defined according to HCl effervescence class values. Sampling points are shown as crosses. Lower: map of estimated clay content at the soil surface using HyMap* $\overline{\mathrm{CR}}_{2206}$.

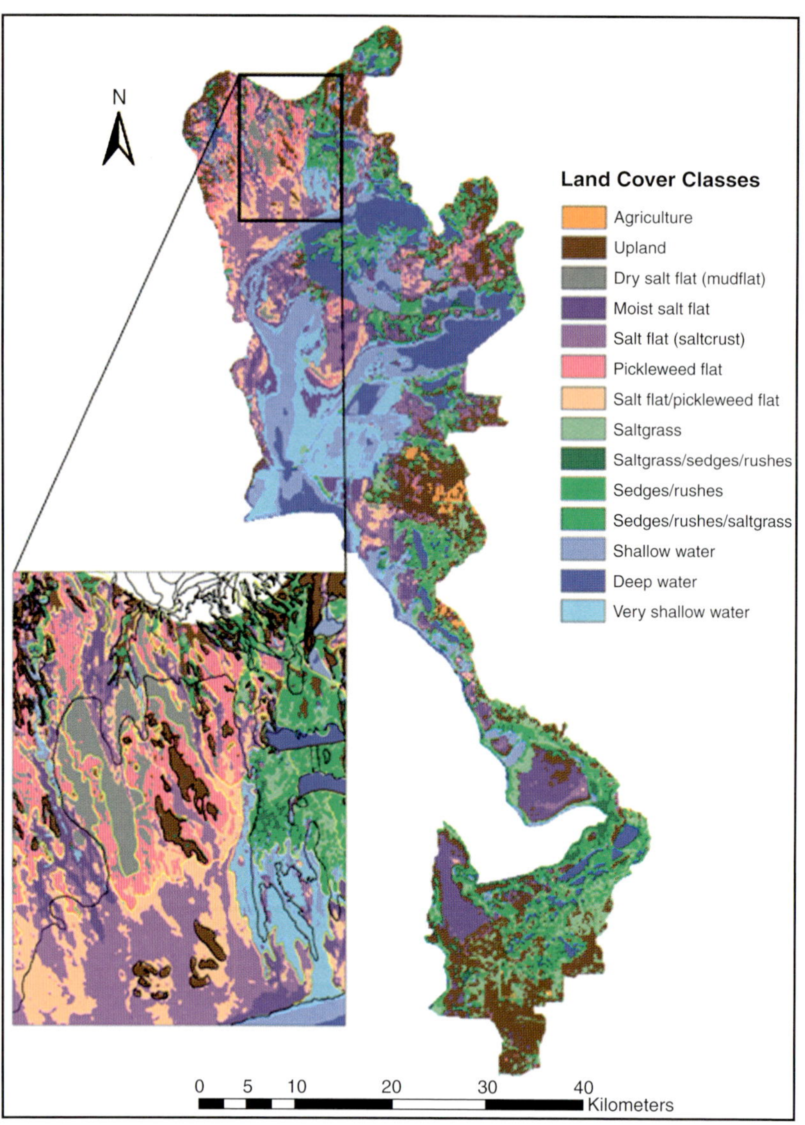

Plate 18. *East Shore Area of the Great Salt Lake, Utah USA. Classified image using supervised fuzzy classification, the minimum distance to means classifier, a fuzzy convolution filter and all six Landsat 7 bands. Inset box shows an example of the original soil survey lines (black, ca. 1975) and the revised lines (yellow) created using the classified image as a guide.*

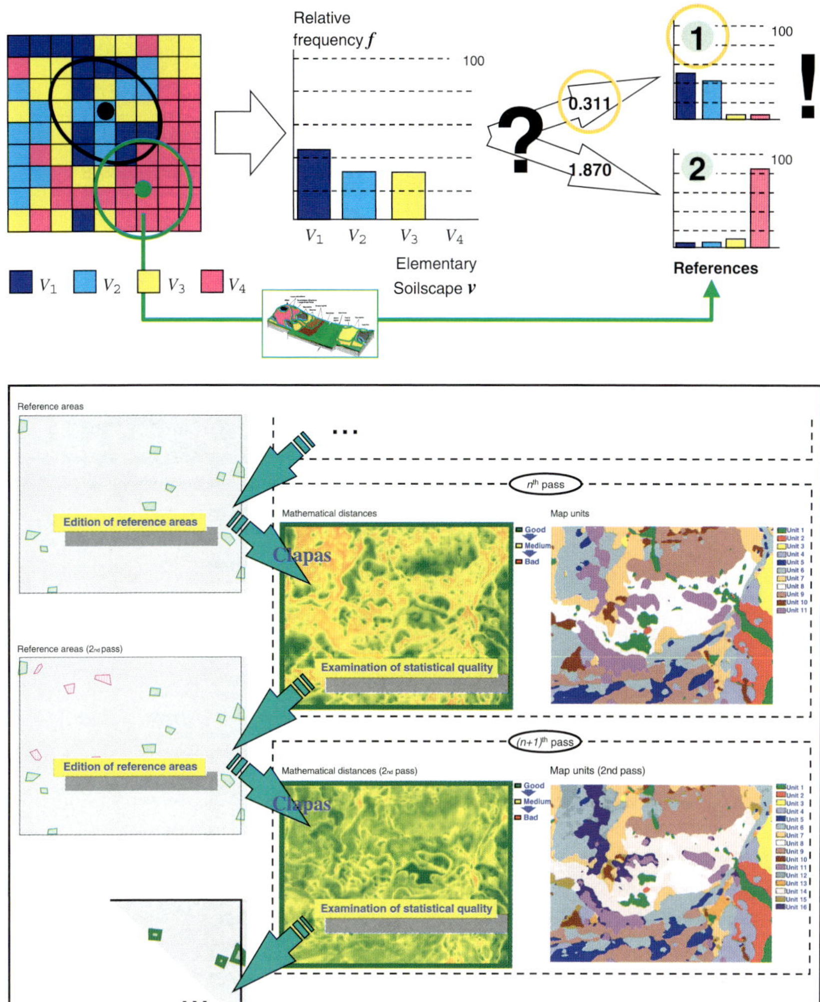

Plate 19. *Upper: computing and using the cover-frequency vectors. (a) In this image containing $p = 4$ classes, from the elliptic neighbourhood of each cell, for example here, the black-dotted cell, (b) one extracts its cover-frequency vector. (c) This is then compared with each of the $q = 2$ reference vectors and by means of a statistical distance d. The cell is assigned to the category whose vector is the most similar, that is the one that minimises d, here with a distance of 0.311. Lower: iterating for an enriched sketch (Caroux Massif, south of France). The sketch is getting more and more precise via enrichment from: (1) examination of statistical and cartographic indicators; (2) field survey. At each step, adding, deleting or merging reference areas allows testing of the hypotheses. It is possible to choose another combination of image themes, descriptors other than cover frequency and other distance metrics.*

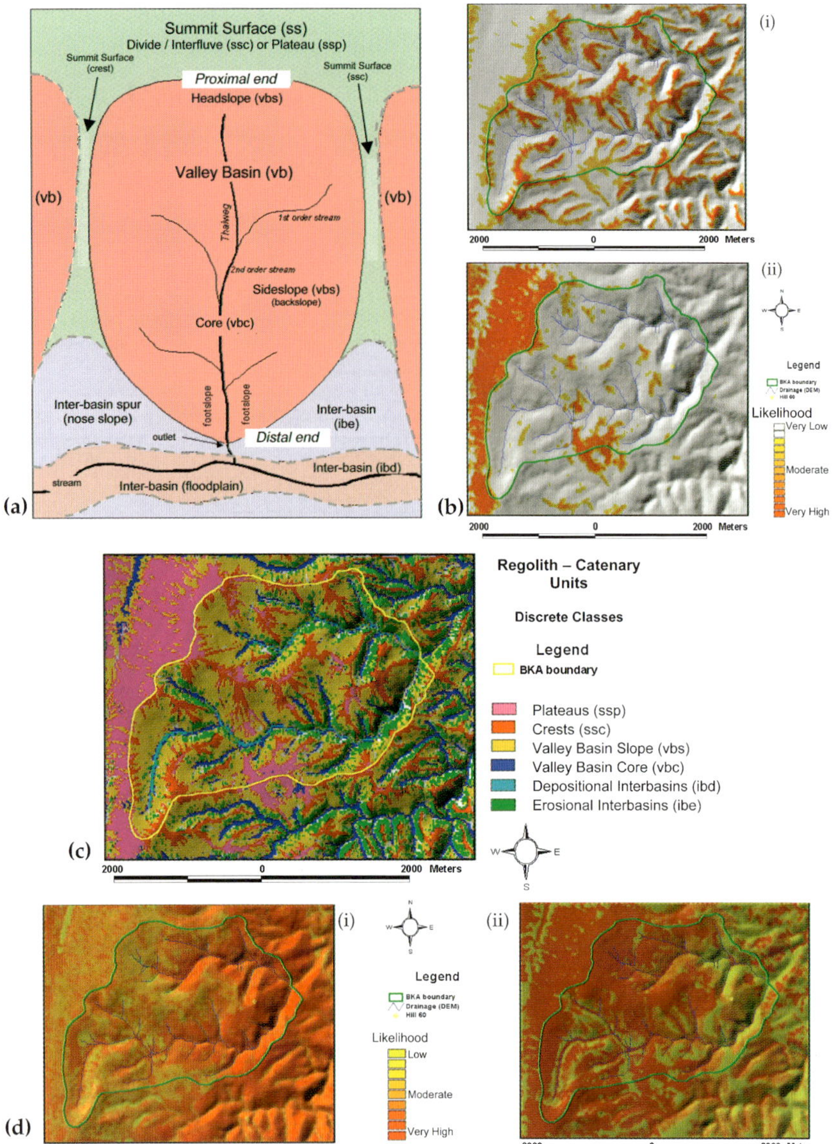

Plate 20. *(a) (Top left) A stylised version of the Regolith-Catenary Units in a simple drainage basin arrangement. Refer to Section 20.3 for a description of the RCU components; (b) (top right) examples of fuzzy classification of RCU components. Hillshading renders the 'Very Low' transparent values as a light grey, (i) likelihood of hillcrests (RCU_{ssc}) and (ii) likelihood of plateaux (RCU_{ssc}); (c) (middle) discretised representation of the RCUs in the BKA. Fuzzy classification of all RCUs cannot be represented in a single image so a 'de-fuzzified' classification is necessary; (d) (bottom) examples of fuzzy classification of regolith–terrain attributes for forestry site-specific management purposes, (i) likelihood of high permeability and (ii) likelihood of deep regolith.*

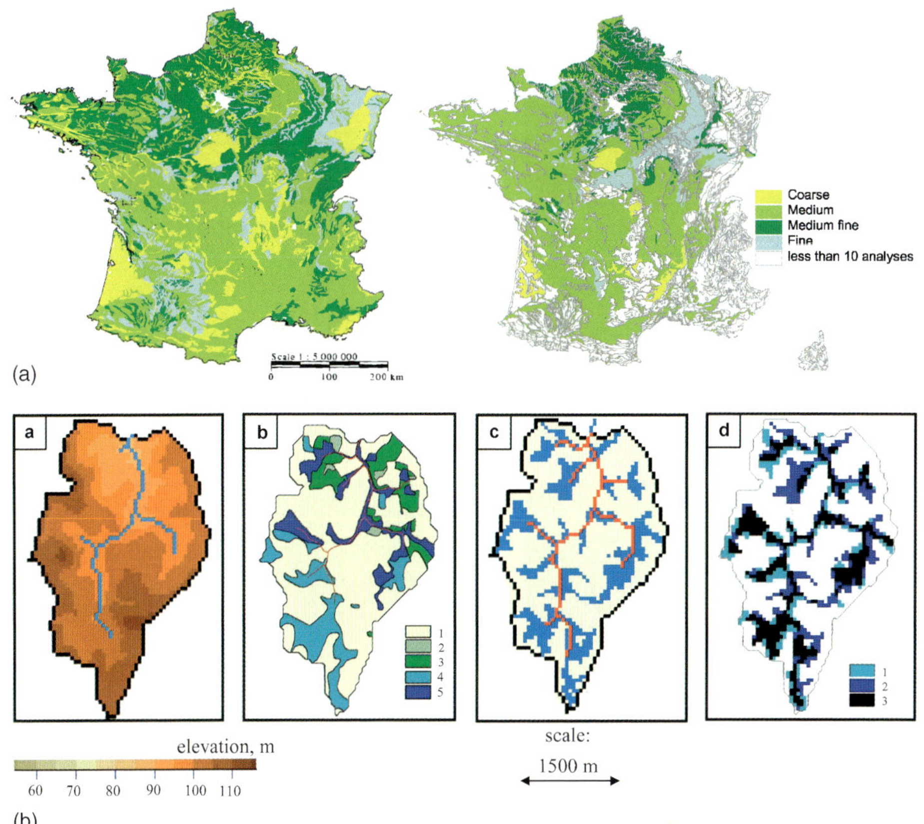

Plate 22. a: *Modal FAO texture class of the 1:1,000,000 soil map of France: (a) estimated by expert knowledge; (b) computed from nearly 30,000 topsoil particle-size analyses (Schvartz et al., 1998). b: Waterlogged soil distribution predicted from a topographic modelling approach in a 250 ha catchment in Brittany, France.: (a) topographic map; (b) soil map (1–3: no or little hydromorphy; 4 and 5: hydromorphy from the soil surface; (c) predicted wetland by topographic index; (d) comparison of the digitized b map and the c map; (1: soil map; 2: topographic index map; 3: superposition of the two units) (after Mérot et al., 2003).*

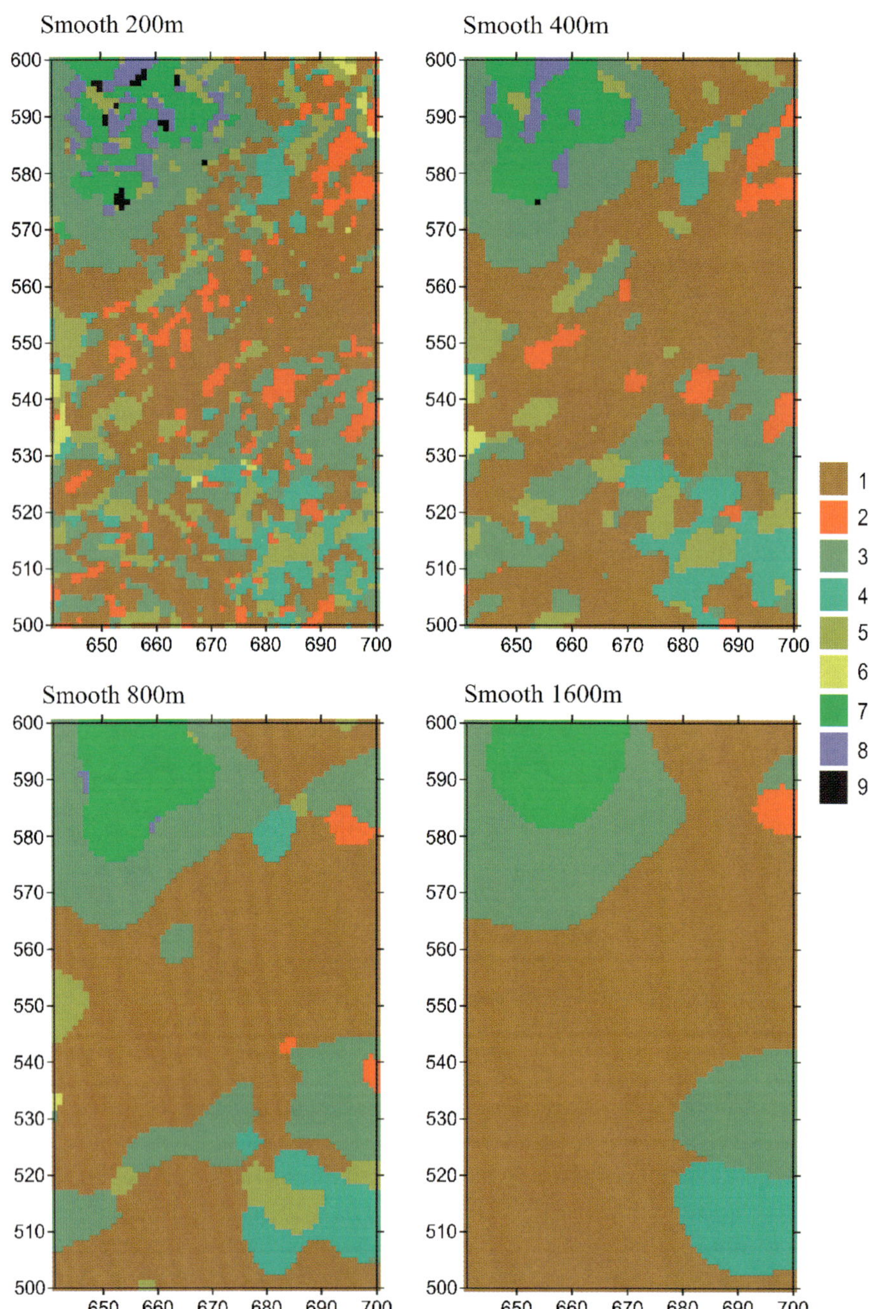

Plate 23. *Smooth components from a two-dimensional (2-D) multiresolution analysis of indicator variables defined on soil classes at Ivybridge, South West England. The class displayed is the one for that the smooth approximation of the indicator is largest at any site. Soil classes are as follows — 1: Typical Brown Earths; 2: Brown Rankers; 3: Brown Podzolic soils; 4: Gleyic Brown Earths; 5: Gleys; 6: Coarse alluvial soil; 7: Humic Gley Stagnopodzol; 8: Stagnopodzol; 9: Blackearth. Nowhere is 'Unclassified' the class with the largest indicator.*

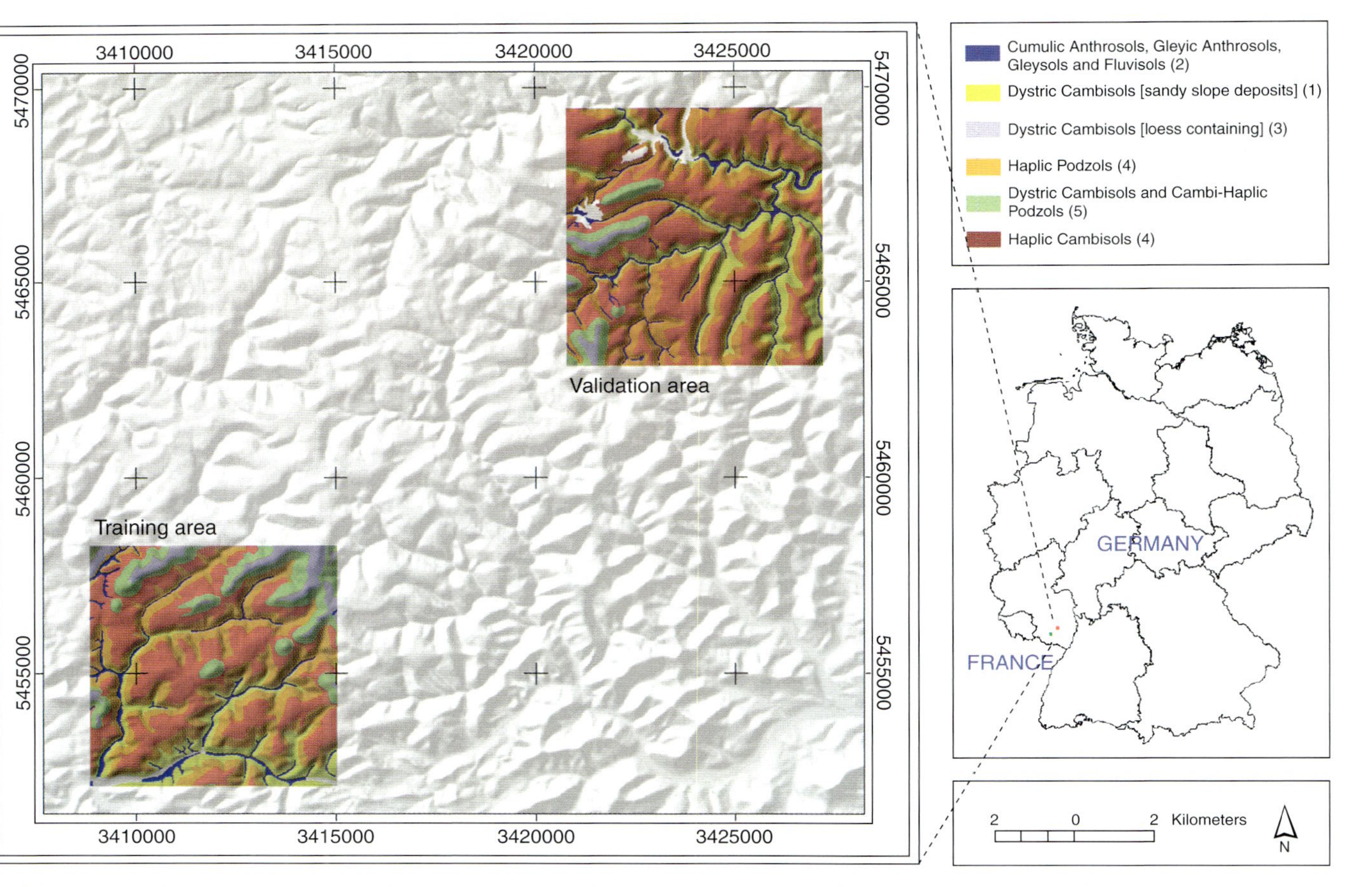

Plate 25. *Study sites in the Palatinate Forest, Rhineland-Palatinate, Germany. The figure shcws the soil maps in the training and the validation area draped on a shaded relief map.*

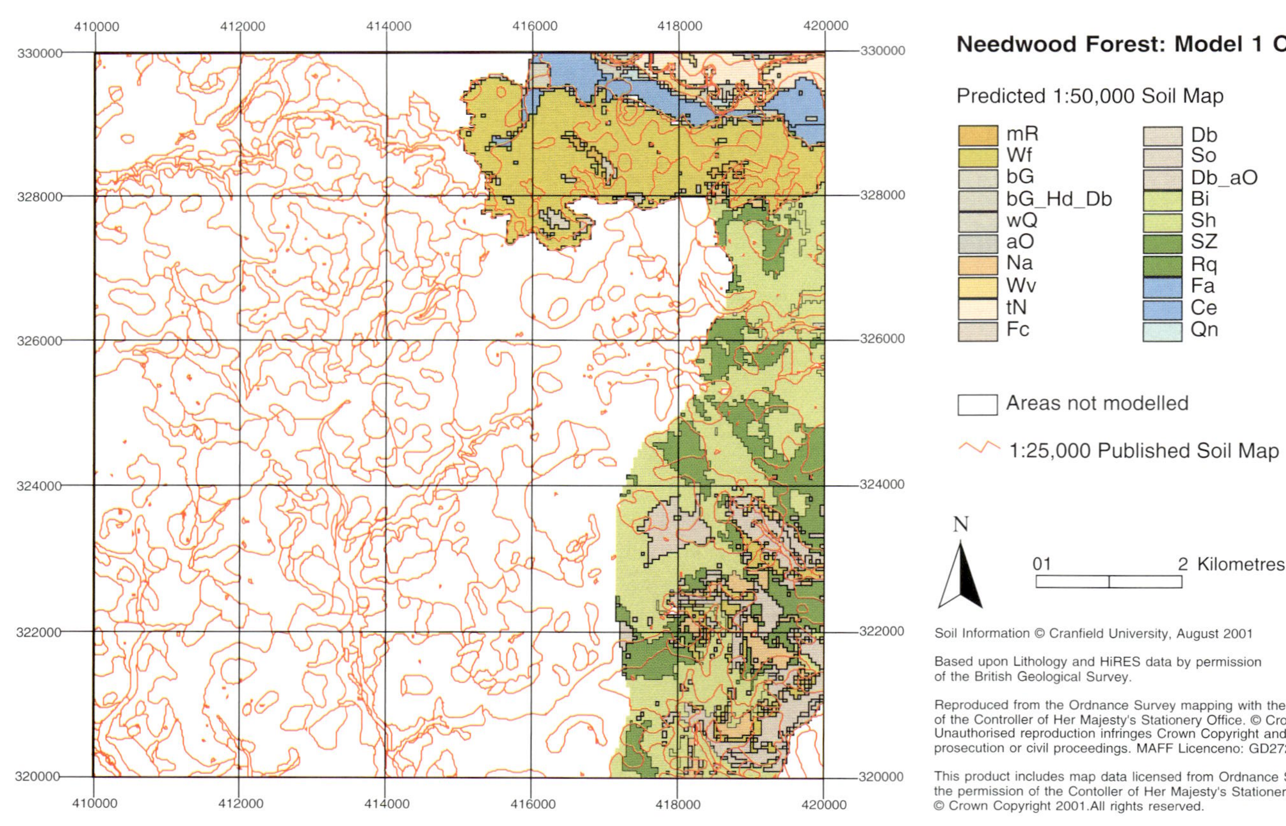

Plate 26. *Extrapolation results for Needwood Forest (England).*

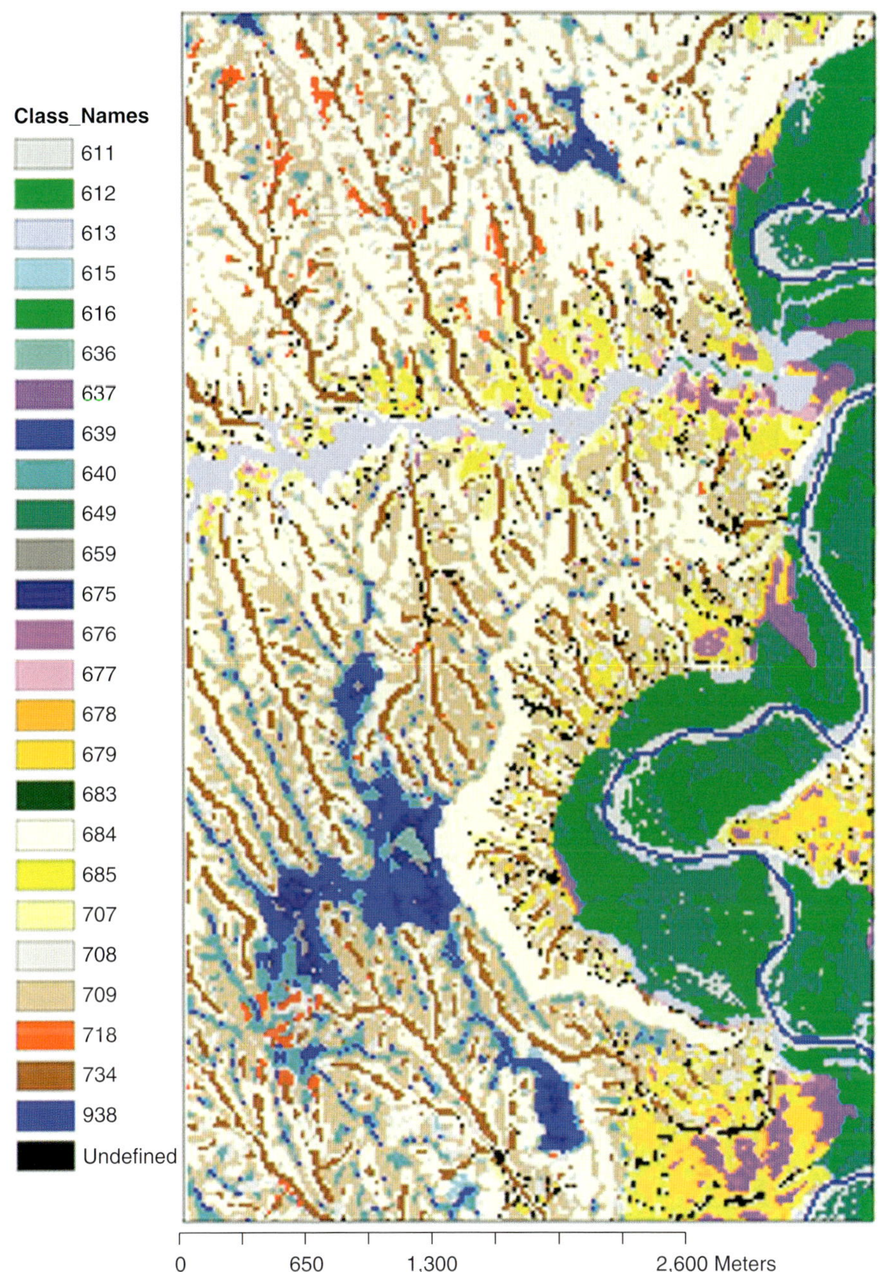

Plate 27. *Final pixel-based map showing map units in the south-eastern quarter of the Juniper Draw 7.5-min quadrangle of the study area (Powder River Basin, Wisconsin, USA).*

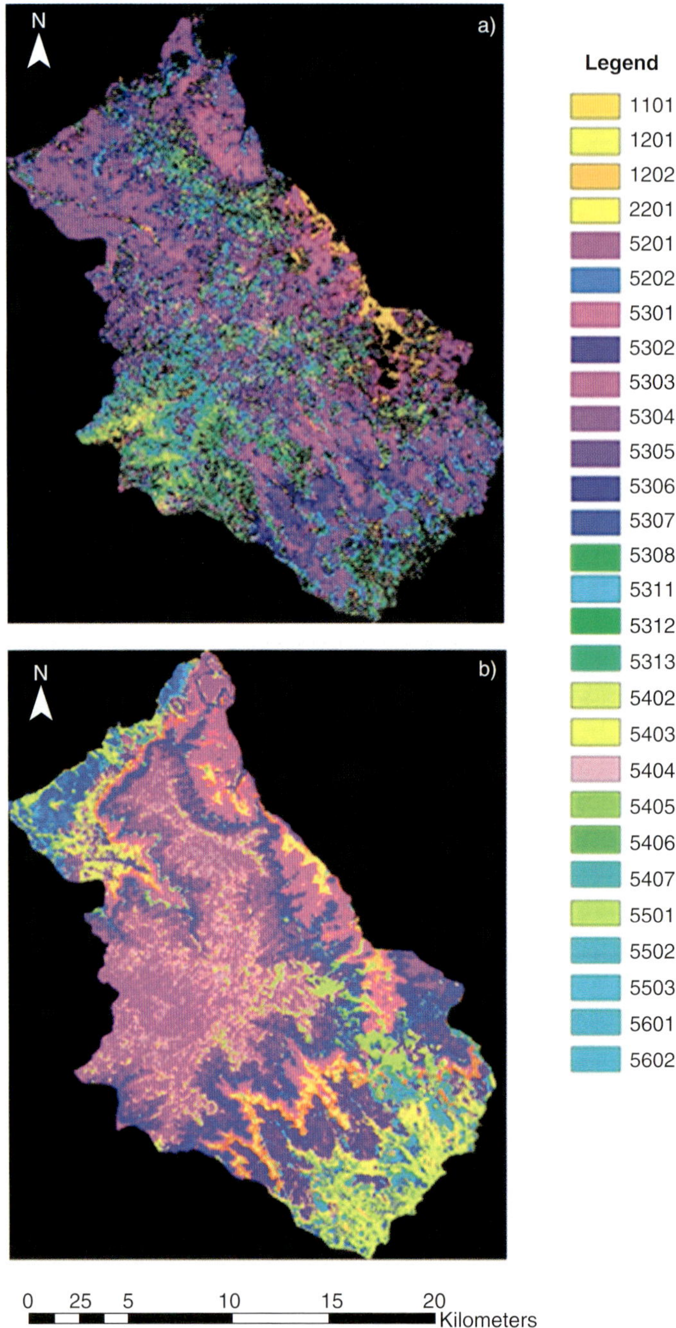

Plate 28. *Green River Basin, Wyoming, USA (a) knowledge-based classification with 28 classes; areas shown in black are unclassified. (b) Classification tree with 28 classes. Key to classes is in Table 28.2.*

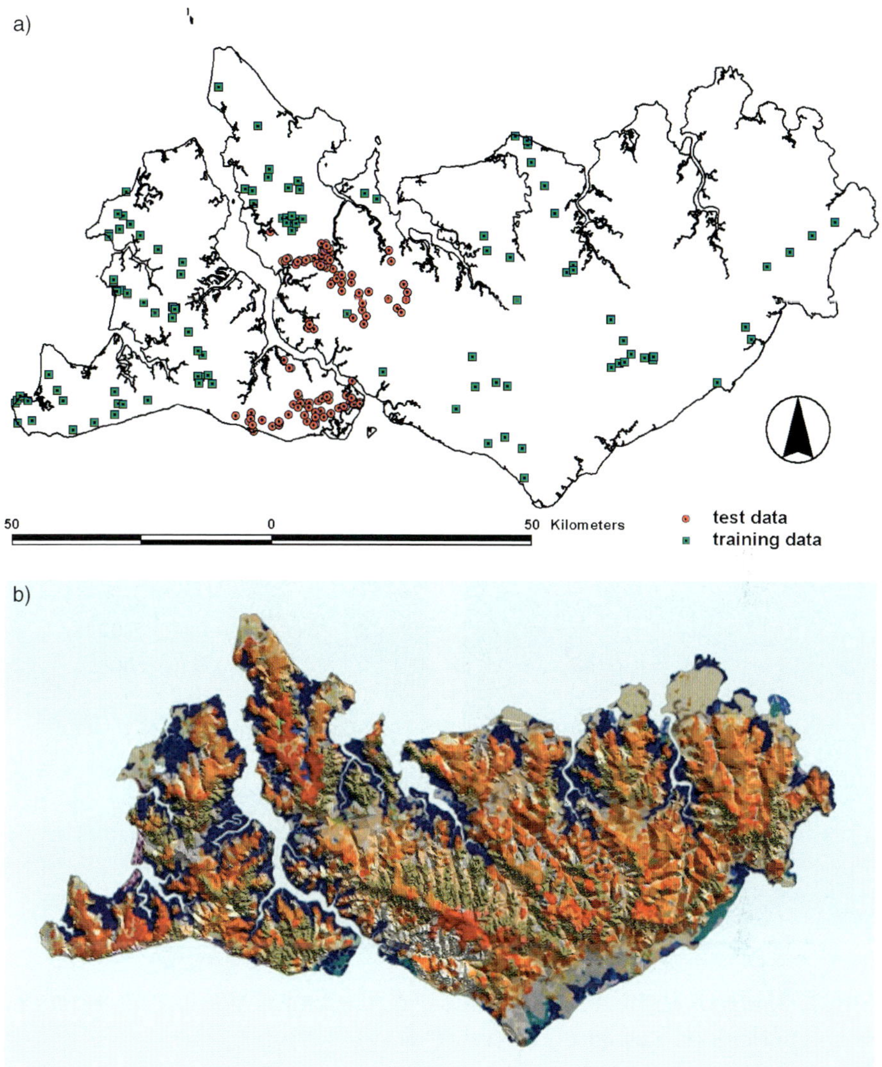

Plate 29. *Tiwi Islands, Northern Territory, Australia (a) Locations of training and test survey sites, (b) Landform units – reddish colours indicating Oxisols, grey Ultisols and blues Aquic soils.*

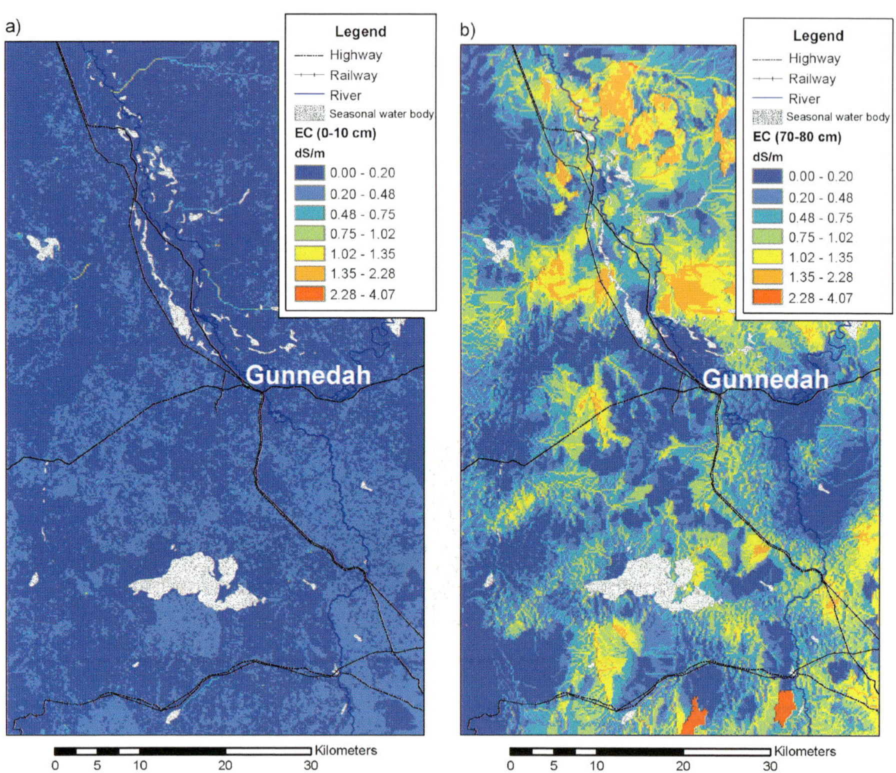

Plate 32. *Northern NSW, Australia. Maps of (a) Topsoil EC (dS/m) 0–10 cm (multiple linear regression), (b) Subsoil EC (dS/m) 70–80 cm (scorpan-kriging).*

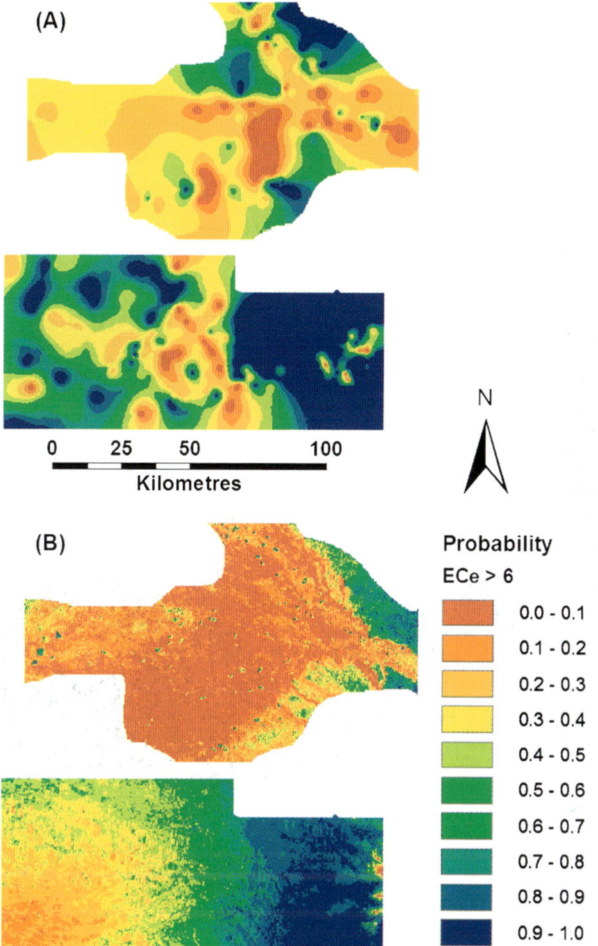

Plate 33. *Northern NSW, Australia. Salinity risk maps indicating the predicted probability of exceeding an ECe threshold of 6 dS/m using (A) Multi-Indicator Kriging (MIK) and (B) discriminant analysis (DA).*

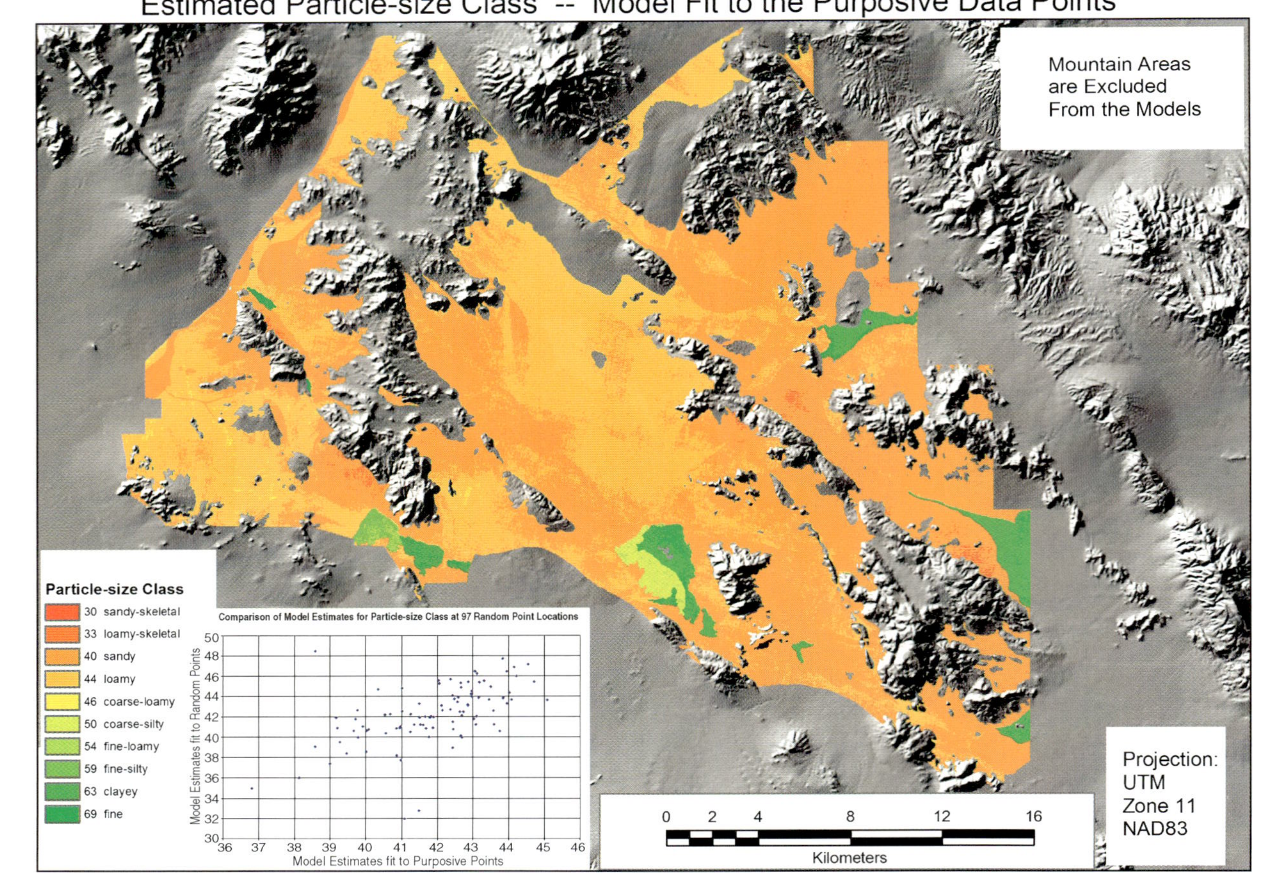

Plate 34. *Model estimate of particle-size class.*

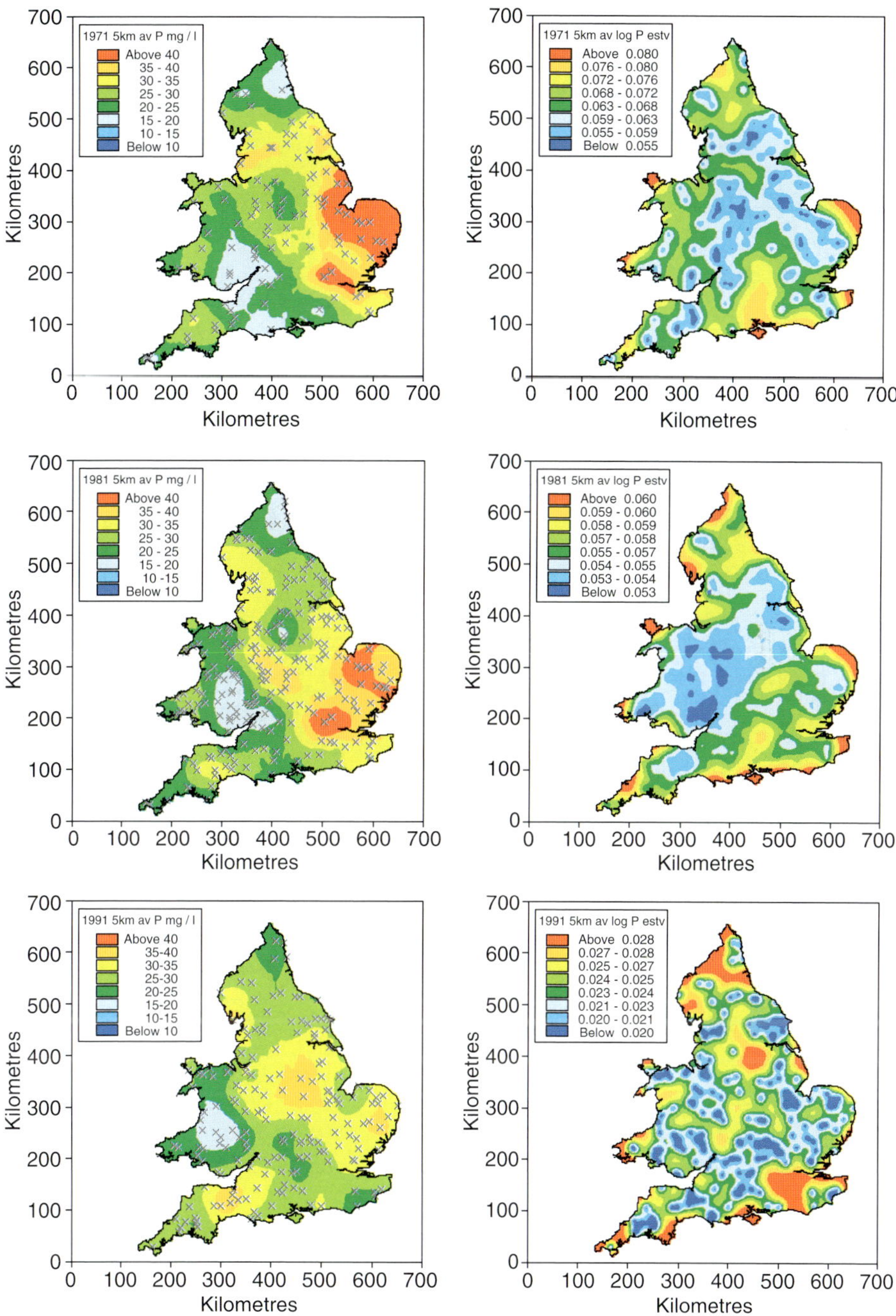

Plate 35a. *Maps of kriged estimates for 1971 (upper), 1981 (middle), 1991 (lower) of P (mg l^{-1}) on the left with the sample locations indicated by the crosses, and maps of $\log_{10}$ P kriging variances on the right.*

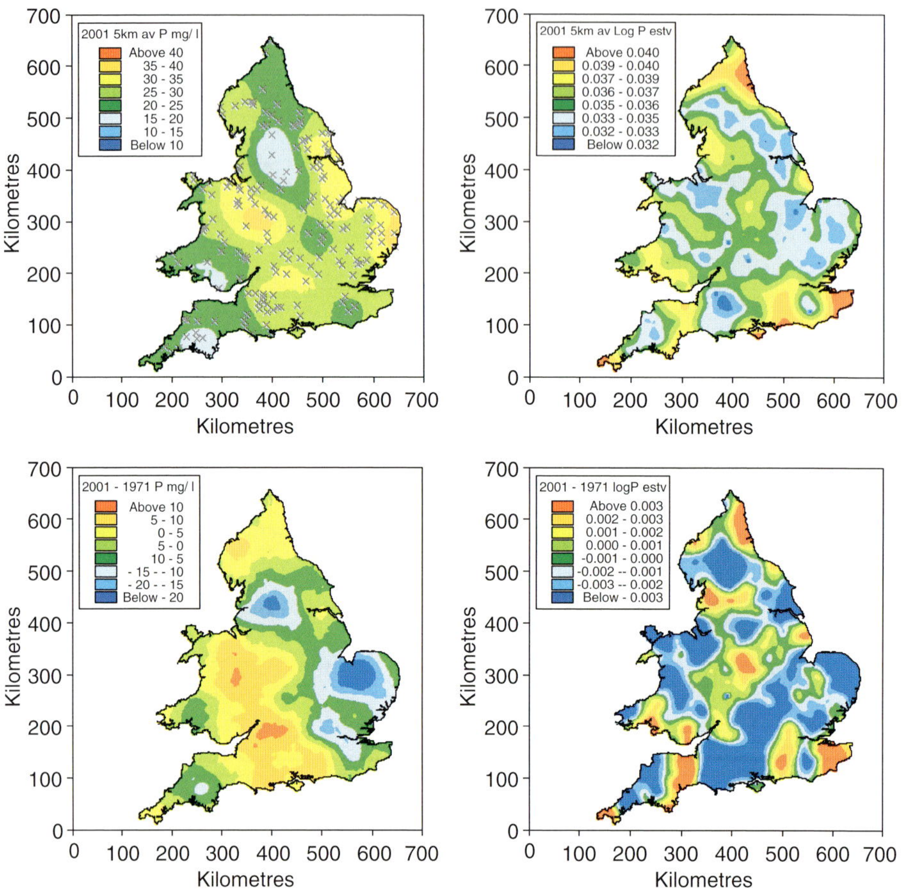

Plate 35b. *Maps of kriged estimates for 2001 of P (mg l^{-1}) (upper) and difference in P between 2001 and 1971 (mg l^{-1}) (lower) on the left with the sample locations indicated by the crosses, and maps of $\log_{10}$ P kriging variances on the right.*

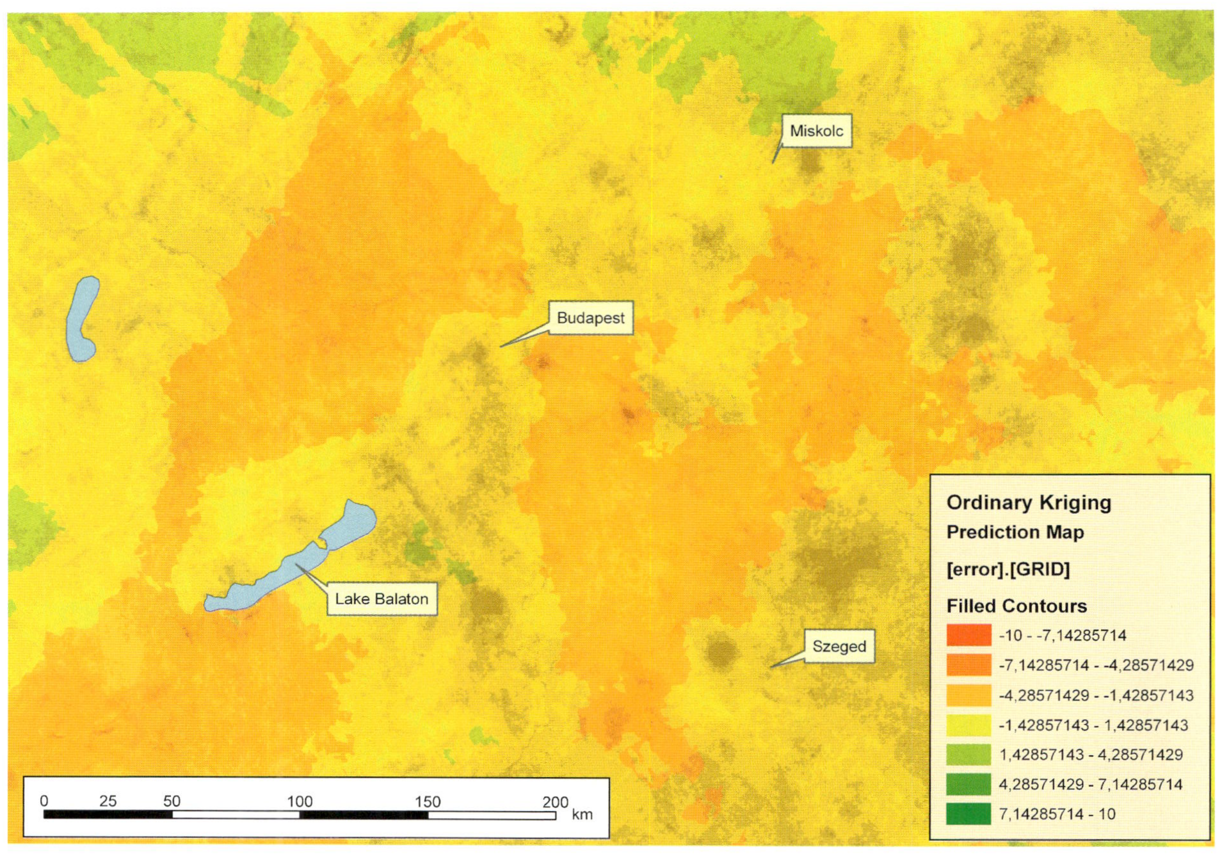

Plate 36a. *Map of organic matter regression model residuals produced by ordinary kriging.*

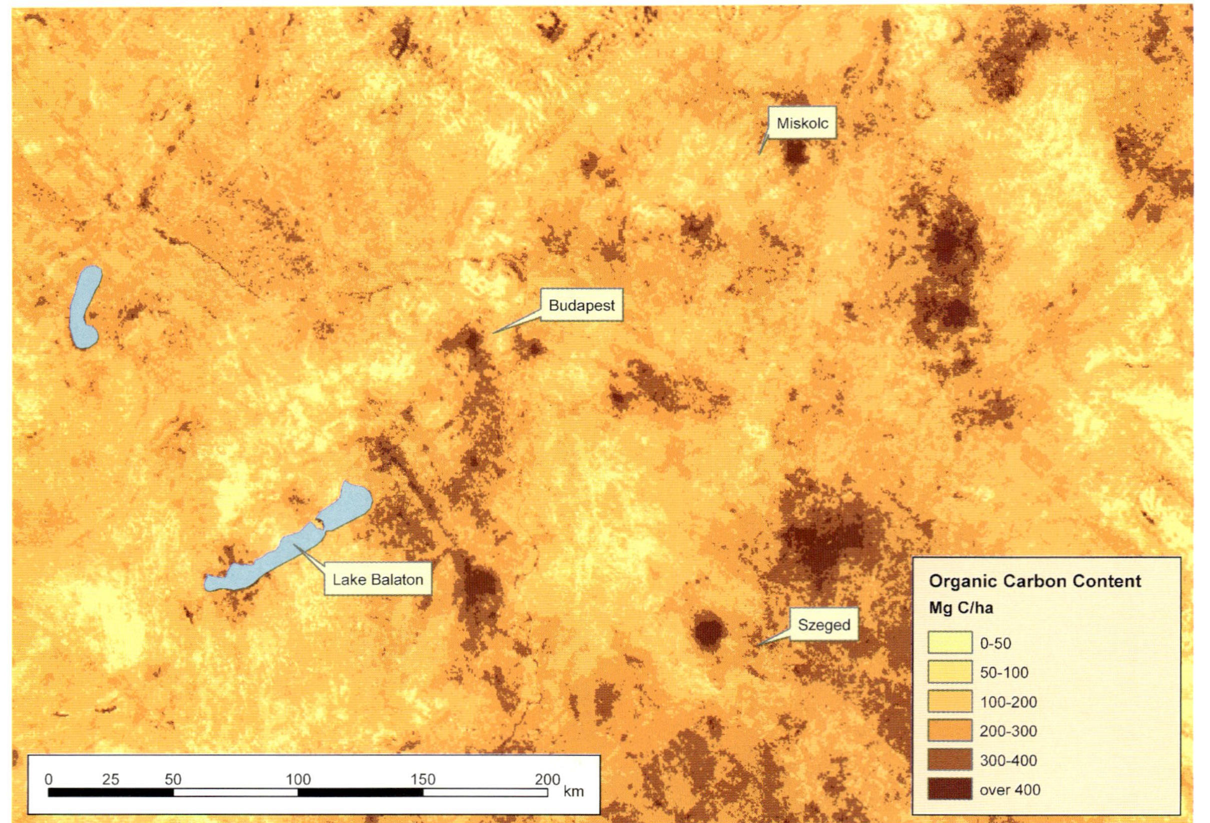

Plate 36b. *Organic matter content of the soils of Hungary derived from MODIS and SRTM-30 data through regression kriging.*

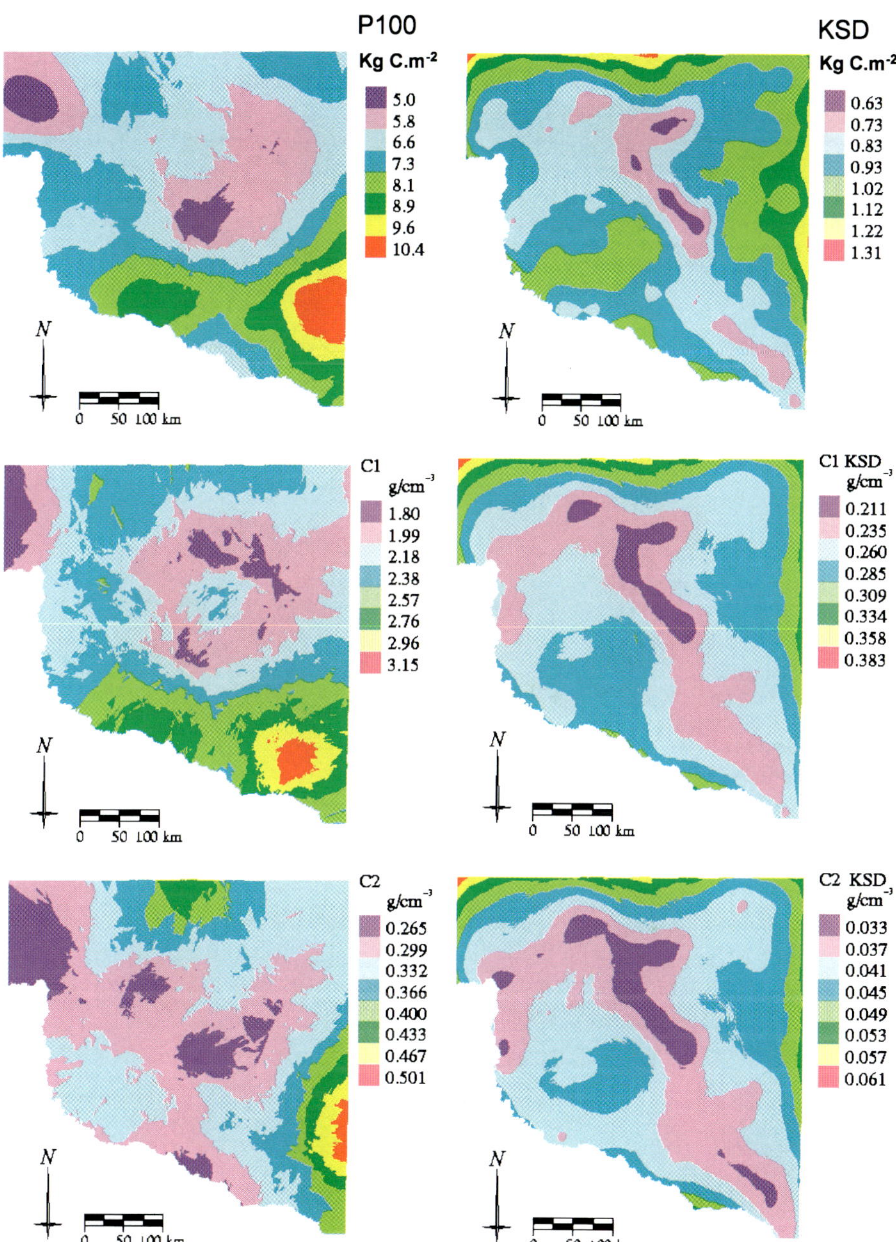

Plate 37. *Soil carbon stocks (P100, 0–100 cm) and vertical model parameters obtained by block kriging and associated errors (KSD).*

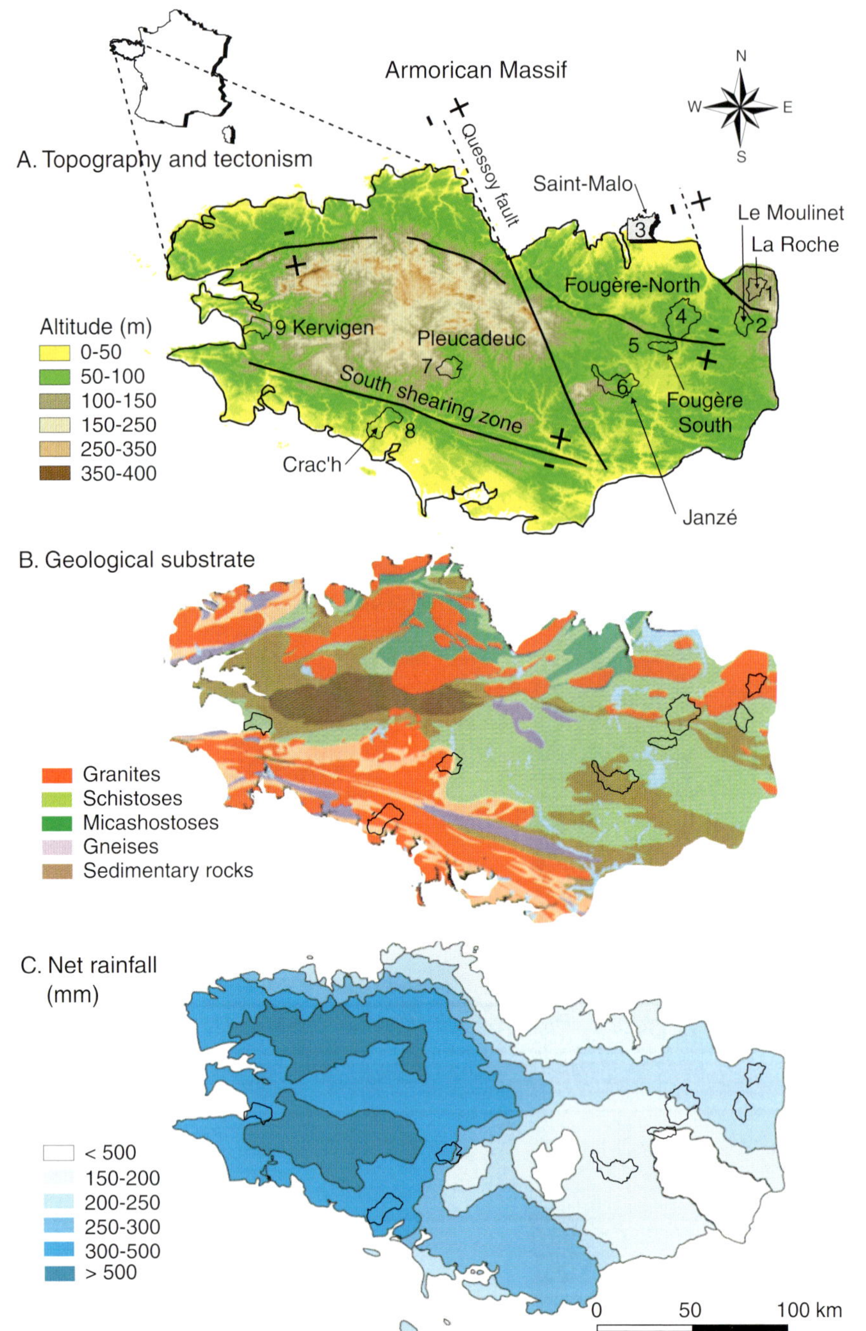

Plate 38. *The Armorican Massif of western France. (a) Location of the study areas, (b) topographic conditions and tectonism; geologic substrates (from Bonnet, 1997) and (c) net rainfall (NR) (difference between precipitation and potential evapotranspiration, PET).*

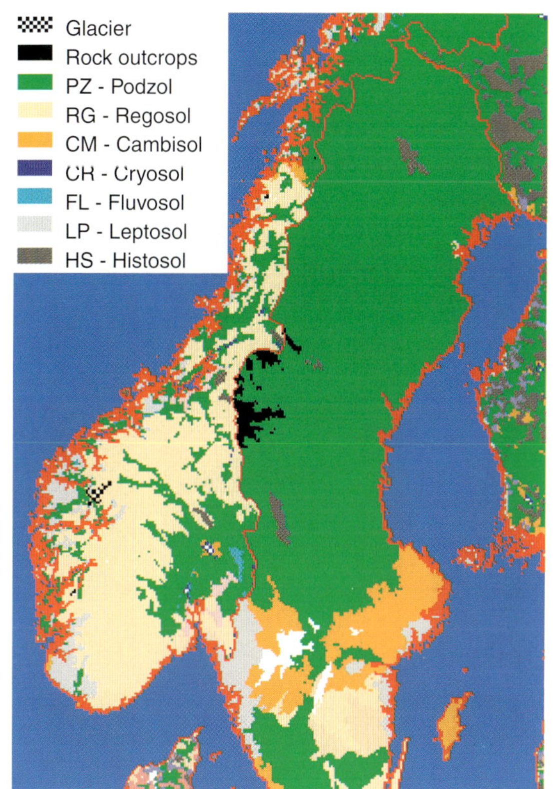

Plate 39. *WRB major soil group delineations in the soil map of Europe 1:1,000,000 (source: http://eusoils.jrc.it/msapps/Soil/SoilDB/SoilDB.phtml).*

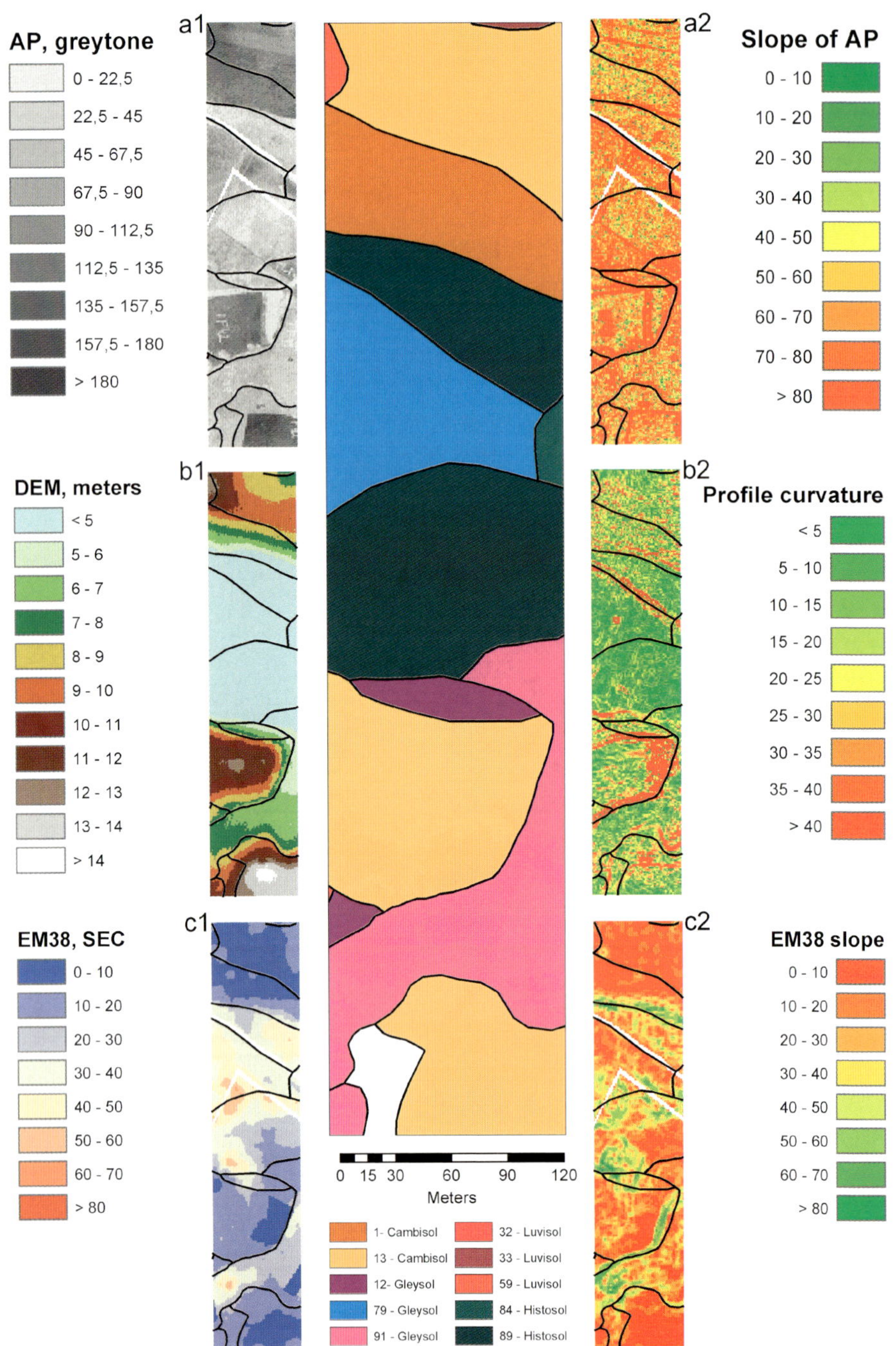

Plate 40. *Different datasets in the 7.4 ha test area situated in the northern part of the island of Funen, Denmark.*

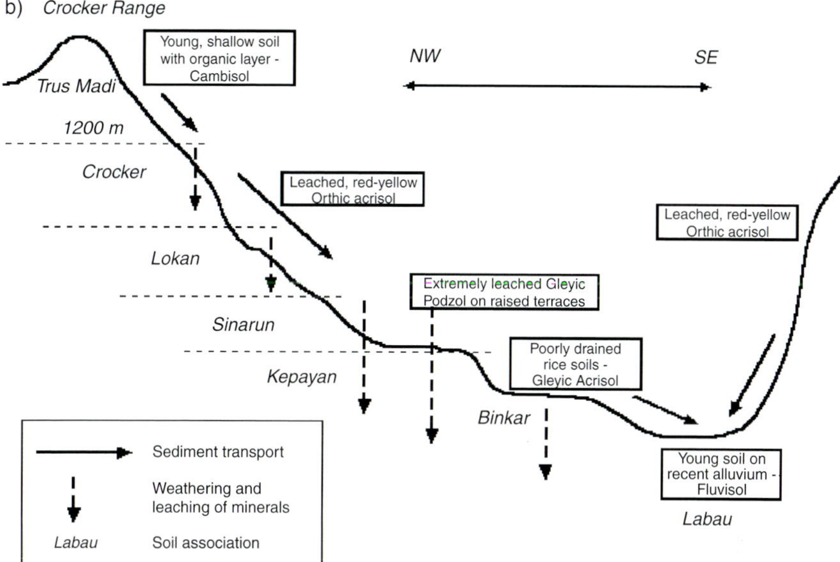

a)

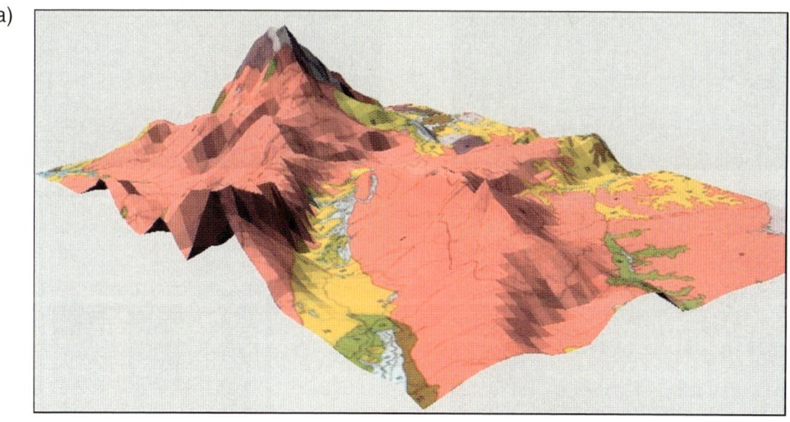

Plate 41. *(a) Catena of soil-landscape relations perpendicular to tectonic structure. (b) Crisp polygon map of soil associations draped over DEM which shows all possible catenas. Original map scale 1:250,000. Vertical exaggeration 3X. Source: Acres et al. (1975).*

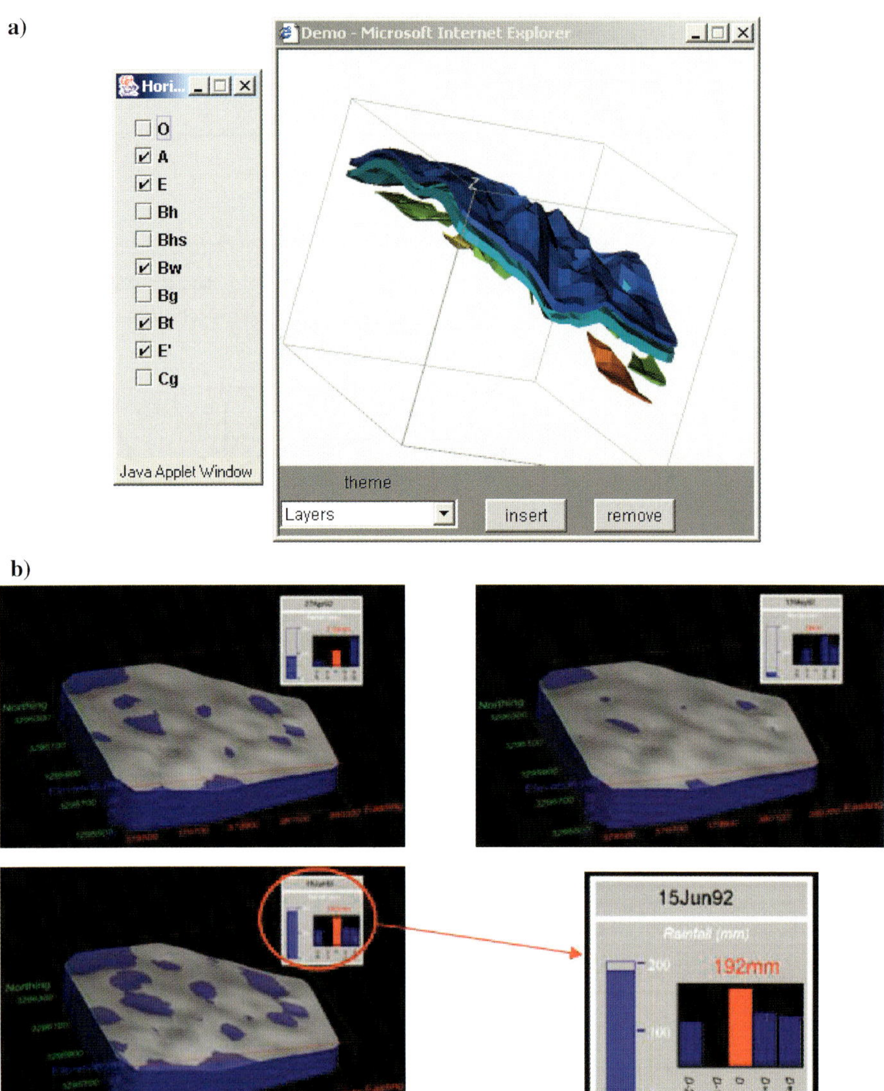

Plate 42. *(a) Soil horizon model infused with a digital elevation model (DEM) representing a Florida flatwood site. (b) Three snapshots of simulations of inundated areas at different time periods.*